CRYPTOLOGY
Classical and Modern
—————SECOND EDITION—————

Chapman & Hall/CRC Cryptography and Network Security Series

Series Editors: Douglas R. Stinson and Jonathan Katz

Algorithmic Cryptanalysis

Antoine Joux

Cryptanalysis of RSA and Its Variants

M. Jason Hinek

Access Control, Security, and Trust: A Logical Approach

Shiu-Kai Chin and Susan Beth Older

Handbook of Financial Cryptography and Security

Burton Rosenberg

Handbook on Soft Computing for Video Surveillance

Sankar K. Pal, Alfredo Petrosino and Lucia Maddalena

Communication System Security

Lidong Chen and Guang Gong

Introduction to Modern Cryptography, Second Edition

Jonathan Katz and Yehuda Lindell

Group Theoretic Cryptography

Maria Isabel Gonzalez Vasco and Rainer Steinwandt

Guide to Pairing-Based Cryptography

Nadia El Mrabet and Marc Joye

Cryptology: Classical and Modern, Second Edition

Richard Klima and Neil Sigmon

For more information about this series, please visit:
https://www.crcpress.com/Chapman--HallCRC-Cryptography-and-Network-Security-Series/book-series/CHCRYNETSEC

CRYPTOLOGY
Classical and Modern
————SECOND EDITION————

Richard Klima

Neil Sigmon

CRC Press
Taylor & Francis Group
Boca Raton London New York

CRC Press is an imprint of the
Taylor & Francis Group, an **informa** business

A CHAPMAN & HALL BOOK

CRC Press
Taylor & Francis Group
6000 Broken Sound Parkway NW, Suite 300
Boca Raton, FL 33487-2742

Printed on acid-free paper

International Standard Book Number-13: 978-1-138-04762-4 (Hardback)

Visit the Taylor & Francis Web site at
http://www.taylorandfrancis.com

and the CRC Press Web site at
http://www.crcpress.com

Contents

Preface

Several years ago we were invited to create a new general education mathematics course for the Honors Academy at Radford University. Wanting to create a multidisciplinary course that would demonstrate some interesting mathematical applications and also be accessible and intriguing to students with a wide variety of interests and backgrounds, we decided on a course in cryptology. Designed for students whose prior experience with mathematics includes only a basic understanding of algebra, statistics, and number theory at the secondary level, the course has been one of the most popular offerings at the Honors Academy.

When deciding on material for the course, since we expected most of the students to come from nontechnical fields, our goal was to choose topics that would be easy to understand, show the importance of cryptology in both cultural and historical contexts, and demonstrate some stimulating but relatively simple mathematical applications. A lesser goal was for students to be motivated to study the subject further and perhaps even consider careers in mathematics or the sciences.

The first edition of this book grew from our experiences teaching this course to students from nontechnical fields at Radford University and Appalachian State University. The first edition contained material that fully served this audience, however when we tried offering the course to students from more technical fields, we found it necessary to supplement the material in the first edition. The second edition of this book is an expanded version of the first edition that, while keeping all of the material that fully served students from nontechnical fields, supplements this material with new content that we believe now allows the book to fully serve students from more technical fields as well. Thus the second edition of this book, while retaining the first edition's ability to reach students at earlier ages, including pre-college, now has the added feature of being able to fully serve students from both technical and nontechnical fields throughout all levels of a collegiate curriculum.

Chapter 1 introduces cryptology, and includes basic terminology as well as some motivation for why the subject is worth studying. Chapters 2–3 introduce several elementary cryptologic methods and techniques, through substitution ciphers in Chapter 2, and transposition ciphers in Chapter 3. Although Chapters 2–3 are elementary in nature, they are not trivial, and include presentations of three specific types of ciphers that are well known and celebrated in history—Playfair ciphers, the Navajo code, and ADFGVX ciphers. Chapter 4 includes a fully developed presentation of the Enigma cipher machine that was used as a German field cipher during World War II, as well as an introduction to the mathematical field of combinatorics, which is used in Chapter 4 to analyze the security of the Enigma. Chapter 5 is completely new to the second edition of this book, and includes a fully developed presentation of the Turing bombe machine that was used in the cryptanalysis of the Enigma during World War II. Chapter 6 includes an introduction to modular arithmetic, which is used in Chapter 6 to create shift and affine ciphers. Chapter 7 introduces polyalphabetic ciphers, through Alberti and Vigenère ciphers, and includes an introduction to probability, which is used in Chapter 7 in the cryptanalysis of Vigenère ciphers. Chapter 8 includes an introduction to matrix algebra, which is used in Chapter 8 to create Hill ciphers. Chapters 9–10 introduce public-key cryptography, through RSA ciphers in Chapter 9, and the Diffie-Hellman key exchange and ElGamal ciphers in Chapter 10. Chapters 9–10 also introduce a variety of mathematical topics, including the Euclidean algorithm, binary exponentiation, primality testing, integer factorization, and discrete logarithms, with each connected to appropriate places within the cryptologic methods and techniques presented in Chapters 9–10. Chapter 11 introduces binary and hexadecimal representations of numbers, which are used in Chapter 11 to create stream ciphers and within a fully developed presentation of the Advanced Encryption Standard. Chapter 12 considers message authenticity, through digital signatures, hash functions, the man-in-the-middle attack, and certificates. Exercises of varying levels of difficulty are included at the end of every section in Chapters 2–12. For many sections, some exercises require online or library research. Instructors teaching from this book should be able to use these research exercises as a springboard for student projects that would greatly enhance their course. Hints and answers for selected exercises are included at the end of the book, and a complete solutions manual is available through the publisher.

This book includes plenty of material for a one-semester course on cryptology, and depending on the audience may contain enough material for a two-semester sequence if supplemented with significant student projects. Some parts of the book are more appropriate for collegiate juniors and seniors, others for collegiate freshman and sophomores, and yet others for advanced secondary students. Depending on the depth to which the topics are covered, Chapters 1–8 would make a nice complete course on classical cryptology. Chapter 9 could be included if a taste of public-key cryptography is desired, as well as some or all of Chapters 10–12 in more advanced courses. Parts of the book could also be used in courses designed to be free of traditional mathematics. Specifically, Chapters 1–3, Section 4.1, Chapter 5, and Sections 7.1–7.2 are essentially free of traditional mathematics.

The first edition of this book included detailed instructions for the use of a technology resource that we have found to be very useful with certain parts of the book. More specifically, in order to include more substantive examples and exercises with the techniques presented in this book, the first edition of the book incorporated Maplets, a technology resource much like Java applets, but which use the engine of the mathematics software package Maple. For the second edition of this book we have moved these instructions, examples, and exercises online, posting them along with all of our Maplets at https://www.radford.edu/npsigmon/cryptobook.html. The reason for this move is threefold. First, it continues to be easy to use the printed second edition of this book without an advanced technology resource. In short, every example and exercise in the printed second edition of this book can be completed with either no technology or a simple hand-held calculator. Second, although most of our Maplets can still be used even by users who have not purchased or downloaded Maple, the ever-increasing availability of free technology resources online for many of the techniques presented in this book means it continues to become easier for users of the book to find and choose their own technology resources. Third, although our Maplets continue to work, Maplesoft, Inc. no longer provides updates or active support for them. Moving our Maplet materials online allows us to provide our own support to users of this book by being able to more easily modify our Maplets and instructions for their use.

We wish to thank Dr. Joe King, former Director of the Honors Academy at Radford University, for his strong support as we developed the course and this book. We are also grateful to Willis Tsosie of Diné College, whose uncle Kenneth was a Navajo code talker, for his inspiration as we developed our section on the Navajo code, and for providing the spoken code words for the Navajo code Maplet that is included in our online materials. We also wish to thank our mentor and friend, Dr. Ernie Stitzinger of North Carolina State University, for his encouragement, interest in our projects, and guidance in our own education and careers. Finally, we wish to thank our families, especially Vicky and Mandy, for their patience and support.

We welcome comments, questions, corrections, and suggestions for future editions of this book, and sincerely hope that you enjoy using it.

Rick Klima
klimare@appstate.edu

Neil Sigmon
npsigmon@radford.edu

Chapter 1

Introduction to Cryptology

Throughout the history of human communication, the practice of keeping information secret by disguising it, known as *cryptography*, has been of great importance. Many important historical figures—for example, Julius Caesar, Francis Bacon, and Thomas Jefferson—have used cryptography to protect sensitive information. Before becoming the "Father of the Modern Computer," Alan Turing played an integral role during World War II in the successful attacks by the Allies on the Enigma machine, which was used by the Germans to disguise information. Important literary figures have also included cryptography in their writings. In fact, William Friedman, called the "Dean of American Cryptology" on a bust at the U.S. National Cryptologic Museum, was first inspired to study the subject through reading Edgar Allan Poe's short story "The Gold Bug." Cryptography is also at the heart of some remarkably fascinating accounts from human history, such as the successful attacks on the Enigma machine by the Allies, and the Allies' own effective use of Navajo code talkers during World War II. There are numerous books devoted exclusively to the history of cryptography, including excellent accounts by David Kahn [13] and Simon Singh [21]. However, cryptography is not just a historical subject. Most of us use cryptographic methods quite frequently, often without knowing or thinking about it, for example, when we purchase items using a credit card or send information using email.

The main purpose of this chapter is to introduce some terminology and concepts involved with studying cryptography, and to preview what lies ahead in this book. We also give a brief description of some of the benefits to learning about cryptography.

1.1　Basic Terminology

In the field of information security, the terms *cryptography*, *cryptanalysis*, and *cryptology* have subtly different meanings. The process of developing a system for disguising information so that ideally it cannot be understood by anyone but the intended recipient of the information is called *cryptography*, and a method designed to perform this process is called a *cryptosystem* or a *cipher*. *Cryptanalysis* refers to the process of an unintended recipient of disguised information attempting to remove the disguise and understand the information, and successful cryptanalysis is sometimes called *breaking* or *cracking* a cipher. *Cryptology* is an all-inclusive term that includes cryptography, cryptanalysis, and the interaction between them.

When a cipher is used by two parties to exchange information, the undisguised information (in this book, usually a message written in ordinary English) is called the *plaintext*, and the disguised information is called the *ciphertext*. The process of converting from plaintext to ciphertext is called *encryption* or *encipherment*. Upon receiving a ciphertext, the recipient must remove the disguise, a process called *decryption* or *decipherment*. To be able to effectively encrypt and decrypt messages, two correspondents must typically share knowledge of a secret *key*, which is used in applying the agreed-upon cipher. More specifically, the key for a cipher is information usually known only to the originator and intended recipient of a message, which the originator uses to encrypt the plaintext, and the recipient uses to decrypt the ciphertext.

Often confused with cryptography is the subject of *coding theory* or *codes*. Unlike with cryptography, in which the concern is primarily concealing information, with codes the concern is usually transmitting information reliably and efficiently over a communications medium. For example, Morse code is not a cipher. On the other hand, cryptologists do sometimes refer to ciphers as codes, for instance, the Navajo code, which we consider a cipher since it primarily existed to conceal information. Determining the proper use of the word *code* is ordinarily easy to derive from context. To minimize confusion, the only cipher that we will refer to as a code is the Navajo code, which we will study in Chapter 2.

1.2　Cryptology in Practice

Throughout this book we will demonstrate many different types of ciphers. In practice, it is usually assumed that when a pair of correspondents use a cipher to communicate a message confidentially, the type of cipher used is known by any adversaries wishing to discover the contents of the message. Thus, the *security* of a cipher, which is simply a measure of how difficult it would be for an adversary to break the cipher, depends only on how difficult it would be for an adversary to find the key for the cipher. The benefit to this is that by correspondents choosing a cipher with an acceptable level of security, they would not have to worry about keeping the type of cipher secret from adversaries.

The various types of ciphers that have been and are used in practice split into two broad categories—*symmetric-key* and *public-key*. Symmetric-key ciphers, the only kind that existed before the 1970s, are also sometimes called *private-key* ciphers. When using a symmetric-key cipher, the originator and intended recipient of a message must keep the key secret from adversaries. In Chapters 2–8 of this book, we will see a variety of different types of symmetric-key ciphers that have been used throughout history. These types of ciphers are more commonly called *classical* ciphers, since they are not typically useful in communicating sensitive information in modern society. They are still fascinating and fun to study, though. In Chapters 2, 3, and 7, we will see some types of ciphers for which the keys are formed using English words called *keywords*. For Enigma machine ciphers, which we will study in Chapter 4, the keys are the initial settings of the machine. In Chapters 6 and 8, we will see some types of ciphers for which the keys are mathematical quantities such as numbers or matrices. A deficiency in symmetric-key ciphers is that correspondents must have a way to identify keys in secret, while the very need for a cipher indicates that they have no secret way to communicate.

The invention of public-key ciphers in the 1970s revolutionized the science of cryptology. Public-key ciphers use a pair of keys, one for encryption and one for decryption. When using a public-key cipher, the intended recipient of a message creates both the encryption and decryption keys, publicizes the encryption key so that anyone can know it, but keeps the decryption key secret. That way, the originator of the message can know the encryption key, which he or she needs to encrypt the plaintext, but only the recipient knows the decryption key. It would seem to be a deficiency in public-key ciphers that adversaries can know encryption keys. However, as we will see when we study the two most common types of public-key ciphers in Chapters 9 and 10, although encryption and decryption keys are obviously related, it usually is not realistically possible to find decryption keys from the knowledge of encryption keys.

The development of public-key ciphers did not lead to the demise of symmetric-key ciphers, though. A major reason for this is the fact that public-key ciphers typically operate much more slowly than symmetric-key ciphers. Thus, for correspondents wishing to use a cipher in communicating a large amount of information, it is often most prudent to use a public-key cipher to exchange the key for a symmetric-key cipher, and then use the symmetric-key cipher to actually communicate the information. In Chapter

11, we will see some types of symmetric-key ciphers that are useful in communicating sensitive information in modern society.

Many fascinating historical accounts of cryptology involve successful cryptanalysis. In Chapter 5, we will study in detail one celebrated such account, the attack on the German Enigma machine by Allied cryptanalysts at Bletchley Park near London, England, during World War II. The goal in cryptanalysis is often to determine the key for a cipher. The most obvious method for accomplishing this, known as a *brute force attack*, involves testing every possible key until finding one that works. Some types of ciphers have a relatively small number of possible keys, and thus can be attacked by brute force. However, brute force is not a legitimate method of attack against most ciphers, even in our technologically advanced society. For example, for the Advanced Encryption Standard, a type of symmetric-key cipher that we will study in Chapter 11, the minimum number of possible keys is 3.4×10^{38}, which would take trillions of years to test even using the most advanced current technology.

The security of a cipher is not always tied directly to the number of possible keys, though. For example, although the number of possible keys for a substitution cipher is more than 4×10^{26}, we will see in Chapter 2 that substitution ciphers can sometimes be broken relatively easily through a technique called *frequency analysis*. Also, as we will see in Chapters 3 and 8, there are other types of ciphers against which both a brute force attack and frequency analysis may be pointless, but which can sometimes still be broken relatively easily by adversaries who know a small part of the plaintext, called a *crib*. In addition, any cipher, no matter how theoretically secure, is always susceptible to being broken due to human error on the part of the users of the cipher. For example, the types of public-key ciphers that we will study in Chapters 9 and 10 are essentially unbreakable, but only provided certain initial parameters are chosen correctly.

The final cryptologic issues we will consider in this book relate to *message authentication*, specifically verifying that a ciphertext received electronically was really sent by the person claiming to have sent it, and that keys identified electronically really belong to the person claiming to own them. Especially in our digital age, confirming that one is communicating with whom he or she believes to be communicating can be as important as what is actually communicated. We will address these issues in Chapter 12, through the ideas of *digital signatures* and *public-key infrastructures*.

1.3   Why Study Cryptology?

An obvious question, especially for individuals with limited experience or natural interest in technical fields, is why would cryptology be worthwhile to study? For that matter, why is the subject of cryptology even important in our society?

One answer to these questions is that due to the ever-increasing dependence of our society upon technology in the communication of information, for instance through ATM transactions and credit card purchases, effective cryptography is essential for commerce that is both private and reliable. Effective cryptography is also essential for personal privacy by individuals who use cell phones or email, or who even just have personal information such as Social Security or driver license numbers stored in government databases. In fact, the dependence of our government and military upon cryptology to ensure secure and authentic communication is so profound that it led to the formation of an entire federal agency, the National Security Agency, whose primary purpose is to create and analyze cryptologic methods, and whose published vision includes "global cryptologic dominance." In the near future, our society will also likely see an increased dependence upon devices such as *smart cards*, which are pocket-size cards with integrated computer circuits embedded with cryptographic methods, for identification and financial transactions.

Cryptology is also a multidisciplinary science. As we have noted, the subject is rich with fascinating historical accounts, several of which we will comment on in this book. As we will see in the earlier chapters of this book,

The National Security Agency

The National Security Agency (NSA) is the primary agency for cryptology in the U.S. It is responsible for collecting and analyzing communications between foreign entities, and developing methods for protecting communications originating from U.S. entities. Created in 1952 by President Harry Truman, the NSA specializes in foreign signals intelligence (SIGINT). SIGINT is information from electronic signals and targets, and can be derived from sources such as communications systems, electronic signals, and weapons systems. Research is also a vital component of the operations of the NSA. Its research goals include dominating global computing and communications networks, coping with information overload, providing methods for secure collaboration within the U.S. government and its partners, and penetrating targets that threaten the U.S.

To achieve its goals, the NSA employs a very large number of mathematicians. Computer scientists, engineers, and linguists are also in high demand at the NSA.

knowledge of letter frequencies is important in cryptanalyzing some types of ciphers. Linguistics thus plays a role in cryptology, since letter frequencies naturally vary in different languages. Sociology and culture are evident in cryptology as well. For instance, the Navajo culture and societal beliefs were critical in the development and success of the Navajo code. As we will see in the later chapters of this book, the design and engineering required to construct computers capable of generating the parameters needed for implementing modern ciphers securely and efficiently also play a role in cryptology.

The discipline that plays the most integral and important role in cryptology, though, is mathematics. Cryptology provides numerous applications of mathematical topics ranging from elementary arithmetic to advanced collegiate mathematics. In Chapter 4, we will see how combinatorics can be used to analyze the difficulty of breaking Enigma machine ciphers. In Chapter 7, we will see how probability and statistics can be used in the cryptanalysis of Vigenère ciphers. Beginning in Chapter 6, we will explore how modular arithmetic can be used in the implementation and cryptanalysis of several types of ciphers. In Chapters 8 and 11, we will see how matrices can be used in the implementation of classical Hill ciphers and the modern Advanced Encryption Standard. In Chapters 9 and 10, we will see how number theory, specifically division, exponentiation, primality, and factorization, is useful in the implementation and cryptanalysis of RSA and ElGamal public-key ciphers. The topics presented in this book should easily convince readers of the importance of mathematics in our society. In addition, since this book is not designed to go too deeply into the mathematical theory involved with studying cryptology, it will hopefully also provide motivation for readers to further explore the mathematics topics in the book, and perhaps even lead to a more purposeful understanding of such areas of mathematics as linear algebra, combinatorics, probability and statistics, number theory, and abstract algebra.

Finally, and we realize most importantly to some, learning about cryptology can be fun and entertaining. Cryptology is a subject that often finds its way into modern popular culture. The television show *NCIS*, which has been voted as America's all-time favorite show, featured cryptology in numerous episodes. Recent Hollywood blockbusters involving cryptology include *The Imitation Game*, *Zodiac*, *Windtalkers*, *U-571*, and Disney's *National Treasure* franchise. Cryptology is also involved in numerous works of literature. For example, Dan Brown's mystery-detective novel *The Da Vinci Code*, which topped national bestseller lists for years and spawned its own blockbuster Hollywood franchise, includes several references to encrypted messages that are essential to the story. Just knowing the basics, which we provide in this book, should give readers the ability to better enjoy and appreciate such examples of cryptology.

Chapter 2

Substitution Ciphers

One common and popular type of cipher for newspaper games and puzzle books is a *substitution* cipher. In simple substitution ciphers, users agree upon a rearrangement, or *permutation*, of the alphabet letters, yielding a collection of correspondences to be used for converting plaintext letters into ciphertext letters. This rearrangement of the alphabet letters is often called the *cipher alphabet*. To say that the cipher alphabet is a permutation means that each possible plaintext letter in the original alphabet is paired with one and only one possible ciphertext letter, and vice versa. With more sophisticated substitution ciphers, messages and cipher alphabets can include numbers, punctuation marks, or mixtures of multiple characters, and substitutions can be made for entire words or phrases.

As we will see in this chapter, simple substitution ciphers are not very secure (meaning they are easy to break). Despite this, substitution ciphers have a rich history of being used. One of the earliest known ciphers, the Hebrew Atbash, was a substitution cipher, as was a cipher developed and used by Julius Caesar that we will consider in Chapter 6, and a cipher incorporated in Edgar Allan Poe's short story "The Gold Bug." Not all substitution ciphers are easy to break, though. The cipher famously created during World War II by the Navajo code talkers, which we will consider in this chapter, and which was not known to have ever been broken during a period of use that extended through the Korean War and into the early stages of the Vietnam War, was essentially a substitution cipher.

2.1   Keyword Substitution Ciphers

One way to form a substitution cipher is to just use a random cipher alphabet.

Example 2.1 Consider a substitution cipher with the following cipher alphabet.

> **Plain:**   A B C D E F G H I J K L M N O P Q R S T U V W X Y Z
> **Cipher:** T V X Z U W Y A D G K N Q B E H R O S C F J M P I L

Using this cipher alphabet, the plaintext YOUTH IS WASTED ON THE YOUNG[1] encrypts to the ciphertext IEFCA DS MTSCUZ EB CAU IEFBY. This ciphertext can be decrypted by using the same cipher alphabet but with the correspondences viewed in the reverse order. □

A problem with using a random cipher alphabet is that it may be inconvenient or cumbersome for users to keep a record of. For instance, users wishing to use a substitution cipher with the cipher alphabet in Example 2.1 would most likely have to keep a written record of this alphabet. One solution to this problem would be for users to use a keyword in forming the cipher alphabet. In this section, we will demonstrate two methods for doing this.

2.1.1   Simple Keyword Substitution Ciphers

For simple keyword substitution ciphers, users agree upon one or more keywords for the cipher. Spaces and duplicate letters in the keyword(s) are removed, and the resulting letters are then listed in order as the ciphertext letters that correspond to the first plaintext letters in alphabetical order. The remaining alphabet letters not included in the keyword(s) are then listed in alphabetical order to correspond to the remaining plaintext letters in alphabetical order.

Example 2.2 Consider a simple keyword substitution cipher with the keywords WILL ROGERS. Removing the space as well as the duplicate letters in these keywords gives WILROGES, and results in the following cipher alphabet.

> **Plain:**   A B C D E F G H I J K L M N O P Q R S T U V W X Y Z
> **Cipher:** W I L R O G E S A B C D F H J K M N P Q T U V X Y Z

Using this cipher alphabet, the plaintext EVERYTHING IS FUNNY AS LONG AS IT IS HAPPENING TO SOMEBODY ELSE[2] encrypts to OUONYQSAHE AP GTHHY WP DJHE WP AQ AP SWKKOHAHE QJ PJFOIJRY ODPO. To decrypt a ciphertext that was formed using this cipher alphabet, we can use the same cipher alphabet but with the characters considered in the reverse order. For example, the ciphertext QSO VADD NJEONP AHPQAQTQO OHEWEOP AH LWNRAJKTDFJHWNY

---

[1]George Bernard Shaw (1856–1950), quote.
[2]Will Rogers (1879–1935), quote.

FORALWD NOPOWNLS decrypts to THE WILL ROGERS INSTITUTE ENGAGES IN CARDIOPULMONARY MEDICAL RESEARCH. □

Example 2.2 reveals a problem with simple keyword substitution ciphers. Often with this type of cipher, notably when the keyword doesn't contain any letters near the end of the alphabet, the last several correspondences in the cipher alphabet are letters corresponding to themselves. Such correspondences are called *collisions*, and can make a cipher more vulnerable to cryptanalysis. Keyword columnar substitution ciphers can help to alleviate this problem.

2.1.2 Keyword Columnar Substitution Ciphers

For keyword columnar substitution ciphers, users again agree upon one or more keywords, and remove spaces and duplicate letters in the keyword(s). The resulting letters are then listed in order in a row, with the alphabet letters not included in the keyword(s) listed in order in successive rows of the same size beneath the keyword letters. The cipher alphabet is then obtained by taking the columns of the resulting array of letters in order starting from the left, and placing these columns as rows under the plaintext letters.

Example 2.3 Consider a keyword columnar substitution cipher with keywords ABE LINCOLN. Removing the space and duplicate letters gives ABELINCO, and placing these letters in a row, with the remaining alphabet letters listed in order in successive rows, yields the following array.

```
A  B  E  L  I  N  C  O
D  F  G  H  J  K  M  P
Q  R  S  T  U  V  W  X
Y  Z
```

Transcribing this array by columns starting from the left yields the following cipher alphabet.

Plain: A B C D E F G H I J K L M N O P Q R S T U V W X Y Z
Cipher: A D Q Y B F R Z E G S L H T I J U N K V C M W O P X

Using this cipher alphabet, the plaintext IF I WERE TWO-FACED WOULD I BE WEARING THIS ONE[3] encrypts to EF E WBNB VWI-FAQBY WICLY E DB WBANETR VZEK ITB, and the ciphertext FICN KQINB ATY KBMBT PBANK ARI decrypts to FOUR SCORE AND SEVEN YEARS AGO.[4] □

---

[3] Abraham Lincoln (1809–1865), quote.
[4] Abraham Lincoln, from the Gettysburg Address.

2.1.3 Exercises

1. Consider a substitution cipher with the following cipher alphabet.

   Plain: A B C D E F G H I J K L M N O P Q R S T U V W X Y Z
   Cipher: X Q K M D B P S E T C L O R U J V A F W Z G H N I Y

   (a)* Use this cipher to encrypt QUOTH THE RAVEN NEVERMORE.[5]

   (b) Use this cipher to encrypt A FEW WORDS ON SECRET WRITING.

   (c) Decrypt WSD BXLL UB WSD SUZFD UB ZFSDA, which was formed using this cipher.

2. Create a substitution cipher with a random cipher alphabet and use it to encrypt a plaintext of your choice with at least 20 letters.

3. Decrypt X PUUM OXR EF SXAM WU BERM QZW BERMERP X PUUM MUP EF DXFI, which was formed using a substitution cipher with the cipher alphabet in Exercise 1.

4. Consider a simple keyword substitution cipher with the keyword GILLIGAN.

   (a)* Use this cipher to encrypt A THREE HOUR TOUR.

   (b) Use this cipher to encrypt A TALE OF A FATEFUL TRIP.

   (c) Decrypt RGQUNJ WNJLD GUAETEOMNA BOR KGRY GMM, which was formed using this cipher.

5. Create a simple keyword substitution cipher and use it to encrypt a plaintext of your choice with at least 20 letters.

6. Decrypt TGLDUCHQ HOTL LSUOLS AGC QRHKSHG E CUPGHO AQ ADUFGT, which was formed using a simple keyword substitution cipher with the keywords APPALACHIAN STATE.

7. Consider a keyword columnar substitution cipher with the keywords MARSHAL DILLON.

   (a)* Use this cipher to encrypt SATURDAY NIGHT IN DODGE CITY.

   (b) Use this cipher to encrypt CAPTAIN KIRK WAS IN ONE EPISODE.

   (c) Decrypt MGA HWDKCZ DLYUF MLLCMZCA VYIZ KWHCD, which was formed using this cipher.

---

*Throughout this book, exercises with hints or answers included at the end of the book are notated with this footnote symbol.
[5]Edgar Allan Poe (1809–1849), from "The Raven."

8. Create a keyword columnar substitution cipher and use it to encrypt a plaintext of your choice with at least 20 letters.

9. Decrypt NZVAHX XHQZVCD PNQF SNZXWXNHNGXI NHC IQELLY AQQKDDKY, which was formed using a keyword columnar substitution cipher with the keywords NC STATE UNIVERSITY.

10. Create and describe a method different from those illustrated in this section for using a keyword to form the cipher alphabet for a substitution cipher. Give at least one example of your method.

11. Find some information about the Hebrew Atbash cipher, including how it worked, and write a summary of your findings.

12. Find some information about Edgar Allan Poe's interest in cryptography, including the various cryptographic challenges he offered, and write a summary of your findings.

13. Find a copy of Edgar Allan Poe's short story "The Gold Bug," and write a summary of how a substitution cipher is integrated into the story.

14. Find a copy of Sir Arthur Conan Doyle's Sherlock Holmes mystery *The Adventure of the Dancing Men*, and write a summary of how a substitution cipher is integrated into the story and how the cipher worked.

2.2 Cryptanalysis of Substitution Ciphers

Considering the number of possible cipher alphabets, substitution ciphers seem impossible to break. With 26 letters, there are more than 4×10^{26} possible cipher alphabets. To test them all would be infeasible. However, as it turns out, most simple substitution ciphers are fairly easy to break through the use of *frequency analysis*. In fact, inadequate security of substitution ciphers has even altered the course of history. For example, the breaking of a substitution cipher led to the execution of Mary, Queen of Scots in 1587.

In languages like English, it is known that certain letters and combinations of letters occur more often than others. In ordinary English, the letters that naturally occur the most often are, in order, E, T, A, O, I, N, and S. The frequency with which each of the 26 letters in our alphabet occurs in ordinary English is shown in Table 2.1 on page 12.

Common digraphs (letter pairs), trigraphs (letter triples), and repeated letters in ordinary English are also known. The most common digraphs are

Letter	Frequency	Letter	Frequency
A	8.17%	N	6.75%
B	1.49%	O	7.51%
C	2.78%	P	1.93%
D	4.25%	Q	0.10%
E	12.70%	R	5.99%
F	2.23%	S	6.33%
G	2.02%	T	9.06%
H	6.09%	U	2.76%
I	6.97%	V	0.98%
J	0.15%	W	2.36%
K	0.77%	X	0.15%
L	4.03%	Y	1.97%
M	2.41%	Z	0.07%

Table 2.1   Letter frequencies in ordinary English.

TH, ER, ON, AN, RE, HE, IN, ED, and ND. The most common trigraphs are
THE, AND, THA, ENT, ION, TIO, FOR, NDE, HAS, and NCE. The most common
repeated letters are LL, EE, SS, TT, OO, MM, and FF. For a thorough analysis
of common letter sequences in ordinary English, see [25].

   For a ciphertext that has been formed using a substitution cipher, with
a sufficient number of ciphertext letters and the spacing between words in
the plaintext preserved, frequency analysis can usually be used to break
the cipher.

Example 2.4 Consider the following ciphertext, which was formed using
a substitution cipher.

```
WZIS VZIL VRRQ VZI CRAEVZ TGISYGISV, M WTJ JMFISV BIOTAJI
M YRS'V YITF YEAPJ. WZIS VZIL VRRQ VZI JMNVZ TGISYGISV, M
QIHV UAMIV BIOTAJI M QSRW M'G MSSROISV. WZIS VZIL VRRQ VZI
JIORSY TGISYGISV, M JTMY SRVZMSP BIOTAJI M YRS'V RWS T
PAS. SRW VZIL'KI ORGI CRE VZI CMEJV TGISYGISV TSY M OTS'V
JTL TSLVZMSP TV TFF.
```

The frequency with which each letter occurs in this ciphertext is shown in
the following table.

Letter:	I	S	V	R	T	M	Z	J	Y	G	A	O	W
Count:	33	28	28	17	17	16	15	11	11	10	7	7	7
Letter:	L	Q	E	F	P	B	C	H	K	N	U	D	X
Count:	6	5	4	4	4	3	3	1	1	1	1	0	0

Based on the frequency and locations of the letter I in the ciphertext, it seems likely that this letter corresponds to E in the plaintext. In addition, the trigraph VZI occurs in the ciphertext eight times, four of these as a single word. Since the trigraph that occurs with the highest frequency in ordinary English is THE, it seems reasonable to suppose that VZI in the ciphertext corresponds to THE in the plaintext. The validity of this is reinforced by the fact that it causes the second most common letter in the ciphertext, V, to correspond to the second most common letter in ordinary English, T. Note also that the one-letter words M and T both occur in the ciphertext. In ordinary English, the most common one-letter words are A and I. The fact that the ciphertext also contains the word M'G suggests it is likely that the ciphertext letter M corresponds to I in the plaintext, and consequently that the ciphertext letter T corresponds to A in the plaintext. The following shows the complete ciphertext, with the part of the plaintext given by the plain/cipher letter correspondences that we have determined provided above the ciphertext letters.

```
HE THE T    THE    TH A E   ET, I A   I E T  E A E
WZIS VZIL VRRQ VZI CRAEVZ TGISYGISV, M WTJ JMFISV BIOTAJI

I   'T EA      .  HE THE T   THE I TH A E   ET, I
M YRS'V YITF YEAPJ. WZIS VZIL VRRQ VZI JMNVZ TGISYGISV, M

  ET  IET E A E I       I'  I    E T. HE THE T    THE
QIHV UAMIV BIOTAJI M QSRW M'G MSSROISV. WZIS VZIL VRRQ VZI

  E    A E  ET, I AI    THI   E A E I    'T      A
JIORSY TGISYGISV, M JTMY SRVZMSP BIOTAJI M YRS'V RWS T

      .    THE ' E   E    THE I T AE  ETA   I A 'T
PAS. SRW VZIL'KI ORGI CRE VZI CMEJV TGISYGISV TSY M OTS'V

  A A THI   AT A  .
JTL TSLVZMSP TV TFF.
```

Next, note that the repeated letters RR occur three times in the ciphertext, each in the middle of the word VRRQ, which suggests that the ciphertext letter R corresponds to a vowel in the plaintext. The repeated vowels most likely to occur in ordinary English are EE and OO, and since a ciphertext letter has already been assigned to the plaintext letter E, it seems reasonable that the ciphertext letter R corresponds to the plaintext letter O. In addition, the fact that each time VRRQ occurs in the ciphertext it is followed by THE in the plaintext suggests that the ciphertext word VRRQ corresponds to TOOK in the plaintext. Thus, we will assign the ciphertext letter Q to the plaintext letter K. Also, note that the third most common letter in the ciphertext is S. Based on the positions of S in the ciphertext, it appears likely that the ciphertext letter S corresponds to a consonant in the plaintext. Since we have already assigned a ciphertext letter to the most common consonant in ordinary English, T, it seems reasonable to assume that the ciphertext letter S corresponds to the second most common consonant in ordinary English, N. The following shows the part of the cipher alphabet that we have assigned so far.

```
Plain:  A B C D E F G H I J K L M N O P Q R S T U V W X Y Z
Cipher: T _ _ _ I _ _ Z M _ Q _ _ S R _ _ _ _ V _ _ _ _ _ _
```

The following again shows the complete ciphertext, along with the part of the plaintext given by our expanded plain/cipher letter assignments.

```
HEN THE  TOOK THE  O  TH A  EN ENT, I  A   I ENT  E A  E
WZIS VZIL VRRQ VZI CRAEVZ TGISYGISV, M WTJ JMFISV BIOTAJI

I  ON'T  EA         .  HEN THE  TOOK THE  I TH A  EN  ENT, I
M YRS'V YITF YEAPJ. WZIS VZIL VRRQ VZI JMNVZ TGISYGISV, M

KE T   IET  E A  E I KNO  I'  INNO ENT.  HEN THE  TOOK THE
QIHV UAMIV BIOTAJI M QSRW M'G MSSROISV. WZIS VZIL VRRQ VZI

 E ON  A  EN ENT, I  AI  NOTHIN  E A  E I  ON'T O N A
JIORSY TGISYGISV, M JTMY SRVZMSP BIOTAJI M YRS'V RWS T

  N.  NO  THE '  E  O E  O  THE  I  T A EN  ENT AN  I  AN'T
PAS. SRW VZIL'KI ORGI CRE VZI CMEJV TGISYGISV TSY M OTS'V

  A  AN THIN  AT A  .
JTL TSLVZMSP TV TFF.
```

The first two words in the plaintext now appear to be WHEN and THEY. Other resulting apparent words in the plaintext are DON'T, INNOCENT, NOTHING, and ANYTHING. The corresponding plain/cipher letter assignments give the following expanded cipher alphabet.

```
Plain:  A B C D E F G H I J K L M N O P Q R S T U V W X Y Z
Cipher: T _ O Y I _ P Z M _ Q _ _ S R _ _ _ _ V _ _ W _ L _
```

The following again shows the complete ciphertext, along with the part of the plaintext given by our expanded cipher alphabet.

```
WHEN THEY TOOK THE  O  TH A  END ENT, I WA   I ENT  ECA  E
WZIS VZIL VRRQ VZI CRAEVZ TGISYGISV, M WTJ JMFISV BIOTAJI

I DON'T DEA  D  G . WHEN THEY TOOK THE  I TH A  END  ENT, I
M YRS'V YITF YEAPJ. WZIS VZIL VRRQ VZI JMNVZ TGISYGISV, M
```

```
KE T   IET  ECA  E I KNOW I'  INNOCENT. WHEN THEY TOOK THE
QIHV UAMIV BIOTAJI M QSRW M'G MSSROISV. WZIS VZIL VRRQ VZI

 ECOND A END ENT, I  AID NOTHING  ECA  E I DON'T OWN A
JIORSY TGISYGISV, M JTMY SRVZMSP BIOTAJI M YRS'V RWS T

G N. NOW THEY' E CO E  O  THE  I  T A END ENT AND I CAN'T
PAS. SRW VZIL'KI ORGI CRE VZI CMEJV TGISYGISV TSY M OTS'V

 AY ANYTHING AT A  .
JTL TSLVZMSP TV TFF.
```

With just a little more thought, the plain/cipher letter assignments can be completed, yielding the following full plaintext.

```
WHEN THEY TOOK THE FOURTH AMENDMENT, I WAS SILENT BECAUSE
I DON'T DEAL DRUGS. WHEN THEY TOOK THE SIXTH AMENDMENT, I
KEPT QUIET BECAUSE I KNOW I'M INNOCENT. WHEN THEY TOOK THE
SECOND AMENDMENT, I SAID NOTHING BECAUSE I DON'T OWN A
GUN. NOW THEY'VE COME FOR THE FIRST AMENDMENT AND I CAN'T
SAY ANYTHING AT ALL.⁶
```

The following is the resulting cipher alphabet, with two ciphertext letters excluded because there were two letters that did not appear in the plaintext.

```
Plain:  A B C D E F G H I J K L M N O P Q R S T U V W X Y Z
Cipher: T B O Y I C P Z M _ Q F G S R H U E J V A K W N L _
```

This cipher alphabet does not seem to be the result of a simple keyword substitution cipher. For a keyword columnar substitution cipher, the second letter in the keyword would most likely be either the O that appears third in the cipher alphabet, or the I that appears fifth in the cipher alphabet. If it were I, then the array of letters that produced the cipher alphabet would begin as follows.

```
T  I  _  _  _  _  _  _
B  _  _  _  _  _  _  _
O  _  _  _  _  _  _  _
Y  _
```

Continuing to fill columns in this array using the cipher alphabet yields the following.

```
T  I  M  F  R  E  A  N
B  C  _  G  H  J  K  L
O  P  Q  S  U  V  W  _
Y  Z
```

---

⁶Tim Freeman, quote.

Thus, the keyword, with spaces and duplicate letters removed, is `TIMFREAN`, which indeed is the result of removing the space and duplicate letters from `TIM FREEMAN`. Filling in the two missing letters gives the following.

```
T  I  M  F  R  E  A  N
B  C  D  G  H  J  K  L
O  P  Q  S  U  V  W  X
Y  Z
```

This allows the full cipher alphabet to be completed.   □

The reason why we were able to break the cipher in Example 2.4 relatively easily is because a sufficient number of ciphertext letters corresponded to plaintext letters that occur frequently in ordinary English. Having the punctuation and spacing between words preserved made it easier to break as well. A ciphertext with a smaller number of letters or in which punctuation and spacing had been removed could have been much more difficult to cryptanalyze. Substitution ciphers in which entire plaintext words are replaced with numbers or words (known as *nomenclators*) can also be more difficult to break, as can ciphers in languages in which letter frequencies are different from those in English. However, history has shown that most substitution ciphers are insecure and can be broken through persistence.

The Beale Ciphers: Riches to Be Discovered or a Hoax?

A real-life story of buried treasure protected by a cipher centers around the adventures of a man named Thomas Beale. Beale reportedly stayed at a hotel in Lynchburg, Virginia, in 1822, and upon departing left a locked box with the hotel's owner, Robert Morriss. After not hearing from Beale for more than two decades, Morriss broke the box open and found a note along with three ciphertexts. The note told how Beale and 29 other men had discovered a large cache of gold in New Mexico. To keep the treasure safe, Beale transported it to Virginia and buried it. Decrypting the first ciphertext would reveal the treasure's location, the second its contents, and the third a list of relatives who were to share in it. After trying to break the ciphers for years, Morriss shared them with an unknown friend, who broke the second cipher using a key formed from the Declaration of Independence. This revealed not only the treasure's value, at more than $20 million in today's standards, but also that it was buried somewhere near Bedford, Virginia. In 1885 the unknown friend published an anonymous pamphlet disclosing the story.

Beale's first and third ciphers remain unbroken, despite being attacked in earnest by some of the world's greatest cryptanalysts. They were even included in a training program for new recruits at the U.S. Signals Intelligence Service, a precursor to the National Security Agency. Because of this, many believe the treasure is a hoax and that Thomas Beale may have never even existed.

2.2.1 Exercises

1.* Cryptanalyze the following ciphertexts, which were formed using substitution ciphers.

(a) QDXU BUDPHKC HK DUCTDVTANR KPAUD, SUDR XJMDTCTQHG. HQ
HK JPQ QBU XDEU QP KXDWTKK TNN PQBUDK TQ FBTQUSUD
GPKQ, AXQ QBU XDEU QP KUDSU PQBUDK TQ FBTQUSUD
GPKQ.[7]

(b) ZVR XIELR PD QALLRQQ EQ VGIU SPIO, URUELGZEPH ZP ZVR
FPC GZ VGHU, GHU ZVR URZRINEHGZEPH ZVGZ SVRZVRI SR
SEH PI WPQR, SR VGJR GXXWERU ZVR CRQZ PD PAIQRWJRQ ZP
ZVR ZGQO GZ VGHU.[8]

(c) IES QJDEVFMRF MVY FIY WQTLMC MVFT FE TEHQYFN? Q UYYC
TFVERBCN FIMF FIY WQTLMC MVFT MVY EU WMTF MRO
QRHMCHLCMACY QJDEVFMRHY. EU HELVTY Q HELCU AY
DVYZLOQHYO. Q MJ M WQTLMC MVF.[9]

2.* The following ciphertexts were formed using keyword substitution ciphers. For each, cryptanalyze the ciphertext, and find the keyword.

(a) RLW RWQR KD KSP MPKAPWQQ BQ JKR VLWRLWP VW TNN RK RLW
THSJNTJEW KD RLKQW VLK LTUW ISEL. BR BQ VLWRLWP VW
MPKUBNW WJKSAL RK RLKQW VLK LTUW GBRRGW.[10]

(b) COJSHOIOJ, WO'HH AO RTLLORRBTH SDER YOVQ EB YKT LVJ
BKLTR KJ SDQOO SDEJCR, VJN SDQOO SDEJCR KJHY: YKTQ
BVIEHY, YKTQ QOHECEKJ, VJN SDO CQOOJ AVY MVLGOQR.[11]

(c) T BRXTRJR ZWAZ AZ ZWR RGV PD ZWR MRGZCHK ZWR CQR PD
SPHVQ AGV NRGRHAX RVCMAZRV PYTGTPG STXX WAJR AXZRHRV
QP ICMW ZWAZ PGR STXX BR ABXR ZP QYRAO PD IAMWTGRQ
ZWTGOTGN STZWPCZ RLYRMZTGN ZP BR MPGZHAVTMZRV.[12]

(d) DUQT QFBL JUT TAS ORCE, DRCE, HRES R ISW LFRI, QTRI,
YJU KJI'T ISSK TJ OS CJY, PJY, DUQT GST YJUPQSFN
NPSS. AJL JI TAS OUQ, GUQ, YJU KJI'T ISSK TJ KBQCUQQ
HUCA, DUQT KPJL JNN TAS ESY, FSS, RIK GST YJUPQSFN
NPSS.[13]

---

[7] Arthur Ashe (1943–1993), quote.
[8] Vince Lombardi (1913–1970), quote.
[9] Kermit the Frog, quote.
[10] Franklin Delano Roosevelt (1882–1945), quote.
[11] Jim Valvano (1946–1993), quote.
[12] Alan Turing (1912–1954), from *Computing Machinery and Intelligence*.
[13] Paul Simon, from "Fifty Ways to Leave Your Lover."

3.* Cryptanalyze the following ciphertext, which was formed using a substitution cipher, with word divisions (punctuation and spaces) removed from the plaintext during encryption.

```
HTLYHKYVYNVYHKVYYHKRUDAEYNFFAKTOFQQTYWYVVPQNPEYAKTOFQQ
THKRAYWYVPQNPTCTOVFRWYAUOLBYVTUYRAFNODKPTOAKTOFQFNODK
YWYVPQNPUOLBYVHETRAHKRUSPTOAKTOFQAGYUQATLYHRLYRUHKTOD
KHNUQUOLBYVHKVYYRAPTOAKTOFQKNWYPTOVYLTHRTUALTWYQHTHYN
VAJTOFQBYKNGGRUYAATVZTPBOHHKRUSNBTOHRHRCPTOFNODKPTOHK
RUSNUQPTOJVPHKNHANCOFFQNPHKNHANKYJSTCNQNPPTOQTHKNHAYW
YUQNPANEYYSPTOVYDTRUDHTKNWYATLYHKRUDAGYJRNF¹⁴
```

4.* The following ciphertext is from the original version of Edgar Allan Poe's short story "The Gold Bug." It was formed using a substitution cipher, with word divisions (punctuation and spaces) removed from the plaintext during encryption, and non-letter characters used in the ciphertext. Cryptanalyze the ciphertext.

```
53++!305))6*;4826)4+.)4+);806*;48!8'60))85;1+(;:+*8!8
3(88)5*!;46(;88*96*?;8)*+(;485);5*!2:*+(;4956*2(5*-4)
8'8*;4069285);)6!8)4++;1(+9;48081;8:8+1;48!85;4)485!5
28806*81(+9;48;(88;4(+?34;48)4+;161;:188;+?;
```

5. Suppose a plaintext P is encrypted, yielding ciphertext M, and then M is encrypted, yielding a new ciphertext C. Encrypting M is called *superencrypting* P.

(a)* Use the ciphers in Examples 2.2 on page 8 and 2.3 on page 9 (in that order) to superencrypt I WAS IN THE POOL.¹⁵

(b) Decrypt O VWC OQ UZI BAAD, which was superencrypted using the ciphers in Examples 2.3 and 2.2 (in that order).

(c) Does superencryption by two substitution ciphers yield more security than encryption by one substitution cipher? In other words, if a plaintext P is encrypted using a substitution cipher, yielding M, and then M is encrypted using another substitution cipher, yielding C, would C be harder in general to cryptanalyze than M? Explain your answer completely, and be as specific as possible.

6. Find some additional information about the Beale ciphers, and write a summary of your findings.

---

¹⁴Jim Valvano, quote.
¹⁵George Costanza, quote.

7. Find some information about the role of the Babington plot and crypt-analysis in the life and death of Mary, Queen of Scots, and write a summary of your findings.

8. Find some information about the encrypted messages sent by the Zodiac killer to the San Francisco Bay area press in 1969–1970, and write a summary of your findings.

2.3 Playfair Ciphers

Substitution ciphers would be less susceptible to attack by frequency analysis if plaintext characters were encrypted in pairs (i.e., digraphs). This is the basis for *Playfair* ciphers. Playfair ciphers were first described in 1854 by English scientist and inventor Sir Charles Wheatstone, but are named for Scottish scientist and politician Baron Lyon Playfair, Wheatstone's friend, who argued for their use by the British government. Although initially rejected because of their perceived complexity, Playfair ciphers were eventually used by the British military during the Second Boer War and World War I, and by British intelligence and the militaries of several countries, including both the United States and Germany, during World War II.

Playfair ciphers use one or more keywords. Spaces and duplicate letters in the keyword(s) are removed, and the resulting letters are then used to form an array of letters, similar to the array used in keyword columnar substitution ciphers, except that for Playfair ciphers this array must always have exactly five letters per row. Also, I and J are considered to be the same letter in Playfair arrays, so J is not included. The reason for this is so that Playfair arrays will always form perfect squares of size 5 × 5. (That is, Playfair arrays will always have exactly five rows and five columns.)

Example 2.5 Consider a Playfair cipher with keyword WHEATSTONE. Removing the duplicate letters in this keyword gives WHEATSON, and using these letters to form the array for a Playfair cipher yields the following.

```
W H E A T
S O N B C
D F G I K
L M P Q R
U V X Y Z
```

We will use a Playfair cipher with this array to encrypt a message in Example 2.7. □

To encrypt a message using a Playfair cipher, spaces are removed from the plaintext, and the plaintext is then split into digraphs. If any digraphs contain repeated letters, an X is inserted in the plaintext between the first pair of repeated letters that were grouped together in a digraph, and the plaintext is again split into digraphs. This process is repeated if necessary and as many times as necessary until no digraphs contain repeated letters. Finally, if necessary, an X is inserted at the end of the plaintext so that the last letter is in a digraph.

Example 2.6 Consider the message IDIOCY OFTEN LOOKS LIKE INTELLI-GENCE. To encrypt this message using a Playfair cipher, we begin by splitting the plaintext into digraphs. This yields the following.

ID IO CY OF TE NL OO KS LI KE IN TE LL IG EN CE

The seventh digraph is the first one that contains repeated letters. Thus, we insert an X between these letters and again split the plaintext into digraphs. This yields the following.

ID IO CY OF TE NL OX OK SL IK EI NT EL LI GE NC E

None of these digraphs contain repeated letters, but now we must insert an X at the end of the plaintext so that the last letter will be in a digraph. This yields the following.

ID IO CY OF TE NL OX OK SL IK EI NT EL LI GE NC EX

We are now ready to encrypt this message using a Playfair cipher, which we will do in Example 2.7.   □

In a Playfair cipher, the 5 × 5 array of letters is used to convert plaintext digraphs into ciphertext digraphs according to the following rules.

- If the letters in a plaintext digraph are in the same row of the array, then the ciphertext digraph is formed by replacing each plaintext letter with the letter in the array in the same row but one position to the right, wrapping from the end of the row to the start if necessary. For example, using the array in Example 2.5, the plaintext digraph ID encrypts to the ciphertext digraph KF, and IK encrypts to KD.

- If the letters in a plaintext digraph are in the same column of the array, then the ciphertext digraph is formed by replacing each plaintext letter with the letter in the array in the same column but one position down, wrapping from the bottom of the column to the top if necessary. For example, using the array in Example 2.5, OF encrypts to FM, and EX encrypts to NE.

- If the letters in a plaintext digraph are not in the same row or column of the array, then the ciphertext digraph is formed by replacing the first plaintext letter with the letter in the array in the same row as the first plaintext letter and the same column as the second plaintext letter, and replacing the second plaintext letter with the letter in the array in the same row as the second plaintext letter and the same column as the first plaintext letter. For example, using the array in Example 2.5, IO encrypts to FB, and OX encrypts to NV.

Example 2.7 The Playfair cipher with keyword WHEATSTONE (for which the array is given in Example 2.5) encrypts the plaintext IDIOCY OFTEN LOOKS LIKE INTELLIGENCE as follows.

> **Plain:** ID IO CY OF TE NL OX OK SL IK EI NT EL LI GE NC EX
> **Cipher:** KF FB BZ FM WA SP NV CF DU KD AG CE WP QD PN BS NE

For decryption, the rules for encryption are reversed. (The first decryption rule is identical to the first encryption rule except letters one position to the left are chosen, wrapping from the start of the row to the end. The second decryption rule is identical to the second encryption rule except letters one position up are chosen, wrapping from the top of the column to the bottom. The third decryption rule is identical to the third encryption rule.)   □

For cryptanalysis, because Playfair ciphers encrypt digraphs, single-letter frequency analysis is in general not helpful. (Note in Example 2.7 the plaintext letters I and O both correspond to four different ciphertext letters.) However, when used to encrypt long messages, it is sometimes possible to break Playfair ciphers using frequency analysis on digraphs, since identical plaintext digraphs will always encrypt to identical ciphertext digraphs. Other weaknesses are that a plaintext digraph and its reverse (e.g., AB and BA) will always encrypt to a ciphertext digraph and its reverse, and that for short keywords the bottom rows of the array may be predictable.

2.3.1   Exercises

1. Consider a Playfair cipher with keyword SEINFELD.

   (a)* Use this cipher to encrypt THE SMELLY CAR.

   (b) Use this cipher to encrypt THE BIZARRO JERRY.

   (c) Decrypt QMSHKZHCILKBXARBIY, which was formed using this cipher.

2. Consider a Playfair cipher with keywords CLINT EASTWOOD.

   (a)* Use this cipher to encrypt DIRTY HARRY IS A CLASSIC.

    (b) Use this cipher to encrypt A FISTFUL OF DOLLARS IS GOOD TOO.

    (c) Decrypt ORAEZCABSNEWWUOSCAFSAFOCCOQZOC, which was formed using this cipher.

3. Create a Playfair cipher and use it to encrypt a plaintext of your choice with at least 20 letters.

4. In Walt Disney Pictures' 2007 movie *National Treasure: Book of Secrets*, a man named Thomas Gates (the great-great-grandfather of treasure hunter Benjamin Franklin Gates, the main character in the movie) is asked by John Wilkes Booth and a colleague to decrypt the ciphertext MEIKQOTXCQTEZXCOMWQCTEHNFBIKMEHAKRQCUNGIKMAV, which was formed using a Playfair cipher with keyword DEATH. Decrypt this ciphertext.

5. On August 2, 1943, the Japanese destroyer *Amagiri* rammed and sank the American patrol boat *PT-109*, which was under the command of U.S. Naval Reserve Lieutenant and future President John F. Kennedy. After reaching shore, Kennedy sent the following ciphertext, which was formed using a Playfair cipher with keywords ROYAL NEW ZEALAND NAVY. Decrypt this ciphertext.

    KXIEYUREBEZWEHEWRYTUHEYFSKREHEGOYFIWUQUTQYOMUQYCAIPOB
    OTEIZONTXBYBNTGONEYCUZWRGDSONSXBOUYWRHEBAAHYUSEDQ

6. Find a copy of Dorothy Sayers' novel *Have His Carcase*, and write a summary of how a Playfair cipher is integrated into the story and the steps described in it for breaking a Playfair cipher.

7. A description of cryptanalysis of Playfair ciphers can be found in U.S. Army Field Manual 34-40-2 [24]. Find a copy of this manual, and write a summary of how it describes Playfair cipher cryptanalysis.

8. Find some information about two-square and four-square ciphers, and write a summary of your findings.

2.4   The Navajo Code

While simple substitution ciphers are not very secure, and even ciphers such as Playfair that substitute for digraphs can be broken through a type of frequency analysis on longer ciphertexts, not all ciphers based on substitution alone are easy to break. The Navajo code, a cipher famously created by Native Americans, primarily from the Navajo Nation that occupies a large region of Utah, Arizona, and New Mexico, and used effectively by the Americans throughout the Pacific Campaign during World War II, was essentially a substitution cipher. The Navajo language was at the time exclusively oral, very complex, and unknown to virtually everyone outside the Navajo Nation. The idea of using Navajos basically speaking their native language as a means for encrypting messages originated in 1942 with a man named Philip Johnston. Having grown up the son of a missionary to the Navajo, Johnston was very familiar with the Navajo culture, and was one of only a handful of non-Navajos who spoke the Navajo language fluently.

Johnston was a veteran of World War I, where he may have seen Native Americans, specifically from the Choctaw Nation, encrypting messages for the U.S. Army basically by speaking their native language. More likely, after the attack on Pearl Harbor, which thrust the United States into World War II, Johnston read of the use of Choctaw by the U.S. Army. Whatever the origin of his idea, Johnston recruited four Navajos to demonstrate to a group of U.S. Marine officers how they could quickly and flawlessly translate English messages into the Navajo language, communicate these messages to each other via radio, and then translate these messages back into English. Convinced of the potential of the Navajo language, the Marines ordered a pilot program in which an eight-week communications training course was completed by a group of 29 Navajos, who became the original Navajo code talkers. A graduation picture from this training course is shown in Figure 2.1.

Figure 2.1   Graduation picture of the original Navajo code talkers.

Before this training course could commence, the Marines had to figure out a way to overcome a problem that had plagued attempts at using Native American languages as a means for encrypting messages during World War I—many military words, for example, SUBMARINE and DIVE BOMBER, had no translatable equivalent in Native American languages. To overcome this problem, the Navajo trainees decided that they would indicate such military words using literal English translations of things in the natural world for which they had Navajo translations. For example, the word SUBMARINE was given the literal English translation IRON FISH, which was translatable in the Navajo language as BESH-LO. A few examples of military words and some other words, the literal English translations of these words used by the Navajo code talkers, and the Navajo translations, or *code words*, of these literal translations are shown in Table 2.2.

English word	Literal translation	Navajo code word
ABANDON	RUN AWAY FROM	YE-TSAN
AMERICA	OUR MOTHER	NE-HE-MAH
ASSAULT	FIRST STRIKER	ALTSEH-E-JAH-HE
BATTALION	RED SOIL	TACHEENE
BRITAIN	BETWEEN WATERS	TOH-TA
CAPTAIN	TWO SILVER BARS	BESH-LEGAI-NAH-KIH
DIVE BOMBER	CHICKEN HAWK	GINI
GERMANY	IRON HAT	BESH-BE-CHA-HE
ORDER	ORDER	BE-EH-HO-ZINI
SAILORS	WHITE CAPS	CHA-LE-GAI
SUBMARINE	IRON FISH	BESH-LO
THE	BLUE JAY	CHA-GEE

Table 2.2   Navajo code words for selected English words.

An encoded phonetic alphabet was also created so less common English words could be translated one letter at a time. Individual letters in such words were also indicated by literal English translations of things for which translations existed in the Navajo language, and then the words were encoded one letter at a time using these translations. The individual letters, the literal English translations of these letters used by the Navajo code talkers, and the Navajo translations of these literal translations are shown in Table 2.3 on page 25. Multiple translations were used for most letters to increase the difficulty of attacking the code using frequency analysis.

By the conclusion of World War II, the full Navajo code included approximately 800 code words. A list of the code words in the full code can be found at https://www.history.navy.mil/research/library/online-reading-room/title-list-alphabetically/n/navajo-code-talker-dictionary.html [17].

	Literal	Code word		Literal	Code word
A	ANT	WOL-LA-CHEE	K	KID	KLIZZIE-YAZZIE
A	APPLE	BE-LA-SANA	L	LAMB	DIBEH-YAZZIE
A	AXE	TSE-NILL	L	LEG	AH-JAD
B	BADGER	NA-HASH-CHID	L	LION	NASH-DOIE-TSO
B	BARREL	TOISH-JEH	M	MATCH	TSIN-TLITI
B	BEAR	SHUSH	M	MIRROR	BE-TAS-TNI
C	CAT	MOASI	M	MOUSE	NA-AS-TSO-SI
C	COAL	TLA-GIN	N	NEEDLE	TSAH
C	COW	BA-GOSHI	N	NOSE	A-CHIN
D	DEER	BE	N	NUT	NESH-CHEE
D	DEVIL	CHINDI	O	OIL	A-KIIA
D	DOG	LHA-CHA-EH	O	ONION	TLO-CHIN
E	EAR	AH-JAH	O	OWL	NE-AHS-JAH
E	ELK	DZEH	P	PANT	CLA-GI-AIH
E	EYE	AH-NAH	P	PIG	BI-SO-DIH
F	FIR	CHUO	P	PRETTY	NE-ZHONI
F	FLY	TSA-E-DONIN-EE	Q	QUIVER	CA-YEILTH
F	FOX	MA-E	R	RABBIT	GAH
G	GIRL	AH-TAD	R	RAM	DAH-NES-TSA
G	GOAT	KLIZZIE	R	RICE	AH-LOSZ
G	GUM	JEHA	S	SHEEP	DIBEH
H	HAIR	TSE-GAH	S	SNAKE	KLESH
H	HAT	CHA	T	TEA	D-AH
H	HORSE	LIN	T	TOOTH	A-WOH
I	ICE	TKIN	T	TURKEY	THAN-ZIE
I	INTESTINE	A-CHI	U	UNCLE	SHI-DA
I	ITCH	YEH-HES	U	UTE	NO-DA-IH
J	JACKASS	TKELE-CHO-GI	V	VICTOR	A-KEH-DI-GLINI
J	JAW	AH-YA-TSINNE	W	WEASEL	GLOE-IH
J	JERK	YIL-DOI	X	CROSS	AL-NA-AS-DZOH
K	KETTLE	JAD-HO-LONI	Y	YUCCA	TSAH-AS-ZIH
K	KEY	BA-AH-NE-DI-TININ	Z	ZINC	BESH-DO-TLIZ

Table 2.3 Navajo code words for alphabet letters.

Example 2.8 Consider the plaintext THE DIVE BOMBER SANK THE SUB-MARINE. The English word SANK was translated in Navajo one letter at a time. Thus, by Tables 2.2 and 2.3, one possible ciphertext of Navajo code words for this message is CHA-GEE GINI DIBEH TSE-NILL A-CHIN KLIZZIE-YAZZIE CHA-GEE BESH-LO. The literal English translation of this ciphertext is BLUE JAY CHICKEN HAWK SHEEP AXE NOSE KID BLUE JAY IRON FISH.   □

Example 2.9 Consider the ciphertext of Navajo code words BESH-LEGAI-NAH-KIH WOL-LA-CHEE LIN BE-LA-SANA TOISH-JEH KLIZZIE BE-LA-SANA A-KEH-DI-GLINI AH-NAH CHA-GEE BE-EH-HO-ZINI. The literal English translation of this ciphertext is TWO SILVER BARS ANT HORSE APPLE BARREL GOAT APPLE VICTOR EYE BLUE JAY ORDER. By Tables 2.2 and 2.3, the plaintext for this message is CAPTAIN AHAB GAVE THE ORDER.   □

In the field while being used as a cipher by the U.S. Marine Corps, the Navajo code was completely oral and never written down. As a result, each code talker, of which there were more than 400 by the end of World War II, had to know every code word by memory. This was not difficult for the Navajos, though, since their language had never had a written script. William McCabe, one of the original 29 code talkers, noted: "In Navajo, everything is in the memory—songs, prayers, everything. That's the way we were raised."

Carl and Zonnie Gorman: Navajo Code Hero and Historian

(Photo courtesy Carl N. Gorman Family collection.)   (Photo courtesy Anthony Anaya-Gorman.)

Dr. Carl Nelson Gorman was a respected artist and educator, and one of the original 29 Navajo code talkers. Like many Native Americans in the first half of the 20th century, Gorman was punished severely as a child for speaking his native language, including being chained to an iron pipe for a week for doing so on the grounds of his mission school. After the war Gorman attended the Otis Art Institute, and became a founding faculty of Native American Studies at the University of California, Davis.

To share her father's legacy and that of all the Navajo code talkers, Gorman's daughter Zonnie has lectured extensively throughout the United States and Canada. Ms. Gorman has appeared in documentaries on both The History Channel and PBS about these famous Navajo communicators. She is recognized for conducting the first extensive interviews with the original 29 code talkers, and her research into the code continues, as she looks for new and important insights about the men and their code.

The speed, accuracy, and security of the Navajo code proved it highly successful. Messages that would have taken hours to encrypt or decrypt using rotor machines like the Enigma were encrypted or decrypted in just minutes using the Navajo code. Just as importantly, the Navajo code is not known to have ever been broken, and it played a critical role in the American success in the Pacific Campaign during World War II. U.S. Major Howard Connor, 5th Marine Division signal officer at Iwo Jima, noted: "Were it not for the Navajos, the Marines would never have taken Iwo Jima." The Navajo code continued to be used successfully by the Americans in the Korean War and in the early stages of the Vietnam War.

The dedication and loyalty of the Navajo code talkers was remarkable, from their tireless work in making the code a success to their humility in keeping their role in its success a secret. Not until years after the code was declassified in 1968 did the code talkers begin to receive the recognition they deserved. In 1982, President Ronald Reagan signed a resolution declaring August 14 National Navajo Code Talkers Day. On July 26, 2001, President George W. Bush presented the original 29 code talkers the Congressional Gold Medal, the highest civilian award in the United States. Four of the five surviving original code talkers at the time were in attendance. The last surviving original code talker, Chester Nez, received his B.F.A. degree from the University of Kansas on November 12, 2012, more than 60 years after having to drop out when his GI Bill funds expired. Nez, who is the author of *Code Talker: The First and Only Memoir by One of the Original Navajo Code Talkers of WWII* [18], passed away on June 4, 2014.

2.4.1 Exercises

1.* For the plaintext in Example 2.8, use the part of the Navajo code in Tables 2.2 and 2.3 to find two ciphertexts different from the one in this example, and give the literal English translation of each.

2.* Repeat Exercise 1 using the plaintext in Example 2.9.

3. For the following plaintexts, use the part of the Navajo code in Tables 2.2 and 2.3 to encrypt the plaintext, and give the literal English translation of the resulting ciphertext.

   (a)* ABANDON HOPE.

   (b) AMERICA THE BEAUTIFUL

4. For the following ciphertexts, which were formed using the Navajo code, use Tables 2.2 and 2.3 to give the literal English translation, and decrypt the ciphertext.

   (a)* NA-AS-TSO-SI BE-LA-SANA BE DZEH YEH-HES TSAH TOH-TA

(b) TSIN-TLITI TSE-NILL CHINDI AH-NAH TKIN A-CHIN BESH-BE-
    CHA-HE

5. For the following plaintexts, use the full Navajo code given in [17] to
   encrypt the plaintext, and give the literal English translation of the
   resulting ciphertext.

   (a)* TANK AMMUNITION DEPLETED.

   (b) POSITIVE ON AERIAL RECONNAISSANCE

6. For the following ciphertexts, which were formed using the Navajo
   code, use the full Navajo code given in [17] to give the literal English
   translation, and decrypt the ciphertext.

   (a)* AL-TAH-JE-JAY DIBEH SHI-DA GAH TKIN NA-HASH-CHID WOL-
        LA-CHEE MOASI TSE-GAH YEH-HES AH-DI HA-YELI-KAHN

   (b) KLESH NO-DA-IH AH-LOSZ A-CHI SHUSH BE-LA-SANA TLA-GIN
       LIN TKIN BIN-KIE-JINH-JIH-DEZ-JAY UT-ZAH-HA-DEZ-BIN

7. Use the full Navajo code in [17] to encrypt a plaintext of your choice
   with at least four words for which Navajo code words existed and at
   least two words that were translated in Navajo one letter at a time.

8. Find some information about the efforts by Carl and Zonnie Gorman
   to obtain recognition for the Navajo code talkers, and summarize your
   findings.

9. The pilot program completed by the original 29 Navajo code talkers
   was designed for 30 recruits, however one of the 30 Navajos recruited
   for the program never entered the program and did not become a
   code talker. The identity and fate of this 30th recruit, as well as his
   reasons for not entering the program, were lost to history until very
   recently, when they were discovered by Zonnie Gorman during a war
   archive search. Find some information about this 30th recruit, and
   summarize your findings.

10. Find some information about the Cherokee, Choctaw, and Comanche
    code talkers used by the U.S. military, and summarize your findings.

Chapter 3

Transposition Ciphers

In substitution ciphers, plaintext letters are encrypted by being replaced by other letters given by correspondences in a cipher alphabet. Transposition ciphers differ from substitution ciphers in that plaintext letters are not encrypted by being replaced by other letters, but rather by being rearranged according to some rule agreed upon by the two parties wishing to exchange the message. That is, to form the ciphertext for a transposition cipher, the plaintext letters are rearranged in some manner, as opposed to being replaced by other letters.

Transposition ciphers, like substitution ciphers, are not very secure, but have a rich history of being used. The scytale cipher used in ancient Greece was a transposition cipher. Transposition ciphers have also been included as parts of larger ciphers, such as the ADFGX and ADFGVX ciphers used by Germany during World War I, which we will consider in this chapter, and the Data and Advanced Encryption Standards, both of which were selected in recent years by the National Institute of Standards and Technology to serve as Federal Information Processing Standards.

3.1   Columnar Transposition Ciphers

For columnar transposition ciphers, users agree upon some prescribed number of columns, and then the actual plaintext letters (with spaces and punctuation removed) are used to form an array of letters, similar to the array used in keyword columnar substitution ciphers, with this number of columns. The ciphertext is obtained by taking the columns of the resulting array in some specified order, and placing the letters in these columns in a row. In the rest of this section, we will demonstrate two methods for doing this.

3.1.1　Simple Columnar Transposition Ciphers

For simple columnar transposition ciphers, the ciphertext is obtained by taking the columns of the array in order, starting from the left, and placing the letters in these columns in a row.

Example 3.1 Consider a simple columnar transposition cipher with six columns. To use this cipher to encrypt the plaintext YOU CANNOT SIMULTA-NEOUSLY PREVENT AND PREPARE FOR WAR,[1] we begin by using these plaintext letters to form the following array.

```
Y   O   U   C   A   N
N   O   T   S   I   M
U   L   T   A   N   E
O   U   S   L   Y   P
R   E   V   E   N   T
A   N   D   P   R   E
P   A   R   E   F   O
R   W   A   R
```

To form the ciphertext, we transcribe this array by columns starting from the left. Thus, the ciphertext is YNUORAPROOLUENAWUTTSVDRACSALEPERAINY NRFNMEPTEO.　　　　　　　　　　　　　　　　　　　　　　□

To decrypt a ciphertext that was formed using a transposition cipher, it helps if the ciphertext is expressed in blocks of equal sizes. For example, the ciphertext in Example 3.1 expressed in blocks of five letters each would be YNUOR APROO LUENA WUTTS VDRAC SALEP ERAIN YNRFN MEPTE O (with the final block shorter than the rest because the end of the message was reached). The reason why expressing a ciphertext in blocks of equal sizes helps with decryption is because the number of letters per column in the array must be determined, and to find this the total number of letters in the ciphertext must be known. Expressing the ciphertext in blocks of equal sizes allows the total number of letters in the ciphertext to be found quickly and reliably. Expressing the ciphertext in blocks of equal sizes can also help to reduce errors in copying messages. In the following example, we illustrate how the number of letters per column in the array for a transposition cipher can be determined.

Example 3.2 Consider the ciphertext DAIST SEROB FIIUA ENOFN CRRAO UIMSE ETTTC AIYSE WYUTC OTROO LHAUI RUTEN MLRRI MAILY DEASN G, which was formed using a simple columnar transposition cipher with 10 columns. Note that there are 76 letters in this message. Dividing the

---

[1] Albert Einstein (1879–1955), quote.

number of columns $c = 10$ into the number of letters $n = 76$ yields the following.

$$10 \overline{)\ 76} \atop \displaystyle \frac{-70}{6} \atop \displaystyle ^7}$$

In this division, the quotient is $q = 7$ and the remainder is $r = 6$. This remainder indicates that of the 10 columns in the array, the first six (the value of r) will be one letter longer than the last four. Also, the value of this quotient is the number of letters in the last four columns. That is, the last $c - r = 10 - 6 = 4$ columns will contain $q = 7$ letters, and thus the first $r = 6$ columns will contain $q + 1 = 8$ letters. So to decrypt this message, we should use the first eight letters in the ciphertext to form the first column of the array, the next eight ciphertext letters to form the second column, the next eight letters to form the third column, and so forth through the sixth column. After six columns have been formed, we should then change to using only seven letters per column for the rest of the array. This yields the following.

```
D  O  N  O  T  W  O  R  R  Y
A  B  O  U  T  Y  O  U  R  D
I  F  F  I  C  U  L  T  I  E
S  I  N  M  A  T  H  E  M  A
T  I  C  S  I  C  A  N  A  S
S  U  R  E  Y  O  U  M  I  N
E  A  R  E  S  T  I  L  L  G
R  E  A  T  E  R
```

Transcribing this array by rows starting from the top gives the plaintext DO NOT WORRY ABOUT YOUR DIFFICULTIES IN MATHEMATICS; I CAN ASSURE YOU MINE ARE STILL GREATER.[2]   □

In general, for a simple columnar transposition cipher with c columns and a ciphertext with n letters, to find the number of letters per column in the array, we divide c into n, obtaining quotient q and remainder r.

$$c \overline{)\ n} \atop \displaystyle \frac{-q \cdot c}{r} \atop \displaystyle ^q}$$

Then the first r columns of the array will contain $q + 1$ letters, and the remaining $c - r$ columns will contain q letters.

---

[2] Albert Einstein, quote.

Example 3.3 Consider a ciphertext with 47 letters formed using a simple columnar transposition cipher with nine columns. Dividing 9 into 47 yields the following.

$$
\begin{array}{r}
5 \\
9 \overline{)\ 47} \\
-45 \\
\hline
2
\end{array}
$$

Thus, the first two columns of the array will contain six letters, and the remaining seven columns will contain five letters.   □

One problem with simple columnar transposition ciphers is that ciphertexts are always obtained by taking the columns of the array in order starting from the left. This can make a cipher more vulnerable to cryptanalysis. Keyword columnar transposition ciphers can help to alleviate this problem.

3.1.2   Keyword Columnar Transposition Ciphers

For keyword columnar transposition ciphers, users agree upon one or more keywords, and remove spaces in the keyword(s). However, unlike for keyword substitution ciphers, for keyword transposition ciphers duplicate letters are not removed from the keyword(s). The number of columns in the array is then equal to the number of keyword letters, with the keyword letters placed in order as labels on the columns, and the ciphertext obtained by taking the columns of the array (not including the keyword letter labels) in alphabetical order by the keyword letter labels, and placing the letters in these columns in a row. If the keyword(s) contain any duplicate letters, then columns with identical keyword letter labels are taken in order starting from the left. Also, to make decryption easier, users can choose to include extra characters at the end of a message so that each column in the array will contain the same number of letters. This is called *padding* the message.

Example 3.4 Consider a keyword columnar transposition cipher with keyword CHURCHILL. To use this cipher to encrypt the plaintext ENDING A SENTENCE WITH A PREPOSITION IS SOMETHING UP WITH WHICH I WILL NOT PUT,[3] we begin by using these keyword and plaintext letters to form the following array. The numbers above the keyword letter labels indicate the order in which the columns should be taken to form the ciphertext. Also, we have padded the plaintext with Xs so that each column in the array will contain the same number of letters.

---

[3]Winston Churchill (1874–1965), quote.

```
1   3   9   8   2   4   5   6   7
C   H   U   R   C   H   I   L   L
E   N   D   I   N   G   A   S   E
N   T   E   N   C   E   W   I   T
H   A   P   R   E   P   O   S   I
T   I   O   N   I   S   S   O   M
E   T   H   I   N   G   U   P   W
I   T   H   W   H   I   C   H   I
W   I   L   L   N   O   T   P   U
T   X   X   X   X   X   X   X   X
```

To form the ciphertext, we transcribe this array by columns chosen in alphabetical order by the keyword letter labels (or, equivalently, in numerical order by the numbers above the keyword letter labels). Thus, the ciphertext expressed in blocks of five letters each is ENHTE IWTNC EINHN XNTAI TTIXG EPSGI OXAWO SUCTX SISOP HPXET IMWIU XINRN IWLXD EPOHH LX. □

Example 3.5 Consider the ciphertext TILTE ICTSE HSFED TNAGE XUIIM IRXIA TAAFX TLHTB EXAET GKDN, which was formed using a keyword columnar transposition cipher with keyword WINSTON. Since there are 49 letters in this message and seven letters in this keyword, each column in the array will contain $49/7 = 7$ letters. Thus, we split the ciphertext into the following blocks of length seven. The numbers under the blocks indicate the keyword letter positions (when ordered alphabetically) under which the blocks should be placed as columns in the array.

TILTEIC TSEHSFE DTNAGEX UIIMIRX IATAAFX TLHTBEX AETGKDN
   1   2   3   4   5   6   7

These blocks and the keyword letters yield the following array.

```
7   1   2   5   6   4   3
W   I   N   S   T   O   N
A   T   T   I   T   U   D
E   I   S   A   L   I   T
T   L   E   T   H   I   N
G   T   H   A   T   M   A
K   E   S   A   B   I   G
D   I   F   F   E   R   E
N   C   E   X   X   X   X
```

Thus, the plaintext is ATTITUDE IS A LITTLE THING THAT MAKES A BIG DIFFERENCE.[4]

---

[4] Winston Churchill, quote.

3.1.3 Exercises

1. Consider a simple columnar transposition cipher with five columns.

   (a)* Use this cipher to encrypt LAKE PLACID IS IN UPSTATE NEW YORK.

   (b) Decrypt IGOIA UNNOU RCNST RSOKT HGANM AEEDD OI, which was formed using this cipher.

2. Consider a simple columnar transposition cipher with seven columns.

   (a) Use this cipher to encrypt USA OVER USSR IN HOCKEY IN LAKE PLACID WAS A STUNNING UPSET, padded with Xs (if necessary) so each column in the array will have the same number of letters.

   (b) Decrypt SSNNH DLIRB GIAIN HYORY ECASA TSRED ELEEX UWVSE AXSOE TNRX, which was formed using this cipher.

3. Consider a simple columnar transposition cipher with nine columns.

   (a) Use this cipher to encrypt I THINK THERE IS A SOLUTION TO ALL THESE PROBLEMS; IT'S JUST ONE, AND IT'S EDUCATION.[5]

   (b) Decrypt IAAOT RMEHD TWHTY DAYHT IICUV SAHSN ACEEN ASGUA AEDTU MSTLN IDPEE IRDKE PIOOE ENAON FN,[6] which was formed using this cipher.

4. Create a simple columnar transposition cipher and use it to encrypt a plaintext of your choice with at least 30 letters.

5. Consider a keyword columnar transposition cipher with the keyword BARNEY.

   (a) Use this cipher to encrypt BARNEY FIFE WAS A VERY INEPT DEPUTY.

   (b) Decrypt ERAEH NHALN TUSCY WSAIL TIWTR PHPUI IG, which was formed using this cipher.

6. Consider a keyword columnar transposition cipher with the keyword MAYBERRY.

   (a)* Use this cipher to encrypt THE ANDY GRIFFITH SHOW WAS SET IN RURAL MAYBERRY NORTH CAROLINA, padded with Xs (if necessary) so each column in the array will have the same number of letters.

   (b) Decrypt NTUAP FRYWT TRRYS NAHAM XAEOW SNEHO IETAX OFRII YXDON SIORM MYNOB X, which was formed using this cipher.

---

[5] Malala Yousafzai, quote.
[6] Malala Yousafzai, quote.

7. Consider a keyword columnar transposition cipher with the keyword EDUCATION.

   (a) Use this cipher to encrypt YOU KNOW THERE IS A PROBLEM WITH THE EDUCATION SYSTEM WHEN YOU REALIZE THAT OUT OF THE THREE R'S, ONLY ONE BEGINS WITH AN R.[7]

   (b) Decrypt DFAOT OSTRU EAOCI IYILP OIHGS AAESE ISCTA SCHLE WTTAA CEINT TTVRI TAUTB DEANM GNLDS AOHHE ONVAL NEUEH EFVEE ITTSP ATYH,[8] which was formed using this cipher.

8. Create a keyword columnar transposition cipher and use it to encrypt a plaintext of your choice with at least 30 letters.

9. Create and describe a method different from those illustrated in this section for forming a transposition cipher. Give at least one example of your method.

10. With the *rail fence* cipher, a plaintext is written across a page in a zigzag pattern, and the ciphertext is formed by transcribing the letters in the resulting pair of rows starting from the top. For example, to encrypt the plaintext NC STATE WOLFPACK using the rail fence cipher, we would use the following zigzag pattern.

```
    N     S     A     E     O     F     A     K
       C     T     T     W     L     P     C     X
```

Thus, the ciphertext is NSAEO FAKCT TWLPC X. To decrypt a ciphertext that was formed using the rail fence cipher, the ciphertext can be split in half, with the letters in each half staggered in a zigzag pattern, and then the plaintext found by reading across the zigzag pattern.

   (a) Use the rail fence cipher to encrypt CHUCK NORRIS IS A TOUGH GUY.

   (b) Decrypt BTLNE SWOMY EOGEU CITAT ODABT UHR, which was formed using the rail fence cipher.

   (c) Explain why the rail fence cipher is a special case of a simple columnar transposition cipher.

11. Find some information about the scytale cipher that was used in ancient Greece, including how it worked, and write a summary of your findings.

---

[7]Dennis Miller, quote.
[8]Russell Green (1933–2012), quote.

12. Find some information about how route ciphers work, and the Union route cipher that was used during the American Civil War, and write a summary of your findings.

13. Find some information about how double transposition ciphers work, and the double transposition ciphers that were used during World Wars I and II, and write a summary of your findings.

3.2   Cryptanalysis of Transposition Ciphers

3.2.1   Cryptanalysis of Simple Columnar Ciphers

For a simple columnar transposition cipher, the only key is the number of columns in the array. Thus, for a ciphertext formed using a simple columnar transposition cipher, the cipher can usually be broken by a brute force attack, meaning we would try arrays with various numbers of columns (in some systematic fashion) until obtaining the correct plaintext.

Example 3.6 Consider the ciphertext TANHY IHWGA MEEAR TYNSS YDFDE ATARS, which was formed using a simple columnar transposition cipher. To decrypt this message without knowledge of the number of columns in the array, we will first try an array with two columns. Since there are 30 letters in this message, for an array with two columns there will be $30/2 = 15$ letters per column. This yields the following array.

$$
\begin{array}{cc}
T & T \\
A & Y \\
N & N \\
H & S \\
Y & S \\
I & Y \\
H & D \\
W & F \\
G & D \\
A & E \\
M & A \\
E & T \\
E & A \\
A & R \\
R & S \\
\end{array}
$$

Transcribing this array by rows gives TTAYNNHSYSIYHDWFGDAEMAETEAARRS, which is clearly not the correct plaintext. So next we will try an array with three columns. For an array with three columns there will be $30/3 = 10$ letters per column. This yields the following array.

```
T  M  Y
A  E  D
N  E  F
H  A  D
Y  R  E
I  T  A
H  Y  T
W  N  A
G  S  R
A  S  S
```

Transcribing this array by rows gives TMYAEDNEFHADYREITAHYTWNAGSRASS, which is also clearly not the correct plaintext. So we will try an array with four columns. For an array with four columns, dividing the number of letters by the number of columns yields the following.

$$4 \overline{)\begin{array}{r} 7 \\ 30 \\ -28 \\ \hline 2 \end{array}}$$

Thus, the first two columns of the array will contain eight letters, and the remaining two columns will contain seven. This yields the following array.

```
T  G  Y  D
A  A  N  E
N  M  S  A
H  E  S  T
Y  E  Y  A
I  A  D  R
H  R  F  S
W  T
```

Transcribing this array by rows gives TGYDAANENMSAHESTYEYAIADRHRFSWT, which is still clearly not the correct plaintext. So we will try an array with five columns, and $30/5 = 6$ letters per column.

```
T  H  E  S  E
A  W  A  S  A
N  G  R  Y  T
H  A  T  D  A
Y  M  Y  F  R
I  E  N  D  S
```

Transcribing this array by rows finally gives the correct plaintext THE SEA WAS ANGRY THAT DAY MY FRIENDS.[9]   □

---
[9] George Costanza, quote.

3.2.2   Cryptanalysis of Keyword Columnar Ciphers

Because keyword columnar transposition ciphers do not necessarily take the columns of the cipher array in order, cryptanalysis can be more difficult than it is for simple columnar transposition ciphers. To break a keyword columnar transposition cipher by a brute force attack, not only must arrays with various numbers of columns be considered, but various ways to order the columns of these arrays must be considered as well. The cryptanalysis process can be simplified, however, if a crib (i.e., a part of the plaintext) longer than the keyword(s) is known.

Example 3.7 Consider the ciphertext AHLCC MSOAO NMSSS MTSSI AASDI NRVLF WANTO ETTIA IOERI HLEYL AECVL W, which was formed using a keyword columnar transposition cipher, and suppose we have the crib THE FAMILY. (That is, suppose we know THE FAMILY is part of the corresponding plaintext.) To try to decrypt this message, in the hope that our crib is longer than the keyword(s) for the cipher, we will start by assuming there are exactly eight letters in the keyword(s). If there are eight letters in the keyword(s), then the array will have eight columns, and the crib would appear in these columns in the following form.

```
        T  H  E  F  A  M  I  L
        Y
```

Thus, the digraph TY would have to appear in the ciphertext. However, TY does not appear in the ciphertext, and so there are not exactly eight letters in the keyword(s). So next we will assume there are exactly seven letters in the keyword(s). If there are seven letters in the keyword(s), then the array will have seven columns, and the crib would appear in these columns in the following form.

```
        T  H  E  F  A  M  I
        L  Y
```

However, the digraphs TL and HY do not both appear in the ciphertext, and so there are not exactly seven letters in the keyword(s). (Although neither digraph appears in the ciphertext, either one failing to appear would be enough to indicate this.) So we will assume there are exactly six letters in the keyword(s), in which case the array will have six columns, and the crib would appear in these columns in the following form.

```
        T  H  E  F  A  M
        I  L  Y
```

Since the digraphs TI, HL, and EY all appear in the ciphertext, it is likely that there are exactly six letters in the keyword(s) and six columns in the array. Dividing the number of letters in the ciphertext by this number of columns yields the following.

$$
\begin{array}{r}
9 \\
6\overline{)\,56} \\
-54 \\
\hline
2
\end{array}
$$

Thus, the first two columns of the array will contain 10 letters, and the remaining four columns will contain nine letters. So we will split the ciphertext into blocks of nine letters each, which we label as follows.

AHLCCMSOA ONMSSSMTS SIAASDINR VLFWANTOE TTIAIOERI HLEYLAECV LW
   1   2   3   4   5   6

Next, we will arrange these blocks as columns in an array in the only way in which the known crib and digraphs TI, HL, and EY all line up correctly. This yields the following.

```
5  1  6  4  3  2
───────────────
   H  V  S  O
T  A  L  L  I  N
T  H  E  F  A  M
I  L  Y  W  A  S
A  C  L  A  S  S
I  C  A  N  D  S
O  M  E  T  I  M
E  S  C  O  N  T
R  O  V  E  R  S
I  A
```

When reading across the rows of this array from the top, the letters begin to form sensible English starting with the columns labeled 1 and 6. Thus, it is likely that these columns are the first two in the original cipher array, and would therefore be the two columns that contain 10 letters instead of nine. So we will split the ciphertext into blocks again, using 10 letters in the blocks labeled 1 and 6, and nine letters in the rest.

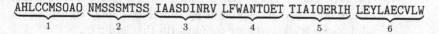

AHLCCMSOAO NMSSSMTSS IAASDINRV LFWANTOET TIAIOERIH LEYLAECVLW
   1   2   3   4   5   6

Arranging these blocks as columns in the same order as in the previous array, with the block labeled 5 moved from the front of the array to the end, yields the following.

```
1   6   4   3   2   5
A   L   L   I   N   T
H   E   F   A   M   I
L   Y   W   A   S   A
C   L   A   S   S   I
C   A   N   D   S   O
M   E   T   I   M   E
S   C   O   N   T   R
O   V   E   R   S   I
A   L   T   V   S   H
O   W
```

Thus, the plaintext is ALL IN THE FAMILY WAS A CLASSIC AND SOMETIMES
CONTROVERSIAL TV SHOW.   □

3.2.3   Exercises

1. Cryptanalyze the following ciphertexts, which were formed using simple columnar transposition ciphers.

   (a) AOANS BUYTE NBIEB ELNDA REVBL DDEAL

   (b) DMAIN TATLR EITVE SBXJS SHEDK AXANM INBEL X

   (c) PTIEO OGTBI NEYRA SICEY AYTRR DOISA FKFRL NGVWE GITOC
       APIHO EILCT RLIOO EDIEH DNNIR TNPNE NMEIS HTONR UFITR
       EIOGN RLDLG WGES[10]

   (d) FMGEO DFKOY AAOYL HIOIE ITUOY EFTBN UOLAN RNTNH LIOAM
       HODDU WTGKD HLAON ULOAN IOSOE TXSII TVGMO NTUFN TNGOX
       ONFYE IATES SYTIT OWX[11]

2. Recall from Exercise 5 in Section 2.2 the concept of *superencryption*.

   (a)* Use simple columnar transposition ciphers with four and six
        columns (in that order) to superencrypt LIKE AN OLD MAN TRYING
        TO SEND BACK SOUP IN A DELI.[12]

   (b) Decrypt LBLPY IEAAD OURLS EKNNO TEOKC MNNND TIADI IASG,
       which was superencrypted using simple columnar transposition
       ciphers with six and four columns (in that order).

---

[10] Groucho Marx (1890–1977), quote.
[11] Jeff Foxworthy, quote.
[12] George Costanza, quote.

(c) Does superencryption by two simple columnar transposition ciphers yield more security than encryption by one simple columnar transposition cipher? In other words, if a plaintext P is encrypted using a simple columnar transposition cipher, yielding M, and then M is encrypted using another simple columnar transposition cipher, yielding C, would C be harder in general to cryptanalyze than M? Explain your answer completely, and be as specific as possible.

3. The following ciphertexts were formed using keyword columnar transposition ciphers. Cryptanalyze each with the given crib.

(a)* UAODI HRNNI AODSE FSOUI CWLAI HSTHO HIBYF TROTI TVRDE LRETF ENEL, with the crib CIVIL WAR

(b) IDHTE NCLEX MECEH ACLHX AHPAO OAROA NTABF HDEFB SSAKT POATL IUESR OSBRL, with the crib PEACH BASKET

(c)* RSELS UEIOT EEINC HYBAG UFETF EEATL RHETE IXWRI VNSRE TFOHI SEIEO BPSUE EULIN SRCEG IIUFI EOETE BONHA LRSEL RTINC TEEEE EEBOG OIVEI TPHRS ECTIN OTTLA TRLRO XITTE,[13] with the crib INTELLIGENCE

(d)* RSRDI HILGS LDRGL GBHTS WLOTA SIDAD SGGTA NDNHD ORSET ROIEH ATUJT GIREB EENAA OTRUY LHATC MEJDD NHORD HHIYD JMAAE ADSRT TKYNI IWEEN CTGEI DOTCH EOEAI MUYME NEEAA IITLO FEBEE GKH, with the crib AT CHRISTMAS

4. Find a copy of Herbert Yardley's book *The American Black Chamber*, and write a summary of the description of the cryptanalysis of a German transposition cipher that can be found in Chapter 7 of it.

3.3 ADFGX and ADFGVX Ciphers

Toward the end of World War I, while most of the rest of the world was using either substitution ciphers or transposition ciphers, Germany began using a new type of cipher that combined features of both. These new ciphers, called *ADFGX* ciphers, are named for the only five letters that can appear in ciphertexts. These five letters were chosen because they sound very different from one another in Morse code, thus minimizing transmission errors. ADFGX ciphers were created by German Signals Officer Colonel Fritz Nebel, and first used by the German military in March 1918.

ADFGX ciphers involve two steps. The first step is a substitution cipher applied to the plaintext (after spaces and punctuation have been removed)

---

[13] Ernest Hemingway (1899–1961), quote.

that, like Playfair ciphers, uses a 5×5 array of letters agreed upon by the users. Also, like Playfair ciphers, I and J are considered to be the same letter in ADFGX ciphers. However, unlike Playfair ciphers, ADFGX ciphers do not encrypt plaintext letters as digraphs. Rather, the rows and columns of ADFGX arrays are each labeled with the letters A, D, F, G, and X, in order, and for the first step in an ADFGX cipher, each plaintext letter is replaced with the pair of row and column (in that order) letter labels of the position that the plaintext letter occupies in the array. This yields a preliminary ciphertext that is twice as long as the plaintext and contains only the letters A, D, F, G, and X. The second step in an ADFGX cipher is a keyword columnar transposition cipher applied to the preliminary ciphertext, using one or more keywords agreed upon by the users. This yields the final ciphertext.

Example 3.8 Consider an ADFGX cipher with the following random 5×5 array of letters and keywords KARL MARX.

	A	D	F	G	X
A	P	G	C	E	N
D	B	Q	O	Z	R
F	S	L	A	F	T
G	M	D	V	I	W
X	K	U	Y	X	H

To encrypt the plaintext I AM NOT A MARXIST[14] using this cipher, for the first step we replace each plaintext letter with the pair of row and column letter labels of the position that the plaintext letter occupies in this array. These pairs of letters are as follows.

Plain:	I	A	M	N	O	T	A	M	A	R	X	I
Cipher:	GG	FF	GA	AX	DF	FX	FF	GA	FF	DX	XG	GG

Plain:	S	T
Cipher:	FA	FX

Thus, the preliminary ciphertext is GGFFG AAXDF FXFFG AFFDX XGGGF AFX. For the second step, we use a keyword columnar transposition cipher with the given keywords.

3	1	6	4	5	2	7	8
K	A	R	L	M	A	R	X
G	G	F	F	G	A	A	X
D	F	F	X	F	F	G	A
F	F	D	X	X	G	G	G
F	A	F	X				

---

[14]Karl Marx (1818–1883), quote.

Thus, the final ciphertext is GFFAA FGGDF FFXXX GFXFF DFAGG XAG. For decryption, the encryption steps must be undone in the opposite order. That is, the transposition cipher must be undone (i.e., decrypted normally) first, and the substitution cipher undone second. □

In June 1918, the Germans increased the size of the array for ADFGX ciphers to 6×6. This allowed for the digits 0–9 to be included in plaintexts, as well as for I and J to be distinguished. It also required that an additional letter be used in ciphertexts, of course. This additional letter was V, thus creating the more commonly known *ADFGVX* ciphers. In *The Codebreakers* [13], David Kahn's masterful encyclopedic history of cryptology, Kahn refers to the ADFGVX system as "probably the most famous field cipher in all cryptology." Kahn later notes that when it was introduced, the ADFGVX cipher system was "the toughest field cipher the world had yet seen."

The Germans believed that ADFGX and ADFGVX ciphers were unbreakable. However, they were indeed broken through the extraordinary cryptanalytic efforts of a French Army Lieutenant named Georges Painvin, whose success is widely considered to be a primary reason why the French were able to stop Germany's ill-fated 1918 Spring Offensive. It was not an easy success for Painvin, though. The slender Painvin, doing nothing more

Arthur Zimmermann: Impact of the Zimmermann Telegram

Cryptology has often played a role in shaping world history. One example of this is the Zimmermann Telegram. In early 1917 during World War I, to counter a British naval blockade, Germany concluded that a total submarine offensive was necessary, which would include destroying civilian U.S. ships bound to and from Britain. Fearing this would draw the neutral U.S. into the war on the side of the Allies, the Germans had their Foreign Secretary, Arthur Zimmermann, encrypt and send a telegram to the Mexican government outlining a possible military alliance. This telegram explained that if the U.S. entered the war, Germany would provide military support to help Mexico reconquer lost territory in Texas, New Mexico, and Arizona. The telegram was intercepted and cryptanalyzed by the British, who then revealed its contents to the U.S. government. This enraged the American people towards Germany, expediting America's entry into the war, and hastening the war's end.

than sitting at his desk breaking ADFGX and ADFGVX ciphers, managed to lose 33 pounds in a little more than three months, and required a long period of rehabilitation. After the war, Painvin embarked on a very successful career in business, becoming president of several important corporations and, eventually, the Chamber of Commerce of Paris. Despite all of these accomplishments, late in life Painvin stated that his cryptanalysis of ADFGX and ADFGVX ciphers left "an indelible mark on my spirit, and remain for me one of the brightest and most outstanding memories of my existence."

The cryptanalysis of ADFGX and ADFGVX ciphers was typical of other cryptanalytic successes during World War I. Most of the ciphers of the time were based on nineteenth-century techniques, and thus could be broken with the right combination of ingenuity and perseverance. This led to the development of more secure ciphers that were used during World War II, one of which we will consider in the next chapter.

3.3.1   Exercises

1. Consider an ADFGX cipher with the array given in Example 3.8 and keyword NIETZSCHE.

   (a)* Use this cipher to encrypt PLATO WAS A BORE.[15]

   (b) Use this cipher to encrypt THE DOER ALONE LEARNETH.[16]

   (c) Decrypt FXGFG GAFXA DDFDX XAAXA XGGXX FGGAF AXFXF XFXAX FGFXD XGDFD XFFFX GFFXG XX,[17] which was formed using this cipher.

2. Consider an ADFGX cipher with the array given in Example 3.8 and keywords FAMOUS GERMANS.

   (a)* Use this cipher to encrypt COMPOSERS INCLUDING BEETHOVEN, BACH, HANDEL, MOZART, AND WAGNER.

   (b) Use this cipher to encrypt SCIENTISTS INCLUDING EINSTEIN, CANTOR, GAUSS, HERTZ, KEPLER, AND OHM.

   (c) Decrypt AAXAG FDGXG XDAXG GAFFA DFADD XAXFD GDXAG AGFAD GGFDX FDFAF GGGFG GFXDA, which was formed using this cipher.

3. Create an ADFGX cipher and use it to encrypt a plaintext of your choice with at least 20 letters.

---

[15] Friedrich Nietzsche (1844–1900), quote.

[16] Friedrich Nietzsche, quote.

[17] Friedrich Nietzsche, quote.

4. Consider an ADFGVX cipher with the following array and keyword SONGS.

	A	D	F	G	V	X
A	8	P	3	D	1	N
D	L	T	4	0	A	H
F	7	K	B	C	5	Z
G	J	U	6	W	G	M
V	X	S	V	I	R	2
X	9	E	Y	0	F	Q

(a)* Use this cipher to encrypt JENNY'S NUMBER WAS 867-5309.

(b) Use this cipher to encrypt 1999 REACHED #2 IN THE UK IN 1985.

(c) Decrypt VGGDG DGDGA AFVDG DGFVV XAXVD GDFGG GAVVG DFVXV VVVAD FDVAX DGDG, which was formed using this cipher.

5. Consider an ADFGVX cipher with the array given in Exercise 4 and keywords GEORGES PAINVIN.

(a)* Use this cipher to encrypt POOR PAINVIN, THIS TIME I DON'T THINK YOU'LL GET IT.[18]

(b) Use this cipher to encrypt BY VIRTUE OF MY JOB I AM THE BEST INFORMED MAN IN FRANCE, AND AT THIS MOMENT I NO LONGER KNOW WHERE THE GERMANS ARE.[19]

(c) Decrypt VVDDG GGVGV GVGDV DDDVD VVAVG DDDDV VVXGD XFAGD AXXDG VXGGD ADXAA VAVDD XGDDV DXGXG XGAGV XDDGD VD,[20] which was formed using this cipher.

6. Create an ADFGVX cipher and use it to encrypt a plaintext of your choice with at least 20 characters.

7. Find some information about the cryptanalysis of the Zimmermann Telegram by the British, and write a summary of your findings.

8. Find some information about how Georges Painvin broke the ADFGX and ADFGVX ciphers, and write a summary of your findings.

9. Find some information about the role of cryptanalysis in France's stopping of Germany's 1918 Spring Offensive, and write a summary of your findings.

---

[18] François Cartier (1862–1953), French military cryptologic bureau chief, March, 1918.
[19] Head of intelligence at the French general headquarters, March 24, 1918.
[20] Head of intelligence at the French general headquarters, March 24, 1918.

Chapter 4

The Enigma Machine

This book is a survey of cryptology, not a history book. However, any survey of cryptology should include or lead readers to discover for themselves at least a little history of the subject, as this book does, not just because of how the subject has been formed by its history, but also because the idea of writing secret messages has led to historical matters, both fictional (e.g., a cipher in a Sherlock Holmes mystery) and actual (e.g., encrypted messages sent by the Zodiac killer), that are (at least to us) inherently interesting. We have placed many of our historical references in exercises to prompt readers to search for this information themselves, because, this being a survey book rather than a history book, what we could include here would not do the information justice relative to what is already readily available in more specialized books and articles in print and, usually, online.

A sizable part of the history of cryptology is related to government communications, especially military during times of conflict. Humans usually function optimally out of necessity, not convenience, and while the Zodiac killer did not need to send encrypted messages, governments and militaries do, especially during times of conflict. So although this is not a history book, we feel it is appropriate to go into some detail about the Enigma cipher machine, the cryptanalysis of which by the Allies during World War II is one of the greatest achievements of the human intellect. This feat, born out of necessity, directly contributed to a swifter end to the greatest war in history, and saved many lives on both sides.

4.1   The Enigma Cipher Machine

In 1918, German electrical engineer Arthur Scherbius applied for a patent for a mechanical cipher machine. This machine, later marketed commercially under the name *Enigma*, was designed with electric current running through revolving wired wheels, called *rotors*. Scherbius offered his machine to the German military, and while they did not find any deficiencies in it, they did not choose at that time to purchase any. Only years later, after learning that their World War I ciphers had routinely been broken, did the Germans adopt various models of the Enigma, which they used as their primary resource for encrypted field communications throughout World War II. In this section, we will present some technical details of two of these models, the *Wehrmacht* Enigma, used by the German Army, and the *Kriegsmarine M4* Enigma, used by the German Navy.

Before presenting any technical details of an Enigma, we should note that descriptions of these details, to the extent they are included in this book, are rarely found in literature aimed at nontechnical audiences. On the other hand, images of the various components of an Enigma abound, and are readily available through a simple Internet search. As such, in this book we will not include images of the various components of an Enigma. However, we do very strongly encourage our readers to do each part of Exercise 1 at the end of this section as the component of an Enigma given in the exercise is described in this section. The components of an Enigma are listed in Exercise 1 at the end of this section in basically the same order in which they are described in this section.

An Enigma consisted of four components: a 26-letter keyboard for entering input letters (either plaintext or ciphertext), a plugboard resembling a miniature old telephone switchboard, a system of rotors, and a 26-letter lampboard for displaying output letters. Pressing an input letter on the keyboard sent an electric current through the plugboard and rotors, where the encryption or decryption took place, and the current ended at the lampboard where a small bulb was illuminated to indicate the output letter. The layout of letters on the keyboard and lampboard was similar, although not identical, to the layout on a modern keyboard.

More specifically, pressing an input letter on the keyboard on an Enigma sent current designating the letter first to the plugboard. The plugboard was situated on the front of an Enigma, and had 26 sockets representing the 26 letters in the alphabet. Each plugboard socket could either be left open or connected to another socket by a short cable. If the sockets in the plugboard representing a pair of letters were connected by a cable, then current designating either letter would be converted at the plugboard to designate the other letter. If the socket in the plugboard representing a letter was left open, then current designating the letter would leave the plugboard still designating the same letter.

Example 4.1 Consider an Enigma plugboard wired with the sockets representing M and Z connected, the sockets representing N and S connected, and all other sockets left open. Then, current designating M will be converted at the plugboard to designate Z. Similarly, current designating S will be converted at the plugboard to designate N. On the other hand, current designating E will leave the plugboard still designating E. □

There were many different choices for which plugboard sockets could be connected in an Enigma, with anywhere from 0 to 13 cables used, and usually a very large number of possibilities for which sockets could be connected by each cable. Varying the number of cables would have maximized security, but standard German operating procedure was to use a fixed number of cables. With a fixed number of cables, 11 cables would have maximized security (as we will verify in Section 4.3), but for most of the war, standard German operating procedure was to use 10 cables. Each Enigma provided for use in the field came with 12 cables, with 2 held in reserve in case any of the 10 in use became faulty.

After leaving the plugboard, current went through a system of rotors that was situated in the back of an Enigma. Each individual rotor was a circular disk about the size of a hockey puck. We will call the flat sides of a rotor the *right* and *left* sides, since rotors could only be placed in an Enigma standing on end with each side facing in a particular direction. Both flat sides of a rotor contained 26 contact points, one to represent each letter, with the letters considered in alphabetical order around both sides of the rotor clockwise (when the rotor was viewed from the right). The contacts on the right side of a rotor were wired to the contact points on the left, but not usually straight across. The idea was that current could enter one side of a rotor at one of the contact positions, representing a letter, and pass through and exit the rotor on the other side at most likely a different contact position, representing a different letter.

Example 4.2 Consider an Enigma rotor wired with the contacts connected as listed in the following table (i.e., with each right contact listed in the first row wired to the left contact below it in the second row).

Right contact:	A B C D E F G H I J K L M N O P Q R S T U V W X Y Z
Left contact:	E K M F L G D Q V Z N T O W Y H X U S P A I B R C J

Then, current designating M that enters the rotor on the right will exit the rotor on the left designating O. Similarly, current designating M that enters the rotor on the left will exit the rotor on the right designating C. Also, current designating S that enters the rotor on either side will exit the rotor on the other side still designating S. □

Wehrmacht Enigmas could accommodate three rotors placed side-by-side, while Kriegsmarine M4 Enigmas could accommodate four rotors. Although rotors could only be situated with each side facing in a particular direction, current could pass through the rotors in either direction. The reason for this is that while current always initially passed through the rotors from right to left, to the left of the rotor slots was a reflector which sent the current back through the rotors from left to right. In addition, the reflector was itself like half a rotor in the sense that on its right side there were 26 contact points, one to represent each possible letter, but on its left side there were no contacts. The contacts on the right side of a reflector were wired to each other in 13 pairs. Unlike plugboard sockets, reflector contacts were always fully connected. Also, unlike rotor contacts, reflector contacts could not be connected in a way such that a letter was connected to itself.

Example 4.3 Consider an Enigma reflector wired with the contacts connected as listed in the following table (i.e., with each contact listed in the first row wired to the contact below it in the second row).

Contact:	A	B	C	D	E	F	G	I	J	K	M	T	V
Paired contact:	Y	R	U	H	Q	S	L	P	X	N	O	Z	W

Then, current designating M that enters the reflector will exit the reflector designating O. Similarly, current designating O that enters the reflector will exit the reflector designating M. □

There were many different choices for how rotor and reflector contacts could be connected in an Enigma, but because rotors and reflectors had to be hard-wired and changing the wiring was very difficult, rotors and reflectors with only a very small number of different wirings were ever produced and used in the field. Rotors with only five different wirings were produced for Wehrmacht Enigmas. These rotors were labeled with the Roman numerals **I–V**, and the contacts connected in each are listed in Table 4.1. Of these five rotors, three were used at a time in Wehrmacht Enigmas. Any three could be used, and they could be arranged in any order.

Right contact:	A B C D E F G H I J K L M N O P Q R S T U V W X Y Z
I left contact:	E K M F L G D Q V Z N T O W Y H X U S P A I B R C J
II left contact:	A J D K S I R U X B L H W T M C Q G Z N P Y F V O E
III left contact:	B D F H J L C P R T X V Z N Y E I W G A K M U S Q O
IV left contact:	E S O V P Z J A Y Q U I R H X L N F T G K D C M W B
V left contact:	V Z B R G I T Y U P S D N H L X A W M J Q O F E C K

Table 4.1   Contacts connected in Wehrmacht rotors.

Recall that while Wehrmacht Enigmas could only accommodate three rotors, Kriegsmarine M4 Enigmas could accommodate four. Kriegsmarine M4

Enigmas could actually hold four rotors in the same space that Wehrmacht Enigmas had for three. This was accomplished by using a thinner reflector, which allowed for a thinner fourth rotor to be inserted between the leftmost full-size rotor and the reflector. For the three full-size rotors in Kriegsmarine M4 Enigmas, any of the Wehrmacht rotors **I–V** could be used, as well as any of three additional rotors with different wirings. These three additional full-size rotors were labeled with the Roman numerals **VI–VIII**, and the contacts connected in each are listed in Table 4.2.

Right contact:	A B C D E F G H I J K L M N O P Q R S T U V W X Y Z
VI left contact:	J P G V O U M F Y Q B E N H Z R D K A S X L I C T W
VII left contact:	N Z J H G R C X M Y S W B O U F A I V L P E K Q D T
VIII left contact:	F K Q H T L X O C B J S P D Z R A M E W N I U Y G V

Table 4.2 Contacts connected in additional Kriegsmarine M4 rotors.

The eight full-size rotors **I–VIII** were too wide to fit into the space available for the thinner fourth rotor in Kriegsmarine M4 Enigmas. For this thinner fourth rotor, rotors with only two different wirings were produced. These rotors were labeled with the Greek letters β (beta) and γ (gamma), and the contacts connected in each are listed in Table 4.3.

Right contact:	A B C D E F G H I J K L M N O P Q R S T U V W X Y Z
β left contact:	L E Y J V C N I X W P B Q M D R T A K Z G F U H O S
γ left contact:	F S O K A N U E R H M B T I Y C W L Q P Z X V G J D

Table 4.3 Contacts connected in thinner Kriegsmarine M4 rotors.

Reflectors with only two different wirings were produced for Wehrmacht Enigmas. These reflectors were labeled with the letters **B** and **C**, and the contacts connected in each are listed in Table 4.4.

Contact:	A	B	C	D	E	F	G	I	J	K	M	T	V
B paired contact:	Y	R	U	H	Q	S	L	P	X	N	O	Z	W
Contact:	A	B	C	D	E	G	H	K	L	M	N	Q	S
C paired contact:	F	V	P	J	I	O	Y	R	Z	X	W	T	U

Table 4.4 Contacts connected in Wehrmacht reflectors.

Because Kriegsmarine M4 Enigmas were modified to hold four rotors instead of three, reflectors produced for Wehrmacht Enigmas were too wide to fit in them. As a result, different thinner reflectors had to be produced for Kriegsmarine M4 Enigmas. Reflectors with only two different wirings were produced for Kriegsmarine M4 Enigmas. These thinner reflectors were

also labeled with the letters **B** and **C**, and the contacts connected in each
are listed in Table 4.5.

Contact:	A	B	C	D	F	G	H	I	L	M	R	S	T
B paired contact:	E	N	K	Q	U	Y	W	J	O	P	X	Z	V
Contact:	A	B	C	E	F	G	H	I	L	P	Q	S	U
C paired contact:	R	D	O	J	N	T	K	V	M	W	Z	X	Y

Table 4.5   Contacts connected in Kriegsmarine M4 reflectors.

As we have noted, pressing an input letter on an Enigma keyboard sent an
electric current through the plugboard and rotors, and the current ended at
the lampboard where a small bulb was illuminated to indicate the output
letter. To be more precise, pressing an input letter on the keyboard sent an
electric current first through the plugboard, then through the rotors (either
three or four depending on the Enigma model) from right to left, through
the reflector, back through the rotors from left to right, and then through
the plugboard a second time. After leaving the plugboard for the second
time, the current went to the lampboard where a bulb was illuminated to
indicate the output letter.

What we have presented so far already gives a very large number of
possible configurations for an Enigma. However, we are not done. Before a
rotor was placed in an Enigma, it could be rotated into any of 26 possible
orientations. The orientation of a rotor in an Enigma dictates the path
that current follows through the rotor.

Example 4.4 Consider an Enigma rotor rotated from its original orientation (for which current designating a letter will travel through the rotor along the path we originally indicated in Tables 4.1–4.3) five positions
counterclockwise (when the rotor is viewed from the right). Then, current
designating B entering the rotor will not travel through the rotor along the
path we originally indicated for current designating B, but rather along the
path we originally indicated for current designating G, since the contact for
G will have rotated into the position originally occupied by the contact for
B. Similarly, current designating any letter entering the rotor will travel
through the rotor along the path we originally indicated for current designating the letter five positions later in the alphabet, wrapping from the end
of the alphabet to the start if necessary. For instance, current designating X
entering the rotor will travel through the rotor along the path we originally
indicated for current designating C.   □

To assist Enigma operators with orienting rotors correctly, etched in a ring
around the edge of each rotor were the letters A–Z (or sometimes the numbers 01–26), listed in order clockwise (when the rotor was viewed from the

right). For each rotor slot in an Enigma, a small window was cut to show the letter (or number) at a particular location on the ring. We will call this letter the *window letter*. The window letter for a rotor indicates the orientation of the rotor.

A number called the *rotor offset* also indicates the orientation of a rotor in an Enigma. The rotor offset for a rotor is a whole number between 0 and 25, with 0 meaning the rotor is in its original orientation (for which the window letter will be A), 1 meaning the rotor has been rotated 1 position counterclockwise (when viewed from the right), 2 meaning the rotor has been rotated 2 positions counterclockwise, and so on, through 25 meaning the rotor has been rotated 25 positions counterclockwise. There is no need to consider rotor offsets larger than 25, since rotating a rotor 26 positions counterclockwise will take the rotor back to its original orientation with rotor offset 0.

The etched ring around the edge of an Enigma rotor was also movable, and could be rotated into any of 26 different positions while the wired part of the rotor was held fixed. This is a complication, because rotating the ring changes the window letter without changing the rotor offset. A number called the *ring setting* indicates the position of the ring on a rotor. The ring setting for a rotor is a whole number between 1 and 26, with 1 meaning the ring is in its original position (for which with rotor offset 0 the window letter is A), 2 meaning the ring has been rotated 1 position counterclockwise (when the rotor is viewed from the right), 3 meaning the ring has been rotated 2 positions counterclockwise, and so on, through 26 meaning the ring has been rotated 25 positions counterclockwise. There is no need to consider ring settings larger than 26, since rotating a ring 26 positions counterclockwise will take the ring back to its original position with ring setting 1.

For a rotor in an Enigma for which the rotor and its ring have been rotated some number of positions, the window letter gives a number of positions it can be assumed that the rotor and its ring have been rotated in total.

Example 4.5 Consider a rotor in an Enigma for which the rotor and its ring have been rotated some number of positions. If the window letter is U, then it can be assumed that the rotor and its ring have been rotated 20 positions counterclockwise (when the rotor is viewed from the right) in total, since U is 20 positions after A in the alphabet. As a result, if the ring setting is known to be 4 (i.e., if the ring is known to have been rotated 3 positions counterclockwise), then the rotor offset will be 17 (i.e., it can be assumed that the rotor has been rotated 17 positions counterclockwise). Put another way, the rotor offset can be obtained by subtracting the ring setting from the number of the position of the window letter in the alphabet.

Since U is the 21st letter in the alphabet and the ring setting is 4, the rotor offset is $21 - 4 = 17$.   □

Example 4.6 Consider a rotor in an Enigma with ring setting 12 and window letter Q. Since Q is the 17th letter in the alphabet, the rotor offset is $17 - 12 = 5$.   □

For a rotor in an Enigma, when the ring setting is subtracted from the number of the position of the window letter in the alphabet, it is possible to obtain a negative result. We show how to account for this in the next example.

Example 4.7 Consider a rotor in an Enigma with ring setting 22 and window letter D. Since D is the 4th letter in the alphabet, subtracting the ring setting from the number of the position of the window letter in the alphabet gives a rotor offset of $4 - 22 = -18$. This means it can be assumed that the rotor has been rotated *negative* 18 positions counterclockwise (when the rotor is viewed from the right). However, recall that rotor offsets are whole numbers between 0 and 25, and so cannot be negative. This negative result is easy to remedy, though, since rotating a rotor negative 18 positions counterclockwise gives the same orientation as rotating the rotor *positive* 8 positions counterclockwise. Thus, the rotor offset is actually 8, a number that can be obtained by simply adding 26 to -18. That is, if subtracting the ring setting from the number of the position of the window letter in the alphabet gives a negative result, the rotor offset can be obtained by adding 26 to the result.   □

Example 4.8 Consider a rotor in an Enigma with ring setting 16 and window letter K. Since K is the 11th letter in the alphabet, subtracting the ring setting from the number of the position of the window letter in the alphabet gives $11 - 16 = -5$. Thus, the rotor offset is $-5 + 26 = 21$.   □

As we have noted, the rotor offset for an Enigma rotor dictates the path that current follows through the rotor.

Example 4.9 Consider Enigma rotor **V** with rotor offset 5, and current designating B entering the rotor on the right. With rotor offset 5, current designating B entering the rotor will not travel through the rotor along the path we originally indicated for current designating B, but rather along the path we originally indicated for current designating G, since the contact for G will have rotated into the position originally occupied by the contact for B. Note that B is the 2nd letter in the alphabet, and if we add the rotor offset to 2, the result is $2 + 5 = 7$, which corresponds to the fact that G is the 7th letter in the alphabet. According to Table 4.1, current designating G that entered rotor **V** on the right would exit the rotor on the left designating T. However, when the current enters the rotor it is supposed to designate B, not G. Since T is the 20th letter in the alphabet, we must only subtract the rotor offset from 20 to obtain $20 - 5 = 15$. Since the 15th letter in the alphabet is O, the current will actually exit the rotor on the left designating O.   □

Example 4.10 Consider Enigma rotor **III** with rotor offset 17, and current designating G entering the rotor on the right. Since G is the 7th letter in the alphabet, and $7 + 17 = 24$, the current will travel through the rotor along the path we originally indicated for current designating the 24th letter in the alphabet, X. According to Table 4.1, current designating X that entered rotor **III** on the right would exit the rotor on the left designating S. Since S is the 19th letter in the alphabet, and $19 - 17 = 2$, the current will actually exit the rotor on the left designating the 2nd letter in the alphabet, B.   □

For a rotor in an Enigma, when the rotor offset is added to or subtracted from the number of the position of a letter in the alphabet, it is possible to obtain a result that is larger than 26 or smaller than 0. We show how to account for this in the next example.

Example 4.11 Consider Enigma rotor **III** with rotor offset 17, and current designating Y entering the rotor on the left. Since Y is the 25th letter in the alphabet, and $25 + 17 = 42$, the current will travel through the rotor along the path we originally indicated for current designating the 42nd letter in the alphabet. To remedy the fact that 42 is outside the range of allowed results (i.e., from 1 through 26), we can subtract 26 from 42 to obtain $42 - 26 = 16$. Thus, the current will travel through the rotor along the path we originally indicated for current designating the 16th letter in the alphabet, P. According to Table 4.1, current designating P that entered rotor **III** on the left would exit the rotor on the right designating H. Since H is the 8th letter in the alphabet, and $8 - 17 = -9$, the current will exit the rotor on the right designating the -9th letter in the alphabet. To remedy the fact that -9 is outside the range of allowed results, we can add 26 to -9 to obtain $-9 + 26 = 17$. Thus, the current will actually exit the rotor on the right designating the 17th letter in the alphabet, Q.   □

Example 4.12 Consider Enigma rotor **V** with rotor offset 5, and current designating M entering the rotor on the left. Since M is the 13th letter in the alphabet, and $13 + 5 = 18$, the current will travel through the rotor along the path we originally indicated for current designating the 18th letter in the alphabet, R. According to Table 4.1, current designating R that entered rotor **V** on the left would exit the rotor on the right designating D. Subtracting the rotor offset from the number of the position of D in the alphabet gives $4 - 5 = -1$. Since $-1 + 26 = 25$, the current will actually exit the rotor on the right designating the 25th letter in the alphabet, Y.　　　　　　　　　　　　　　　　　　　　　　　　　□

The various rotor offsets and ring settings increase the number of possible configurations for an Enigma to an astronomically large number. Even so, everything we have presented so far would have ultimately made for nothing more than a glorified substitution cipher had it not been for one final feature that we have not yet mentioned—the rotors revolved within the machine during the actual encryption and decryption processes.

Encrypting and decrypting messages with an Enigma was done one letter at a time, and each time an input letter was pressed on the keyboard, the rightmost rotor would immediately (before the current reached the rotors) rotate one position counterclockwise (when the rotor was viewed from the right). In addition, for each Enigma rotor **I**–**VIII**, there was either one or two notches on the ring around the rotor. Since each notch was on the ring, its position in the rotor slot at any time could be identified solely by the window letter. For each notch, there was one particular position in the rotor slot, identified by a window letter called the *notch letter*, for which if the rotor rotated one position counterclockwise, the notch would cause the rotor to the left, if it were one of the rotors **I**–**VIII**, to also rotate one position counterclockwise. That is, for each notch on the ring on the rightmost rotor, once every 26 times the rotor rotated one position counterclockwise the notch would cause the middle full-size rotor to also rotate one position counterclockwise, and for each notch on the ring on the middle full-size rotor, once every 26 times the rotor rotated one position counterclockwise the notch would cause the leftmost full-size rotor to also rotate one position counterclockwise. Additionally, for the middle full-size rotor only, if a notch letter was showing in the window when an input letter was pressed, the middle full-size rotor would itself rotate one position counterclockwise, regardless of whether a notch on the ring on the rightmost rotor would have caused it to rotate.

To clarify, during the actual encryption and decryption processes, only Enigma rotors rotated, not the rings around the rotors. Once a ring had been set in the initial configuration of the machine, its location around its rotor was fixed, and during the encryption and decryption processes, the rotation of the rotor alone changed the window letter. Also, only the full-size rotors **I**–**V** used in Wehrmacht Enigmas and **I**–**VIII** used for the rightmost three rotors in Kriegsmarine M4 Enigmas rotated. The thinner rotors β and γ used for the leftmost rotor in Kriegsmarine M4 Enigmas never rotated, although they could be set in the initial configuration of the machine with a nonzero rotor offset. On the other hand, in both Enigma models the reflectors **B** and **C**, which also never rotated, were always set with a zero offset. Finally, after an input letter was pressed on the keyboard, all rotation of the rotors occurred before the current reached the rotors, and no additional rotation occurred until the next input letter was pressed.

The notch letters for each of the Enigma rotors **I–VIII** are listed in Table 4.6.

Rotor	Notch letters	Rotor I–VIII to left rotates when window letter changes (from) → (to)
I	Q	Q → R
II	E	E → F
III	V	V → W
IV	J	J → K
V	Z	Z → A
VI	M, Z	M → N, Z → A
VII	M, Z	M → N, Z → A
VIII	M, Z	M → N, Z → A

Table 4.6   Notch letters for Enigma rotors **I–VIII**.

Example 4.13 Consider a Wehrmacht Enigma initially configured with rotors **V**, **III**, and **I**, in order from left to right, and corresponding window letters QUO. Then, during an actual encryption or decryption with this initial configuration, the window letters would change according to the following sequence: QUO → QUP → QUQ → QVR → RWS → RWT → RWU → ...   □

Example 4.14 Consider a Kriegsmarine M4 Enigma initially configured with rotors β, **IV**, **VII**, and **VI**, in order from left to right, and corresponding window letters DJYL. Then, during an actual encryption or decryption with this initial configuration, the window letters would change according to the following sequence: DJYL → DJYM → DJZN → DKAO → DKAP → DKAQ → DKAR → ...   □

We have now presented all of the details of the operation of an Enigma, and are ready to see the full encryption process.

Example 4.15 Consider a Wehrmacht Enigma initially configured with the plugboard wired with the sockets representing M and Z connected, the sockets representing N and S connected, and all other sockets left open, rotors **V**, **III**, and **I**, in order from left to right, with corresponding ring settings 12, 4, and 8 and initial window letters QUO, and reflector **B**. We will determine the result of using an Enigma with this initial configuration to encrypt the plaintext ENIGMA. The first thing we do is press the key for the initial input plaintext letter E on the keyboard. This changes the window letters to QUP, and sends current designating E to the plugboard. Since the socket representing E in the plugboard is open, the current leaves the plugboard still designating E. Next, the current goes to the rotors. The rotor offsets are given in the following table. (We determined these rotor offsets for rotors **III** and **V** in Examples 4.5 and 4.6.)

Rotor	Ring setting	Window letter	Rotor offset
I	8	P	$16 - 8 = 8$
III	4	U	$21 - 4 = 17$
V	12	Q	$17 - 12 = 5$

The following bulleted items follow the current through the system of rotors:

- First, current designating E enters rotor **I** on the right. Since E is the 5th letter in the alphabet, and $5 + 8 = 13$, the current will travel through the rotor along the path we originally indicated for current designating the 13th letter in the alphabet, M. According to Table 4.1 on page 50, current designating M that entered rotor **I** on the right would exit the rotor on the left designating O. Since O is the 15th letter in the alphabet, and $15 - 8 = 7$, the current will actually exit the rotor on the left designating the 7th letter in the alphabet, G.

- Next, current designating G enters rotor **III** on the right. In Example 4.10, we determined that this current will exit rotor **III** on the left designating B.

- Next, current designating B enters rotor **V** on the right. In Example 4.9, we determined that this current will exit rotor **V** on the left designating O.

- Next, current designating O enters reflector **B**. According to Table 4.4 on page 51, current designating O that enters reflector **B** will exit the reflector designating M.

- Next, current designating M enters rotor **V** on the left. In Example 4.12, we determined that this current will exit rotor **V** on the right designating Y.

- Next, current designating Y enters rotor **III** on the left. In Example 4.11, we determined that this current will exit rotor **III** on the right designating Q.

- Next, current designating Q enters rotor **I** on the left. Since Q is the 17th letter in the alphabet, and $17 + 8 = 25$, the current will travel through the rotor along the path we originally indicated for current designating the 25th letter in the alphabet, Y. According to Table 4.1 on page 50, current designating Y that entered rotor **I** on the left would exit the rotor on the right designating O. Since O is the 15th letter in the alphabet, and $15 - 8 = 7$, the current will actually exit the rotor on the right designating the 7th letter in the alphabet, G.

Now through the system of rotors, current designating G goes back to the plugboard. Since the socket representing G in the plugboard is open, the current leaves the plugboard still designating G. The current then goes to the lampboard, where a bulb is illuminated to indicate G as the initial output ciphertext letter. For the encryption of the first plaintext letter E that results in the ciphertext letter G, a diagram of the flow of current through the entire machine is shown in Figure 4.1.

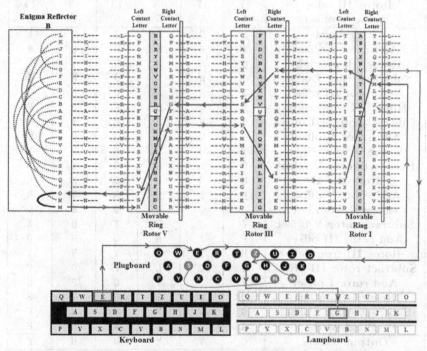

Figure 4.1 Flow of current through a Wehrmacht Enigma.

Next we press the key for the second input plaintext letter N on the keyboard. This changes the window letters to QUQ (which changes the rotor offset for the rightmost rotor), and begins the journey through the machine for current initially designating N. A summary of the full encryption process for the complete plaintext ENIGMA is shown in Table 4.7 on page 60. The last line in Table 4.7 shows the complete ciphertext, GVYWHF. □

Summary of encryption of plaintext ENIGMA using a Wehrmacht Enigma with plugboard connections MZ and NS, left-to-right rotors V, III, I with ring settings 12, 4, 8 and initial window letters QUO, and reflector B.						
Input letters	E	N	I	G	M	A
Window letters	QUP	QUQ	QVR	RWS	RWT	RWU
Rotor I offset	8	9	10	11	12	13
Rotor III offset	17	17	18	19	19	19
Rotor V offset	5	5	5	6	6	6
Plugboard	E	S	I	G	Z	A
Add rotor I offset	M	B	S	R	L	N
Rotor I from right	O	K	S	U	T	W
Subtract rotor I offset	G	B	I	J	H	J
Add rotor III offset	X	S	A	C	A	C
Rotor III from right	S	G	B	F	B	F
Subtract rotor III offset	B	P	J	M	I	M
Add rotor V offset	G	U	O	S	O	S
Rotor V from right	T	Q	L	M	L	M
Subtract rotor V offset	O	L	G	G	F	G
Reflector B	M	G	L	L	S	L
Add rotor V offset	R	L	Q	R	Y	R
Rotor V from left	D	O	U	D	H	D
Subtract rotor V offset	Y	J	P	X	B	X
Add rotor III offset	P	A	H	Q	U	Q
Rotor III from left	H	T	D	Y	W	Y
Subtract rotor III offset	Q	C	L	F	D	F
Add rotor I offset	Y	L	V	Q	P	S
Rotor I from left	O	E	I	H	T	S
Subtract rotor I offset	G	V	Y	W	H	F
Plugboard	G	V	Y	W	H	F
Output letters	G	V	Y	W	H	F

Table 4.7   Summary of encryption using a Wehrmacht Enigma.

Finally, for current traveling through an Enigma, since the current went through the plugboard at the start of its journey and then again at the end, and through the rotors from right to left before going through the reflector and then again from left to right after, and reflector contacts were always connected in pairs, the machine would always produce input/output letters in pairs. That is, for example, for identical configurations of an Enigma, if entering input letter E would yield output letter G, then entering input letter G would yield output letter E.

What is important about this is that for a ciphertext formed using an Enigma, the ciphertext could be decrypted by initially configuring the machine identically to how it had been initially configured during the encryption of the message, and then inputting the ciphertext letters. That is, for example, for a Wehrmacht Enigma initially configured identically to the initial configuration of the machine in Example 4.15, if the ciphertext letters GVYWHF were input into the machine, the resulting output would be the plaintext letters ENIGMA.

In the field, a ciphertext formed using an Enigma was decrypted by initially configuring a different Enigma identically to how the machine used to encrypt the message had been initially configured, and then inputting the ciphertext letters. Thus, the key for an Enigma cipher was the complete initial configuration of the machine used to encrypt the message, including how the plugboard had been wired, which rotors had been used in order with ring settings and initial window letters, and which reflector had been used. Despite this, from a technical perspective the Enigma was not difficult for operators to use in the field, since they did not have to understand the encryption or decryption processes, but only how to configure the machine.

4.1.1 Exercises

1. Find one or more images of the following.

    (a) Full view of an Enigma

    (b) Enigma in use in the field

    (c) Enigma keyboard

    (d) Enigma lampboard

    (e) Enigma plugboard

    (f) Enigma rotor

    (g) Enigma rotor taken apart to show wiring

    (h) Enigma reflector

    (i) Enigma reflector taken apart to show wiring

    (j) Enigma rotor ring etched with letters

(k) Enigma rotor ring etched with numbers

(l) Enigma with top panel removed to show rotors in slots

(m) Enigma with window letters (or numbers) showing through windows

(n) Enigma rotor ring showing notch

2. Consider an Enigma rotor wired with the contacts connected as listed in the following table.

Right contact:	A B C D E F G H I J K L M N O P Q R S T U V W X Y Z
Left contact:	E S O V P Z J A Y Q U I R H X L N F T G K D C M W B

(a)* Suppose current designating B enters the rotor on the right. Find the letter designated by the current when it exits the rotor on the left.

(b) Suppose current designating I enters the rotor on the right. Find the letter designated by the current when it exits the rotor on the left.

(c)* Suppose current designating D enters the rotor on the left. Find the letter designated by the current when it exits the rotor on the right.

(d) Suppose current designating T enters the rotor on the left. Find the letter designated by the current when it exits the rotor on the right.

3. Consider an Enigma reflector wired with the contacts connected as listed in the following table.

Contact:	A	B	C	E	F	G	H	I	L	P	Q	S	U
Paired contact:	R	D	O	J	N	T	K	V	M	W	Z	X	Y

(a)* Suppose current designating D enters the reflector. Find the letter designated by the current when it exits the reflector.

(b) Suppose current designating N enters the reflector. Find the letter designated by the current when it exits the reflector.

4. Find the rotor offset for the following Enigma rotors.

(a)* A rotor with ring setting 7 and window letter J

(b) A rotor with ring setting 2 and window letter B

(c)* A rotor with ring setting 22 and window letter E

(d) A rotor with ring setting 16 and window letter L

5. Consider Enigma rotor **IV** with rotor offset 4.

   (a)* Suppose current designating E enters the rotor on the right. Find the letter designated by the current when it exits the rotor on the left.

   (b) Suppose current designating P enters the rotor on the left. Find the letter designated by the current when it exits the rotor on the right.

6. Consider Enigma rotor **II** with rotor offset 9.

   (a)* Suppose current designating L enters the rotor on the left. Find the letter designated by the current when it exits the rotor on the right.

   (b) Suppose current designating W enters the rotor on the right. Find the letter designated by the current when it exits the rotor on the left.

7. Consider Enigma rotor **VIII** with rotor offset 22.

   (a)* Suppose current designating P enters the rotor on the right. Find the letter designated by the current when it exits the rotor on the left.

   (b) Suppose current designating Y enters the rotor on the left. Find the letter designated by the current when it exits the rotor on the right.

8. Consider a Kriegsmarine M4 Enigma initially configured with rotors γ, **II**, **VIII**, and **IV**, in order from left to right, and corresponding window letters BELI. Suppose this Enigma is used to encrypt a plaintext.

   (a)* Find the window letters after each of the first and second plaintext letters are entered.

   (b) Find the window letters after each of the third, fourth, fifth, and sixth plaintext letters are entered.

9.* Consider a Kriegsmarine M4 Enigma initially configured with the plugboard wired with the sockets representing G and Y connected, the sockets representing N and S connected, the sockets representing R and T connected, and all other sockets left open, rotors γ, **II**, **VIII**, and **IV**, in order from left to right, with corresponding ring settings 2, 22, 16, and 7 and initial window letters BELI, and reflector **C**. Determine the result of using an Enigma with this initial configuration to encrypt the plaintext GERMAN, and fill in each of the entries in Table 4.8 on page 64 to summarize the full encryption process.

Summary of encryption of plaintext GERMAN using a Kriegsmarine M4 Enigma with plugboard connections GY, NS, and RT, left-to-right rotors γ, II, VIII, IV with ring settings 2, 22, 16, 7 and initial window letters BELI, and reflector C.						
Input letters	G	E	R	M	A	N
Window letters						
Rotor IV offset Rotor VIII offset Rotor II offset Rotor γ offset						
Plugboard Add rotor IV offset Rotor IV from right Subtract rotor IV offset Add rotor VIII offset Rotor VIII from right Subtract rotor VIII offset Add rotor II offset Rotor II from right Subtract rotor II offset Add rotor γ offset Rotor γ from right Subtract rotor γ offset Reflector C Add rotor γ offset Rotor γ from left Subtract rotor γ offset Add rotor II offset Rotor II from left Subtract rotor II offset Add rotor VIII offset Rotor VIII from left Subtract rotor VIII offset Add rotor IV offset Rotor IV from left Subtract rotor IV offset Plugboard						
Output letters						

Table 4.8   Summary of encryption using a Kriegsmarine M4 Enigma.

10. Find some information about the American *M-209* cipher machine, and write a summary of your findings.

11. Find some information about the American *SIGABA* cipher machine, and write a summary of your findings.

12. Find some information about the British *Typex* cipher machine, and write a summary of your findings.

13. Find some information about the German *Lorenz* cipher machine, and write a summary of your findings.

14. Find a copy of Wolfgang Petersen's 1981 movie *Das Boot*, and write a summary of how the Enigma is presented in it as being used in day-to-day German military operations during World War II.

4.2   Combinatorics

German cryptologists during World War II were confident that the Enigma cipher machine was unbreakable, due to the astronomically large number of initial configurations of the machine. One of our goals in Section 4.3 will be to count this number of possible initial configurations. To do this, we first need to briefly review some basics about combinatorics.

In simplest terms, combinatorics is the study of methods for counting the number of possible outcomes for an experiment. Often counting the number of possible outcomes for an experiment is trivial. For example, for the experiment of being dealt a single card from a deck of 52 standard playing cards, there are 52 possible outcomes. On the other hand, sometimes counting the number of possible outcomes for an experiment can be more difficult. For example, for the experiment of being dealt five cards from a deck of 52 standard playing cards, it is more difficult to count the number of possible outcomes. In this section, we will consider some mathematical tools designed to help with counting the number of possible outcomes for an experiment.

4.2.1   The Multiplication Principle

Consider the experiment of rolling a pair of dice, one red and one white, and observing the numbers showing on the top faces of the dice after the roll. One way to count the number of possible outcomes for this experiment is to list all of the outcomes separately, as we have done in the following array. Each ordered pair in this array represents a possible outcome of the experiment, with the first number indicating the number showing on the top face of the red die after the roll, and the second number indicating the number showing on the top face of the white die after the roll.

$$(1,1) \quad (1,2) \quad (1,3) \quad (1,4) \quad (1,5) \quad (1,6)$$
$$(2,1) \quad (2,2) \quad (2,3) \quad (2,4) \quad (2,5) \quad (2,6)$$
$$(3,1) \quad (3,2) \quad (3,3) \quad (3,4) \quad (3,5) \quad (3,6)$$
$$(4,1) \quad (4,2) \quad (4,3) \quad (4,4) \quad (4,5) \quad (4,6)$$
$$(5,1) \quad (5,2) \quad (5,3) \quad (5,4) \quad (5,5) \quad (5,6)$$
$$(6,1) \quad (6,2) \quad (6,3) \quad (6,4) \quad (6,5) \quad (6,6)$$

Since there are 36 ordered pairs in this array, there are 36 possible outcomes for this experiment. The reason there are exactly 36 possible outcomes is because with six potential numbers showing on the top face of the red die after the roll, and also six potential numbers showing on top face of the white die after the roll, the number of possible outcomes is $6 \cdot 6 = 36$.

As another example, suppose that after allowing your dirty laundry to build up for a while, you open your closet one morning to discover only the following clean items.

Shirts	Socks	Pants
White	Dress	Jeans
Blue	Athletic	Slacks
Red		

Consider the experiment of choosing a shirt, socks, and pants from these clean items. One way to count the number of possible outcomes for this experiment is to draw the following *tree diagram*.

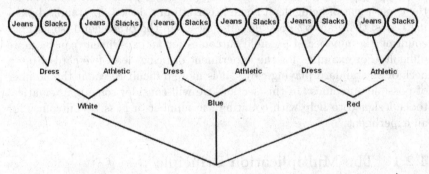

The bottom of this tree diagram is called the *root*, and the lines are called the *branches*. The pants choices at the top are circled to indicate that each represents a different outcome for the experiment. To determine the actual outcome from a circled item at the top, we start with the circled pants choice and follow the branches down through the socks and shirt choices to the root. The reason there are exactly 12 possible outcomes is because with three shirt choices, two socks choices, and two pants choices, the number of possible outcomes is $3 \cdot 2 \cdot 2 = 12$.

The previous examples of rolling a pair of dice and choosing clothes from your closet both illustrate the following *multiplication principle*.

Theorem 4.1 *(The Multiplication Principle) For an experiment involving two operations, if the first operation can be completed in s different ways and the second in t different ways, then the number of possible outcomes for the experiment is s · t. More generally, for an experiment involving k operations, if the operations can be completed in $s_1, s_2, \ldots, s_k$ ways, respectively, then the number of possible outcomes for the experiment is $s_1 \cdot s_2 \cdots s_k$.*

Example 4.16 Consider a restaurant offering pizzas with three possible types of crust, four possible types of sauce, and 20 possible toppings. The number of different one-topping pizzas with a single type of crust and sauce available at the restaurant is $3 \cdot 4 \cdot 20 = 240$. □

Example 4.17 Basic license plates for vehicles in North Carolina consist of three capital letters followed by four digits. Assuming no restrictions on the choices of letters and digits, since there are 26 possible choices for each letter and 10 for each digit, the number of different license plates available in North Carolina is $26 \cdot 26 \cdot 26 \cdot 10 \cdot 10 \cdot 10 \cdot 10 = 175,760,000$. □

4.2.2 Permutations

Suppose you have three extra tickets to a football game to give to your friends Allie, Sophie, and Trixie. The first ticket is for a seat in a luxury box, the second is for a seat in the first row, and the third is for a seat in the last row. The different ways in which you could distribute the three tickets to your friends are shown in Table 4.9.

Luxury Box	First Row	Last Row
Allie	Sophie	Trixie
Allie	Trixie	Sophie
Sophie	Allie	Trixie
Sophie	Trixie	Allie
Trixie	Allie	Sophie
Trixie	Sophie	Allie

Table 4.9 Ways to distribute three types of tickets to three friends.

The reason there are exactly six different ways in which you could distribute the tickets to your friends is the following. If we think of your identifying the recipient of your luxury box ticket first, you have all three of your friends from whom to choose. Then, after the luxury box ticket has been designated to someone, if we think of you identifying the recipient of your first row ticket, only two of your friends are left from whom to choose. Finally, if we think of you identifying the recipient of your last row ticket, only one friend is left to choose. Using the multiplication principle, this gives $3 \cdot 2 \cdot 1 = 6$ different ways in which you could distribute the tickets to your friends.

This method for determining the number of ways in which you could distribute tickets to your friends could easily be extended if there were more tickets and friends. For example, if you had six different types of tickets to give to six friends, then the number of ways in which you could distribute the tickets would be $6 \cdot 5 \cdot 4 \cdot 3 \cdot 2 \cdot 1 = 720$.

In both of the previous examples of distributing tickets to your friends, the final calculation can be expressed using a *factorial*.

Definition 4.2 *For a positive integer n, the factorial of n, denoted $n!$ ("n-factorial"), is defined as follows.*

$$n! = n \cdot (n - 1) \cdot (n - 2) \cdots 2 \cdot 1$$

In addition, $0!$ is defined to be 1.

For example, using a factorial we can express the number of ways in which you could distribute three different types of tickets to three friends as $3! = 3 \cdot 2 \cdot 1 = 6$. Similarly, we can express the number of ways in which you could distribute six different types of tickets to six friends as $6! = 6 \cdot 5 \cdot 4 \cdot 3 \cdot 2 \cdot 1 = 720$. In both of these examples, the order in which the tickets are distributed matters, since the tickets are all of different types. As such, both of these examples illustrate the more general mathematical concept of a *permutation*.

Definition 4.3 *A permutation of a collection of objects is an arrangement of the objects in an ordered list.*

For example, for the collection of objects {Allie, Sophie, Trixie}, there are $3! = 6$ different permutations, each of which is shown in Table 4.9. More generally, for a collection of n objects, there are $n!$ different permutations.

Example 4.18 The starting lineup for a basketball team consists of players in five different positions (point guard, shooting guard, small forward, power forward, and center). Suppose the coach of a particular team has decided upon the five players who will form the starting lineup, but not the positions these players will occupy. If each of the players could be used in each of the positions, and we consider the same players in different positions as a different starting lineup, then the number of different starting lineups the coach could form would be the number of permutations of the five players. That is, the number of different starting lineups the coach could form would be $5! = 5 \cdot 4 \cdot 3 \cdot 2 \cdot 1 = 120$.   □

Example 4.19 Referring to the basketball team in Example 4.18, suppose there are 12 total players on the team, and the coach has not yet decided upon which five of the 12 players will form the starting lineup. If each of the 12 players could be used in each of the five positions, and we consider the same players in different positions as a different starting lineup, then the number of different starting lineups the coach could form would be $12 \cdot 11 \cdot 10 \cdot 9 \cdot 8 = 95{,}040$.   □

Note that in Example 4.19, each of the permutations being counted only includes five of the objects in the original collection of 12 players. More specifically, in Example 4.19, there are 12 players from which the coach can choose a starting lineup, but only five of these 12 players will actually be used in the starting lineup. This is why in the calculation in Example 4.19, the factors being multiplied decreased from 12 only to 8, instead of all the way down to 1, as they would have in a factorial. This example is generalized as the following theorem.

Theorem 4.4 *From a collection of n objects, the number of different ways to choose t of the objects (for any $1 \le t \le n$) and arrange these objects in an ordered list is given by the following quantity.*[1]

$$P(n,t) = n \cdot (n-1) \cdot (n-2) \cdots (n-t+2) \cdot (n-t+1)$$

That is, $P(n,t)$ gives the number of different permutations of t objects chosen from n objects (for any $1 \le t \le n$).

For example, using Theorem 4.4 we can express the number of different starting lineups in Example 4.19 as $P(12,5) = 12 \cdot 11 \cdot 10 \cdot 9 \cdot 8 = 95{,}040$.

Example 4.20 From a full collection of all 26 capital letters in our alphabet, the number of different ways to choose three of the letters and arrange these letters in an ordered list is $P(26,3) = 26 \cdot 25 \cdot 24 = 15{,}600$. Similarly, from a full collection of all 10 single digits, the number of different ways to choose four of the digits and arrange these digits in an ordered list is

---

[1]The formula for $P(n,t)$ given in Theorem 4.4 can be expressed as $P(n,t) = \frac{n!}{(n-t)!}$. This equivalent expression for $P(n,t)$ is often used in literature, since not only is it more concise, but it also allows for $t = 0$ in Theorem 4.4.

$P(10, 4) = 10 \cdot 9 \cdot 8 \cdot 7 = 5,040$. Referring to Example 4.17, the multiplication principle gives that the number of different license plates available in North Carolina that have no duplicate letters and no duplicate digits is $P(26, 3) \cdot P(10, 4) = 15,600 \cdot 5,040 = 78,624,000$.   □

4.2.3   Combinations

Suppose you are going to a baseball game, and you have two extra general admission tickets to give to two of your friends. Since the tickets are for general admission, they are identical, and it does not matter if a particular friend receives one ticket or the other. Giving the first ticket to a first friend and the second ticket to a second would be no different than giving the first ticket to the second friend and the second ticket to the first. Suppose also that you have five friends from whom to choose to give the tickets—David, Horto, Jeff, Mike, and Tony. The different choices for which of your friends will receive the tickets are shown in Table 4.10.

General admission
David, Horto
David, Jeff
David, Mike
David, Tony
Horto, Jeff
Horto, Mike
Horto, Tony
Jeff, Mike
Jeff, Tony
Mike, Tony

Table 4.10   Ways to distribute two identical tickets among five friends.

Note that since the tickets are identical, we have not listed "David, Horto" and "Horto, David" separately in Table 4.10 as different choices for which of your friends will receive the tickets. In this example, the order in which the tickets are distributed does not matter. As such, this example illustrates the general mathematical concept of a *combination*.

Definition 4.5 *From a collection of n objects, a combination is an unordered subset of t of the objects (for any $1 \leq t \leq n$).*

For example, for the collection of objects {David, Horto, Jeff, Mike, Tony}, there are 10 different combinations, each of which is shown in Table 4.10. This number of combinations can also be determined without actually listing them all by using the formula in the following theorem.

Theorem 4.6 *The number of different combinations of t objects chosen from n objects (for any $1 \leq t \leq n$) is given by the following quantity.*[2]

$$C(n,t) = \frac{P(n,t)}{t!}$$

For example, using Theorem 4.6 we can express the number of different ways in which you can choose two of five friends to receive a pair of identical tickets as $C(5,2) = \frac{P(5,2)}{2!} = \frac{5 \cdot 4}{2 \cdot 1} = 10$.

Example 4.21 Consider a restaurant offering pizzas with 20 possible toppings. The number of different three-topping pizzas available at the restaurant is $C(20,3) = \frac{P(20,3)}{3!} = \frac{20 \cdot 19 \cdot 18}{3 \cdot 2 \cdot 1} = 1140$. □

Example 4.22 As a continuation of Example 4.21, suppose that of the 20 toppings offered at the restaurant, 12 are vegetables and eight are meats. For a three-topping pizza on which two of the toppings are to be vegetables and one is to be a meat, the vegetables can be chosen in $C(12,2)$ ways and the meat in $C(8,1)$ ways. The multiplication principle gives that the number of different three-topping pizzas with two vegetables and one meat available at the restaurant is $C(12,2) \cdot C(8,1) = \frac{12 \cdot 11}{2 \cdot 1} \cdot \frac{8}{1} = 66 \cdot 8 = 528$. □

Example 4.23 Suppose 30 students try out for the basketball team at a school, but the coach can only keep 12 players on the team. The number of different ways in which the coach can choose 12 of the 30 students to be on the team is $C(30,12) = \frac{P(30,12)}{12!} = \frac{30 \cdot 29 \cdot 28 \cdots 20 \cdot 19}{12 \cdot 11 \cdot 10 \cdots 2 \cdot 1} = 86{,}493{,}225$. Interestingly, we can also find this by determining the number of different ways in which the coach can choose 18 of the 30 students to *not* be on the team:[3] $C(30,18) = \frac{P(30,18)}{18!} = \frac{30 \cdot 29 \cdot 28 \cdots 14 \cdot 13}{18 \cdot 17 \cdot 16 \cdots 2 \cdot 1} = 86{,}493{,}225$. □

Example 4.24 As a continuation of Example 4.23, recall from Example 4.19 that after the coach chooses the 12 players to be on the team, the number of different starting lineups the coach could form would be $P(12,5) = 12 \cdot 11 \cdot 10 \cdot 9 \cdot 8 = 95{,}040$. Thus, beginning with the full collection of 30 students who try out for the team, for the experiment of creating a press release listing the 12 players on the team in alphabetical order, with the five players in the starting lineup identified by position, the multiplication principle gives that the number of possible outcomes is $C(30,12) \cdot P(12,5) = 86{,}493{,}225 \cdot 95{,}040 = 8{,}220{,}316{,}104{,}000$. Interestingly,

---

[2] An explanation for why the formula for $C(n,t)$ in Theorem 4.6 is correct is left as an exercise (Exercise 25 at the end of this section). This formula can also be expressed as $C(n,t) = \frac{n!}{t!(n-t)!}$. This equivalent expression for $C(n,t)$ is often used in literature, since it allows for $t = 0$ in Definition 4.5 and Theorem 4.6.

[3] It is true in general, for any collection of n objects, $C(n,t) = C(n,n-t)$ for any integer $1 \leq t \leq n-1$.

we can also find this by considering the coach first choosing the five players for the starting lineup from the 30 students who try out for the team,
and then choosing the seven other players from the remaining 25 students:
$P(30,5) \cdot C(25,7) = 17,100,720 \cdot 480,700 = 8,220,316,104,000.$ □

4.2.4   Exercises

1. An appliance manufacturer makes four types of refrigerators, five
   types of ovens, and six types of dishwashers.

   (a)* For the experiment of choosing a refrigerator and an oven from
        the manufacturer, how many different outcomes are possible?

   (b) For the experiment of choosing a refrigerator, an oven, and a
       dishwasher from the manufacturer, how many different outcomes
       are possible?

2. To access a bank account using an ATM, a four-digit personal identification number (PIN) is required.

   (a)* How many different PINs are possible?

   (b) How many different PINs are possible if the first digit cannot be
       0?

   (c) How many different PINs are possible if the first digit cannot be
       0 and duplicate digits are not allowed?

3. (a)* How many different seven-digit telephone numbers are possible?

   (b) How many different seven-digit telephone numbers are possible
       if the first digit cannot be 0 or 1?

   (c) How many different seven-digit telephone numbers are possible
       if the first digit cannot be 0 or 1 and duplicate digits are not
       allowed?

4. Twelve horses enter a race for which prizes are awarded for first,
   second, and third place. Assuming the prizes are all different, in how
   many different ways can the prizes be awarded?

5. A quiz is automatically generated by randomly choosing eight questions from a bank of 24 questions.

   (a)* How many different quizzes are possible if the same eight ques-
        tions in different orders are considered different quizzes?

   (b) How many different quizzes are possible if the same eight ques-
       tions in different orders are considered the same quiz?

6. From a collection of 15 books, students in an English class must read exactly three.

   (a)* In how many different ways can a student choose three of the books to read?

   (b) The students in the class must turn in a list of the names of the books they will read in the order in which they will read them. How many different lists are possible?

7. The starting lineup for a baseball team consists of players in eight different regular positions (catcher, first base, second base, third base, shortstop, right field, center field, and left field). Consider the same players in different regular positions as a different starting lineup.

   (a)* Suppose the coach of a baseball team has decided upon the eight players for the starting lineup, but not the actual positions these players will occupy. If each of the players could be used in each of the positions, how many different starting lineups could the coach form?

   (b) Suppose there are 15 total players on a baseball team that could be used in regular positions, and the coach of the team has not yet decided upon which eight of the 15 players will form the starting lineup. If each of the 15 players could be used in each of the eight regular positions, how many different starting lineups could the coach form?

   (c) Suppose 20 students try out for the regular positions on the baseball team at a school, but the coach can only keep 15 of the players on the team. For the experiment of creating a press release listing the 15 players on the team in alphabetical order, with the eight players in the starting lineup identified by position, how many different outcomes are possible?

8. For the experiment of being dealt five cards from a deck of 52 standard playing cards, how many different outcomes are possible?

9. Consider a restaurant offering burritos with two possible types of tortilla, 12 possible ingredients, and four possible types of salsa.

   (a)* How many different burritos are available at the restaurant with four ingredients and a single type of tortilla and salsa?

   (b) How many different burritos are available at the restaurant with four ingredients, a single type of tortilla, and two types of salsa?

10. An animal shelter contains 30 cats and 25 dogs.

    (a)* In how many different ways can a family select three cats from the shelter.

    (b) In how many different ways can a family select three pets from the shelter.

    (c) In how many different ways can a family select two cats and a dog from the shelter.

11. An advertisement is posted inviting applications for five equal positions within a company. Suppose 100 applications are received, 40 from men and 60 from women, and all 100 applicants are equally qualified for the positions.

    (a)* In how many different ways can the company fill the positions from the 100 applicants?

    (b) In how many different ways can the company fill the positions from the 100 applicants if they will only hire women?

    (c) In how many different ways can the company fill the positions from the 100 applicants if they want to hire exactly two men and three women?

12. For a substitution cipher with a 26-letter alphabet, how many different cipher alphabets are possible?

13.* For an ADFGX cipher, how many different 5×5 arrays are possible?

14. For an ADFGVX cipher, how many different 6×6 arrays are possible?

15.* From an empty Enigma plugboard, how many different ways are there to choose the sockets for exactly two letters to connect with a single cable?

16. From an empty Enigma plugboard, how many different ways are there to choose the sockets for exactly six letters to leave unconnected?

17.* How many different wirings are theoretically possible for an Enigma rotor?

18.* How many different wirings are theoretically possible for an Enigma reflector?

19.* From a collection of one copy each of the five rotors that were actually produced for Wehrmacht Enigmas, in how many different ways can rotors be arranged from left to right in a Wehrmacht Enigma?

20.* From a collection of one copy each of the five rotors and two reflectors that were actually produced for Wehrmacht Enigmas, in how many different ways can a reflector and rotors be arranged from left to right in a Wehrmacht Enigma?

21. From a collection of one copy each of the 10 rotors that were actually produced for Kriegsmarine M4 Enigmas, in how many different ways can rotors be arranged from left to right in a Kriegsmarine M4 Enigma?

22. From a collection of one copy each of the 10 rotors and two reflectors that were actually produced for Kriegsmarine M4 Enigmas, in how many different ways can a reflector and rotors be arranged from left to right in a Kriegsmarine M4 Enigma?

23.* How many different sequences of window letters are possible in a Wehrmacht Enigma?

24. How many different sequences of window letters are possible in a Kriegsmarine M4 Enigma?

25.* Explain why the formula for $C(n,t)$ given in Theorem 4.6 is correct.

4.3 Security of the Enigma Machine

German cryptologists during World War II were confident that the Enigma machine was unbreakable, given the fact that due to the astronomically large number of initial configurations of the machine, it would have been impossible for an Enigma cipher to be broken by a brute force attack. It is shown in [16] that the theoretical number of initial configurations of a Wehrmacht Enigma is more than 3×10^{114}, and of a Kriegsmarine M4 Enigma is more than 2×10^{145}, numbers so large that they are beyond basic human comprehension. However, recall that the Germans only actually used rotors and reflectors with a very small number of different wirings, and for most of the war used a fixed number of plugboard cables. This dramatically reduced the actual number of possible initial configurations.

4.3.1 Number of Initial Configurations

In order to determine the actual number of possible initial configurations of an Enigma, we will consider separately the five variable components of the machine—the plugboard, arrangement of rotors, ring settings, initial window letters, and choice of reflector. More specifically, we will consider separately the number of initial configurations of each of these variable components, and then combine them using the multiplication principle to find the number of possible initial configurations of the full machine. We will illustrate using the Wehrmacht Enigma, and leave similar calculations for the Kriegsmarine M4 Enigma as exercises at the end of this section.

Plugboard

Recall that an Enigma plugboard consisted of 26 sockets representing the 26 letters in the alphabet, and each socket could either be left open or connected to another socket by a short cable. The number of cables used in the machine could thus range from 0 to 13, with exactly twice as many sockets connected. As long as at least one cable was used, then the number of ways to choose the sockets to be connected can be found using the formula for $C(n, t)$ given in Theorem 4.6. More specifically, if p cables were used, then $2p$ sockets would be connected, and (as long as p is at least 1) the number of ways to choose the $2p$ sockets to be connected can be found using the following formula.

$$C(26, 2p) = \frac{P(26, 2p)}{(2p)!} = \frac{26 \cdot 25 \cdots (26 - 2p + 1)}{(2p)!}$$

For example, recall that for most of the war standard German operating procedure was to use a fixed number of 10 cables. Using this formula for $C(26, 2p)$, we can find that the number of ways to choose 20 sockets to connect using these cables is $C(26, 20) = \frac{P(26,20)}{20!} = \frac{26 \cdot 25 \cdot 24 \cdots 8 \cdot 7}{20 \cdot 19 \cdot 18 \cdots 2 \cdot 1} = 230{,}230$.

After the $2p$ sockets have been chosen, they must actually be connected using the p cables. Suppose one end of a first cable is plugged into one of the sockets. This leaves $2p - 1$ choices for the socket into which the other end of the first cable will be plugged. After this is done, suppose one end of a second cable is plugged into one of the remaining sockets. This leaves $2p - 3$ choices for the socket into which the other end of the second cable will be plugged. Continuing in this manner, the multiplication principle gives that the number of ways to connect all $2p$ sockets using p cables (as long as p is at least 1) is given by the following quantity.

$$S_p = (2p - 1) \cdot (2p - 3) \cdot (2p - 5) \cdots 3 \cdot 1$$

For example, with 20 sockets chosen to be connected using 10 cables, this formula for S_p gives that the number of ways to actually connect the sockets using the cables is $S_{10} = 19 \cdot 17 \cdot 15 \cdots 3 \cdot 1 = 654{,}729{,}075$.

Combining the formulas for $C(26, 2p)$ and S_p using the multiplication principle, we find that as long as p is at least 1, the number of ways to first choose $2p$ sockets to be connected, and then actually connect these sockets using p cables, is given by the following quantity.

$$N_p = C(26, 2p) \cdot S_p$$

This quantity N_p gives the number of possible initial configurations of an Enigma plugboard, assuming exactly p cables are used, for any value of p from 1 to 13. For example, this formula for N_p gives that the number of possible initial configurations of an Enigma plugboard using exactly 10 cables is $N_{10} = C(26, 20) \cdot S_{10} = 230{,}230 \cdot 654{,}729{,}075 = 150{,}738{,}274{,}937{,}250$.

Arrangement of Rotors

Rotors with five different wirings were produced for Wehrmacht Enigmas, with three in use in the machine at a time. Using the formula for $P(n, t)$ given in Theorem 4.4 on page 69, we can find that the number of different ways in which rotors can be arranged from left to right in a Wehrmacht Enigma is $P(5, 3) = 5 \cdot 4 \cdot 3 = 60$.

Ring Settings

Around each Enigma rotor was a movable ring that could be rotated into any of 26 different positions while the wired part of the rotor was held fixed. With three rotors in use in the machine at a time, the multiplication principle gives that the number of possible ring settings for all of the rotors in a Wehrmacht Enigma is $26 \cdot 26 \cdot 26 = 17{,}576$.

Initial Window Letters

Before a rotor was placed in an Enigma, it could be rotated into any of 26 possible orientations, each yielding a unique window letter. With three rotors in use in the machine at a time, the multiplication principle gives that the number of possible initial window letters for all of the rotors in a Wehrmacht Enigma is $26 \cdot 26 \cdot 26 = 17{,}576$.

Choice of Reflector

Reflectors with two different wirings were produced for Wehrmacht Enigmas, with one in use in the machine at a time. Thus, the number of different ways in which a reflector can be chosen for a Wehrmacht Enigma is 2.

The Full Machine

Combining the number of initial configurations of each of the variable components of a Wehrmacht Enigma, the multiplication principle gives that the number of possible initial configurations of the full machine using exactly 10 plugboard cables is the following.[4]

$$N_{10} \cdot 60 \cdot 17{,}576 \cdot 17{,}576 \cdot 2 = 5{,}587{,}851{,}741{,}017{,}032{,}206{,}720{,}000$$

This number, which is approximately 5.5879×10^{24}, or more than five million billion billion, was still much too large for a brute force attack on the Enigma to have been possible during World War II. However, it is also very much less than the theoretical number of initial configurations of the full machine. This fact along with other general mistakes in the overall implementation of the Enigma by the Germans as well as specific mistakes by German operators in the field allowed the Allies to successfully break the machine during World War II.

4.3.2 Background on Cryptanalysis

Even before World War II began, a small team of Polish mathematicians led by Marian Rejewski succeeded in breaking early versions of the Enigma machine. Typically, German Enigma operators were provided with a code-book that gave an initial configuration to be used for all messages encrypted on any particular day. However, so that identical initial rotor offsets would not be used for all messages encrypted on any particular day, operators were instructed to choose their own initial window letters each time they encrypted a message. The operators at this time were supposed to use the initial configuration given in the codebook, which included initial window letters, to encrypt their own initial window letters, and then change the window letters on their machine to their initial window letters and encrypt their message. They were then to transmit their encrypted initial window letters followed immediately by their encrypted message. Upon receipt, intended recipients would use the initial configuration given in the codebook to decrypt the operator's initial window letters, and then change the window letters on their machine to the operator's initial window letters and decrypt the message.

Some specific mistakes made occasionally by German Enigma operators in the field included choosing predictable initial window letters, such as ABC or short names, and using the same initial window letters when encrypting different messages. More importantly, a general mistake in the

---

[4]From the perspective of someone actually trying to break an Enigma cipher, this number can be reduced by a factor of 26. Because the notch on the ring on the leftmost rotor in an Enigma had no effect on the operation of the machine, for each possible orientation of the leftmost rotor (considering both the ring setting and initial window letter), there were 26 different combinations of a ring setting and initial window letter for which the operation of the machine was identical. For someone trying to break an Enigma cipher, only one of these 26 combinations would need to be considered.

overall implementation of the machine was that to reduce inaccuracies resulting from transmission error, when operators encrypted their initial window letters, they were supposed to repeat their initial window letters. For example, if the initial window letters given in the codebook for a particular day were NPS and an operator's initial window letters were REK, the operator was supposed to initially configure the machine with window letters NPS, encrypt REKREK, and then change the window letters on the machine to REK and encrypt the message. Upon receipt, the intended recipient would initially configure the machine with the window letters NPS given in the codebook, decrypt the first six letters of the received message to determine that the operator's initial window letters were REK, and then change the window letters on the machine to REK and decrypt the rest of the message.

Rejewski and his team exploited the fact that German Enigma operators repeated their initial window letters before encrypting their messages. By identifying patterns caused by these repetitions, Rejewski and his team reduced the number of possible initial configurations to 105,456. Cataloging these initial configurations made it possible to check them all through the use of an electromechanical machine designed by Rejewski called a *bomba*. At its peak effectiveness, the bomba allowed the Poles to decrypt and read much of the German Enigma traffic for any particular day in about two

---

Agnes Meyer Driscoll: First Lady of Naval Cryptology

Agnes Meyer Driscoll was a cryptanalyst for the U.S. Navy responsible for breaking several types of Japanese ciphers before and during World War II. After obtaining degrees in mathematics and physics with proficiencies in French, German, Japanese, and Latin, Driscoll joined the Navy at the highest possible rank of Chief Yeoman. As a cryptanalyst with the Navy, Driscoll broke several Japanese ciphers, including the Red Book code in the 1920s and the Blue Book code in 1930. In 1940, she made critical contributions towards breaking the JN-25 code, the most secure cipher used by the Japanese Imperial Navy before and during World War II. After the war, Driscoll was part of the naval contingent that joined the nation's new cryptologic agencies, including the Armed Forces Agency in 1949 and the National Security Agency in 1952. Driscoll was laid to rest in 1971 at Arlington National Cemetery, and in 2000 was inducted into the National Security Agency's Hall of Honor.

hours. This continued until December 1938, when German cryptographers made modifications to the Enigma, most notably increasing the number of different rotors produced from three to five, which increased its security.

This increased security proved too much for the Poles to account for in a timely manner, and so in July 1939, just five weeks before Poland fell to the Germans, Rejewski and his team shared their work with cryptologists from Britain and France. Building upon their work, British and French cryptanalysts working at Bletchley Park near London were able to make further advances in breaking the Enigma. In particular, British mathematician Alan Turing, a bona fide genius considered the "Father of the Modern Computer," identified weaknesses in the Enigma encryption process using patterns generated by cribs, which were made easier to find through the frequent mistaken use of standard salutations, titles, and addresses by German operators. Inspired by Rejewski's bomba, Turing designed a more powerful electromechanical machine called a *bombe* to search for these patterns. This machine, like Rejewski's, dramatically reduced the time required to check a collection of initial configurations. At its peak effectiveness, attacking a more complex Enigma than the Poles had attacked with the bomba, the bombe allowed the British to decrypt and read much of the German Enigma traffic for any particular day in about three hours. We will give a detailed description of how the bombe was designed in Chapter 5.

Not all of the work to break the Enigma was done by mathematicians, though. For example, a French intelligence officer purchased information about the Enigma from an employee of the German cryptologic agency. In addition, some stories of daring attempts, both successful and unsuccessful, by the Allies to capture Enigma machines and codebooks are legendary. David Kahn's 1991 book *Seizing the Enigma: The Race to Break the German U-Boat Codes, 1939–1943* [12] is a remarkable history of not only some of these attempts, but also the development of the machine and its cryptanalysis. The attempts by the Allies to capture Enigma machines and codebooks also inspired the 2000 fictional Hollywood movie *U-571*. The cryptanalysis of the Enigma at Bletchley Park inspired Robert Harris's 1995 novel *Enigma*, upon which the 2001 movie of the same name is based. The movie *Enigma* was co-produced by Mick Jagger, who along with providing funds, also provided access to his own personal Enigma machine.

The cryptanalysis of the Enigma at Bletchley Park also provided the inspiration for the fact-based 2014 Hollywood blockbuster *The Imitation Game*, which was filmed in part at Bletchley Park. After Bletchley Park fell into disrepair and was considered for demolition, the Bletchley Park Trust was formed and tasked to restore the area as a museum. Bletchley Park opened to visitors in this capacity in 1993. As of this writing, the most recent major renovations were completed in 2014, although the restoration of Bletchley Park continues in earnest to this day.

4.3.3 Exercises

1. Find the number of initial configurations of an Enigma plugboard, assuming p cables are used, for the following values of p.

   (a)* 8

   (b) 9

   (c)* 11

   (d) 12

2. Assuming p plugboard cables are used, for what value of p is the number of configurations of an Enigma plugboard the largest?

3. Find the number of different ways in which rotors can be arranged from left to right in a Kriegsmarine M4 Enigma.

4. Find the number of possible ring settings for all of the rotors in a Kriegsmarine M4 Enigma.

5. Find the number of possible window letters for all of the rotors in a Kriegsmarine M4 Enigma.

6. Find the number of initial configurations of a full Wehrmacht Enigma (considering the plugboard, arrangement of rotors, ring settings, initial window letters, and choice of reflector), assuming p plugboard cables are used, for the following values of p.

   (a)* 8

   (b) 9

   (c)* 11

   (d) 12

7. Find the number of initial configurations of a full Kriegsmarine M4 Enigma (considering the plugboard, arrangement of rotors, ring settings, initial window letters, and choice of reflector), assuming p plugboard cables are used, for the following values of p.

   (a)* 8

   (b) 9

   (c) 10

   (d)* 11

   (e) 12

8.* Another general mistake in the overall implementation of the Enigma by the Germans that was exploited by Allied cryptanalysts was that no matter how the machine was configured, it was impossible for a ciphertext output letter to ever match the corresponding plaintext input letter. Explain why, given the description of the components of a German Enigma in Section 4.1, it is impossible for a ciphertext output letter to ever match the corresponding plaintext input letter.

9. Find some information about the small team of Polish mathematicians led by Marian Rejewski that succeeded in breaking early versions of the Enigma machine, and write a summary of your findings.

10. Find some information about the bomba, the electromechanical machine designed by Marian Rejewski that Polish cryptanalysts used in breaking early versions of the Enigma machine, and write a summary of your findings.

11. Find some information about Alan Turing, the British mathematician who played a decisive role in breaking later versions of the Enigma machine, and write a summary of your findings.

12. Find some information about Bletchley Park and its role as the Allied headquarters for cryptanalysis during World War II, and write a summary of your findings.

13. Find some additional information about the career in cryptology of Agnes Meyer Driscoll, and write a summary of your findings.

Chapter 5

The Turing Bombe

As we have noted, Germany's "unbreakable" Enigma ciphers were in fact broken. Early versions of the Enigma were broken by a team of Polish mathematicians led by Marian Rejewski using a machine designed by Rejewski called a *bomba*. More complex versions of the Enigma were later broken at Bletchley Park by a team of Allied mathematicians that included Alan Turing using a machine designed by Turing called a *bombe*. Books that describe some of the technical details regarding the cryptanalysis of the Enigma tend to focus more on Rejewski's bomba; for example, see [2], a comprehensive history of cryptology written by a colleague whose interest in the subject was inspired in part by the same person who inspired ours. In this chapter, we will describe some of the technical details regarding the cryptanalysis of the Enigma, however our focus will be on Turing's bombe. This will be a challenge, but one well worth the effort, as the cryptanalysis of the Enigma at Bletchley Park is not only one of the supreme achievements of the human intellect, but it also saved many lives on both sides during World War II by directly contributing to a swifter end to the war. To make this discussion as accessible as possible, we will limit our focus to the cryptanalysis of the three-rotor Wehrmacht Enigma at Bletchley Park.

5.1   Cribs and Menus

Turing had no interest in replicating or extending Rejewski's methods for breaking early versions of the Enigma, because he believed that the Germans would make improvements to the security of the machine that would render Rejewski's methods obsolete. This is indeed exactly what happened in December 1938, when the Germans made modifications to the Enigma, most notably increasing the number of different rotors produced from three to five. Turing's attack on the machine was thus different from Rejewski's, and began with the use of cribs.

Recall that when trying to break any type of cipher, we call a known part of a plaintext a *crib*. In their attack on the Enigma, the cryptanalysts at Bletchley Park were able to consistently obtain reliable cribs because German Enigma operators often included standard salutations, titles, addresses, and introductory or concluding remarks in their messages. Also, an Enigma could never encrypt a plaintext letter as itself, because of the fully wired reflector in the middle of the encryption process that did not allow current to follow the same path away from the reflector as it had followed toward it. The cryptanalysts at Bletchley Park were well aware of this fact, and used it to help them find possible positions where the encrypted letters resulting from a known crib might exist within a given ciphertext. For example, suppose the crib FOLLOW ORDERS TO was known to encrypt to some part of the ciphertext NUENT ZERLO HHBTD SHLHI YWEAB HTQKC. To determine a possible position where the encrypted letters resulting from the crib might be, we can slide the crib along the ciphertext until we find a position where no plaintext letter is encrypted as itself. Consider first the following alignment.

Crib: F O L L O W O **R** D E R S T O
Cipher: N U E N T Z E **R** L O H H B T D S H L H I Y W E A B H T Q K C

Since this alignment between the crib and ciphertext results in the first R in the crib being encrypted as itself, it cannot be the correct alignment. So we slide the crib one position to the right, which results in the following second alignment.

Crib: F O L L O W O R D E R S **T** O
Cipher: N U E N T Z E R L O H H B **T** D S H L H I Y W E A B H T Q K C

Since this second alignment also results in a letter in the crib being encrypted as itself, it also cannot be the correct alignment. So we again slide the crib one position to the right, resulting in the following third alignment.

Crib: F O L L O W O R D E R S T O
Cipher: N U E N T Z E R L O H H B T D S H L H I Y W E A B H T Q K C

Since this third alignment results in no letter in the crib being encrypted as itself, it could be the correct alignment between the crib and ciphertext.

In our analysis of this possible correct alignment, we will start by labeling the crib and ciphertext letters with position numbers as follows.

Position:	1	2	3	4	5	6	7	8	9	10	11	12	13	14
Crib:	F	O	L	L	O	W	O	R	D	E	R	S	T	O
Cipher:	E	N	T	Z	E	R	L	O	H	H	B	T	D	S

Since an Enigma used identical settings for encryption and decryption, we can see that for an Enigma in position 1 with plaintext letter F encrypted as ciphertext letter E, it would also be true that plaintext letter E would be encrypted as ciphertext letter F. Similarly, for the machine in position 2 with plaintext letter O encrypted as ciphertext letter N, it would also be true that plaintext letter N would be encrypted as ciphertext letter O. These and the rest of the crib/ciphertext pairs are expressed in Figure 5.1 in what the cryptanalysts at Bletchley Park called a *menu*.

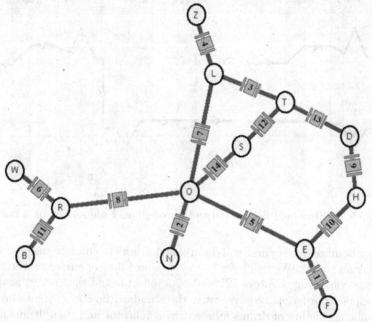

Figure 5.1   A crib/ciphertext menu.

Links connecting letters in menus represented actual physical cables in a Turing bombe. These cables contained 26 individual wires, one for each of the 26 letters in the alphabet.

Recall that when a letter was encrypted or decrypted by a Wehrmacht Enigma, current designating the letter first went to the plugboard, which could change the current to designate a different letter. The current then passed from right to left through three rotors, each of which could change the current to designate a different letter. After passing through a reflector which definitely changed the current to designate a different letter, the current passed back through the rotors from left to right, again went to the plugboard, and then finally arrived at the lampboard where it lit the bulb of the decrypted or encrypted letter. In a Turing bombe, the plugboard,

rotors, and reflector were connected in a *double-ended* fashion. The effect
on current of traveling through a double-ended bombe as opposed to an
Enigma was the same, but in a bombe the current did not change direction
at the reflector, instead traveling in the same direction through the entire
machine. A diagram illustrating this is shown in Figure 5.2, comparing the
standard flow of current through an Enigma (on the left) and the standard
flow for the same path through a double-ended bombe (on the right).

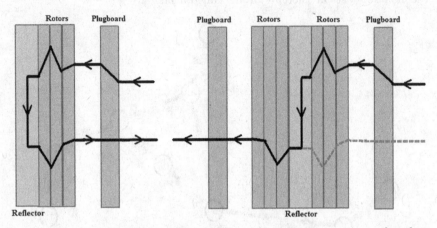

Figure 5.2   Standard flow of current through an Enigma versus a bombe.

Turing bombes used cylindrical disks called *drums* to emulate the operation
of Enigma rotors. We will refer to the standard flow of current through an
Enigma consisting of rotors followed by a reflector and then rotors again in
the opposite order, or, equivalently, the standard flow of current through
a bombe consisting of drums followed by a reflector and then drums again
in the opposite order, as a *double scrambler*. In a bombe, drums were
mounted vertically on shafts, with the drum emulating the rightmost rotor
in an Enigma at the top, the drum emulating the middle rotor in the
middle, and the drum emulating the leftmost rotor at the bottom. Before
drums were loaded onto shafts, a plate was fixed onto each shaft. Each
plate contained four concentric circles of 26 contact points, for a total of
104 contacts on each plate. The back of each drum contained 104 small
wire brushes, which were placed to touch the contact points on the plate
when the drum was loaded onto a shaft. An image showing two columns
of shafts fixed with plates, with one also loaded with drums, is shown in
Figure 5.3 on page 87. In this image, the column of shafts on the left is
fixed with plates but not loaded with drums, while the column on the right
is fixed with plates and loaded with drums. An image showing some of the
wire brushes on the back of a drum is shown in Figure 5.4 on page 87. Each

Figure 5.3   Two columns of plates, one loaded with drums.

Figure 5.4   Back of a drum showing wire brushes.

drum was color coded to indicate which of the five Wehrmacht Enigma rotors it emulated, and also contained two sets of contact points, one for current traveling toward the emulator of the reflector, and the other for current traveling away from the emulator of the reflector. The drums and the reflector were connected into double scramblers using wiring that was situated behind the plates, and there were sockets at each end of the double scrambler into which a cable containing 26 individual wires, one for each of the 26 letters in the alphabet, could be plugged.

When a bombe was in operation, current designating a letter would enter a double scrambler through the 26-wire cable plugged into the outermost ring of contacts on the top plate. It would pass through the top plate and then via the outermost ring of wire brushes into the top drum, where it could be changed to designate a different letter. It would then leave the top drum via the wire brushes one ring inside the outermost, and go back through the top plate via the contacts one ring inside the outermost. These contacts were wired to the outermost ring of contacts on the middle plate, where the current would go next. It would travel via the outermost rings through the middle plate and into the middle drum, where it could again be changed to designate a different letter, and then out of the middle drum and back through the middle plate via the wire brushes and contacts one ring inside the outermost. These contacts were wired to the outermost ring of contacts on the bottom plate, where the current would go next. It would travel via the outermost rings through the bottom plate and into the bottom drum, where it could once again be changed to designate a different letter, and then out of the bottom drum and back through the bottom plate via the wire brushes and contacts one ring inside the outermost. These contacts were connected to an emulator of the chosen reflector, where the current would definitely be changed to designate a different letter. The current would then travel back through the drums and plates in the reverse order, possibly being changed to designate a different letter by each drum, entering each drum via the wire brushes one ring outside the innermost and leaving each via the innermost ring of wire brushes. The current finally exited the double scrambler through the 26-wire cable plugged into the innermost ring of contacts on the top plate.

5.1.1 Exercises

1. The following ciphertexts were formed using an Enigma. Using only the fact that an Enigma could never encrypt a plaintext letter as itself, find all possible positions where the encrypted letters formed from the given crib could be.

   (a)* The crib **ATTACK AT TWELVE**, which is known to encrypt to some part of the ciphertext PARLH EWCNI CGIRG KR

(b) The crib THE WRATH OF KHAN AND, which is known to encrypt to
some part of the ciphertext HTHFX BXNHI QRCEE VLYEL

(c)* The crib GORDON WELCHMAN, which is known to encrypt to some
part of the ciphertext CWGBT GHGDD THDHJ TI

(d) The crib FOLLOW ORDERS TO, which is known to encrypt to some
part of the ciphertext NUENT ZERLO HHBTD SHLHI YWEAB HTQKC

2. Suppose the crib GLOBAL MISSION is known to encrypt to some part
of the ciphertext GLBCJ QQQYL QJOZM, which was formed using an
Enigma.

(a) Find the only possible position in the ciphertext where the encrypted letters formed from the crib could be.

(b) Draw a menu that expresses the crib/ciphertext pairs resulting
from part (a).

3.* Suppose the crib GOOD PET OFTEN IS is known to encrypt to some part
of the ciphertext KMUTV MUUPW FFTSF SGK, which was formed using an
Enigma.

(a) Find the only possible position in the ciphertext where the encrypted letters formed from the crib could be.

(b) Draw a menu that expresses the crib/ciphertext pairs resulting
from part (a).

4. Suppose the crib THE WRATH OF KHAN AND is known to encrypt to the
ciphertext WPMOI NNMUM GNGEL UR, which was formed using an Enigma.
Draw a menu that expresses the crib/ciphertext pairs.

5.* Suppose the crib GORDON WELCHMAN is known to encrypt to the ciphertext EMHSS MBNSX NLWW, which was formed using an Enigma. Draw a
menu that expresses the crib/ciphertext pairs.

6. Suppose the crib THE SOLAR SYSTEM IS is known to encrypt to the
ciphertext AJJRJ EHMRI OFURZ J, which was formed using an Enigma.
Draw a menu that expresses the crib/ciphertext pairs.

7. Find some information about a replica Turing bombe in operation at
Bletchley Park, and write a summary of your findings.

8. Find some information about the actual Enigma cribs that the cryptanalysts at Bletchley Park were able to obtain, and write a summary
of your findings.

5.2   Loops and Logical Inconsistencies

Recall that, of the astronomical number of possible initial configurations
of an Enigma, the plugboard contributed by far the largest factor to this
number. It is ironic then that Turing figured out a way to use the plugboard
in his attack on the machine. This attack began with the choice of a menu
letter, normally one with a large number of links connected to it, called the
central letter. For example, from the menu in Figure 5.1, a natural choice
for the central letter would be O, since this letter has the largest number of
links connected to it.

Once a central letter was selected, a possible plugboard partner for it
was chosen. For example, with the menu in Figure 5.1, for the central
letter O suppose we choose a plugboard partner, which for now we will just
label with the Greek letter α (alpha). Physically in a bombe, α would
be activated by having a voltage applied to its relay in a 26-relay device
known as the *indicator unit*, which itself was connected to the central letter
by a 26-wire cable. For the menu in Figure 5.1, consider specifically the
encryption of the letter O with the machine in position 14 that results in the
letter S. Recall that during this encryption, the plugboard would be applied
both before and after the double scrambler. Thus, since α is the plugboard
partner of O, the double scrambler in position 14 actually transforms α into
the plugboard partner of S, which for now we will just label with the Greek
letter β (beta). This is emphasized in the menu in Figure 5.5 on page 91.

In the menu in Figure 5.5, note that there are three sequences of links
that form closed loops. For example, starting with the central letter O, one
such closed loop is the menu letters O $\rightarrow$ S $\rightarrow$ T $\rightarrow$ L $\rightarrow$ O, which result with
the machine in the sequence of positions 14, 12, 3, 7. As we have noted, for
the first of these positions, 14, the double scrambler actually transforms α,
the plugboard partner of O, into β, the plugboard partner of S. Similarly,
for the second of these positions, 12, the double scrambler transforms β into
the plugboard partner of T, which for now we will just label with the Greek
letter δ (delta). Likewise, for the third of these positions, 3, the double
scrambler transforms δ into the plugboard partner of L, which for now we
will just label with the Greek letter μ (mu). This closed loop is separated
from the rest of the menu and shown in Figure 5.6 on page 91, with the
plugboard partners of the menu letters noted. It is important to note that
because each double scrambler was applied only to the plugboard partner
of a menu letter, whatever β, δ, and μ turn out to be depends only on the
choice of α and the double scrambler positions.

From the letters in closed menu loops and their plugboard partners, Turing formulated a working hypothesis about the reflector, rotor order, and
window letters that had been used when a given ciphertext was formed.
Recall that when a ciphertext was formed using an Enigma, although the

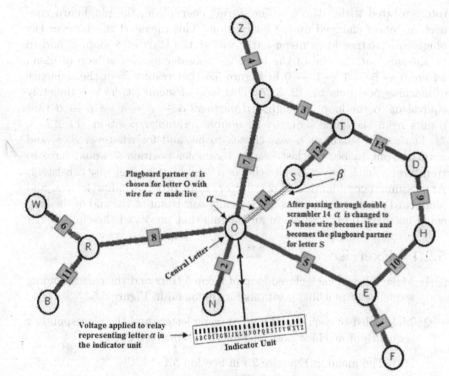

Figure 5.5   Menu emphasizing plugboard partners.

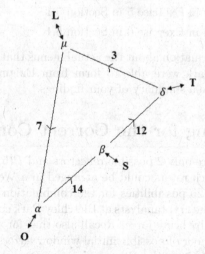

Figure 5.6   Menu loop emphasizing plugboard partners.

rotors rotated within the machine during encryption, the plugboard connections never changed during encryption. This meant that whatever the plugboard partner for a menu letter was at the start of a loop, it had to be the same at the end of the loop. So consider again the loop of menu letters $O \rightarrow S \rightarrow T \rightarrow L \rightarrow O$ in Figure 5.5 that results from the sequence of machine positions 14, 12, 3, 7. This loop of menu letters is completely equivalent to the loop of plugboard partners $\alpha \rightarrow \beta \rightarrow \delta \rightarrow \mu \rightarrow \alpha$ that results from the same sequence of double scrambler positions 14, 12, 3, 7. Thus, with whatever α was chosen to be, and for whatever β, δ, and μ turned out to be, the last double scrambler position 7 would have to transform μ back into α, or else the loop would not be logically consistent. Any assumed combination of a reflector, rotor order, window letters, and plugboard partner α that was not logically consistent at the end of the loop could not have been possible in an Enigma that produced the ciphertext.

5.2.1   Exercises

1. Make a list of each closed loop of menu letters and the corresponding sequence of machine positions for the menu in Figure 5.5.

2. Make a list of each closed loop of menu letters and the corresponding sequence of machine positions for the following menus.

   (a) The menu in Exercise 2b in Section 5.1

   (b)* The menu in Exercise 3b in Section 5.1

   (c) The menu in Exercise 4 in Section 5.1

   (d)* The menu in Exercise 5 in Section 5.1

   (e) The menu in Exercise 6 in Section 5.1

3. Find some information about the actual menus that the cryptanalysts at Bletchley Park were able to form from Enigma crib/ciphertext pairs, and write a summary of your findings.

5.3   Searching for the Correct Configuration

Recall that there were only 2 possible reflectors and $P(5,3) = 5 \cdot 4 \cdot 3 = 60$ possible ways in which rotors could be arranged in a Wehrmacht Enigma. With only $2 \cdot 60 = 120$ possibilities for the combination of a reflector and rotor arrangement, the cryptanalysts at Bletchley Park could determine the correct combination by brute force. Recall also that for an arrangement of three rotors, the number of possible initial window letters and ring settings was $26 \cdot 26 \cdot 26 = 17{,}576$ each. However, when attempting to decrypt an individual message, the cryptanalysts at Bletchley Park found it unnecessary to determine both initial window letters and ring settings. To see why, consider an arrangement of rotors with initial window letters TRA, which are the alphabet letters in positions 20, 18, 1, and ring settings 10, 15, 21. Then as we saw in Section 4.1, the initial rotor offsets for the machine would be 10, 3, 6, since $20 - 10 = 10$, $18 - 15 = 3$, and $1 - 21 = -20$ with $-20 + 26 = 6$. However, for an arrangement of rotors with initial window letters ZZZ, or 26, 26, 26, and ring settings 16, 23, 20, the initial rotor offsets would also be 10, 3, 6, since $26 - 16 = 10$, $26 - 23 = 3$, and $26 - 20 = 6$. Thus, assuming the same reflector and rotor arrangement, a Wehrmacht Enigma configured with initial window letters TRA and ring settings 10, 15, 21 would operate exactly the same way as one configured with initial window letters ZZZ and ring settings 16, 23, 20, at least until there was a rotation of the middle rotor in one but not the other. As a result, the cryptanalysts at Bletchley Park could assume any initial window letters, which often were indeed assumed to be ZZZ, and then search from only 17,576 possibilities for ring settings that gave the correct initial rotor offsets, which they called the *rotor core starting positions*.

Of course, each of the 17,576 possibilities for ring settings had to be considered for each of the 120 possible combinations of a reflector and rotor arrangement, giving a total of $17{,}576 \cdot 120 = 2{,}109{,}120$ configurations that could require some level of testing. Although this number is obviously significant, the cryptanalysts at Bletchley Park found it manageable.

To test one of these 2,109,120 configurations on a given crib/ciphertext alignment, a crib/ciphertext menu was formed, a central letter chosen from the menu, a bombe set up with an indicator unit, and then a voltage applied to one of the 26 relays in the indicator unit, specifically the relay corresponding to a chosen potential plugboard partner of the central letter. If the menu contained a loop, it was checked for a logical inconsistency at the central letter.

Example 5.1 Consider the Wehrmacht Enigma configuration with reflector **C**, rotor order **I, V, III**, initial window letters ZZZ, and ring settings 16, 23, 18, to be tested on the crib/ciphertext alignment at the bottom of page 84. Given the fact that pressing a key on an Enigma keyboard caused the rightmost rotor to rotate one position before the encryption or decryption of the letter, and assuming no rotation of the middle or leftmost rotors, we can notate this alignment with window letters as follows.

Position:	1	2	3	4	5	6	7	8	9	10	11	12	13	14
Window:	ZZA	ZZB	ZZC	ZZD	ZZE	ZZF	ZZG	ZZH	ZZI	ZZJ	ZZK	ZZL	ZZM	ZZN
Crib:	F	O	L	L	O	W	O	R	D	E	R	S	T	O
Cipher:	E	N	T	Z	E	R	L	O	H	H	B	T	D	S

The menu in Figure 5.1 on page 85 is reproduced in Figure 5.7 on page 94, with Figure 5.7 also showing the window letters along each link. Also, for

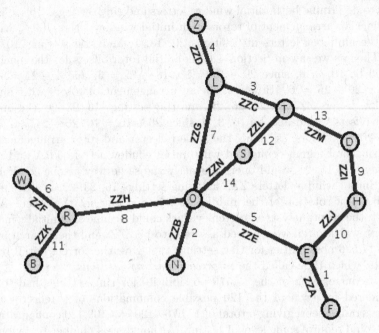

Figure 5.7 Menu showing window letters.

	A	B	C	D	E	F	G	H	I	J	K	L	M	N	O	P	Q	R	S	T	U	V	W	X	Y	Z
ZZA	I	C	B	E	D	Q	J	P	A	G	R	W	V	X	U	H	F	K	Z	Y	O	M	L	N	T	S
ZZB	M	R	X	K	H	N	Z	E	O	P	D	V	A	F	I	J	W	B	T	S	Y	L	Q	C	U	G
ZZC	D	Z	K	A	G	H	E	F	N	M	C	T	J	I	U	W	V	Y	X	L	O	Q	P	S	R	B
ZZD	F	N	H	P	L	A	K	C	Z	V	G	E	U	B	W	D	Y	S	R	X	M	J	O	T	Q	I
ZZE	L	F	P	H	S	B	X	D	J	I	T	A	U	Z	V	C	Y	W	E	K	M	O	R	G	Q	N
ZZF	M	J	D	C	O	S	I	W	G	B	T	U	A	V	E	Y	R	Q	F	K	L	N	H	Z	P	X
ZZG	H	E	K	O	B	G	F	A	S	T	C	M	L	U	D	V	W	X	I	J	N	P	Q	R	Z	Y
ZZH	N	F	G	L	Y	B	C	S	T	W	O	D	V	A	K	X	Z	U	H	I	R	M	J	P	E	Q
ZZI	B	A	Q	P	I	R	S	X	E	Y	L	K	U	Z	V	D	C	F	G	W	M	O	T	H	J	N
ZZJ	X	D	M	B	R	T	I	W	G	N	O	S	C	J	K	V	U	E	L	F	Q	P	H	A	Z	Y
ZZK	E	M	W	S	A	G	F	U	Z	K	J	R	B	P	T	N	Y	L	D	O	H	X	C	V	Q	I
ZZL	U	L	Q	H	Z	M	P	D	T	V	S	B	F	O	N	G	C	X	K	I	A	J	Y	R	W	E
ZZM	S	L	D	C	X	N	M	V	J	I	R	B	G	F	U	W	T	K	A	Q	O	H	P	E	Z	Y
ZZN	G	K	T	R	J	P	A	L	W	E	B	H	N	M	S	F	V	D	O	C	Y	Q	I	Z	U	X

Table 5.1 Double scrambler outputs for the Wehrmacht Enigma configuration in Example 5.1 (with reflector **C**, rotor order **I**, **V**, **III**, and ring settings 16, 23, 18).

a Wehrmacht Enigma with reflector **C**, rotor order **I**, **V**, **III**, and ring settings 16, 23, 18, the outputs from the double scrambler for each possible input and all window letters in the menu are shown in Table 5.1 on page 94. Each letter inside this table is the output that would result from the input labeling the column with the window letters labeling the row. Consider the loop O → S → T → L → O in the menu, and suppose we choose A as the plugboard partner of the central letter O. If a voltage were applied to the relay corresponding to A in the indicator unit, and current designating this letter entered double scrambler 14 in a bombe with drums set for window letters ZZN, then according to Table 5.1 the output from the double scrambler would be current designating G, which would thus be the plugboard partner of S. Next, if G entered double scrambler 12 with drum setting ZZL, then according to Table 5.1 the output would be P, which would thus be the plugboard partner of T. Next, if P entered double scrambler 3 with drum setting ZZC, the output would be W, which would thus be the plugboard partner of L. Finally, if W entered double scrambler 7 with drum setting ZZG, the output would be Q, which would thus be the plugboard partner of O. In summary, this machine configuration with plugboard partner A of O at the start of the loop would result in the following plugboard partners of the menu letters in the loop.

$$
\begin{array}{llllll}
\textbf{Drum Setting:} & \text{ZZN} & \text{ZZL} & \text{ZZC} & \text{ZZG} \\
\textbf{Menu Letter:} & \text{O} \longrightarrow & \text{S} \longrightarrow & \text{T} \longrightarrow & \text{L} \longrightarrow & \text{O} \\
\textbf{Plug Partner:} & \text{A} \longrightarrow & \text{G} \longrightarrow & \text{P} \longrightarrow & \text{W} \longrightarrow & \text{Q}
\end{array}
$$

Note that this gives different plugboard partners of the central letter at the start and end of the loop. This logical inconsistency indicates that the assumed settings, including the reflector, rotor order, initial window letters, ring settings, and plugboard partner of the central letter, cannot all be correct. Continuing with the same reflector, rotor order, initial window letters, and ring settings, suppose we now choose S as the plugboard partner of O. This would result in the following plugboard partners of the menu letters in the loop.

$$
\begin{array}{llllll}
\textbf{Drum Setting:} & \text{ZZN} & \text{ZZL} & \text{ZZC} & \text{ZZG} \\
\textbf{Menu Letter:} & \text{O} \longrightarrow & \text{S} \longrightarrow & \text{T} \longrightarrow & \text{L} \longrightarrow & \text{O} \\
\textbf{Plug Partner:} & \text{S} \longrightarrow & \text{O} \longrightarrow & \text{N} \longrightarrow & \text{I} \longrightarrow & \text{S}
\end{array}
$$

Since this gives the same plugboard partner of the central letter at the start and end of the loop, the assumed settings could be correct.   □

As we have noted, drum positions were set to correspond to particular window letters. Drums were loaded in groups of three, and each drum was color coded to indicate which of the five Wehrmacht Enigma rotors it emulated. Each drum was also labeled with the 26 letters of the alphabet in a circular pattern, which could be rotated and set to any of 26 positions. Each group of drums was loaded in a vertical sequence, with the drum emulating the rightmost rotor in an Enigma at the top, the drum emulating the middle rotor in the middle, and the drum emulating the leftmost rotor at the bottom. When a bombe was fully loaded with drums and viewed from the front, an observer would see three rows of drum groups, with 12 groups in the top and bottom rows, and 13 in the middle row, as shown in Figure 5.8.

Figure 5.8   Front view of a replica Turing bombe.

Within each group, drums were initially rotated to positions that would emulate a double scrambler with window letters corresponding to a particular link in a menu. More specifically, with assumed initial window letters ZZZ, the drums in the leftmost group on the top row of a bombe would be rotated to emulate a double scrambler for the encryption in position 1 with window letters ZZA, the group to its right would be rotated to emulate a double scrambler for the encryption in position 2 with window letters ZZB, the group to its right would be rotated to emulate a double scrambler for the encryption in position 3 with window letters ZZC, and so forth across the top row. A diagram illustrating this is shown in Figure 5.9 on page 97.

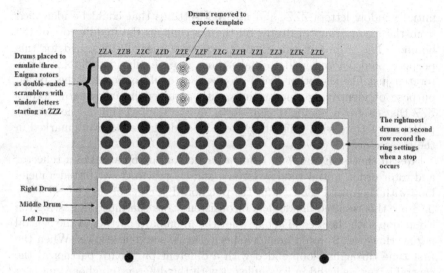

Figure 5.9 Front view of a Turing bombe.

The first 12 drum groups in the middle row of a bombe and all 12 groups in the bottom row were constructed to operate in the same way as the 12 groups in the top row, but could be loaded with drums that corresponded to different rotor orders than the top row. That is, because the bombe included three rows of drum groups, three different rotor orders could sometimes (depending on the menu) be tested simultaneously.[1] The extra drum group on the far right in the middle row served a different purpose that we will describe shortly.

When a bombe was ready to be run, a switch on the side was flipped, which sent voltage to the relay in the indicator unit corresponding to a potential plugboard partner of the central letter. As it ran, the drums would rotate to emulate advancing window letters in an Enigma, with the top drum in each drum group rotating the fastest. If a combination of a reflector and rotor order, initial window letters, ring settings, and a plugboard partner of the central letter were tested for which no logical inconsistency occurred at the central letter, the bombe would shut down so the settings could be recorded to be checked further using a different process.

Initial window letters and ring settings that caused a bombe to shut down could be recorded in two different ways. One way was to assume ring settings and use the positions to which the drums had rotated to indicate initial window letters. However, a more convenient way was to assume

---

[1] The bombe thus had the added distinction of serving as a very early example of parallel computing.

initial window letters ZZZ, and use ring settings that could be identified from the extra group of drums on the far right in the middle row of the bombe. The drums in this extra group were specially designed for this purpose, and colored gold to make them more easily identifiable. They rotated just like the other drum groups, but could only serve the special purpose of identifying ring settings with assumed initial window letters ZZZ that caused a bombe to shut down, due to the fact that in the circular pattern of letters with which they were labeled, the letters were marked in the reverse order.

Recall that in Example 5.1, for a particular combination of a reflector and rotor order, initial window letters, and ring settings, we tested a menu loop with the initial choice A for the plugboard partner of the central letter O. Since this resulted in a logical inconsistency, we continued by testing the same loop with the second choice S for the plugboard partner of the central letter. However, S would not have been Turing's second choice. When the first time through a loop ended with a different plugboard partner of the central letter, as it did in Example 5.1 with the different plugboard partner Q of O, Turing found it more efficient to continue testing the same loop with this different plugboard partner as the second choice. That is, Turing's second choice for the plugboard partner of the central letter in Example 5.1 would have been Q. When the second time through the loop resulted in another different plugboard partner of the central letter, Turing would have again tested the same loop with this other different plugboard partner of the central letter, carrying on in this manner until eventually cycling back around to the original choice for the plugboard partner of the central letter. Turing designed the bombe to run continuously in this manner, with each choice for the plugboard partner of the central letter that led to a logical inconsistency at the central letter recorded in the indicator unit.

Example 5.2 In Example 5.1, we first tested the menu loop O → S → T → L → O with the initial choice A for the plugboard partner of the central letter O. This resulted in the following plugboard partners for the menu letters in the loop, which gave a logical inconsistency.

$$\begin{array}{llcccc}\textbf{Drum Setting:} & \text{ZZN} & \text{ZZL} & \text{ZZC} & \text{ZZG} \\ \textbf{Menu Letter:} & \text{O} \longrightarrow & \text{S} \longrightarrow & \text{T} \longrightarrow & \text{L} \longrightarrow & \text{O} \\ \textbf{Plug Partner:} & \text{A} \longrightarrow & \text{G} \longrightarrow & \text{P} \longrightarrow & \text{W} \longrightarrow & \text{Q}\end{array}$$

Using Turing's scheme, next we would test the same loop again but with Q as the plugboard partner of O. This choice results in the following plugboard partners for the menu letters, which again gives a logical inconsistency.

$$\begin{array}{llcccc}\textbf{Drum Setting:} & \text{ZZN} & \text{ZZL} & \text{ZZC} & \text{ZZG} \\ \textbf{Menu Letter:} & \text{O} \longrightarrow & \text{S} \longrightarrow & \text{T} \longrightarrow & \text{L} \longrightarrow & \text{O} \\ \textbf{Plug Partner:} & \text{Q} \longrightarrow & \text{V} \longrightarrow & \text{J} \longrightarrow & \text{M} \longrightarrow & \text{L}\end{array}$$

Again using Turing's scheme, next we would test the same loop again but with L as the plugboard partner of O. This choice results in the following plugboard partners for the menu letters, which again gives a logical inconsistency at the central letter.

Drum Setting: ZZN ZZL ZZC ZZG
Menu Letter: O $\longrightarrow$ S $\longrightarrow$ T $\longrightarrow$ L $\longrightarrow$ O
Plug Partner: L $\longrightarrow$ H $\longrightarrow$ D $\longrightarrow$ A $\longrightarrow$ H

Continuing Turing's scheme would involve testing the same loop in this manner until eventually cycling back around to the original choice A for the plugboard partner of O. This would require testing the loop nine more times, and the choices for the plugboard partner of O would be the following letters in order.

$$A \mapsto Q \mapsto L \mapsto H \mapsto Y \mapsto O \mapsto K \mapsto J \mapsto E \mapsto W \mapsto M \mapsto N \mapsto A$$

For convenience we can represent these letters as the *cycle* (AQLHYOKJEWMN). For the reflector, rotor order, initial window letters, and ring settings assumed in Example 5.1, the letters in this cycle can all be eliminated as possible plugboard partners of O. Of course, we could have started the whole process with an initial choice for the plugboard partner of O that was not any of the letters in this cycle. For example, the initial choice B would have resulted in the cycle (BRGDIXF), the initial choice C in the cycle (CVU), and the initial choice P in the cycle (PT). The letters in these three cycles can thus also be eliminated as possible plugboard partners of O for the reflector, rotor order, initial window letters, and ring settings assumed in Example 5.1. The only letters not eliminated by any of these cycles are S and Z, which as we verified in Example 5.1 for S, and which turns out to be true for Z as well, do not result in a logical inconsistency at the central letter when chosen as the plugboard partner of O. These letters could thus be represented as the cycles (S) and (Z), giving the following complete list of cycles for the loop O → S → T → L → O.

$$(AQLHYOKJEWMN)(BRGDIXF)(CVU)(PT)(S)(Z) \qquad \square$$

In Example 5.2, for a particular combination of a reflector and rotor order, initial window letters, and ring settings, we tested a menu loop with initial choice A for the plugboard partner of the central letter O. This resulted in the cycle (AQLHYOKJEWMN), all letters in which could be eliminated as possible plugboard partners of O. We went on in Example 5.2 to note some other letters that could be eliminated as possible plugboard partners of O by testing the same menu loop with other initial choices for the plugboard partner of O. However, since the menu in this example (which is shown in

Figure 5.7 on page 94) had multiple loops, this is not what Turing would have done. For menus with multiple loops, Turing chose to test all of the loops with the same initial choice for the plugboard partner of the central letter, and used the resulting cycles together to eliminate possible plugboard partners of the central letter more quickly.

Example 5.3 The menu in Figure 5.7 on page 94 has three loops, which can be represented as follows.

1. $0 \to S \to T \to L \to 0$

2. $0 \to S \to T \to D \to H \to E \to 0$

3. $0 \to L \to T \to D \to H \to E \to 0$

In Example 5.2, we tested loop 1 with various choices for the plugboard partner of the central letter, which resulted in the following complete list of cycles for loop 1.

$$\text{(AQLHYOKJEWMN)(BRGDIXF)(CVU)(PT)(S)(Z)}$$

Testing loops 2 and 3 would result in different cycles, of course. The complete list of cycles that would result for each loop is given in the following table, with the cycles labeled for later reference.

Loop	Cycles
1	$C_1 = \text{(AQLHYOKJEWMN)}$, $C_2 = \text{(BRGDIXF)}$, $C_3 = \text{(CVU)}$, $C_4 = \text{(PT)}$, $C_5 = \text{(S)}$, $C_6 = \text{(Z)}$
2	$D_1 = \text{(ABHVOKGCNQWUFYJZELMP)}$, $D_2 = \text{(DXRT)}$, $D_3 = \text{(I)}$, $D_4 = \text{(S)}$
3	$E_1 = \text{(AQBYVFRHMUNPDCOJGT)}$, $E_2 = \text{(EZ)}$, $E_3 = \text{(IX)}$, $E_4 = \text{(K)}$, $E_5 = \text{(LW)}$, $E_6 = \text{(S)}$

In Example 5.2, from the cycles C_1–C_6 for loop 1, we were able to eliminate all letters except S and Z as possible plugboard partners of the central letter, since every letter except S and Z appears in a cycle with at least one other letter. From the cycle D_1 for loop 2, we can also eliminate Z. Further, since S also appears in a cycle by itself for both loops 2 and 3, we see that considering the complete list of cycles for all three loops, S is the only possible plugboard partner of the central letter. Recall though that eliminating possible plugboard partners by forming the complete list of cycles for each loop was not Turing's approach. Instead, Turing would have begun by testing all three loops with the same initial choice A for the plugboard partner of the central letter. That is, after forming the cycle C_1, Turing would have next formed D_1 and E_1. These three cycles eliminate all letters except I, S, and X as possible plugboard partners of the central letter. Next, given that B was eliminated by D_1 but not C_1, Turing could have gone back to the first loop with B as the plugboard partner of the central letter, and formed C_2. This further eliminates I and X, again leaving S as the only possible plugboard partner of the central letter. The letters, as they are eliminated by the cycles C_1, D_1, E_1, and C_2 with initial choice A for the plugboard partner of the central letter, are summarized in the following table.

Test letter	A	B	C	D	E	F	G	H	I	J	K	L	M	N	O	P	Q	R	S	T	U	V	W	X	Y	Z
A activates C_1	A				E			H		J	K	L	M	N	O		Q						W		Y	
A activates D_1	A	B	C		E	F	G	H		J	K	L	M	N	O	P	Q				U	V	W		Y	Z
A activates E_1	A	B	C	D		F	G	H		J			M	N	O	P	Q	R		T	U	V			Y	
Join C_1, D_1, E_1	A	B	C	D	E	F	G	H		J	K	L	M	N	O	P	Q	R		T	U	V	W		Y	Z
B activates C_2		B		D		F	G		I									R						X		
Join C_2	A	B	C	D	E	F	G	H	I	J	K	L	M	N	O	P	Q	R		T	U	V	W	X	Y	Z

When this testing was done using a bombe, the fact that S was the only possible plugboard partner of the central letter would have been noted from the indicator unit, where every relay except the one corresponding to S would have received a voltage. □

In Example 5.3, for a particular combination of a reflector and rotor order, initial window letters, and ring settings, we showed that S was the only possible plugboard partner of the central letter in a menu that resulted from a given crib/ciphertext alignment. However, this possible plugboard partner is tied to that particular combination of a reflector and rotor order, initial window letters, and ring settings, which may or may not actually be the correct settings of the Enigma that produced the crib/ciphertext alignment. All we really showed in Example 5.3 was that *if* a Wehrmacht Enigma were configured as noted at the start of Example 5.1 on page 93 (with reflector **C**, rotor order **I, V, III**, initial window letters ZZZ, and ring settings 16, 23, 18), then S would be the only possible plugboard partner of the central letter O in the menu in Figure 5.7 on page 94 that results from the crib/ciphertext alignment in Example 5.1. Different combinations of a reflector and rotor order, initial window letters, and ring settings might lead to other possible plugboard partners of the central letter though.

Example 5.4 Consider the Wehrmacht Enigma configuration with reflector **C**, rotor order **I, V, III**, initial window letters ZZZ, and ring settings 13, 2, 6, to be tested on the crib/ciphertext alignment in Example 5.1 on page 93, with menu shown in Figure 5.7 on page 94 and central letter O. For a Wehrmacht Enigma with reflector **C**, rotor order **I, V, III**, and ring settings 13, 2, 6, the outputs from the double scrambler for each possible input and all window letters in the menu are shown in Table 5.2 on page

103. With the menu loops labeled 1–3 as in Example 5.3, the complete list of cycles that would result for each loop is given in the following table.

Loop	Cycles
1	$C_1 = \text{(AGWLK)}$, $C_2 = \text{(B)}$, $C_3 = \text{(CYSVJTEU)}$, $C_4 = \text{(DRMQOFNPI)}$, $C_5 = \text{(HZ)}$, $C_6 = \text{(X)}$
2	$D_1 = \text{(AOE)}$, $D_2 = \text{(B)}$, $D_3 = \text{(CMUJT)}$, $D_4 = \text{(DKYWSRVX)}$, $D_5 = \text{(FIGLNQZH)}$, $D_6 = \text{(P)}$
3	$E_1 = \text{(AFD)}$, $E_2 = \text{(B)}$, $E_3 = \text{(CQHNOUTYLPIWVXRJEGKSM)}$, $E_4 = \text{(Z)}$

Letters, as they are eliminated as possible plugboard partners of the central letter by these cycles with initial choice A for the plugboard partner of the central letter, are summarized in the following table.

Test letter	A	B	C	D	E	F	G	H	I	J	K	L	M	N	O	P	Q	R	S	T	U	V	W	X	Y	Z
A activates C_1	A						G				K	L											W			
A activates D_1	A				E										O											
A activates E_1	A			D		F																				
Join C_1,D_1,E_1	A			D	E	F	G				K	L			O								W			
D activates C_4				D		F			I				M	N	O	P	Q	R								
D activates D_4				D							K							R	S			V	W	X	Y	
Join C_4,D_4	A			D	E	F	G		I		K	L	M	N	O	P	Q	R	S			V	W	X	Y	
E activates C_3			C		E					J									S	T	U	V			Y	
E activates E_3			C		E		G	H	I	J	K	L	M	N	O	P	Q	R	S	T	U	V	W	X	Y	
Join C_3,E_3	A		C	D	E	F	G	H	I	J	K	L	M	N	O	P	Q	R	S	T	U	V	W	X	Y	
F activates D_5						F	G	H	I			L		N			Q									Z
Join D_5	A		C	D	E	F	G	H	I	J	K	L	M	N	O	P	Q	R	S	T	U	V	W	X	Y	Z

This indicates that if a Wehrmacht Enigma were configured with reflector **C**, rotor order **I, V, III**, initial window letters ZZZ, and ring settings 13, 2, 6, then B would be the only possible plugboard partner of the central letter O in the menu in Figure 5.7 that results from the crib/ciphertext alignment in Example 5.1.   □

For a particular combination of a reflector and rotor order, initial window letters, and ring settings, it could also occur that all 26 letters would result in a logical inconsistency at the central letter in at least one loop when tested as the plugboard partner of the central letter in a menu that resulted from a given crib/ciphertext alignment. This would mean that regardless of how the plugboard was connected, that particular combination of a reflector and rotor order, initial window letters, and ring settings in a Wehrmacht Enigma could not have produced the crib/ciphertext alignment.

Example 5.5 Consider the Wehrmacht Enigma configuration with reflector **C**, rotor order **I, V, III**, initial window letters ZZZ, and ring settings

	A	B	C	D	E	F	G	H	I	J	K	L	M	N	O	P	Q	R	S	T	U	V	W	X	Y	Z
ZZA	F	D	S	B	Z	A	L	M	U	T	R	G	H	Y	W	V	X	K	C	J	I	P	O	Q	N	E
ZZB	O	E	W	N	B	T	U	V	S	L	R	J	P	D	A	M	Y	K	I	F	G	H	C	Z	Q	X
ZZC	U	Y	E	Q	C	S	L	V	W	M	Z	G	J	T	R	X	D	O	F	N	A	H	I	P	B	K
ZZD	N	I	X	Y	U	W	Q	V	B	R	P	M	L	A	T	K	G	J	Z	O	E	H	F	C	D	S
ZZE	Y	W	I	Z	T	K	R	O	C	S	F	U	Q	V	H	X	M	G	J	E	L	N	B	P	A	D
ZZF	O	Z	R	Q	V	T	N	J	S	H	X	M	L	G	A	Y	D	C	I	F	W	E	U	K	P	B
ZZG	Y	H	W	P	K	L	O	B	N	R	E	F	T	I	G	D	S	J	Q	M	Z	X	C	V	A	U
ZZH	W	M	S	X	L	N	J	O	Y	G	U	E	B	F	H	T	V	Z	C	P	K	Q	A	D	I	R
ZZI	G	T	R	L	S	P	A	K	Q	N	H	D	X	J	U	F	I	C	E	B	O	Z	Y	M	W	V
ZZJ	P	R	Y	L	O	G	F	Q	X	U	V	D	S	W	E	A	H	B	M	Z	J	K	N	I	C	T
ZZK	M	F	H	U	L	B	Z	C	Q	R	O	E	A	W	K	V	I	J	Y	X	D	P	N	T	S	G
ZZL	Z	R	M	K	P	G	F	T	V	X	D	O	C	S	L	E	Y	B	N	H	W	I	U	J	Q	A
ZZM	S	T	R	Y	N	I	H	G	F	V	M	Z	K	E	U	X	W	C	A	B	O	J	Q	P	D	L
ZZN	O	I	S	M	R	H	N	F	B	X	W	U	D	G	A	Y	Z	E	C	V	L	T	K	J	P	Q

Table 5.2 Double scrambler outputs for the Wehrmacht Enigma configuration in Example 5.4 (with reflector **C**, rotor order **I**, **V**, **III**, and ring settings 13, 2, 6).

	A	B	C	D	E	F	G	H	I	J	K	L	M	N	O	P	Q	R	S	T	U	V	W	X	Y	Z
ZZA	X	J	M	Z	W	V	K	O	U	B	G	P	C	S	H	L	T	Y	N	Q	I	F	E	A	R	D
ZZB	C	Y	A	I	W	L	K	T	D	O	G	F	Q	X	J	S	M	V	P	H	Z	R	E	N	B	U
ZZC	T	M	H	L	O	V	Z	C	N	U	R	D	B	I	E	S	Y	K	P	A	J	F	X	W	Q	G
ZZD	I	W	D	C	H	M	V	E	A	P	X	N	F	L	R	J	U	O	Z	Y	Q	G	B	K	T	S
ZZE	S	U	T	G	J	I	D	Q	F	E	M	R	K	Z	Y	V	H	L	A	C	B	P	X	W	O	N
ZZF	J	D	W	B	Y	V	O	U	M	A	S	Q	I	X	G	T	L	Z	K	P	H	F	C	N	E	R
ZZG	W	R	M	U	F	E	P	L	Z	T	Y	H	C	Q	V	G	N	B	X	J	D	O	A	S	K	I
ZZH	Q	D	G	B	U	L	C	S	Y	R	X	F	P	Z	W	M	A	J	H	V	E	T	O	K	I	N
ZZI	S	F	R	W	K	B	Y	U	T	Z	E	N	P	L	V	M	X	C	A	I	H	O	D	Q	G	J
ZZJ	H	D	K	B	S	M	X	A	R	Y	C	O	F	Q	L	W	N	I	E	U	T	Z	P	G	J	V
ZZK	C	J	A	T	R	H	M	F	V	B	U	P	G	Y	S	L	Z	E	O	D	K	I	X	W	N	Q
ZZL	D	E	T	A	B	S	Z	O	U	V	W	X	Q	R	H	Y	M	N	F	C	I	J	K	L	P	G
ZZM	W	D	Z	B	G	L	E	X	U	M	O	F	J	Y	K	S	V	T	P	R	I	Q	A	H	N	C
ZZN	T	F	L	Y	V	B	O	W	K	N	I	C	X	J	G	Q	P	S	R	A	Z	E	H	M	D	U

Table 5.3 Double scrambler outputs for the Wehrmacht Enigma configuration in Example 5.5 (with reflector **C**, rotor order **I**, **V**, **III**, and ring settings 12, 6, 16).

12, 6, 16, to be tested on the crib/ciphertext alignment in Example 5.1 on page 93, with menu shown in Figure 5.7 on page 94 and central letter O. For a Wehrmacht Enigma with reflector **C**, rotor order **I, V, III**, and ring settings 12, 6, 16, the outputs from the double scrambler for each possible input and all window letters in the menu are shown in Table 5.3 on page 103. With the menu loops labeled 1–3 as in Example 5.3, the complete list of cycles that would result for each loop is given in the following table.

Loop	Cycles
1	$C_1 = $ (ACVEDXQSJUGITBNYLOHKRPF), $C_2 = $ (MZW)
2	$D_1 = $ (ALWFVC), $D_2 = $ (BGMUISZQNEDXKTH), $D_3 = $ (JY), $D_4 = $ (OPR)
3	$E_1 = $ (AOFEXRHNDQYUTKBIJLMGZSW), $E_2 = $ (C), $E_3 = $ (P), $E_4 = $ (V)

Letters, as they are eliminated as possible plugboard partners of the central letter by these cycles with initial choice A for the plugboard partner of the central letter, are summarized in the following table.

Test letter	A	B	C	D	E	F	G	H	I	J	K	L	M	N	O	P	Q	R	S	T	U	V	W	X	Y	Z
A activates C_1	A	B	C	D	E	F	G	H	I	J	K	L		N	O	P	Q	R	S	T	U	V		X	Y	
A activates D_1	A		C			F						L										V	W			
A activates E_1	A	B		D	E	F	G	H	I	J	K	L	M	N	O		Q	R	S	T	U		W	X	Y	Z
Join C_1, D_1, E_1	A	B	C	D	E	F	G	H	I	J	K	L	M	N	O	P	Q	R	S	T	U	V	W	X	Y	Z

Since this results in no possible plugboard partner of the central letter O, it indicates that a Wehrmacht Enigma configured with reflector **C**, rotor order **I, V, III**, initial window letters ZZZ, and ring settings 12, 6, 16 could not have produced the crib/ciphertext alignment in Example 5.1. When this testing was done using a bombe, this fact would have been noted from the indicator unit, where every relay would have received a voltage.   □

For a particular combination of a reflector and rotor order, initial window letters, and ring settings, when a bombe was run as we have described to test an initial choice for the plugboard partner of the central letter in a menu formed from a given crib/ciphertext alignment, the result was almost always one of the following three outcomes.

1. No logical inconsistency occurred at the central letter for any menu loop. An example of this would be if B were chosen as the initial plugboard partner of the central letter O in Example 5.4. This fact would be noted from the indicator unit, where only the relay corresponding to B would have received a voltage. If this happened, the bombe would shut down so this information could be recorded for further analysis.

2. A logical inconsistency occurred at the central letter for exactly 25 letters, thus eliminating exactly 25 letters as possible plugboard partners of the central letter, but leaving one letter still as a possible plugboard partner of the central letter. An example of this would be if A were chosen as the initial plugboard partner of the central letter O in Example 5.4, with every letter but B eventually being eliminated. This fact would also be noted from the indicator unit, where every relay except the one corresponding to B would have received a voltage. If this happened, the bombe would also shut down so this information could be recorded for further analysis.

3. A logical inconsistency occurred at the central letter for all 26 letters, thus eliminating all 26 letters as possible plugboard partners of the central letter. An example of this would be if A were chosen as the initial plugboard partner of the central letter O in Example 5.5. In this case, every relay in the indicator unit would receive a voltage. This would cause at least one drum in each group to rotate one position, and the bombe to automatically continue running to test this new configuration.

Bombe operators referred to the machine shutting down due to outcomes 1 or 2 as a *stop*. Not every stop gave a correct Enigma configuration for a given crib/ciphertext alignment though. Stops that gave correct configurations were called *good* stops, and others were called *false*. However, whether good or false, stops due to outcomes 1 or 2 always occurred because a particular letter appeared in a cycle by itself for every loop in a menu. For example, because B appears in a cycle by itself for all three menu loops in Example 5.4, an outcome 1 stop would result from the initial choice B for the plugboard partner of the central letter O, but for the same reason an outcome 2 stop would result from the initial choice A for the plugboard partner of O. It could just be by chance though (as opposed to because the configuration was correct) that B appears in a cycle by itself for all three menu loops in Example 5.4. The reason any particular letter ever appeared in a cycle by itself for a menu loop was because when that letter was chosen as the initial plugboard partner of the central letter at the start of the loop, the same letter came back out at the end of the loop. This would always happen with a correct configuration, of course, but it would also happen on average $\frac{1}{26}$ of the time with an incorrect configuration, since what comes out at the end of a loop has to be one of the 26 letters in the alphabet.

An important observation from this is that the likelihood of a stop being false was in direct correspondence with the number of loops in the menu being tested. An outcome 1 or 2 false stop would only occur if a letter appeared by chance in a cycle by itself for every loop in the menu. For each loop, there was a $\frac{1}{26}$ chance that a letter would appear in a cycle by itself. Thus, for a menu with n loops, there would only be a $\frac{1}{26^n}$ chance that a letter would appear in a cycle by itself for every loop. This means that for a crib/ciphertext alignment that resulted in a menu with n loops, and a particular combination of a reflector and rotor order and assumed initial window letters, given that $26^3 = 17,576$ ring settings were possible in a Wehrmacht Enigma, the expected number of stops in a bombe run would be $\frac{17,576}{26^n}$. For example, for menus with 3 loops, there would be $\frac{17,576}{26^3} = 1$ stop on average for every combination of a reflector and rotor order and assumed initial window letters. The cryptanalysts at Bletchley Park found this manageable. Menus with fewer loops did result in an unmanageable number of stops, however. Fortunately, the crib/ciphertext menus they considered did at least sometimes have three or more loops.

5.3.1   Exercises

1. For a Wehrmacht Enigma with the following combinations of initial window letters and ring settings, find the rotor core starting positions.

    (a)* Initial window letters PRT, ring settings 10, 17, 5

    (b) Initial window letters SAM, ring settings 2, 9, 13

    (c)* Initial window letters ZME, ring settings 12, 20, 26

    (d) Initial window letters DIX, ring settings 5, 10, 25

2.* For a Wehrmacht Enigma with initial window letters ZZZ, find ring settings that give the same rotor core starting positions as those that result from each part of Exercise 1.

3. Each part of this exercise gives one or more cycles that result for the loops in a menu formed from a Wehrmacht Enigma crib/ciphertext alignment, and an initial choice for the plugboard partner of the central letter in the menu. For each part, from only the information given, find a letter that cannot be eliminated as a possible plugboard partner of the central letter. (The menus, crib/ciphertext alignments, and central letters are neither given nor necessary.)

    (a)* The following cycles, with initial choice D for the plugboard partner of the central letter:

Loop	Cycles
1	$C_1 = $ (AGY), $C_2 = $ (BHXEQUOZMKPNLRWTCFVSJ), $C_3 = $ (D)
2	$D_1 = $ (ARZEI), $D_2 = $ (BPVSOHGFQXLNUJ), $D_3 = $ (D)
3	$E_1 = $ (AIQVNRGXULEMK), $E_2 = $ (B), $E_3 = $ (D)

(b) The following cycles, with initial choice A for the plugboard partner of the central letter:

Loop	Cycles
1	$C_1 = $ (AZMRTVGBKPSHLUJIDNFXE)
2	$D_1 = $ (ACRLFPVSHY)
3	$E_1 = $ (AZUINLGJMCWHRBKFXEYTPVDQ)

(c)* The following cycles, with initial choice A for the plugboard partner of the central letter:

Loop	Cycles
1	$C_1 = $ (ANKYZ), $C_2 = $ (BHPXVJLFCTQIORW)
2	$D_1 = $ (AJWMSOYCI), $D_2 = $ (BZQVLNXHF)
3	$E_1 = $ (ALHUKOPNRJV), $E_2 = $ (BXQCDGTZISMWFY)

(d) The following cycles, with initial choice A for the plugboard partner of the central letter:

Loop	Cycles
1	$C_1 = $ (AZQIHOLFTBYRCNS), $C_2 = $ (DWJVP)
2	$D_1 = $ (ALJPF), $D_2 = $ (BOYRNMDVEKQTIWZSHUX)
3	$E_1 = $ (AFKJZCTYL), $E_2 = $ (BVNADQUHYSLCFP)

4. Suppose the crib GLOBAL MISSION is known to encrypt to the ciphertext BCJQQ QYLQJ OZM, which was formed using a Wehrmacht Enigma.

  (a) Assuming initial window letters ZZZ, draw a menu that expresses the crib/ciphertext pairs with the window letters shown along each link. (We considered this crib and ciphertext previously in Exercise 2 in Section 5.1, in which we drew a menu that expresses the crib/ciphertext pairs.)

  (b) Consider the Wehrmacht Enigma configuration with reflector **B**, rotor order **IV, III, I**, initial window letters ZZZ, and ring settings 9, 6, 25, to be tested on the crib/ciphertext alignment and menu from part (a) with central letter Q. For a Wehrmacht Enigma with reflector **B**, rotor order **IV, III, I**, and ring settings 9, 6, 25, the outputs from the double scrambler for each possible input and all window letters in the menu are shown in Table 5.4 on page 108. Show that for the only loop in the menu, the choice S for the plugboard partner of the central letter results in no logical inconsistency at the central letter.

  (c) Consider the Wehrmacht Enigma configuration with reflector **B**, rotor order **II, IV, I**, initial window letters ZZZ, and ring settings 9, 6, 25, to be tested on the crib/ciphertext alignment and menu

	A	B	C	D	E	F	G	H	I	J	K	L	M	N	O	P	Q	R	S	T	U	V	W	X	Y	Z
ZZA	T	J	R	S	Q	P	I	M	G	B	X	Y	H	O	N	F	E	C	D	A	V	U	Z	K	L	W
ZZB	N	D	O	B	Z	U	R	W	X	K	J	S	T	A	C	V	Y	G	L	M	F	P	H	I	Q	E
ZZC	D	K	H	A	L	V	T	C	X	O	B	E	P	Z	J	M	S	U	Q	G	R	F	Y	I	W	N
ZZD	D	U	J	A	K	Y	L	N	Q	C	E	G	V	H	Z	W	I	T	X	R	B	M	P	S	F	O
ZZE	S	W	X	R	M	N	Z	J	L	H	T	I	E	F	P	O	Y	D	A	K	V	U	B	C	Q	G
ZZF	K	C	B	Q	F	E	Y	U	M	O	A	N	I	L	J	S	D	V	P	Z	H	R	X	W	G	T
ZZG	P	W	Y	O	K	M	R	I	H	L	E	J	F	Z	D	A	T	G	V	Q	X	S	B	U	C	N
ZZH	S	G	K	R	J	M	B	N	P	E	C	T	F	H	Q	I	O	D	A	L	W	X	U	V	Z	Y
ZZI	M	G	L	N	O	V	B	X	R	T	U	C	A	D	E	Z	S	I	Q	J	K	F	Y	H	W	P
ZZJ	T	N	P	G	K	H	D	F	O	Q	E	R	X	B	I	C	J	L	Z	A	W	Y	U	M	V	S
ZZK	T	G	Z	L	R	Q	B	U	O	N	S	D	Y	J	I	V	F	E	K	A	H	P	X	W	M	C
ZZL	H	S	G	Y	W	I	C	A	F	Q	R	P	O	U	M	L	J	K	B	Z	N	X	E	V	D	T
ZZM	D	V	I	A	O	M	U	T	C	P	R	X	F	Q	E	J	N	K	W	H	G	B	S	L	Z	Y

Table 5.4 Double scrambler outputs for the Wehrmacht Enigma configuration in Exercise 4b (with reflector **B**, rotor order **IV**, **III**, **I**, and ring settings 9, 6, 25).

	A	B	C	D	E	F	G	H	I	J	K	L	M	N	O	P	Q	R	S	T	U	V	W	X	Y	Z
ZZA	T	Y	Q	P	N	W	J	I	H	G	L	K	V	E	X	D	C	Z	U	A	S	M	F	O	B	R
ZZB	C	E	A	U	B	M	P	W	Z	R	O	Y	F	Q	K	G	N	J	X	V	D	T	H	S	L	I
ZZC	F	L	U	K	M	A	N	Q	X	R	D	B	E	G	W	S	H	J	P	Y	C	Z	O	I	T	V
ZZD	W	V	T	H	S	N	Y	D	J	I	Q	U	R	F	Z	X	K	M	E	C	L	B	A	P	G	O
ZZE	R	W	S	K	Q	G	F	N	V	Y	D	Z	O	H	M	T	E	A	C	P	X	I	B	U	J	L
ZZF	J	I	X	Q	P	R	Z	N	B	A	M	T	K	H	V	E	D	F	U	L	S	O	Y	C	W	G
ZZG	Z	O	R	V	K	N	P	X	S	Q	E	Y	U	F	B	G	J	C	I	W	M	D	T	H	L	A
ZZH	J	N	K	S	O	U	M	Q	W	A	C	P	G	B	E	L	H	Y	D	V	F	T	I	Z	R	X
ZZI	S	K	O	Q	R	Y	I	X	G	U	B	P	W	T	C	L	D	E	A	N	J	Z	M	H	F	V
ZZJ	D	Z	K	A	J	S	L	U	V	E	C	G	X	P	Q	N	O	W	F	Y	H	I	R	M	T	B
ZZK	H	G	T	Z	N	X	B	A	Q	K	J	Y	W	E	V	S	I	U	P	C	R	O	M	F	L	D
ZZL	T	O	W	Y	I	Z	U	V	E	X	Q	M	L	S	B	R	K	P	N	A	G	H	C	J	D	F
ZZM	D	P	M	A	R	X	H	G	Z	T	Y	U	C	S	W	B	V	E	N	J	L	Q	O	F	K	I

Table 5.5 Double scrambler outputs for the Wehrmacht Enigma configuration in Exercise 4c (with reflector **B**, rotor order **II**, **IV**, **I**, and ring settings 9, 6, 25).

from part (a) with central letter Q. For a Wehrmacht Enigma with reflector **B**, rotor order **II**, **IV**, **I**, and ring settings 9, 6, 25, the outputs from the double scrambler for each possible input and all window letters in the menu are shown in Table 5.5 on page 108. Show that for the only loop in the menu, the choice R for the plugboard partner of the central letter results in no logical inconsistency at the central letter.

5.* Suppose the crib GOOD PET OFTEN IS is known to encrypt to the ciphertext MUTVM UUPWF FTSF, which was formed using a Wehrmacht Enigma.

   (a) Assuming initial window letters ZZZ, draw a menu that expresses the crib/ciphertext pairs with the window letters shown along each link. (We considered this crib and ciphertext previously in Exercise 3 in Section 5.1, in which we drew a menu that expresses the crib/ciphertext pairs.)

   (b) Consider the Wehrmacht Enigma configuration with reflector **C**, rotor order **III**, **V**, **II**, initial window letters ZZZ, and ring settings 23, 9, 17, to be tested on the crib/ciphertext alignment and menu from part (a) with central letter T. For a Wehrmacht Enigma with reflector **C**, rotor order **III**, **V**, **II**, and ring settings 23, 9, 17, the outputs from the double scrambler for each possible input and all window letters in the menu are shown in Table 5.6 on page 110. Show that for all loops in the menu, the choice K for the plugboard partner of the central letter results in no logical inconsistencies at the central letter.

   (c) Consider the Wehrmacht Enigma configuration with reflector **C**, rotor order **IV**, **V**, **II**, initial window letters ZZZ, and ring settings 26, 23, 19, to be tested on the crib/ciphertext alignment and menu from part (a) with central letter T. For a Wehrmacht Enigma with reflector **C**, rotor order **IV**, **V**, **II**, and ring settings 26, 23, 19, the outputs from the double scrambler for each possible input and all window letters in the menu are shown in Table 5.7 on page 110. Make a complete list of cycles that result for each loop in the menu, and if possible, find a letter that cannot be eliminated as the plugboard partner of the central letter.

   (d) Consider the Wehrmacht Enigma configuration with reflector **C**, rotor order **IV**, **V**, **II**, initial window letters ZZZ, and ring settings 19, 25, 15, to be tested on the crib/ciphertext alignment and menu from part (a) with central letter T. For a Wehrmacht Enigma with reflector **C**, rotor order **IV**, **V**, **II**, and ring settings

	A	B	C	D	E	F	G	H	I	J	K	L	M	N	O	P	Q	R	S	T	U	V	W	X	Y	Z
ZZA	R	J	I	H	Y	G	F	D	C	B	V	O	Z	U	L	W	X	A	T	S	N	K	P	Q	E	M
ZZB	W	C	B	S	G	Q	E	J	K	H	I	T	Y	U	Z	X	F	V	D	L	N	R	A	P	M	O
ZZC	D	F	K	A	O	B	M	Y	P	X	C	T	G	S	E	I	Z	U	N	L	R	W	V	J	H	Q
ZZD	J	L	F	Q	G	C	E	N	K	A	I	B	S	H	Z	V	D	X	M	W	Y	P	T	R	U	O
ZZE	Z	Q	V	T	Y	G	F	L	N	R	X	H	O	I	M	U	B	J	W	D	P	C	S	K	E	A
ZZF	R	J	O	H	S	Q	V	D	W	B	U	T	X	Y	C	Z	F	A	E	L	K	G	I	M	N	P
ZZG	L	K	Y	X	Q	T	Z	I	H	R	B	A	O	S	M	W	E	J	N	F	V	U	P	D	C	G
ZZH	L	Y	J	V	T	G	F	Q	W	C	R	A	P	Z	X	M	H	K	U	E	S	D	I	O	B	N
ZZI	W	J	F	L	P	C	R	T	K	B	I	D	S	X	V	E	Y	G	M	H	Z	O	A	N	Q	U
ZZJ	H	W	N	M	Y	K	Q	A	J	I	F	O	D	C	L	S	G	Z	P	X	V	U	B	T	E	R
ZZK	Q	L	V	T	P	J	U	X	W	F	Y	B	O	Z	M	E	A	S	R	D	G	C	I	H	K	N
ZZL	J	P	D	C	L	S	X	W	Q	A	N	E	R	K	Y	B	I	M	F	V	Z	T	H	G	O	U
ZZM	I	M	T	P	H	Z	U	E	A	O	X	N	B	L	J	D	Y	W	V	C	G	S	R	K	Q	F
ZZN	S	H	J	K	R	X	V	B	Z	C	D	T	N	M	W	Y	U	E	A	L	Q	G	O	F	P	I

Table 5.6   Double scrambler outputs for the Wehrmacht Enigma configuration in Exercise 5b (with reflector **C**, rotor order **III**, **V**, **II**, and ring settings 23, 9, 17).

	A	B	C	D	E	F	G	H	I	J	K	L	M	N	O	P	Q	R	S	T	U	V	W	X	Y	Z
ZZA	O	E	D	C	B	K	N	Y	S	W	F	R	V	G	A	X	Z	L	I	U	T	M	J	P	H	Q
ZZB	X	H	E	S	C	W	Z	B	M	K	J	O	I	P	L	N	U	Y	D	V	Q	T	F	A	R	G
ZZC	M	I	G	P	W	U	C	K	B	R	H	S	A	Z	Q	D	O	J	L	Y	F	X	E	V	T	N
ZZD	F	E	R	G	B	A	D	O	T	Y	Z	W	Q	S	H	U	M	C	N	I	P	X	L	V	J	K
ZZE	R	M	G	E	D	L	C	Z	Y	S	O	F	B	V	K	X	W	A	J	U	T	N	Q	P	I	H
ZZF	F	D	Q	B	W	A	N	U	M	K	J	S	I	G	R	T	C	O	L	P	H	X	E	V	Z	Y
ZZG	H	X	J	Y	I	S	N	A	E	C	V	Q	T	G	W	Z	L	U	F	M	R	K	O	B	D	P
ZZH	F	O	L	E	D	A	M	Y	U	P	Z	C	G	Q	B	J	N	X	V	W	I	S	T	R	H	K
ZZI	M	R	N	G	X	P	D	Y	K	L	I	J	A	C	S	F	U	B	O	V	Q	T	Z	E	H	W
ZZJ	H	K	V	S	W	M	U	A	P	T	B	R	F	X	Y	I	Z	L	D	J	G	C	E	N	O	Q
ZZK	P	V	T	H	M	O	Q	D	L	K	J	I	E	U	F	A	G	W	Z	C	N	B	R	Y	X	S
ZZL	W	C	B	Z	H	O	K	E	P	N	G	T	X	J	F	I	R	Q	V	L	Y	S	A	M	U	D
ZZM	D	J	Q	A	I	K	X	M	E	B	F	Z	H	R	T	S	C	N	P	O	V	U	Y	G	W	L
ZZN	M	S	K	G	Z	L	D	Y	J	I	C	F	A	U	Q	X	O	V	B	W	N	R	T	P	H	E

Table 5.7   Double scrambler outputs for the Wehrmacht Enigma configuration in Exercise 5c (with reflector **C**, rotor order **IV**, **V**, **II**, and ring settings 26, 23, 19).

19, 25, 15, the outputs from the double scrambler for each possible input and all window letters in the menu are shown in Table 5.8 on page 112. Make a complete list of cycles that result for each loop in the menu, and if possible, find a letter that cannot be eliminated as the plugboard partner of the central letter.

6. Suppose the crib THE WRATH OF KHAN AND is known to encrypt to the ciphertext WPMOI NNMUM GNGEL UR, which was formed using a Wehrmacht Enigma.

   (a) Assuming initial window letters ZZZ, draw a menu that expresses the crib/ciphertext pairs with the window letters shown along each link. (We considered this crib and ciphertext previously in Exercise 4 in Section 5.1, in which we drew a menu that expresses the crib/ciphertext pairs.)

   (b) Consider the Wehrmacht Enigma configuration with reflector C, rotor order I, IV, V, initial window letters ZZZ, and ring settings 19, 22, 24, to be tested on the crib/ciphertext alignment and menu from part (a) with central letter N. For a Wehrmacht Enigma with reflector C, rotor order I, IV, V, and ring settings 19, 22, 24, the outputs from the double scrambler for each possible input and all window letters in the menu are shown in Table 5.9 on page 112. Show that for all loops in the menu, the choice X for the plugboard partner of the central letter results in no logical inconsistencies at the central letter.

   (c) Consider the Wehrmacht Enigma configuration with reflector C, rotor order I, II, III, initial window letters ZZZ, and ring settings 14, 23, 26, to be tested on the crib/ciphertext alignment and menu from part (a) with central letter N. For a Wehrmacht Enigma with reflector C, rotor order I, II, III, and ring settings 14, 23, 26, the outputs from the double scrambler for each possible input and all window letters in the menu are shown in Table 5.10 on page 113. Make a complete list of cycles that result for each loop in the menu, and if possible, find a letter that cannot be eliminated as the plugboard partner of the central letter.

   (d) Consider the Wehrmacht Enigma configuration with reflector C, rotor order I, IV, V, initial window letters ZZZ, and ring settings 20, 23, 24, to be tested on the crib/ciphertext alignment and menu from part (a) with central letter N. For a Wehrmacht Enigma with reflector C, rotor order I, IV, V, and ring settings 20, 23, 24, the outputs from the double scrambler for each possible input and all window letters in the menu are shown in Table

	A	B	C	D	E	F	G	H	I	J	K	L	M	N	O	P	Q	R	S	T	U	V	W	X	Y	Z
ZZA	F	Q	U	G	H	A	D	E	P	M	Z	X	J	S	R	I	B	O	N	V	C	T	Y	L	W	K
ZZB	J	I	Z	Q	S	H	R	F	B	A	O	T	U	P	K	N	D	G	E	L	M	Y	X	W	V	C
ZZC	B	A	M	T	F	E	Q	R	J	I	Z	V	C	X	U	Y	G	H	W	D	O	L	S	N	P	K
ZZD	H	Y	X	M	N	S	K	A	W	L	G	J	D	E	Q	Z	O	U	F	V	R	T	I	C	B	P
ZZE	H	M	G	X	T	Y	C	A	J	I	N	U	B	K	Z	W	R	Q	V	E	L	S	P	D	F	O
ZZF	J	C	B	F	O	D	N	Q	U	A	R	W	X	G	E	V	H	K	T	S	I	P	L	M	Z	Y
ZZG	J	S	F	L	N	C	I	P	G	A	O	D	Q	E	K	H	M	T	B	R	V	U	X	W	Z	Y
ZZH	J	Q	U	E	D	V	L	K	Y	A	H	G	W	S	P	O	B	Z	N	X	C	F	M	T	I	R
ZZI	Z	D	T	B	F	E	U	I	H	R	L	K	O	Y	M	Q	P	J	X	C	G	W	V	S	N	A
ZZJ	J	Z	F	V	I	C	S	L	E	A	P	H	W	Q	Y	K	N	X	G	U	T	D	M	R	O	B
ZZK	O	P	T	I	M	S	X	L	D	Z	Y	H	E	Q	A	B	N	W	F	C	V	U	R	G	K	J
ZZL	N	C	B	K	V	X	W	T	P	O	D	S	Y	A	J	I	Z	U	L	H	R	E	G	F	M	Q
ZZM	M	S	F	K	X	C	H	G	P	T	D	Z	A	V	Q	I	O	U	B	J	R	N	Y	E	W	L
ZZN	T	D	O	B	W	I	Z	X	F	U	L	K	Y	Q	C	V	N	S	R	A	J	P	E	H	M	G

Table 5.8   Double scrambler outputs for the Wehrmacht Enigma configuration in Exercise 5d (with reflector **C**, rotor order **IV**, **V**, **II**, and ring settings 19, 25, 15).

	A	B	C	D	E	F	G	H	I	J	K	L	M	N	O	P	Q	R	S	T	U	V	W	X	Y	Z
ZZA	H	T	E	M	C	Z	L	A	X	W	Y	G	D	V	U	R	S	P	Q	B	O	N	J	I	K	F
ZZB	J	M	G	E	D	V	C	Y	Z	A	R	S	B	T	Q	X	O	K	L	N	W	F	U	P	H	I
ZZC	I	D	Z	B	N	Q	M	Y	A	P	R	U	G	E	X	J	F	K	T	S	L	W	V	O	H	C
ZZD	D	Y	U	A	P	Q	O	V	W	R	L	K	X	Z	G	E	F	J	T	S	C	H	I	M	B	N
ZZE	P	N	Z	F	H	D	U	E	V	Y	T	R	S	B	W	A	X	L	M	K	G	I	O	Q	J	C
ZZF	M	C	B	E	D	O	R	J	X	H	P	U	A	Y	F	K	Z	G	V	W	L	S	T	I	N	Q
ZZG	Z	I	S	N	H	J	Y	E	B	F	T	X	O	D	M	Q	P	U	C	K	R	W	V	L	G	A
ZZH	H	E	I	J	B	W	N	A	C	D	Q	U	Z	G	P	O	K	X	V	Y	L	S	F	R	T	M
ZZI	D	K	Y	A	V	Q	S	M	N	T	B	R	H	I	U	Z	F	L	G	J	O	E	X	W	C	P
ZZJ	K	R	H	Y	Q	U	L	C	T	X	A	G	S	Z	P	O	E	B	M	I	F	W	V	J	D	N
ZZK	Y	K	V	E	D	M	Q	W	R	Z	B	P	F	O	N	L	G	I	X	U	T	C	H	S	A	J
ZZL	Y	O	D	C	I	S	P	Q	E	R	X	W	T	U	B	G	H	J	F	M	N	Z	L	K	A	V
ZZM	N	L	M	G	H	Q	D	E	W	S	T	B	C	A	V	X	F	Z	J	K	Y	O	I	P	U	R
ZZN	Z	M	S	N	P	X	H	G	R	O	Y	V	B	D	J	E	U	I	C	W	Q	L	T	F	K	A
ZZO	B	A	M	E	D	N	Z	Y	T	V	U	Q	C	F	S	R	L	P	O	I	K	J	X	W	H	G
ZZP	N	C	B	Z	K	Y	J	S	Q	G	E	R	V	A	W	T	I	L	H	P	X	M	O	U	F	D
ZZQ	T	R	O	J	I	L	K	Y	E	D	G	F	U	V	C	Z	W	B	X	A	M	N	Q	S	H	P

Table 5.9   Double scrambler outputs for the Wehrmacht Enigma configuration in Exercise 6b (with reflector **C**, rotor order **I**, **IV**, **V**, and ring settings 19, 22, 24).

	A	B	C	D	E	F	G	H	I	J	K	L	M	N	O	P	Q	R	S	T	U	V	W	X	Y	Z
ZZA	I	W	G	K	Q	M	C	S	A	R	D	P	F	U	T	L	E	J	H	O	N	Y	B	Z	V	X
ZZB	D	M	R	A	G	J	E	Q	T	F	L	K	B	U	W	S	H	C	P	I	N	Y	O	Z	V	X
ZZC	T	F	H	G	M	B	D	C	Z	N	L	K	E	J	V	S	W	X	P	A	Y	O	Q	R	U	I
ZZD	P	C	B	W	G	L	E	M	V	Z	R	F	H	O	N	A	T	K	X	Q	Y	I	D	S	U	J
ZZE	E	T	O	N	A	Y	K	I	H	S	G	U	Z	D	C	V	W	X	J	B	L	P	Q	R	F	M
ZZF	Z	Y	V	G	O	M	D	J	K	H	I	W	F	T	E	Q	P	X	U	N	S	C	L	R	B	A
ZZG	T	Z	J	L	G	X	E	Y	Q	C	W	D	P	R	U	M	I	N	V	A	O	S	K	F	H	B
ZZH	X	U	I	S	N	H	V	F	C	R	Z	M	L	E	W	T	Y	J	D	P	B	G	O	A	Q	K
ZZI	G	Y	L	H	K	V	A	D	X	N	E	C	O	J	M	U	T	W	Z	Q	P	F	R	I	B	S
ZZJ	R	T	K	Z	O	N	Y	V	X	U	C	W	S	F	E	Q	P	A	M	B	J	H	L	I	G	D
ZZK	H	X	Q	E	D	G	F	A	U	P	Z	O	T	S	L	J	C	Y	N	M	I	W	V	B	R	K
ZZL	Q	C	B	N	W	R	K	L	X	O	G	H	P	D	J	M	A	F	Y	U	T	Z	E	I	S	V
ZZM	T	H	V	X	L	G	F	B	O	S	R	E	Z	W	I	U	Y	K	J	A	P	C	N	D	Q	M
ZZN	Q	Z	D	C	K	T	V	X	S	O	E	N	P	L	J	M	A	Y	I	F	W	G	U	H	R	B
ZZO	S	R	Y	U	G	X	E	Q	Z	L	N	J	W	K	T	V	H	B	A	O	D	P	M	F	C	I
ZZP	H	Z	R	P	Y	V	N	A	M	U	X	O	I	G	L	D	S	C	Q	W	J	F	T	K	E	B
ZZQ	W	R	F	Y	J	C	K	M	P	E	G	T	H	S	V	I	X	B	N	L	Z	O	A	Q	D	U

Table 5.10 Double scrambler outputs for the Wehrmacht Enigma configuration in Exercise 6c (with reflector **C**, rotor order **I**, **II**, **III**, and ring settings 14, 23, 26).

	A	B	C	D	E	F	G	H	I	J	K	L	M	N	O	P	Q	R	S	T	U	V	W	X	Y	Z
ZZA	Z	C	B	F	Q	D	P	M	L	S	W	I	H	R	Y	G	E	N	J	V	X	T	K	U	O	A
ZZB	H	K	D	C	W	R	U	A	Q	P	B	M	L	S	T	J	I	F	N	O	G	X	E	V	Z	Y
ZZC	U	G	X	W	P	T	B	R	S	L	Z	J	V	O	N	E	Y	H	I	F	A	M	D	C	Q	K
ZZD	F	H	R	E	D	A	T	B	N	L	U	J	W	I	X	Y	Z	C	V	G	K	S	M	O	P	Q
ZZE	I	Q	G	W	J	R	C	O	A	E	V	S	X	Z	H	T	B	F	L	P	Y	K	D	M	U	N
ZZF	V	K	Y	N	G	R	E	Q	L	W	B	I	X	D	S	U	H	F	O	Z	P	A	J	M	C	T
ZZG	O	R	Y	L	G	W	E	S	V	M	U	D	J	P	A	N	T	B	H	Q	K	I	F	Z	C	X
ZZH	P	U	Z	F	O	D	S	W	J	I	M	T	K	R	E	A	Y	N	G	L	B	X	H	V	Q	C
ZZI	H	O	M	V	X	L	P	A	Q	K	J	F	C	R	B	G	I	N	T	S	W	D	U	E	Z	Y
ZZJ	L	Y	J	F	N	D	W	P	U	C	T	A	Z	E	V	H	S	X	Q	K	I	O	G	R	B	M
ZZK	L	P	U	M	O	W	R	X	J	I	Y	A	D	Z	E	B	V	G	T	S	C	Q	F	H	K	N
ZZL	O	C	B	H	K	U	Y	D	J	I	E	Z	W	T	A	S	V	X	P	N	F	Q	M	R	G	L
ZZM	U	M	R	J	I	Y	O	P	E	D	L	K	B	S	G	H	Z	C	N	V	A	T	X	W	F	Q
ZZN	M	Y	D	C	R	H	J	F	L	G	Q	I	A	P	X	N	K	E	Z	U	T	W	V	O	B	S
ZZO	R	Y	T	P	G	S	E	L	V	Q	M	H	K	X	W	D	J	A	F	C	Z	I	O	N	B	U
ZZP	V	N	M	W	R	K	T	O	X	Z	F	Q	C	B	H	U	L	E	Y	G	P	A	D	I	S	J
ZZQ	G	L	I	F	S	D	A	R	C	N	U	B	T	J	Y	V	Z	H	E	M	K	P	X	W	O	Q

Table 5.11 Double scrambler outputs for the Wehrmacht Enigma configuration in Exercise 6d (with reflector **C**, rotor order **I**, **IV**, **V**, and ring settings 20, 23, 24).

5.11 on page 113. Make a complete list of cycles that result for each loop in the menu, and if possible, find a letter that cannot be eliminated as the plugboard partner of the central letter.

7.* Find the expected number of stops in a bombe run for menus with 1 loop and menus with 2 loops.

8. Find some information about how bombes were engineered and wired for particular menus, and write a summary of your findings.

5.4   The Diagonal Board

The cryptanalysts at Bletchley Park initially only had limited success using bombes to decrypt Enigma ciphertexts, due to the dependence of bombes on crib/ciphertext menus with multiple loops, which were not always available. For instance, consider the following example.

Example 5.6 Consider the Wehrmacht Enigma configuration with reflector **B**, rotor order **III**, **IV**, **I**, initial window letters ZZZ, and ring settings 16, 22, 23, to be tested on the following crib/ciphertext alignment.

Position:	1	2	3	4	5	6	7	8	9	10	11	12	13	14
Window:	ZZA	ZZB	ZZC	ZZD	ZZE	ZZF	ZZG	ZZH	ZZI	ZZJ	ZZK	ZZL	ZZM	ZZN
Crib:	F	O	R	E	C	A	S	T	I	S	S	N	O	W
Cipher:	C	C	C	K	G	Z	I	Z	C	L	A	H	T	V

A menu that expresses these crib/ciphertext pairs is shown in Figure 5.10 on page 115. Also, for a Wehrmacht Enigma with reflector **B**, rotor order **III**, **IV**, **I**, and ring settings 16, 22, 23, the outputs from the double scrambler for each possible input and all window letters in the menu are shown in Table 5.12 on page 115. With central letter C, the following is the complete list of cycles that result for the menu loop $C \to O \to T \to Z \to A \to S \to I \to C$.

$$(A)(BYNOLFGXPUVQHJMWTDCS)(EZ)(I)(KR)$$

From these cycles, we can eliminate all letters except A and I as possible plugboard partners of the central letter. However, since the menu contains no other loops, we cannot proceed as in Section 5.3 by checking to see if either A or I produces a logical inconsistency at the central letter when tested as the plugboard partner of the central letter in another loop.   □

British mathematician Gordon Welchman had a very clever idea that helped alleviate some of the early problems presented by menus with too few loops. Welchman's idea led to a new component of the bombe called the *diagonal*

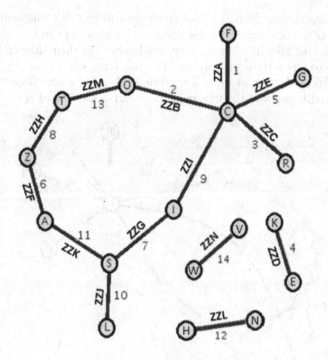

Figure 5.10   Menu with one loop.

	A	B	C	D	E	F	G	H	I	J	K	L	M	N	O	P	Q	R	S	T	U	V	W	X	Y	Z
ZZA	P	W	R	O	I	L	J	Y	E	G	M	F	K	U	D	A	V	C	T	S	N	Q	B	Z	H	X
ZZB	Z	D	X	B	J	M	S	L	O	E	Y	H	F	T	I	Q	P	W	G	N	V	U	R	C	K	A
ZZC	F	K	G	O	R	A	C	U	N	W	B	M	L	I	D	Z	T	E	Y	Q	H	X	J	V	S	P
ZZD	C	E	A	O	B	L	M	Q	S	W	T	F	G	P	D	N	H	U	I	K	R	Z	J	Y	X	V
ZZE	V	K	S	L	R	J	X	U	Z	F	B	D	P	T	Q	M	O	E	C	N	H	A	Y	G	W	I
ZZF	G	M	L	U	W	O	A	Z	Y	N	S	C	B	J	F	X	V	T	K	R	D	Q	E	P	I	H
ZZG	T	O	P	E	D	G	F	M	Z	V	L	K	H	Y	B	C	W	S	R	A	X	J	Q	U	N	I
ZZH	J	U	H	K	W	Y	V	C	Q	A	D	O	R	X	L	S	I	M	P	Z	B	G	E	N	F	T
ZZI	Q	U	I	H	Z	K	Y	D	C	L	F	J	P	X	W	M	A	T	V	R	B	S	O	N	G	E
ZZJ	Q	E	S	N	B	M	X	O	U	W	P	V	F	D	H	K	A	T	C	R	I	L	J	G	Z	Y
ZZK	P	H	R	F	W	D	K	B	Q	L	G	J	O	X	M	A	I	C	Z	Y	V	U	E	N	T	S
ZZL	Z	L	P	M	I	Y	K	R	E	N	G	B	D	J	V	C	U	H	T	S	Q	O	X	W	F	A
ZZM	G	D	R	B	Z	T	A	X	K	L	I	J	S	Y	V	W	U	C	M	F	Q	O	P	H	N	E
ZZN	L	Z	Y	I	F	E	U	X	D	V	N	A	P	K	S	M	R	Q	O	W	G	J	T	H	C	B

Table 5.12   Double scrambler outputs for the Wehrmacht Enigma configuration in Example 5.6 (with reflector **B**, rotor order **III**, **IV**, **I**, and ring
settings 16, 22, 23).

board. The diagonal board, whose shape was in fact not diagonal at all, consisted of 26 terminals, one for each of the 26 letters in the alphabet. A collection of 26-wire cables connected letters via their diagonal board terminals to the letters in a menu. For the large connected part of the menu in Figure 5.10, a diagram illustrating the cable connections between the diagonal board and the letters in the menu is shown in Figure 5.11.

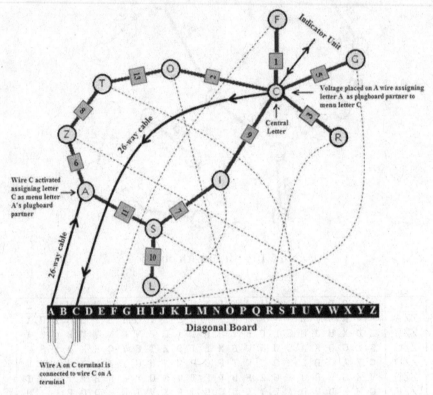

Figure 5.11   Menu with diagonal board.

The contribution of the diagonal board was based on the fact that Enigma plugboards were *reciprocal*, meaning for example that if current entering a plugboard designating A would exit designating C, then current entering the same plugboard designating C would exit designating A. Figure 5.11 illustrates, with initial choice A for the plugboard partner of the central letter C, and voltage placed on the relay corresponding to A in an indicator unit, current ready to begin its journey around the loop C → O → T → Z → A → S → I → C. When it arrived back at the central letter, it would perhaps reveal a logical inconsistency that would eliminate A as a possible plugboard partner of C, although also perhaps not. Welchman's idea was, because A is also in the menu, to have current also travel from the central letter C to the diagonal board, traveling along the A wire in the 26-wire cable connecting the central letter to the terminal corresponding to C in the diagonal board. This current could then travel through the diagonal board from the A wire in the cable connected to the C terminal to the C wire in the cable connected to the A terminal. It could then travel along this wire to A in the menu. Since the initial choice A for the plugboard partner of C dictates that C would have to be the plugboard partner of A, this plugboard partner C of A could then go on its own journey around the remaining part of the loop (A → S → I → C). When it arrived back at the central letter, it would also perhaps reveal a logical inconsistency that would eliminate A as a possible plugboard partner of C. This would thus increase the likelihood that A would be eliminated as a possible plugboard partner of C, if in fact it should be eliminated.

Further, since Enigma plugboards were reciprocal not just for the central letter and its plugboard partner, but for all menu letters and their plugboard partners, the diagonal board could be used in this way not just at the central letter, but anywhere in the menu. Generally, if letters α and β were both part of a menu, and an initial choice for the plugboard partner of the central letter led to β being identified as the plugboard partner of α anywhere in the menu, then current could travel from α in the menu to the diagonal board, traveling along the β wire in the 26-wire cable connecting α in the menu to the terminal corresponding to α in the diagonal board. This current could then travel through the diagonal board from the β wire in the cable connected to the α terminal to the α wire in the cable connected to the β terminal. It could then travel along this wire to β in the menu.

Example 5.7 Consider the configuration, crib/ciphertext alignment, and menu in Example 5.6. Recall that from the complete list of cycles that result for the only loop in this menu, we were not able to eliminate A as the plugboard partner of the central letter C. The initial choice A for the plugboard partner of C results in the following plugboard partners for all of the letters in the large connected part of the menu.

Drum Setting:	ZZB	ZZM	ZZH	ZZF	ZZK	ZZG	ZZI
Menu Letter:	C →	O →	T →	Z →	A →	S →	I → C
Plug Partner:	A →	Z →	E →	W →	E →	W →	Q → A

Drum Setting:	ZZA
Menu Letter:	C → F
Plug Partner:	A → P

Drum Setting:	ZZE
Menu Letter:	C → G
Plug Partner:	A → V

Drum Setting: ZZC
  Menu Letter: C $\longrightarrow$ R
  Plug Partner: A $\longrightarrow$ F

Drum Setting: ZZJ
  Menu Letter: S $\longrightarrow$ ~~L~~
  Plug Partner: W $\longrightarrow$ ~~J~~

The three plugboard pairs that are crossed out in these lists, ~~I/Q~~, ~~F/P~~, and ~~L/J~~, would not activate the diagonal board, since the plugboard partner does not appear in the menu. For five of the plugboard pairs that would activate the diagonal board, T/E, Z/W, A/E, S/W, and G/V, the plugboard partners appear in the menu in a place from which the central letter could not be reached. Even so, useful information might be obtained from having the diagonal board exchange the menu letters and plugboard partners in these pairs. Doing this, and then traveling along the link to which the new menu letters are connected in the menu, results in the following.

Drum Setting: ZZD
  Menu Letter: E $\longrightarrow$ K
  Plug Partner: T $\longrightarrow$ K

Drum Setting: ZZN
  Menu Letter: W $\longrightarrow$ ~~V~~
  Plug Partner: Z $\longrightarrow$ ~~B~~

Drum Setting: ZZD
  Menu Letter: E $\longrightarrow$ K
  Plug Partner: A $\longrightarrow$ C

Drum Setting: ZZN
  Menu Letter: W $\longrightarrow$ V
  Plug Partner: S $\longrightarrow$ O

Drum Setting: ZZN
  Menu Letter: V $\longrightarrow$ ~~W~~
  Plug Partner: G $\longrightarrow$ ~~U~~

This gives three additional plugboard pairs that would activate the diagonal board, K/K, K/C, and V/O. For the pair K/K, the diagonal board would not result in any additional information, since it would not initiate current at a different place in the menu. For the five remaining plugboard pairs that would activate the diagonal board, C/A, O/Z, R/F, K/C, and V/O, the plugboard partners appear in the menu in a place from which the central letter could be reached. For the pair K/C, having the diagonal board exchange the menu letter and plugboard partner would give C/K, and place current directly on the central letter. For the other four pairs, having the diagonal board exchange the menu letters and plugboard partners, and then traveling to the central letter in the menu (in the C → O → T → Z → A → S → I → C direction if going around the loop), results in the following.

Drum Setting: ZZK   ZZG   ZZI
Menu Letter: A ⟶ S ⟶ I ⟶ C
Plug Partner: C ⟶ R ⟶ S ⟶ V

Drum Setting: ZZF   ZZK   ZZG   ZZI
Menu Letter: Z ⟶ A ⟶ S ⟶ I ⟶ C
Plug Partner: O ⟶ F ⟶ D ⟶ E ⟶ Z

Drum Setting: ZZA
Menu Letter: F ⟶ C
Plug Partner: R ⟶ C

Drum Setting: ZZM   ZZH   ZZF   ZZK   ZZG   ZZI
Menu Letter: O ⟶ T ⟶ Z ⟶ A ⟶ S ⟶ I ⟶ C
Plug Partner: V ⟶ O ⟶ L ⟶ C ⟶ R ⟶ S ⟶ V

Recall again that in Example 5.6 we were not able to eliminate A as the plugboard partner of the central letter C. With the diagonal board, though, we are able to eliminate A as the plugboard partner of C. Notably, the diagonal board gives the plugboard pairs C/K, C/V, C/Z, and C/C, all of which are logically inconsistent with the pair C/A with which we started this example. In fact, while this eliminates A as the plugboard partner of C, it also eliminates K, V, Z, and C. This means that using the diagonal board, which we will label as D, within the technique from Section 5.3 with the cycles $C_1 = (A)$, $C_2 = (BYNOLFGXPUVQHJMWTDCS)$, $C_3 = (EZ)$, $C_4 = (I)$, and $C_5 = (KR)$, letters, as they are eliminated as plugboard partners of the central letter with initial choice A for the plugboard partner of the central letter, are summarized in the following table.

Test letter	A	B	C	D	E	F	G	H	I	J	K	L	M	N	O	P	Q	R	S	T	U	V	W	X	Y	Z
A activates D	A		C								K											V				Z
C activates C_2		B	C	D		F	G	H		J		L	M	N	O	P	Q		S	T	U	V	W	X	Y	
Join D, C_2	A	B	C	D		F	G	H		J	K	L	M	N	O	P	Q		S	T	U	V	W	X	Y	Z
K activates C_5											K							R								
Join C_5	A	B	C	D		F	G	H		J	K	L	M	N	O	P	Q	R	S	T	U	V	W	X	Y	Z
Z activates C_3					E																					Z
Join C_3	A	B	C	D	E	F	G	H		J	K	L	M	N	O	P	Q	R	S	T	U	V	W	X	Y	Z

This indicates that if a Wehrmacht Enigma were configured as in Example 5.6, then I would be the only possible plugboard partner of the central letter C in the menu in Figure 5.10 that results from the crib/ciphertext alignment in Example 5.6.   □

5.4.1   Exercises

1. For the crib/ciphertext alignment and menu in Example 5.6 on page 114, determine whether the following plugboard pairs would activate the diagonal board.

   (a)* O/I
   (b) R/Y
   (c)* F/X
   (d) K/V

2.* Suppose the crib GORDON WELCHMAN is known to encrypt to the ciphertext EMHSS MBNSX NLWW, which was formed using a Wehrmacht Enigma.

   (a) Assuming initial window letters ZZZ, draw a menu that expresses the crib/ciphertext pairs with the window letters shown along each link. (We considered this crib and ciphertext previously in Exercise 5 in Section 5.1, in which we drew a menu that expresses the crib/ciphertext pairs.)

   (b) Consider the Wehrmacht Enigma configuration with reflector **B**, rotor order **I, II, III**, initial window letters ZZZ, and ring settings 26, 11, 9, to be tested on the crib/ciphertext alignment and menu from part (a) with central letter M. For a Wehrmacht Enigma with reflector **B**, rotor order **I, II, III**, and ring settings 26, 11, 9, the outputs from the double scrambler for each possible input and all window letters in the menu are shown in Table 5.13 on page 121. Make a complete list of cycles that result for the only loop in the menu.

   (c) For the Wehrmacht Enigma configuration in part (b), and with initial choice A for the plugboard partner of the central letter M in the menu from part (a), find the plugboard partners that result for all of the letters in the large connected part of the menu, and indicate which plugboard pairs would activate the diagonal board.

   (d) For each of the plugboard pairs in part (c) that would activate the diagonal board, exchange the menu letter and plugboard partner, and then follow the links in the menu, to the central letter if possible, identifying the plugboard partners of the menu letters along the way.

   (e) Using your answers to parts (b) and (d), if possible, find a letter that cannot be eliminated as the plugboard partner of the central letter.

	A	B	C	D	E	F	G	H	I	J	K	L	M	N	O	P	Q	R	S	T	U	V	W	X	Y	Z
ZZA	Y	O	G	M	Z	N	C	L	U	W	X	H	D	F	B	S	V	T	P	R	I	Q	J	K	A	E
ZZB	N	U	Q	Y	G	V	E	I	H	Z	L	K	P	A	X	M	C	S	R	W	B	F	T	O	D	J
ZZC	W	D	T	B	K	L	H	G	Z	Y	E	F	V	P	S	N	R	Q	O	C	X	M	A	U	J	I
ZZD	M	D	Y	B	Q	X	H	G	V	K	J	U	A	Z	S	R	E	P	O	W	L	I	T	F	C	N
ZZE	D	J	R	A	N	G	F	K	U	B	H	V	P	E	Q	M	O	C	X	W	I	L	T	S	Z	Y
ZZF	U	N	L	P	T	G	F	X	Q	Z	O	C	Y	B	K	D	I	W	V	E	A	S	R	H	M	J
ZZG	T	X	O	N	F	E	R	W	V	Z	Q	M	L	D	C	Y	K	G	U	A	S	I	H	B	P	J
ZZH	M	C	B	V	F	E	W	U	O	L	Z	J	A	Y	I	X	S	T	Q	R	H	D	G	P	N	K
ZZI	B	A	K	E	D	S	P	R	W	T	C	M	L	V	X	G	U	H	F	J	Q	N	I	O	Z	Y
ZZJ	B	A	U	E	D	O	H	G	L	X	W	I	Y	R	F	T	S	N	Q	P	C	Z	K	J	M	V
ZZK	Z	K	D	C	J	X	M	P	Y	E	B	O	G	Q	L	H	N	U	T	S	R	W	V	F	I	A
ZZL	Z	W	K	L	V	G	F	T	X	U	C	D	P	Q	Y	M	N	S	R	H	J	E	B	I	O	A
ZZM	W	J	N	Q	U	P	L	K	X	B	H	G	O	C	M	F	D	V	T	S	E	R	A	I	Z	Y
ZZN	T	P	I	S	G	W	E	M	C	X	O	N	H	L	K	B	U	V	D	A	Q	R	F	J	Z	Y

Table 5.13 Double scrambler outputs for the Wehrmacht Enigma configuration in Exercise 2b (with reflector **B**, rotor order **I**, **II**, **III**, and ring settings 26, 11, 9).

	A	B	C	D	E	F	G	H	I	J	K	L	M	N	O	P	Q	R	S	T	U	V	W	X	Y	Z
ZZA	C	X	A	L	R	N	M	K	T	W	H	D	G	F	Q	V	O	E	U	I	S	P	J	B	Z	Y
ZZB	I	H	U	T	F	E	W	B	A	Y	L	K	X	Z	R	S	V	O	P	D	C	Q	G	M	J	N
ZZC	Y	V	I	U	O	L	J	T	C	G	X	F	Z	W	E	R	S	P	Q	H	D	B	N	K	A	M
ZZD	J	R	D	C	Y	N	P	L	M	A	T	H	I	F	W	G	U	B	X	K	Q	Z	O	S	E	V
ZZE	G	F	L	U	W	B	A	Q	R	O	M	C	K	Y	J	S	H	I	P	Z	D	X	E	V	N	T
ZZF	U	R	K	W	V	J	Y	M	Z	F	C	O	H	S	L	Q	P	B	N	X	A	E	D	T	G	I
ZZG	F	T	Q	Z	P	A	R	S	K	Y	I	N	U	L	W	E	C	G	H	B	M	X	O	V	J	D
ZZH	V	R	O	L	W	P	Q	I	H	M	S	D	J	Z	C	F	G	B	K	U	T	A	E	Y	X	N
ZZI	Y	O	V	H	Z	Q	M	D	K	L	I	J	G	U	B	W	F	T	X	R	N	C	P	S	A	E
ZZJ	M	T	D	C	J	P	W	V	Q	E	L	K	A	O	N	F	I	Y	Z	B	X	H	G	U	R	S
ZZK	J	F	S	X	V	B	Z	U	T	A	L	K	Y	O	N	R	W	P	C	I	H	E	Q	D	M	G
ZZL	I	L	H	G	T	S	D	C	A	N	P	B	V	J	Z	K	X	Y	F	E	W	M	U	Q	R	O
ZZM	V	F	P	U	N	B	Y	M	J	I	W	Q	H	E	R	C	L	O	X	Z	D	A	K	S	G	T
ZZN	Q	H	R	V	W	K	O	B	M	Z	F	T	I	U	G	S	A	C	P	L	N	D	E	Y	X	J
ZZO	Z	N	Q	U	W	I	O	K	F	V	H	X	S	B	G	R	C	P	M	Y	D	J	E	L	T	A
ZZP	U	G	X	T	V	Z	B	K	R	O	H	S	Y	W	J	Q	P	I	L	D	A	E	N	C	M	F

Table 5.14 Double scrambler outputs for the Wehrmacht Enigma configuration in Exercise 3b (with reflector **C**, rotor order **V**, **III**, **I**, and ring settings 17, 26, 16).

3. Suppose the crib `THE SOLAR SYSTEM IS` is known to encrypt to the ciphertext `AJJRJ EHMRI OFURZ J`, which was formed using a Wehrmacht Enigma.

   (a) Assuming initial window letters ZZZ, draw a menu that expresses the crib/ciphertext pairs with the window letters shown along each link. (We considered this crib and ciphertext previously in Exercise 6 in Section 5.1, in which we drew a menu that expresses the crib/ciphertext pairs.)

   (b) Consider the Wehrmacht Enigma configuration with reflector **C**, rotor order **V**, **III**, **I**, initial window letters ZZZ, and ring settings 17, 26, 16, to be tested on the crib/ciphertext alignment and menu from part (a) with central letter J. For a Wehrmacht Enigma with reflector **C**, rotor order **V**, **III**, **I**, and ring settings 17, 26, 16, the outputs from the double scrambler for each possible input and all window letters in the menu are shown in Table 5.14 on page 121. Make a complete list of cycles that result for the menu loop J → O → S → J.

   (c) For the Wehrmacht Enigma configuration in part (b), and with initial choice **A** for the plugboard partner of the central letter J in the menu from part (a), find the plugboard partners that result for all of the letters in the large connected part of the menu, and indicate which plugboard pairs would activate the diagonal board.

   (d) For each of the plugboard pairs in part (c) that would activate the diagonal board, exchange the menu letter and plugboard partner, and then follow the links in the menu, to the central letter if possible, identifying the plugboard partners of the menu letters along the way.

   (e) Using your answers to parts (b) and (d), if possible, find a letter that cannot be eliminated as the plugboard partner of the central letter.

4. To further illustrate how the diagonal board in a bombe exchanged menu letters and plugboard partners, a diagram of a four-letter diagonal board is shown in Figure 5.12 on page 123. This four-letter diagonal board contains four terminals, which are indicated by capital letters. To each of these terminals a four-wire cable is connected, whose wires are indicated by lowercase letters. Note that for any pair of differing letters α and β, the wire labeled α in the cable connected to the terminal labeled α is not interconnected to any of the wires in the cable connected to the terminal labeled β, while the wire labeled

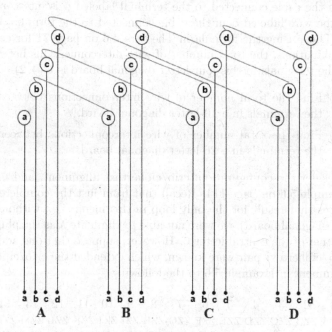

Figure 5.12   A four-letter diagonal board.

	A	B	C	D	E	F	G	H	I	J	K	L	M	N	O	P	Q	R	S	T	U	V	W	X	Y	Z
ZZA	P	W	R	O	I	L	J	Y	E	G	M	F	K	U	D	A	V	C	T	S	N	Q	B	Z	H	X
ZZB	Z	D	X	B	J	M	S	L	O	E	Y	H	F	T	I	Q	P	W	G	N	V	U	R	C	K	A
ZZC	F	K	G	O	R	A	C	U	N	W	B	M	L	I	D	Z	T	E	Y	Q	H	X	J	V	S	P
ZZD	C	E	A	O	B	L	M	Q	S	W	T	F	G	P	D	N	H	U	I	K	R	Z	J	Y	X	V
ZZE	V	K	S	L	R	J	X	U	Z	F	B	D	P	T	Q	M	O	E	C	N	H	A	Y	G	W	I
ZZF	G	M	L	U	W	O	A	Z	Y	N	S	C	B	J	F	X	V	T	K	R	D	Q	E	P	I	H
ZZG	T	O	P	E	D	G	F	M	Z	V	L	K	H	Y	B	C	W	S	R	A	X	J	Q	U	N	I
ZZH	J	U	H	K	W	Y	V	C	Q	A	D	O	R	X	L	S	I	M	P	Z	B	G	E	N	F	T
ZZI	Q	U	I	H	Z	K	Y	D	C	L	F	J	P	X	W	M	A	T	V	R	B	S	O	N	G	E
ZZJ	Q	E	S	N	B	M	X	O	U	W	P	V	F	D	H	K	A	T	C	R	I	L	J	G	Z	Y
ZZK	P	H	R	F	W	D	K	B	Q	L	G	J	O	X	M	A	I	C	Z	Y	V	U	E	N	T	S
ZZL	Z	L	P	M	I	Y	K	R	E	N	G	B	D	J	V	C	U	H	T	S	Q	O	X	W	F	A
ZZM	G	D	R	B	Z	T	A	X	K	L	I	J	S	Y	V	W	U	C	M	F	Q	O	P	H	N	E
ZZN	L	Z	Y	I	F	E	U	X	D	V	N	A	P	K	S	M	R	Q	O	W	G	J	T	H	C	B
ZZO	Z	L	X	P	J	V	H	G	N	E	M	B	K	I	R	D	S	O	Q	W	Y	F	T	C	U	A

Table 5.15   Double scrambler outputs for the Wehrmacht Enigma configuration in Exercise 5 (with reflector **B**, rotor order **III**, **IV**, **I**, and ring
settings 16, 22, 23).

β in the cable connected to the terminal labeled α is interconnected to the wire labeled α in the cable connected to the terminal labeled β. Thus, using the formula in Theorem 4.6 on page 71 for counting combinations, the total number of wire interconnections between all of the terminals in this four-letter diagonal board is $C(4,2) = 6$.

(a)* Find the total number of wire interconnections between all of the terminals in a 10-letter diagonal board.

(b) Find the total number of wire interconnections between all of the terminals in a 26-letter diagonal board.

5. Consider the configuration, crib/ciphertext alignment, and menu in Example 5.6 on page 114. Recall that from just the complete list of cycles that result for the only loop in this menu (i.e., without using the diagonal board), we were not able to eliminate A as the plugboard partner of the central letter C. However, suppose that one additional crib/ciphertext pair were known, which extended the crib/ciphertext alignment in Example 5.6 to the following.

1	2	3	4	5	6	7	8	9	10	11	12	13	14	15
ZZA	ZZB	ZZC	ZZD	ZZE	ZZF	ZZG	ZZH	ZZI	ZZJ	ZZK	ZZL	ZZM	ZZN	ZZO
F	O	R	E	C	A	S	T	I	S	S	N	O	W	R
C	C	C	K	G	Z	I	Z	C	L	A	H	T	V	C

For a Wehrmacht Enigma with the configuration in Example 5.6 (i.e., reflector **B**, rotor order **III, IV, I**, and ring settings 16, 22, 23), the outputs from the double scrambler for each possible input and all window letters in the extended crib/ciphertext alignment are shown in Table 5.15 on page 123. With the one additional crib/ciphertext pair, can A be eliminated as the plugboard partner of the central letter without having to use the diagonal board?

6. Find some information about Gordon Welchman, the British mathematician whose idea led to the diagonal board, and write a summary of your findings.

5.5 The Checking Machine

Recall that when a bombe was used to test Enigma configurations on a given crib/ciphertext alignment, multiple configurations could result in stops. At most one stop could be good though (i.e., give a correct Enigma configuration and plugboard partner of the central letter), meaning at least all but one had to be false (i.e., give an incorrect Enigma configuration and/or plugboard partner of the central letter). To identify a stop as false, and otherwise to continue the cryptanalysis process, the configuration and plugboard partner of the central letter given by the stop were checked by hand using a machine called, naturally, a *checking machine*.

Checking machines were indeed used for two purposes. One purpose was to check for logical inconsistencies away from the central letter in a menu. For instance, consider again Example 5.7 on page 117, specifically near the start where with initial choice A for the plugboard partner of the central letter C, the plugboard partners of the menu letters around the loop $C \to 0 \to T \to Z \to A \to S \to I \to C$ were given as follows.

Drum Setting:	ZZB	ZZM	ZZH	ZZF	ZZK	ZZG	ZZI
Menu Letter:	C $\longrightarrow$	0 $\longrightarrow$	T $\longrightarrow$	Z $\longrightarrow$	A $\longrightarrow$	S $\longrightarrow$	I $\longrightarrow$ C
Plug Partner:	A $\longrightarrow$	Z $\longrightarrow$	E $\longrightarrow$	W $\longrightarrow$	E $\longrightarrow$	W $\longrightarrow$	Q $\longrightarrow$ A

Note that in this loop, although there is no logical inconsistency at the central letter, there are several logical inconsistencies away from the central letter. Notably, W appears as the plugboard partner of both Z and S, E appears as the plugboard partner of both T and A, and Z appears as the plugboard partner of both 0 and W. Any of these logical inconsistencies alone would be enough to eliminate A as the plugboard partner of C for the Enigma configuration in Example 5.7. However, the bombe was not designed to detect logical inconsistencies away from the central letter. Such logical inconsistencies were only able to be identified through the use of a checking machine, since it allowed human operators to see plugboard partners everywhere in a menu, as opposed to just at the central letter. The downside to this, of course, was that checking machines had to be interacted with by human operators, who tend to be slower than machines. Human interaction was inevitable, however, since even for a correct combination of a Wehrmacht Enigma reflector and rotor order, initial window letters, and ring settings, the only additional information that could be obtained directly from the bombe was a single plugboard pair.

This leads to the second purpose for checking machines, which was, in the absence of logical inconsistencies anywhere in a menu, to find some or all of the other plugboard pairs besides the central letter and its partner. That is, for a particular combination of a Wehrmacht Enigma reflector and rotor order, initial window letters, and ring settings being tested on a crib/ciphertext alignment with a given menu, if a choice for the plugboard partner of the central letter resulted in no logical inconsistencies anywhere in the menu, checking machines could be used to identify some or all of the other plugboard pairs used in the Enigma that formed the ciphertext. Recall that for most of the war, standard German operating procedure was to use exactly 10 plugboard cables connecting a total of 20 letters in pairs. Thus, the second purpose of checking machines was to find some or all of the other nine plugboard pairs (besides the central letter and its partner), and by consequence the six letters left unconnected in the plugboard.

Example 5.8 Consider the Wehrmacht Enigma configuration with reflector **C**, rotor order **I, V, III**, initial window letters ZZZ, and ring settings 16, 23, 18, to be tested on the crib/ciphertext alignment in Example 5.1 on page 93, with menu shown in Figure 5.7 on page 94 and central letter O. In Example 5.3 on page 100, we saw that the only possible plugboard partner of O is S. With this plugboard partner of the central letter, traveling along all of the links in the menu results in the following.

Drum Setting:	ZZB		
Menu Letter:	O $\longrightarrow$ N		
Plug Partner:	S $\longrightarrow$ T		

Drum Setting:	ZZH	ZZF	
Menu Letter:	O $\longrightarrow$ R $\longrightarrow$ W		
Plug Partner:	S $\longrightarrow$ H $\longrightarrow$ W		

Drum Setting:	ZZK	
Menu Letter:	R $\longrightarrow$ B	
Plug Partner:	H $\longrightarrow$ U	

Drum Setting:	ZZN	ZZL	ZZC	ZZG
Menu Letter:	O $\longrightarrow$ S $\longrightarrow$ T $\longrightarrow$ L $\longrightarrow$ O			
Plug Partner:	S $\longrightarrow$ O $\longrightarrow$ N $\longrightarrow$ I $\longrightarrow$ S			

Drum Setting:	ZZD
Menu Letter:	L $\longrightarrow$ Z
Plug Partner:	I $\longrightarrow$ Z

Drum Setting:	ZZM	ZZI	ZZJ	ZZE
Menu Letter:	T $\longrightarrow$ D $\longrightarrow$ H $\longrightarrow$ E $\longrightarrow$ O			
Plug Partner:	N $\longrightarrow$ F $\longrightarrow$ R $\longrightarrow$ E $\longrightarrow$ S			

Drum Setting:	ZZA
Menu Letter:	E $\longrightarrow$ F
Plug Partner:	E $\longrightarrow$ D

These results give no logical inconsistencies anywhere in the menu, and indicate, if the Enigma configuration in this example were indeed correct, that in addition to the central letter and its plugboard partner O/S, the letters N/T, R/H, B/U, L/I, and D/F would be plugboard pairs. They also indicate that the letters W, Z, and E would be left unconnected in the plugboard. This does not give the complete setup of the plugboard though, since four plugboard pairs would remain to be determined. To find these

last four plugboard pairs, we can try decrypting other parts of the ciphertext with the plugboard settings we already know. Recall that in Section
5.1 we saw the crib/ciphertext alignment in Example 5.1 as part of the
following expanded plaintext/ciphertext alignment.

> **Plain:** . . F O L L O W O R D E R S T O . . . .
> **Cipher:** N U E N T Z E R L O H H B T D S H L H I Y W E A B H T Q K C

Since the Enigma configuration in this example includes initial window
letters ZZZ for the crib/ciphertext alignment, which would rotate to ZZA
for the first crib/ciphertext pair, the window letters for each position in
this expanded plaintext/ciphertext alignment, assuming no rotation of the
middle or leftmost rotors, would be as follows.

Window: ZZY ZZZ ZZA ZZB ZZC ZZD ZZE ZZF ZZG ZZH ZZI ZZJ ZZK
Plain: .   .   F   O   L   L   O   W   O   R   D   E   R
Cipher: N   U   E   N   T   Z   E   R   L   O   H   H   B

Window: ZZL ZZM ZZN ZZO ZZP ZZQ ZZR ZZS ZZT ZZU ZZV ZZW ZZX
Plain: S   T   O   .   .   .   .
Cipher: T   D   S   H   L   H   I   Y   W   E   A   B   H

Window: ZZY ZZZ ZZA ZZB
Plain: .   .   .   .
Cipher: T   Q   K   C

For a Wehrmacht Enigma with reflector **C**, rotor order **I, V, III**, and ring
settings 16, 23, 18, the outputs from the double scrambler for each possible input and all window letters in the expanded plaintext/ciphertext
alignment are shown in Table 5.16 on page 128. Consider now the first
ciphertext letter N. Since we know the plugboard partner of N is T, and
can find from Table 5.16 that the output from the double scrambler with
window letters ZZY and input T is A, we could decrypt this first ciphertext
letter N if we knew the plugboard partner of A. Unfortunately, we have not
yet determined the plugboard partner of A, so we do not have enough information to decrypt the first ciphertext letter N. However, we do have enough
information to decrypt the second ciphertext letter U. Since we know the
plugboard partner of U is B, can find from Table 5.16 that the output from
the double scrambler with window letters ZZZ and input B is S, and know
the plugboard partner of S is O, then O must be the second plaintext letter.
Further expanding the plaintext/ciphertext alignment with this information and all additional information that we could obtain similarly gives the
following (with **DScram** representing the double scrambler for the window
letters in each position).

	A	B	C	D	E	F	G	H	I	J	K	L	M	N	O	P	Q	R	S	T	U	V	W	X	Y	Z
ZZA	I	C	B	E	D	Q	J	P	A	G	R	W	V	X	U	H	F	K	Z	Y	O	M	L	N	T	S
ZZB	M	R	X	K	H	N	Z	E	O	P	D	V	A	F	I	J	W	B	T	S	Y	L	Q	C	U	G
ZZC	D	Z	K	A	G	H	E	F	N	M	C	T	J	I	U	W	V	Y	X	L	O	Q	P	S	R	B
ZZD	F	N	H	P	L	A	K	C	Z	V	G	E	U	B	W	D	Y	S	R	X	M	J	O	T	Q	I
ZZE	L	F	P	H	S	B	X	D	J	I	T	A	U	Z	V	C	Y	W	E	K	M	O	R	G	Q	N
ZZF	M	J	D	C	O	S	I	W	G	B	T	U	A	V	E	Y	R	Q	F	K	L	N	H	Z	P	X
ZZG	H	E	K	O	B	G	F	A	S	T	C	M	L	U	D	V	W	X	I	J	N	P	Q	R	Z	Y
ZZH	N	F	G	L	Y	B	C	S	T	W	O	D	V	A	K	X	Z	U	H	I	R	M	J	P	E	Q
ZZI	B	A	Q	P	I	R	S	X	E	Y	L	K	U	Z	V	D	C	F	G	W	M	O	T	H	J	N
ZZJ	X	D	M	B	R	T	I	W	G	N	O	S	C	J	K	V	U	E	L	F	Q	P	H	A	Z	Y
ZZK	E	M	W	S	A	G	F	U	Z	K	J	R	B	P	T	N	Y	L	D	O	H	X	C	V	Q	I
ZZL	U	L	Q	H	Z	M	P	D	T	V	S	B	F	O	N	G	C	X	K	I	A	J	Y	R	W	E
ZZM	S	L	D	C	X	N	M	V	J	I	R	B	G	F	U	W	T	K	A	Q	O	H	P	E	Z	Y
ZZN	G	K	T	R	J	P	A	L	W	E	B	H	N	M	S	F	V	D	O	C	Y	Q	I	Z	U	X
ZZO	Z	T	K	Q	M	L	X	I	H	Y	C	F	E	W	V	R	D	P	U	B	S	O	N	G	J	A
ZZP	J	I	Q	P	M	T	K	S	B	A	G	N	E	L	V	D	C	Y	H	F	Z	O	X	W	R	U
ZZQ	Q	U	I	M	K	W	O	N	C	Z	E	R	D	H	G	T	A	L	X	P	B	Y	F	S	V	J
ZZR	G	T	H	M	O	J	A	C	X	F	L	K	D	U	E	Y	W	S	R	B	N	Z	Q	I	P	V
ZZS	F	H	J	K	Y	A	M	B	W	C	D	T	G	Q	Z	S	N	X	P	L	V	U	I	R	E	O
ZZT	E	G	Y	R	A	T	B	U	K	L	I	J	O	Z	M	S	V	D	P	F	H	Q	X	W	C	N
ZZU	G	J	H	R	F	E	A	C	T	B	V	Q	Z	U	W	Y	L	D	X	I	N	K	O	S	P	M
ZZV	F	X	J	G	V	A	D	N	W	C	R	Z	S	H	Y	Q	P	K	M	U	T	E	I	B	O	L
ZZW	V	O	S	E	D	M	Z	I	H	T	Q	X	F	Y	B	R	K	P	C	J	W	A	U	L	N	G
ZZX	P	J	G	S	I	Z	C	W	E	B	X	O	Y	T	L	A	R	Q	D	N	V	U	H	K	M	F
ZZY	T	P	E	Y	C	H	U	F	Z	X	O	S	Q	W	K	B	M	V	L	A	G	R	N	J	D	I
ZZZ	K	S	F	P	L	C	N	R	X	O	A	E	W	G	J	D	Y	H	B	U	T	Z	M	I	Q	V

Table 5.16   Double scrambler outputs for the Wehrmacht Enigma configuration in Example 5.8 (with reflector **C**, rotor order **I**, **V**, **III**, and ring
settings 16, 23, 18).

Window:	ZZY	ZZZ	ZZA	ZZB	ZZC	ZZD	ZZE	ZZF	ZZG	ZZH	ZZI	ZZJ	ZZK
Plain:	.	O	F	O	L	L	O	W	O	R	D	E	R
Plug:	A	S	D	S	I	I	S	W	S	H	F	E	H
DScram:	↕	↕	↕	↕	↕	↕	↕	↕	↕	↕	↕	↕	↕
Plug:	T	B	E	T	N	Z	E	H	I	S	R	R	U
Cipher:	N	U	E	N	T	Z	E	R	L	O	H	H	B

Window:	ZZL	ZZM	ZZN	ZZO	ZZP	ZZQ	ZZR	ZZS	ZZT	ZZU	ZZV	ZZW	ZZX
Plain:	S	T	O	.	U	I	.	.	.	.	.	W	.
Plug:	O	N	S	P	B	L	K	.	X	F	.	W	Q
DScram:	↕	↕	↕	↕	↕	↕	↕	↕	↕	↕	↕	↕	↕
Plug:	N	F	O	R	I	R	L	.	W	E	.	U	R
Cipher:	T	D	S	H	L	H	I	Y	W	E	A	B	H

Window:	ZZY	ZZZ	ZZA	ZZB
Plain:	W	.	.	.
Plug:	W	.	.	.
DScram:	↕	↕	↕	↕
Plug:	N	.	.	.
Cipher:	T	Q	K	C

If we assume that the second plaintext letter O is the second letter in a two-letter word, candidates for this word would include DO, GO, NO, SO, and TO. However, we have already identified plugboard partners for D, N, S, and T that are different from A. Thus the first plaintext letter would have to be G, and G/A would be another plugboard pair. Next, consider the plaintext letters UI occurring at window letter positions ZZP and ZZQ. It seems reasonable that this UI would be preceded in the plaintext at window letter position ZZO by the letter Q, which gives Q/P as another plugboard pair. Further, following these same plaintext letters QUI, the plaintext letter at window letter position ZZR cannot be T or P, since we have already identified plugboard partners for T and P that are different from K. The only other reasonable choice for the plaintext letter at window letter position ZZR is C, which gives C/K as yet another plugboard pair. Finally, following these same plaintext letters QUIC, the only reasonable choice for the plaintext letter at window letter position ZZS is K. In this position, since the plugboard partner of K would be C, and the output from the double scrambler with window letters ZZS and input C is J, the last plugboard pair would be J/Y. We have thus determined the complete setup of the plugboard for this example—O/S, N/T, R/H, B/U, L/I, D/F, G/A, Q/P, C/K, and J/Y are the ten pairs of letters connected in the plugboard, leaving W, Z, E, V, M, and X as the six letters left unconnected in the plugboard. □

5.5.1   Exercises

1. For the following sequences of menu letters and plugboard partners, determine whether there are any logical inconsistencies.

   (a)* **Menu Letter:** M O S D L N E G H R W A B
         Plug Partner: A N E Q D U H L R Q Q D X

   (b)   **Menu Letter:** M O S D L N E G H R W A B
         Plug Partner: C Y J Q F U P T Z K W I X

   (c)* **Menu Letter:** S H A E B C R M G N T J Z
         Plug Partner: S K I F U O X Y W V T L Z

   (d)   **Menu Letter:** M R O I N D L A E H C P F
         Plug Partner: B R U S N I W G Y Q T X F

2. Consider the Wehrmacht Enigma configuration, crib/ciphertext alignment, and menu in Example 5.6 on page 114, for which in Example 5.7 on page 117 we showed that I is the only possible plugboard partner of the central letter C.

   (a) Show that this plugboard partner of the central letter gives no logical inconsistencies anywhere in the menu, and find as many other plugboard pairs and letters left unconnected in the plugboard as possible. (The outputs from the double scrambler for each possible input and all window letters in the menu, which were originally shown in Table 5.12 on page 115, are included in Table 5.17 on page 131.)

   (b)* Suppose the full ciphertext is UCCCK GZIZC LAHTV CTMB. For the Wehrmacht Enigma configuration in this exercise, the outputs from the double scrambler for each possible input and all window letters in an expanded plaintext/ciphertext alignment are shown in Table 5.17 on page 131. By decrypting other parts of the ciphertext with the plugboard settings you found in part (a), find the remaining plugboard pairs and letters left unconnected in the plugboard. Then use this information to decrypt the full ciphertext.

3. Consider the crib/ciphertext alignment and menu in Exercise 4 in Section 5.3, for which in Exercise 4c in Section 5.3 you showed that for a particular Wehrmacht Enigma configuration and choice R for the plugboard partner of the central letter Q, there is no logical inconsistency at the central letter. Show that there is a logical inconsistency somewhere else in the menu.

	A	B	C	D	E	F	G	H	I	J	K	L	M	N	O	P	Q	R	S	T	U	V	W	X	Y	Z
ZZA	P	W	R	O	I	L	J	Y	E	G	M	F	K	U	D	A	V	C	T	S	N	Q	B	Z	H	X
ZZB	Z	D	X	B	J	M	S	L	O	E	Y	H	F	T	I	Q	P	W	G	N	V	U	R	C	K	A
ZZC	F	K	G	O	R	A	C	U	N	W	B	M	L	I	D	Z	T	E	Y	Q	H	X	J	V	S	P
ZZD	C	E	A	O	B	L	M	Q	S	W	T	F	G	P	D	N	H	U	I	K	R	Z	J	Y	X	V
ZZE	V	K	S	L	R	J	X	U	Z	F	B	D	P	T	Q	M	O	E	C	N	H	A	Y	G	W	I
ZZF	G	M	L	U	W	O	A	Z	Y	N	S	C	B	J	F	X	V	T	K	R	D	Q	E	P	I	H
ZZG	T	O	P	E	D	G	F	M	Z	V	L	K	H	Y	B	C	W	S	R	A	X	J	Q	U	N	I
ZZH	J	U	H	K	W	Y	V	C	Q	A	D	O	R	X	L	S	I	M	P	Z	B	G	E	N	F	T
ZZI	Q	U	I	H	Z	K	Y	D	C	L	F	J	P	X	W	M	A	T	V	R	B	S	O	N	G	E
ZZJ	Q	E	S	N	B	M	X	O	U	W	P	V	F	D	H	K	A	T	C	R	I	L	J	G	Z	Y
ZZK	P	H	R	F	W	D	K	B	Q	L	G	J	O	X	M	A	I	C	Z	Y	V	U	E	N	T	S
ZZL	Z	L	P	M	I	Y	K	R	E	N	G	B	D	J	V	C	U	H	T	S	Q	O	X	W	F	A
ZZM	G	D	R	B	Z	T	A	X	K	L	I	J	S	Y	V	W	U	C	M	F	Q	O	P	H	N	E
ZZN	L	Z	Y	I	F	E	U	X	D	V	N	A	P	K	S	M	R	Q	O	W	G	J	T	H	C	B
ZZO	Z	L	X	P	J	V	H	G	N	E	M	B	K	I	R	D	S	O	Q	W	Y	F	T	C	U	A
ZZP	V	H	D	C	G	L	E	B	N	R	W	F	O	I	M	S	Y	J	P	U	T	A	K	Z	Q	X
ZZQ	O	L	J	V	F	E	K	Y	M	C	G	B	I	P	A	N	S	U	Q	X	R	D	Z	T	H	W
ZZR	X	S	Z	K	U	P	L	I	H	Q	D	G	T	V	W	F	J	Y	B	M	E	N	O	A	R	C
ZZS	W	N	U	P	I	Z	V	Q	E	O	T	X	Y	B	J	D	H	S	R	K	C	G	A	L	M	F
ZZT	K	O	H	Y	U	G	F	C	W	R	A	N	T	L	B	S	Z	J	P	M	E	X	I	V	D	Q
ZZU	S	F	L	E	D	B	K	M	T	R	G	C	H	Y	P	O	W	J	A	I	Z	X	Q	V	N	U
ZZV	I	J	G	O	V	L	C	P	A	B	X	F	T	Q	D	H	N	U	Z	M	R	E	Y	K	W	S
ZZW	J	X	M	N	P	W	I	O	G	A	Q	R	C	D	H	E	K	L	Z	V	Y	T	F	B	U	S
ZZX	K	U	T	R	P	V	Y	Q	W	S	A	N	Z	L	X	E	H	D	J	C	B	F	I	O	G	M
ZZY	J	O	P	U	T	K	L	R	X	A	F	G	N	M	B	C	S	H	Q	E	D	Z	Y	I	W	V
ZZZ	D	P	L	A	R	Z	S	O	T	Y	N	C	V	K	H	B	U	E	G	I	Q	M	X	W	J	F

Table 5.17   Double scrambler outputs for the Wehrmacht Enigma configuration in Exercise 2 (with reflector **B**, rotor order **III**, **IV**, **I**, and ring settings 16, 22, 23).

4. Consider the crib/ciphertext alignment and menu in Exercise 4 in Section 5.3, for which in Exercise 4b in Section 5.3 you showed that for a particular Wehrmacht Enigma configuration and choice S for the plugboard partner of the central letter Q, there is no logical inconsistency at the central letter.

   (a) Show that there are no logical inconsistencies anywhere in the menu, and find as many other plugboard pairs and letters left unconnected in the plugboard as possible. (The outputs from the double scrambler for each possible input and all window letters in the menu, which were originally shown in Table 5.4 on page 108, are included in Table 5.18 on page 133.)

   (b)* Suppose the full ciphertext is GLBCJ QQQYL QJOZM BYSGA NA. For the Wehrmacht Enigma configuration in this exercise, the outputs from the double scrambler for each possible input and all window letters in an expanded plaintext/ciphertext alignment are shown in Table 5.18 on page 133. By decrypting other parts of the ciphertext with the plugboard settings you found in part (a), find the remaining plugboard pairs and letters left unconnected in the plugboard. Then use this information to decrypt the full ciphertext.

5.* Consider the crib/ciphertext alignment and menu in Exercise 5 in Section 5.3, for which in Exercise 5b in Section 5.3 you showed that for a particular Wehrmacht Enigma configuration and choice K for the plugboard partner of the central letter T, there is no logical inconsistency at the central letter. Show that there is a logical inconsistency somewhere else in the menu.

6.* Consider the crib/ciphertext alignment and menu in Exercise 5 in Section 5.3, for which in Exercise 5c in Section 5.3 you showed that for a particular Wehrmacht Enigma configuration and choice Q for the plugboard partner of the central letter T, there is no logical inconsistency at the central letter.

   (a) Show that there are no logical inconsistencies anywhere in the menu, and find as many other plugboard pairs and letters left unconnected in the plugboard as possible. (The outputs from the double scrambler for each possible input and all window letters in the menu, which were originally shown in Table 5.7 on page 110, are included in Table 5.19 on page 134.)

   (b) Suppose the full ciphertext is KMUTV MUUPW FFTSF SGKMF U. For the Wehrmacht Enigma configuration in this exercise, the outputs from the double scrambler for each possible input and all

	A	B	C	D	E	F	G	H	I	J	K	L	M	N	O	P	Q	R	S	T	U	V	W	X	Y	Z
ZZA	T	J	R	S	Q	P	I	M	G	B	X	Y	H	O	N	F	E	C	D	A	V	U	Z	K	L	W
ZZB	N	D	O	B	Z	U	R	W	X	K	J	S	T	A	C	V	Y	G	L	M	F	P	H	I	Q	E
ZZC	D	K	H	A	L	V	T	C	X	O	B	E	P	Z	J	M	S	U	Q	G	R	F	Y	I	W	N
ZZD	D	U	J	A	K	Y	L	N	Q	C	E	G	V	H	Z	W	I	T	X	R	B	M	P	S	F	O
ZZE	S	W	X	R	M	N	Z	J	L	H	T	I	E	F	P	O	Y	D	A	K	V	U	B	C	Q	G
ZZF	K	C	B	Q	F	E	Y	U	M	O	A	N	I	L	J	S	D	V	P	Z	H	R	X	W	G	T
ZZG	P	W	Y	O	K	M	R	I	H	L	E	J	F	Z	D	A	T	G	V	Q	X	S	B	U	C	N
ZZH	S	G	K	R	J	M	B	N	P	E	C	T	F	H	Q	I	O	D	A	L	W	X	U	V	Z	Y
ZZI	M	G	L	N	O	V	B	X	R	T	U	C	A	D	E	Z	S	I	Q	J	K	F	Y	H	W	P
ZZJ	T	N	P	G	K	H	D	F	O	Q	E	R	X	B	I	C	J	L	Z	A	W	Y	U	M	V	S
ZZK	T	G	Z	L	R	Q	B	U	O	N	S	D	Y	J	I	V	F	E	K	A	H	P	X	W	M	C
ZZL	H	S	G	Y	W	I	C	A	F	Q	R	P	O	U	M	L	J	K	B	Z	N	X	E	V	D	T
ZZM	D	V	I	A	O	M	U	T	C	P	R	X	F	Q	E	J	N	K	W	H	G	B	S	L	Z	Y
ZZN	R	E	M	Q	B	G	F	Y	K	Z	I	X	C	P	S	N	D	A	O	U	T	W	V	L	H	J
ZZO	I	W	L	N	X	K	J	P	A	G	F	C	U	D	S	H	T	Y	O	Q	M	Z	B	E	R	V
ZZP	E	Y	R	J	A	Z	S	K	M	D	H	W	I	U	Q	X	O	C	G	V	N	T	L	P	B	F
ZZQ	C	U	A	T	L	X	W	Q	R	N	Z	E	O	J	M	V	H	I	Y	D	B	P	G	F	S	K
ZZR	O	K	Z	S	X	M	Y	J	U	H	B	W	F	Q	A	R	N	P	D	V	I	T	L	E	G	C
ZZS	G	I	L	N	X	R	A	O	B	Q	W	C	V	D	H	Z	J	F	T	S	Y	M	K	E	U	P
ZZT	F	I	X	U	P	A	W	T	B	S	R	N	Z	L	Q	E	O	K	J	H	D	Y	G	C	V	M
ZZU	B	A	J	H	R	Q	U	D	M	C	Z	S	I	O	N	W	F	E	L	X	G	Y	P	T	V	K
ZZV	M	V	D	C	I	Q	T	W	E	S	Y	Z	A	R	U	X	F	N	J	G	O	B	H	P	K	L
ZZW	P	Z	X	F	J	D	R	Q	N	E	T	O	S	I	L	A	H	G	M	K	W	Y	U	C	V	B
ZZX	E	S	T	W	A	G	F	R	Z	O	Q	Y	X	P	J	N	K	H	B	C	V	U	D	M	L	I
ZZY	O	F	P	I	L	B	X	W	D	T	Z	E	S	Q	A	C	N	Y	M	J	V	U	H	G	R	K
ZZZ	L	Q	I	O	M	H	T	F	C	U	S	A	E	P	D	N	B	Z	K	G	J	Y	X	W	V	R

Table 5.18 Double scrambler outputs for the Wehrmacht Enigma configuration in Exercise 4 (with reflector **B**, rotor order **IV**, **III**, **I**, and ring settings 9, 6, 25).

	A	B	C	D	E	F	G	H	I	J	K	L	M	N	O	P	Q	R	S	T	U	V	W	X	Y	Z
ZZA	O	E	D	C	B	K	N	Y	S	W	F	R	V	G	A	X	Z	L	I	U	T	M	J	P	H	Q
ZZB	X	H	E	S	C	W	Z	B	M	K	J	O	I	P	L	N	U	Y	D	V	Q	T	F	A	R	G
ZZC	M	I	G	P	W	U	C	K	B	R	H	S	A	Z	Q	D	O	J	L	Y	F	X	E	V	T	N
ZZD	F	E	R	G	B	A	D	O	T	Y	Z	W	Q	S	H	U	M	C	N	I	P	X	L	V	J	K
ZZE	R	M	G	E	D	L	C	Z	Y	S	O	F	B	V	K	X	W	A	J	U	T	N	Q	P	I	H
ZZF	F	D	Q	B	W	A	N	U	M	K	J	S	I	G	R	T	C	O	L	P	H	X	E	V	Z	Y
ZZG	H	X	J	Y	I	S	N	A	E	C	V	Q	T	G	W	Z	L	U	F	M	R	K	O	B	D	P
ZZH	F	O	L	E	D	A	M	Y	U	P	Z	C	G	Q	B	J	N	X	V	W	I	S	T	R	H	K
ZZI	M	R	N	G	X	P	D	Y	K	L	I	J	A	C	S	F	U	B	O	V	Q	T	Z	E	H	W
ZZJ	H	K	V	S	W	M	U	A	P	T	B	R	F	X	Y	I	Z	L	D	J	G	C	E	N	O	Q
ZZK	P	V	T	H	M	O	Q	D	L	K	J	I	E	U	F	A	G	W	Z	C	N	B	R	Y	X	S
ZZL	W	C	B	Z	H	O	K	E	P	N	G	T	X	J	F	I	R	Q	V	L	Y	S	A	M	U	D
ZZM	D	J	Q	A	I	K	X	M	E	B	F	Z	H	R	T	S	C	N	P	O	V	U	Y	G	W	L
ZZN	M	S	K	G	Z	L	D	Y	J	I	C	F	A	U	Q	X	O	V	B	W	N	R	T	P	H	E
ZZO	U	W	D	C	S	Q	Z	Y	N	T	M	V	K	I	R	X	F	O	E	J	A	L	B	P	H	G
ZZP	O	I	V	P	F	E	L	Y	B	U	R	G	W	T	A	D	S	K	Q	N	J	C	M	Z	H	X
ZZQ	R	F	Y	W	J	B	N	O	Q	E	L	K	X	G	H	U	I	A	T	S	P	Z	D	M	C	V
ZZR	Y	I	J	V	R	U	O	M	B	C	S	X	H	Z	G	T	W	E	K	P	F	D	Q	L	A	N
ZZS	T	H	S	V	F	E	K	B	U	X	G	P	W	Y	R	L	Z	O	C	A	I	D	M	J	N	Q
ZZT	M	E	W	N	B	V	T	Q	X	R	S	O	A	D	L	U	H	J	K	G	P	F	C	I	Z	Y
ZZU	G	F	P	W	Q	B	A	U	K	T	I	R	Y	S	V	C	E	L	N	J	H	O	D	Z	M	X
ZZV	T	X	P	J	V	L	Q	U	K	D	I	F	O	Z	M	C	G	Y	W	A	H	E	S	B	R	N
ZZW	E	D	V	B	A	J	N	Y	K	F	I	U	Z	G	S	R	T	P	O	Q	L	C	X	W	H	M
ZZX	U	F	K	O	Y	B	J	L	Q	G	C	H	V	T	D	X	I	Z	W	N	A	M	S	P	E	R
ZZY	C	M	A	F	J	D	X	L	Y	E	T	H	B	Z	U	S	V	W	P	K	O	Q	R	G	I	N
ZZZ	E	D	J	B	A	S	M	V	Y	C	W	N	G	L	X	Z	T	U	F	Q	R	H	K	O	I	P

Table 5.19 Double scrambler outputs for the Wehrmacht Enigma configuration in Exercise 6 (with reflector **C**, rotor order **IV**, **V**, **II**, and ring
settings 26, 23, 19).

window letters in an expanded plaintext/ciphertext alignment are shown in Table 5.19 on page 134. By decrypting other parts of the ciphertext with the plugboard settings you found in part (a), find the remaining plugboard pairs and letters left unconnected in the plugboard. Then use this information to decrypt the full ciphertext.

7. Consider the crib/ciphertext alignment and menu in Exercise 6 in Section 5.3, for which in Exercise 6c in Section 5.3 you showed that for a particular Wehrmacht Enigma configuration and choice B for the plugboard partner of the central letter N, there is no logical inconsistency at the central letter. Show that there is a logical inconsistency somewhere else in the menu.

8. Consider the crib/ciphertext alignment and menu in Exercise 6 in Section 5.3, for which in Exercise 6b in Section 5.3 you showed that for a particular Wehrmacht Enigma configuration and choice X for the plugboard partner of the central letter N, there is no logical inconsistency at the central letter.

   (a) Show that there are no logical inconsistencies anywhere in the menu, and find as many other plugboard pairs and letters left unconnected in the plugboard as possible. (The outputs from the double scrambler for each possible input and all window letters in the menu, which were originally shown in Table 5.9 on page 112, are included in Table 5.20 on page 136.)

   (b) Suppose the full ciphertext is WPMOI NNMUM GNGEL URGXT OY. For the Wehrmacht Enigma configuration in this exercise, the outputs from the double scrambler for each possible input and all window letters in an expanded plaintext/ciphertext alignment are shown in Table 5.20 on page 136. By decrypting other parts of the ciphertext with the plugboard settings you found in part (a), find the remaining plugboard pairs and letters left unconnected in the plugboard. Then use this information to decrypt the full ciphertext.

9.* Consider the crib/ciphertext alignment and menu in Exercise 2 in Section 5.4, for which you showed that for a particular Wehrmacht Enigma configuration and choice B for the plugboard partner of the central letter M, there is no logical inconsistency at the central letter.

   (a) Show that there are no logical inconsistencies anywhere in the menu, and find as many other plugboard pairs and letters left unconnected in the plugboard as possible. (The outputs from the

	A	B	C	D	E	F	G	H	I	J	K	L	M	N	O	P	Q	R	S	T	U	V	W	X	Y	Z
ZZA	H	T	E	M	C	Z	L	A	X	W	Y	G	D	V	U	R	S	P	Q	B	O	N	J	I	K	F
ZZB	J	M	G	E	D	V	C	Y	Z	A	R	S	B	T	Q	X	O	K	L	N	W	F	U	P	H	I
ZZC	I	D	Z	B	N	Q	M	Y	A	P	R	U	G	E	X	J	F	K	T	S	L	W	V	O	H	C
ZZD	D	Y	U	A	P	Q	O	V	W	R	L	K	X	Z	G	E	F	J	T	S	C	H	I	M	B	N
ZZE	P	N	Z	F	H	D	U	E	V	Y	T	R	S	B	W	A	X	L	M	K	G	I	O	Q	J	C
ZZF	M	C	B	E	D	O	R	J	X	H	P	U	A	Y	F	K	Z	G	V	W	L	S	T	I	N	Q
ZZG	Z	I	S	N	H	J	Y	E	B	F	T	X	O	D	M	Q	P	U	C	K	R	W	V	L	G	A
ZZH	H	E	I	J	B	W	N	A	C	D	Q	U	Z	G	P	O	K	X	V	Y	L	S	F	R	T	M
ZZI	D	K	Y	A	V	Q	S	M	N	T	B	R	H	I	U	Z	F	L	G	J	O	E	X	W	C	P
ZZJ	K	R	H	Y	Q	U	L	C	T	X	A	G	S	Z	P	O	E	B	M	I	F	W	V	J	D	N
ZZK	Y	K	V	E	D	M	Q	W	R	Z	B	P	F	O	N	L	G	I	X	U	T	C	H	S	A	J
ZZL	Y	O	D	C	I	S	P	Q	E	R	X	W	T	U	B	G	H	J	F	M	N	Z	L	K	A	V
ZZM	N	L	M	G	H	Q	D	E	W	S	T	B	C	A	V	X	F	Z	J	K	Y	O	I	P	U	R
ZZN	Z	M	S	N	P	X	H	G	R	O	Y	V	B	D	J	E	U	I	C	W	Q	L	T	F	K	A
ZZO	B	A	M	E	D	N	Z	Y	T	V	U	Q	C	F	S	R	L	P	O	I	K	J	X	W	H	G
ZZP	N	C	B	Z	K	Y	J	S	Q	G	E	R	V	A	W	T	I	L	H	P	X	M	O	U	F	D
ZZQ	T	R	O	J	I	L	K	Y	E	D	G	F	U	V	C	Z	W	B	X	A	M	N	Q	S	H	P
ZZR	N	F	W	I	K	B	S	P	D	M	E	Z	J	A	Y	H	U	T	G	R	Q	X	C	V	O	L
ZZS	W	D	K	B	Y	R	I	J	G	H	C	X	U	P	T	N	S	F	Q	O	M	Z	A	L	E	V
ZZT	X	J	R	S	K	L	H	G	P	B	E	F	Q	T	U	I	M	C	D	N	O	Z	Y	A	W	V
ZZU	Q	N	H	Y	T	J	P	C	O	F	X	U	S	B	I	G	A	V	M	E	L	R	Z	K	D	W
ZZV	K	G	R	F	S	D	B	Y	N	W	A	T	X	I	U	Q	P	C	E	L	O	Z	J	M	H	V
ZZW	X	M	E	Z	C	U	T	K	S	Y	H	R	B	W	P	O	V	L	I	G	F	Q	N	A	J	D
ZZX	W	P	G	H	O	S	C	D	U	V	Q	Y	T	Z	E	B	K	X	F	M	I	J	A	R	L	N
ZZY	J	F	Z	Q	N	B	P	X	T	A	W	S	R	E	V	G	D	M	L	I	Y	O	K	H	U	C
ZZZ	Z	S	F	K	O	C	W	M	P	Y	D	N	H	L	E	I	R	Q	B	X	V	U	G	T	J	A

Table 5.20   Double scrambler outputs for the Wehrmacht Enigma configuration in Exercise 8 (with reflector **C**, rotor order **I**, **IV**, **V**, and ring settings 19, 22, 24).

	A	B	C	D	E	F	G	H	I	J	K	L	M	N	O	P	Q	R	S	T	U	V	W	X	Y	Z
ZZA	Y	O	G	M	Z	N	C	L	U	W	X	H	D	F	B	S	V	T	P	R	I	Q	J	K	A	E
ZZB	N	U	Q	Y	G	V	E	I	H	Z	L	K	P	A	X	M	C	S	R	W	B	F	T	O	D	J
ZZC	W	D	T	B	K	L	H	G	Z	Y	E	F	V	P	S	N	R	Q	O	C	X	M	A	U	J	I
ZZD	M	D	Y	B	Q	X	H	G	V	K	J	U	A	Z	S	R	E	P	O	W	L	I	T	F	C	N
ZZE	D	J	R	A	N	G	F	K	U	B	H	V	P	E	Q	M	O	C	X	W	I	L	T	S	Z	Y
ZZF	U	N	L	P	T	G	F	X	Q	Z	O	C	Y	B	K	D	I	W	V	E	A	S	R	H	M	J
ZZG	T	X	O	N	F	E	R	W	V	Z	Q	M	L	D	C	Y	K	G	U	A	S	I	H	B	P	J
ZZH	M	C	B	V	F	E	W	U	O	L	Z	J	A	Y	I	X	S	T	Q	R	H	D	G	P	N	K
ZZI	B	A	K	E	D	S	P	R	W	T	C	M	L	V	X	G	U	H	F	J	Q	N	I	O	Z	Y
ZZJ	B	A	U	E	D	O	H	G	L	X	W	I	Y	R	F	T	S	N	Q	P	C	Z	K	J	M	V
ZZK	Z	K	D	C	J	X	M	P	Y	E	B	O	G	Q	L	H	N	U	T	S	R	W	V	F	I	A
ZZL	Z	W	K	L	V	G	F	T	X	U	C	D	P	Q	Y	M	N	S	R	H	J	E	B	I	O	A
ZZM	W	J	N	Q	U	P	L	K	X	B	H	G	O	C	M	F	D	V	T	S	E	R	A	I	Z	Y
ZZN	T	P	I	S	G	W	E	M	C	X	O	N	H	L	K	B	U	V	D	A	Q	R	F	J	Z	Y
ZZO	F	H	P	R	N	A	M	B	J	I	O	Q	G	E	K	C	L	D	T	S	W	Z	U	Y	X	V
ZZP	B	A	P	E	D	V	Q	T	O	N	U	Z	W	J	I	C	G	S	R	H	K	F	M	Y	X	L
ZZQ	Z	R	H	O	Y	U	J	C	S	G	L	K	N	M	D	Q	P	B	I	V	F	T	X	W	E	A
ZZR	Z	D	I	B	W	T	H	G	C	Q	S	M	L	Y	P	O	J	U	K	F	R	X	E	V	N	A
ZZS	H	O	I	V	J	Q	W	A	C	E	R	P	N	M	B	L	F	K	U	X	S	D	G	T	Z	Y
ZZT	Y	P	W	Z	U	N	X	J	S	H	O	M	L	F	K	B	T	V	I	Q	E	R	C	G	A	D
ZZU	C	V	A	I	O	K	P	U	D	Q	F	Y	N	M	E	G	J	T	W	R	H	B	S	Z	L	X
ZZV	F	W	Q	I	J	A	Z	K	D	E	H	M	L	O	N	S	C	T	P	R	V	U	B	Y	X	G
ZZW	D	S	O	A	H	M	N	E	Y	K	J	P	F	G	C	L	Z	V	B	U	T	R	X	W	I	Q
ZZX	Q	S	Z	P	W	K	H	G	J	I	F	R	N	M	Y	D	A	L	B	U	T	X	E	V	O	C
ZZY	J	K	Z	Y	N	M	H	G	O	A	B	Q	F	E	I	R	L	P	T	S	W	X	U	V	D	C
ZZZ	I	W	X	R	H	G	F	E	A	V	Y	M	L	Q	Z	T	N	D	U	P	S	J	B	C	K	O

Table 5.21 Double scrambler outputs for the Wehrmacht Enigma configuration in Exercise 9 (with reflector **B**, rotor order **I**, **II**, **III**, and ring settings 26, 11, 9).

	A	B	C	D	E	F	G	H	I	J	K	L	M	N	O	P	Q	R	S	T	U	V	W	X	Y	Z
ZZA	C	X	A	L	R	N	M	K	T	W	H	D	G	F	Q	V	O	E	U	I	S	P	J	B	Z	Y
ZZB	I	H	U	T	F	E	W	B	A	Y	L	K	X	Z	R	S	V	O	P	D	C	Q	G	M	J	N
ZZC	Y	V	I	U	O	L	J	T	C	G	X	F	Z	W	E	R	S	P	Q	H	D	B	N	K	A	M
ZZD	J	R	D	C	Y	N	P	L	M	A	T	H	I	F	W	G	U	B	X	K	Q	Z	O	S	E	V
ZZE	G	F	L	U	W	B	A	Q	R	O	M	C	K	Y	J	S	H	I	P	Z	D	X	E	V	N	T
ZZF	U	R	K	W	V	J	Y	M	Z	F	C	O	H	S	L	Q	P	B	N	X	A	E	D	T	G	I
ZZG	F	T	Q	Z	P	A	R	S	K	Y	I	N	U	L	W	E	C	G	H	B	M	X	O	V	J	D
ZZH	V	R	O	L	W	P	Q	I	H	M	S	D	J	Z	C	F	G	B	K	U	T	A	E	Y	X	N
ZZI	Y	O	V	H	Z	Q	M	D	K	L	I	J	G	U	B	W	F	T	X	R	N	C	P	S	A	E
ZZJ	M	T	D	C	J	P	W	V	Q	E	L	K	A	O	N	F	I	Y	Z	B	X	H	G	U	R	S
ZZK	J	F	S	X	V	B	Z	U	T	A	L	K	Y	O	N	R	W	P	C	I	H	E	Q	D	M	G
ZZL	I	L	H	G	T	S	D	C	A	N	P	B	V	J	Z	K	X	Y	F	E	W	M	U	Q	R	O
ZZM	V	F	P	U	N	B	Y	M	J	I	W	Q	H	E	R	C	L	O	X	Z	D	A	K	S	G	T
ZZN	Q	H	R	V	W	K	O	B	M	Z	F	T	I	U	G	S	A	C	P	L	N	D	E	Y	X	J
ZZO	Z	N	Q	U	W	I	O	K	F	V	H	X	S	B	G	R	C	P	M	Y	D	J	E	L	T	A
ZZP	U	G	X	T	V	Z	B	K	R	O	H	S	Y	W	J	Q	P	I	L	D	A	E	N	C	M	F
ZZQ	O	Y	V	Z	G	S	E	M	T	R	Q	W	H	U	A	X	K	J	F	I	N	C	L	P	B	D
ZZR	B	A	T	W	M	Y	P	X	Z	K	J	O	E	V	L	G	R	Q	U	C	S	N	D	H	F	I
ZZS	U	J	H	M	I	K	X	C	E	B	F	N	D	L	Q	S	O	Y	P	V	A	T	Z	G	R	W
ZZT	S	O	U	E	D	K	X	V	L	M	F	I	J	T	B	Y	Z	W	A	N	C	H	R	G	P	Q
ZZU	R	V	F	N	P	C	K	Z	S	L	G	J	W	D	Q	E	O	A	I	X	Y	B	M	T	U	H
ZZV	E	T	Y	I	A	H	R	F	D	S	M	P	K	W	Q	L	O	G	J	B	X	Z	N	U	C	V
ZZW	D	S	N	A	H	T	V	E	X	K	J	M	L	C	Z	R	U	P	B	F	Q	G	Y	I	W	O
ZZX	G	J	I	Z	P	T	A	X	C	B	W	V	O	U	M	E	Y	S	R	F	N	L	K	H	Q	D
ZZY	E	Q	T	F	A	D	M	V	O	K	J	Z	G	P	I	N	B	X	U	C	S	H	Y	R	W	L
ZZZ	H	X	R	M	S	N	K	A	Y	Z	G	P	D	F	W	L	T	C	E	Q	V	U	O	B	I	J

Table 5.22   Double scrambler outputs for the Wehrmacht Enigma configuration in Exercise 10 (with reflector **C**, rotor order **V**, **III**, **I**, and ring settings 17, 26, 16).

double scrambler for each possible input and all window letters in the menu, which were originally shown in Table 5.13 on page 121, are included in Table 5.21 on page 137.)

(b) Suppose the full ciphertext is EMHSS MBNSX NLWWR HISAO. For the Wehrmacht Enigma configuration in this exercise, the outputs from the double scrambler for each possible input and all window letters in an expanded plaintext/ciphertext alignment are shown in Table 5.21 on page 137. By decrypting other parts of the ciphertext with the plugboard settings you found in part (a), find the remaining plugboard pairs and letters left unconnected in the plugboard. Then use this information to decrypt the full ciphertext.

10. Consider the crib/ciphertext alignment and menu in Exercise 3 in Section 5.4, for which you showed that for a particular Wehrmacht Enigma configuration and choice L for the plugboard partner of the central letter J, the menu loop J → O → S → J gives no logical inconsistency at the central letter.

(a) Show that there are no logical inconsistencies anywhere in the menu, and find as many other plugboard pairs and letters left unconnected in the plugboard as possible. (The outputs from the double scrambler for each possible input and all window letters in the menu, which were originally shown in Table 5.14 on page 121, are included in Table 5.22 on page 138.)

(b) Suppose the full ciphertext is AJJRJ EHMRI OFURZ JOTIB TFWH. For the Wehrmacht Enigma configuration in this exercise, the outputs from the double scrambler for each possible input and all window letters in an expanded plaintext/ciphertext alignment are shown in Table 5.22 on page 138. By decrypting other parts of the ciphertext with the plugboard settings you found in part (a), find the remaining plugboard pairs and letters left unconnected in the plugboard. Then use this information to decrypt the full ciphertext.

5.6 Turnovers

In Section 5.5 we saw how, for a given ciphertext formed using a Wehrmacht Enigma, a checking machine could be used to determine the complete setup of the plugboard. However, this would typically only give enough information to accurately decrypt a small number of ciphertext letters on either side of a known crib/ciphertext alignment, because it did not indicate where the middle or leftmost rotors would have rotated in the machine during encryption. The rotation of the rightmost rotor was easy to account for, since it predictably rotated exactly one position each time a letter was pressed on the keyboard. However, recall that in a Wehrmacht Enigma, once every 26 times the rightmost rotor rotated one position, it would cause the middle rotor to rotate one position, and once every 26 times the middle rotor rotated one position, it would cause the leftmost rotor to rotate one position. The cryptanalysts at Bletchley Park referred to rotor rotations as *turnovers*, and while turnovers in middle and leftmost rotors were not hard to identify and account for, they did have to be identified and accounted for each time they occurred.

Example 5.9 The complete setup of the plugboard given at the end of Example 5.8 on page 126 with the double scrambler outputs in Table 5.16 on page 128 results in the following decryption of the full ciphertext in Example 5.8.

Window:	ZZY	ZZZ	ZZA	ZZB	ZZC	ZZD	ZZE	ZZF	ZZG	ZZH	ZZI	ZZJ	ZZK
Plain:	G	O	F	O	L	L	O	W	O	R	D	E	R
Plug:	A	S	D	S	I	I	S	W	S	H	F	E	H
DScram:	↕	↕	↕	↕	↕	↕	↕	↕	↕	↕	↕	↕	↕
Plug:	T	B	E	T	N	Z	E	H	I	S	R	R	U
Cipher:	N	U	E	N	T	Z	E	R	L	O	H	H	B

Window:	ZZL	ZZM	ZZN	ZZO	ZZP	ZZQ	ZZR	ZZS	ZZT	ZZU	ZZV	ZZW	ZZX
Plain:	S	T	O	Q	U	I	C	K	X	D	F	W	P
Plug:	O	N	S	P	B	L	K	C	X	F	D	W	Q
DScram:	↕	↕	↕	↕	↕	↕	↕	↕	↕	↕	↕	↕	↕
Plug:	N	F	O	R	I	R	L	J	W	E	G	U	R
Cipher:	T	D	S	H	L	H	I	Y	W	E	A	B	H

Window:	ZZY	ZZZ	ZZA	ZZB
Plain:	W	F	U	F
Plug:	W	D	B	D
DScram:	↕	↕	↕	↕
Plug:	N	P	C	K
Cipher:	T	Q	K	C

Note that the resulting plaintext is legible until window letter position ZZT. If the middle rotor had rotated at this position, then the window letters at this position would have been ZAT rather than ZZT. Using the new middle window letter A in this and all subsequent positions with the additional double scrambler outputs in Table 5.23 on page 141 results in the following corrected decryption of the full ciphertext in Example 5.8.

	A	B	C	D	E	F	G	H	I	J	K	L	M	N	O	P	Q	R	S	T	U	V	W	X	Y	Z
ZAA	D	V	K	A	Y	O	J	Q	P	G	C	U	N	M	F	I	H	X	W	Z	L	B	S	R	E	T
ZAB	Y	D	Z	B	K	L	V	Q	N	M	E	F	J	I	P	O	H	W	T	S	X	G	R	U	A	C
ZAC	P	U	R	M	Q	H	O	F	L	N	T	I	D	J	G	A	E	C	Y	K	B	Z	X	W	S	V
ZAD	X	I	P	Q	K	U	N	J	B	H	E	M	L	G	Z	C	D	S	R	W	F	Y	T	A	V	O
ZAE	N	L	Z	H	T	P	J	D	K	G	I	B	S	A	X	F	U	Y	M	E	Q	W	V	O	R	C
ZAF	B	A	K	G	P	O	D	T	J	I	C	S	X	W	F	E	R	Q	L	H	Y	Z	N	M	U	V
ZAG	Z	O	R	L	I	G	F	K	E	Y	H	D	S	X	B	T	V	C	M	P	W	Q	U	N	J	A
ZAH	O	G	D	C	H	Y	B	E	Z	S	T	U	W	Q	A	X	N	V	J	K	L	R	M	P	F	I
ZAI	H	C	B	N	I	P	Q	A	E	M	Z	R	J	D	S	F	G	L	O	U	T	W	V	Y	X	K
ZAJ	F	H	N	E	D	A	R	B	L	T	U	I	X	C	Y	V	S	G	Q	J	K	P	Z	M	O	W
ZAK	Z	J	M	P	V	G	F	O	L	B	Q	I	C	X	H	D	K	T	U	R	S	E	Y	N	W	A
ZAL	V	T	W	R	M	X	J	K	Q	G	H	Y	E	O	N	S	I	D	P	B	Z	A	C	F	L	U
ZAM	I	Y	D	C	S	R	N	K	A	Q	H	V	P	G	W	M	J	F	E	X	Z	L	O	T	B	U
ZAN	O	T	S	L	P	I	J	Q	F	G	X	D	N	M	A	E	H	U	C	B	R	Y	Z	K	V	W
ZAO	Z	K	H	E	D	M	Q	C	N	U	B	P	F	I	R	L	G	O	X	W	J	Y	T	S	V	A
ZAP	K	S	G	M	H	P	C	E	R	W	A	N	D	L	T	F	X	I	B	O	Y	Z	J	Q	U	V
ZAQ	R	K	D	C	M	I	Z	V	F	Y	B	U	E	O	N	Q	P	A	X	W	L	H	T	S	J	G
ZAR	O	F	M	N	H	B	W	E	V	X	L	K	C	D	A	S	R	Q	P	Z	Y	I	G	J	U	T
ZAS	K	C	B	J	H	L	Z	E	V	D	A	F	W	Q	U	Y	N	S	R	X	O	I	M	T	P	G
ZAT	L	Y	F	G	W	C	D	X	K	U	I	A	S	P	Q	N	O	Z	M	V	J	T	E	H	B	R
ZAU	B	A	J	N	T	Y	K	Z	O	C	G	Q	X	D	I	W	L	V	U	E	S	R	P	M	F	H
ZAV	X	E	Y	Z	B	J	N	U	W	F	O	T	Q	G	K	R	M	P	V	L	H	S	I	A	C	D
ZAW	Z	O	G	M	X	J	C	Y	V	F	P	T	D	W	B	K	S	U	Q	L	R	I	N	E	H	A
ZAX	T	F	Y	H	R	B	I	D	G	X	N	S	Q	K	P	O	M	E	L	A	V	U	Z	J	C	W
ZAY	U	D	H	B	W	M	Y	C	Z	V	R	O	F	Q	L	S	N	K	P	X	A	J	E	T	G	I
ZAZ	C	X	A	Z	O	V	P	Q	J	I	W	S	Y	R	E	G	H	N	L	U	T	F	K	B	M	D

Table 5.23 Double scrambler outputs for the Wehrmacht Enigma configuration in Example 5.9 (with reflector **C**, rotor order **I**, **V**, **III**, and ring settings 16, 23, 18).

Window:	ZZY	ZZZ	ZZA	ZZB	ZZC	ZZD	ZZE	ZZF	ZZG	ZZH	ZZI	ZZJ	ZZK
Plain:	G	O	F	O	L	L	O	W	O	R	D	E	R
Plug:	A	S	D	S	I	I	S	W	S	H	F	E	H
DScram:	↕	↕	↕	↕	↕	↕	↕	↕	↕	↕	↕	↕	↕
Plug:	T	B	E	T	N	Z	E	H	I	S	R	R	U
Cipher:	N	U	E	N	T	Z	E	R	L	O	H	H	B

Window:	ZZL	ZZM	ZZN	ZZO	ZZP	ZZQ	ZZR	ZZS	ZAT	ZAU	ZAV	ZAW	ZAX
Plain:	S	T	O	Q	U	I	C	K	E	N	T	H	E
Plug:	O	N	S	P	B	L	K	C	E	T	N	R	E
DScram:	↕	↕	↕	↕	↕	↕	↕	↕	↕	↕	↕	↕	↕
Plug:	N	F	O	R	I	R	L	J	W	E	G	U	R
Cipher:	T	D	S	H	L	H	I	Y	W	E	A	B	H

Window:	ZAY	ZAZ	ZAA	ZAB
Plain:	P	A	C	E
Plug:	P	G	K	E
DScram:	↕	↕	↕	↕
Plug:	N	P	C	K
Cipher:	T	Q	K	C

This finally reveals the full plaintext: GO FOLLOW ORDERS TO QUICKEN THE
PACE.   □

We should note and strongly emphasize that in addition to the ingenuity and hard work of Turing, Welchman, and the other cryptanalysts at
Bletchley Park, decrypting Enigma ciphertexts also required some luck.
For example, for the decryption in Example 5.9 to work, it was necessary that there had been no turnover of the middle rotor anywhere in the
crib/ciphertext alignment so that one of the bombe stops could give a correct Enigma configuration and plugboard partner of the central letter, and
it was also necessary that there had been no turnover of the middle rotor
close to the crib in the plaintext/ciphertext alignment so that the checking
machine could give the remaining plugboard pairs. Since a turnover of the
middle rotor always had to occur within each span of 26 letters, for cribs
longer than 13 letters there was a greater than 50% chance that a turnover
of the middle rotor within the crib/ciphertext alignment would ruin the entire cryptanalysis process. Of course, longer cribs also increased the chances
of loops occurring in menus. So the reality for the cryptanalysts at Bletchley Park was that they needed cribs that were long but not too long in
order to maximize the chances of the entire cryptanalysis process working.
Luckily they had a very large volume of encrypted messages to work with.

5.6.1   Exercises

1. Consider the ciphertext UMTZR HUDZB PXGQC, which was formed using a Wehrmacht Enigma configured as in Example 5.9 (i.e., with the plugboard pairs O/S, N/T, R/H, B/U, L/I, D/F, G/A, Q/P, C/K, and J/Y, the letters W, Z, E, G, M, and X left unconnected in the plugboard, and the double scrambler outputs in Table 5.16 on page 128 and Table 5.23 on page 141), with the following preliminary decryption. Find the corrected decryption of the ciphertext.

Window:	ZZL	ZZM	ZZN	ZZO	ZZP	ZZQ	ZZR	ZZS	ZZT	ZZU	ZZV	ZZW
Plain:	I	A	M	G	O	I	N	G	T	T	Q	I
Plug:	L	G	M	A	S	L	T	A	N	N	P	L
DScram:	$\updownarrow$	$\updownarrow$	$\updownarrow$	$\updownarrow$	$\updownarrow$	$\updownarrow$	$\updownarrow$	$\updownarrow$	$\updownarrow$	$\updownarrow$	$\updownarrow$	$\updownarrow$
Plug:	B	M	N	Z	H	R	B	F	Z	U	Q	X
Cipher:	U	M	T	Z	R	H	U	D	Z	B	P	X

Window:	ZZX	ZZY	ZZZ
Plain:	Q	U	G
Plug:	P	B	A
DScram:	$\updownarrow$	$\updownarrow$	$\updownarrow$
Plug:	A	P	K
Cipher:	G	Q	C

2. As a continuation of Exercise 2 in Section 5.5, suppose the ciphertext is extended to UCCCK GZIZC LAHTV CTMBZ YYWFN U. Find the window letter position where the middle rotor rotated during encryption, and use this information with the double scrambler outputs in Table 5.17 on page 131 and additional double scrambler outputs in Table 5.24 on page 144 to decrypt the full extended ciphertext.

3. As a continuation of Exercise 4 in Section 5.5, suppose the ciphertext is extended to GLBCJ QQQYL QJOZM BYSGA NAVKA IKVZX SIXRI. Find the window letter position where the middle rotor rotated during encryption, and use this information with the double scrambler outputs in Table 5.18 on page 133 and additional double scrambler outputs in Table 5.25 on page 145 to decrypt the full extended ciphertext.

4.* As a continuation of Exercise 6 in Section 5.5, suppose the ciphertext is extended to KMUTV MUUPW FFTSF SGKMF USKRC UDAHB P. Find the window letter position where the middle rotor rotated during encryption, and use this information with the double scrambler outputs in Table 5.19 on page 134 and additional double scrambler outputs in Table 5.26 on page 146 to decrypt the full extended ciphertext.

	A	B	C	D	E	F	G	H	I	J	K	L	M	N	O	P	Q	R	S	T	U	V	W	X	Y	Z
ZAA	D	U	I	A	K	L	P	R	C	Z	E	F	W	X	T	G	Y	H	V	O	B	S	M	N	Q	J
ZAB	L	K	H	Z	I	R	N	C	E	Y	B	A	U	G	V	Q	P	F	X	W	M	O	T	S	J	D
ZAC	Y	X	V	Z	M	L	S	W	J	I	O	F	E	R	K	U	T	N	G	Q	P	C	H	B	A	D
ZAD	L	K	O	J	F	E	X	P	N	D	B	A	Z	I	C	H	Y	U	W	V	R	T	S	G	Q	M
ZAE	F	Q	V	P	G	A	E	K	U	L	H	J	R	O	N	D	B	M	Z	X	I	C	Y	T	W	S
ZAF	K	F	M	W	H	B	V	E	J	I	A	X	C	S	U	Q	P	T	N	R	O	G	D	L	Z	Y
ZAG	T	Y	Z	N	G	J	E	P	V	F	X	S	Q	D	U	H	M	W	L	A	O	I	R	K	B	C
ZAH	E	U	G	Y	A	H	C	F	J	I	W	M	L	S	Z	T	X	V	N	P	B	R	K	Q	D	O
ZAI	P	S	Z	H	W	J	M	D	X	F	V	N	G	L	Y	A	R	Q	B	U	T	K	E	I	O	C
ZAJ	B	A	N	P	W	O	R	J	U	H	L	K	V	C	F	D	Z	G	X	Y	I	M	E	S	T	Q
ZAK	W	Y	V	M	S	R	U	P	K	L	I	J	D	Z	T	H	X	F	E	O	G	C	A	Q	B	N
ZAL	S	P	N	M	G	Z	E	Y	V	R	Q	W	D	C	T	B	K	J	A	O	X	I	L	U	H	F
ZAM	I	N	W	E	D	H	T	F	A	R	Q	Z	Y	B	V	X	K	J	U	G	S	O	C	P	M	L
ZAN	L	I	U	H	N	Q	K	D	B	M	G	A	J	E	R	X	F	O	T	S	C	W	V	P	Z	Y
ZAO	M	Y	U	W	O	G	F	K	R	T	H	N	A	L	E	Z	S	I	Q	J	C	X	D	V	B	P
ZAP	D	K	X	A	P	H	Q	F	R	S	B	Y	W	Z	V	E	G	I	J	U	T	O	M	C	L	N
ZAQ	H	V	L	P	N	U	R	A	O	M	S	C	J	E	I	D	X	G	K	Y	F	B	Z	Q	T	W
ZAR	E	J	L	U	A	K	O	Y	W	B	F	C	T	R	G	S	V	N	P	M	D	Q	I	Z	H	X
ZAS	P	N	G	F	W	D	C	T	Y	M	Z	O	J	B	L	A	U	X	V	H	Q	S	E	R	I	K
ZAT	P	U	D	C	H	Z	O	E	Y	N	T	M	L	J	G	A	S	W	Q	K	B	X	R	V	I	F
ZAU	M	R	O	F	G	D	E	T	K	L	I	J	A	Y	C	W	Z	B	X	H	V	U	P	S	N	Q
ZAV	P	O	W	L	U	G	F	N	M	Q	T	D	I	H	B	A	J	X	Z	K	E	Y	C	R	V	S
ZAW	T	J	P	Q	Z	H	I	F	G	B	M	U	K	V	Y	C	D	X	W	A	L	N	S	R	O	E
ZAX	B	A	G	T	N	I	C	R	F	U	X	S	Z	E	W	Y	V	H	L	D	J	Q	O	K	P	M
ZAY	U	O	I	V	P	R	W	N	C	L	X	J	Z	H	B	E	T	F	Y	Q	A	D	G	K	S	M
ZAZ	T	G	L	K	U	X	B	V	W	Z	D	C	P	O	N	M	Y	S	R	A	E	H	I	F	Q	J

Table 5.24   Double scrambler outputs for the Wehrmacht Enigma configuration in Exercise 2 (with reflector **B**, rotor order **III**, **IV**, **I**, and ring settings 16, 22, 23).

	A	B	C	D	E	F	G	H	I	J	K	L	M	N	O	P	Q	R	S	T	U	V	W	X	Y	Z
ZAA	H	O	V	W	N	M	P	A	U	Q	X	T	F	E	B	G	J	Z	Y	L	I	C	D	K	S	R
ZAB	T	E	L	M	B	K	U	R	O	V	F	C	D	W	I	Z	Y	H	X	A	G	J	N	S	Q	P
ZAC	D	U	P	A	N	L	H	G	S	M	X	F	J	E	Q	C	O	W	I	Y	B	Z	R	K	T	V
ZAD	W	K	V	I	Z	N	R	M	D	L	B	J	H	F	T	Q	P	G	X	O	Y	C	A	S	U	E
ZAE	B	A	N	I	M	G	F	K	D	Y	H	S	E	C	W	U	T	X	L	Q	P	Z	O	R	J	V
ZAF	G	Z	N	P	F	E	A	J	X	H	M	Y	K	C	V	D	T	U	W	Q	R	O	S	I	L	B
ZAG	Z	G	R	H	Q	S	B	D	P	L	V	J	O	U	M	I	E	C	F	X	N	K	Y	T	W	A
ZAH	L	Q	W	S	O	J	V	U	M	F	X	A	I	P	E	N	B	T	D	R	H	G	C	K	Z	Y
ZAI	T	Z	R	N	V	H	I	F	G	U	M	S	K	D	W	Y	X	C	L	A	J	E	O	Q	P	B
ZAJ	D	U	Q	A	Y	H	J	F	T	G	X	N	S	L	Z	V	C	W	M	I	B	P	R	K	E	O
ZAK	P	N	Z	H	S	X	T	D	J	I	W	Y	U	B	R	A	V	O	E	G	M	Q	K	F	L	C
ZAL	L	P	Y	O	W	Z	R	V	U	M	X	A	J	T	D	B	S	G	Q	N	I	H	E	K	C	F
ZAM	Q	R	S	H	Y	X	U	D	Z	L	N	J	P	K	T	M	A	B	C	O	G	W	V	F	E	I
ZAN	G	J	S	M	W	X	A	Y	V	B	T	Z	D	U	Q	R	O	P	C	K	N	I	E	F	H	L
ZAO	I	R	W	G	U	J	D	S	A	F	M	X	K	T	V	Z	Y	B	H	N	E	O	C	L	Q	P
ZAP	L	W	U	O	Y	V	R	Q	K	S	I	A	T	Z	D	X	H	G	J	M	C	F	B	P	E	N
ZAQ	V	G	L	H	X	O	B	D	R	M	W	C	J	U	F	Z	S	I	Q	Y	N	A	K	E	T	P
ZAR	W	K	F	M	P	C	I	Z	G	L	B	J	D	O	N	E	R	Q	X	U	T	Y	A	S	V	H
ZAS	S	N	L	G	V	M	D	K	Q	R	H	C	F	B	P	O	I	J	A	Y	X	E	Z	U	T	W
ZAT	L	Y	E	U	C	G	F	J	X	H	W	A	T	O	N	R	V	P	Z	M	D	Q	K	I	B	S
ZAU	U	N	L	E	D	W	O	I	H	P	Z	C	T	B	G	J	X	Y	V	M	A	S	F	Q	R	K
ZAV	J	R	D	C	G	I	E	O	F	A	W	M	L	P	H	N	Y	B	U	Z	S	X	K	V	Q	T
ZAW	Q	S	U	F	T	D	H	G	L	Z	X	I	R	Y	V	W	A	M	B	E	C	O	P	K	N	J
ZAX	O	V	P	Q	M	G	F	K	U	L	H	J	E	W	A	C	D	X	Z	Y	I	B	N	R	T	S
ZAY	U	Q	P	W	J	R	I	X	G	E	V	S	N	M	Y	C	B	F	L	Z	A	K	D	H	O	T
ZAZ	G	T	V	N	X	I	A	O	F	U	Q	Y	Z	D	H	S	K	W	P	B	J	C	R	E	L	M

Table 5.25 Double scrambler outputs for the Wehrmacht Enigma configuration in Exercise 3 (with reflector **B**, rotor order **IV**, **III**, **I**, and ring settings 9, 6, 25).

	A	B	C	D	E	F	G	H	I	J	K	L	M	N	O	P	Q	R	S	T	U	V	W	X	Y	Z
ZAA	V	L	N	E	D	X	H	G	P	Q	S	B	Y	C	U	I	J	W	K	Z	O	A	R	F	M	T
ZAB	R	M	L	X	N	J	P	V	Q	F	T	C	B	E	U	G	I	A	Y	K	O	H	Z	D	S	W
ZAC	C	M	A	W	I	Z	V	X	E	Q	S	R	B	T	P	O	J	L	K	N	Y	G	D	H	U	F
ZAD	M	Z	K	X	R	O	U	S	P	L	C	J	A	Y	F	I	V	E	H	W	G	Q	T	D	N	B
ZAE	T	L	W	I	F	E	V	J	D	H	U	B	X	Q	Y	Z	N	S	R	A	K	G	C	M	O	P
ZAF	X	W	H	P	R	L	I	C	G	M	N	F	J	K	V	D	S	E	Q	Y	Z	O	B	A	T	U
ZAG	T	F	O	I	K	B	M	P	D	U	E	S	G	W	C	H	R	Q	L	A	J	Y	N	Z	V	X
ZAH	T	U	W	V	I	L	X	P	E	N	R	F	Y	J	Z	H	S	K	Q	A	B	D	C	G	M	O
ZAI	T	X	S	U	J	M	O	L	W	E	Y	H	F	R	G	V	Z	N	C	A	D	P	I	B	K	Q
ZAJ	W	I	S	R	J	Z	T	X	B	E	O	N	V	L	K	Y	U	D	C	G	Q	M	A	H	P	F
ZAK	R	C	B	Y	I	N	X	V	E	L	M	J	K	F	Z	T	S	A	Q	P	W	H	U	G	D	O
ZAL	P	J	V	O	M	X	U	W	T	B	R	S	E	Z	D	A	Y	K	L	I	G	C	H	F	Q	N
ZAM	M	E	N	X	B	W	K	S	P	O	G	T	A	C	J	I	U	V	H	L	Q	R	F	D	Z	Y
ZAN	P	L	W	F	U	D	M	K	N	V	H	B	G	I	S	A	R	Q	O	Y	E	J	C	Z	T	X
ZAO	G	H	X	F	U	D	A	B	V	M	R	T	J	Y	Z	Q	P	K	W	L	E	I	S	C	N	O
ZAP	F	X	U	W	Y	A	Z	K	L	M	H	I	J	P	Q	N	O	V	T	S	C	R	D	B	E	G
ZAQ	L	O	V	X	I	U	R	J	E	H	Q	A	T	Y	B	Z	K	G	W	M	F	C	S	D	N	P
ZAR	F	L	Y	O	T	A	S	R	N	X	Z	B	P	I	D	M	V	H	G	E	W	Q	U	J	C	K
ZAS	K	L	Q	M	G	T	E	Z	X	O	A	B	D	W	J	U	C	V	Y	F	P	R	N	I	S	H
ZAT	X	P	E	M	C	Y	Q	J	N	H	R	U	D	I	S	B	G	K	O	Z	L	W	V	A	F	T
ZAU	W	S	U	Y	X	L	Q	O	Z	V	P	F	T	R	H	K	G	N	B	M	C	J	A	E	D	I
ZAV	L	T	J	M	U	I	Y	N	F	C	O	A	D	H	K	Q	P	W	X	B	E	Z	R	S	G	V
ZAW	Y	R	K	H	X	O	P	D	T	N	C	Z	S	J	F	G	V	B	M	I	W	Q	U	E	A	L
ZAX	S	M	G	H	P	X	C	D	U	T	V	R	B	W	Z	E	Y	L	A	J	I	K	N	F	Q	O
ZAY	L	Q	P	W	M	Y	V	S	T	K	J	A	E	O	N	C	B	Z	H	I	X	G	D	U	F	R
ZAZ	N	W	D	C	X	U	Y	T	S	R	Q	O	P	A	L	M	K	J	I	H	F	Z	B	E	G	V

Table 5.26   Double scrambler outputs for the Wehrmacht Enigma configuration in Exercise 4 (with reflector **C**, rotor order **IV**, **V**, **II**, and ring settings 26, 23, 19).

	A	B	C	D	E	F	G	H	I	J	K	L	M	N	O	P	Q	R	S	T	U	V	W	X	Y	Z
ZAA	Q	U	L	K	V	W	I	M	G	P	D	C	H	Y	R	J	A	O	T	S	B	E	F	Z	N	X
ZAB	F	E	D	C	B	A	V	X	R	S	W	Z	Y	U	P	O	T	I	J	Q	N	G	K	H	M	L
ZAC	Z	J	F	W	Q	C	K	U	P	B	G	S	N	M	V	I	E	Y	L	X	H	O	D	T	R	A
ZAD	V	R	M	O	Y	S	J	N	P	G	U	T	C	H	D	I	W	B	F	L	K	A	Q	Z	E	X
ZAE	O	J	L	M	Q	T	Z	I	H	B	V	C	D	W	A	U	E	X	Y	F	P	K	N	R	S	G
ZAF	X	T	V	F	M	D	L	Q	W	N	Z	G	E	J	P	O	H	Y	U	B	S	C	I	A	R	K
ZAG	G	X	Y	S	T	Z	A	L	P	Q	R	H	U	V	W	I	J	K	D	E	M	N	O	B	C	F
ZAH	Z	F	E	U	C	B	L	P	Q	M	X	G	J	R	V	H	I	N	T	S	D	O	Y	K	W	A
ZAI	I	G	U	E	D	L	B	K	A	V	H	F	W	Z	T	X	S	Y	Q	O	C	J	M	P	R	N
ZAJ	X	C	B	J	Y	O	V	T	M	D	P	N	I	L	F	K	S	W	Q	H	Z	G	R	A	E	U
ZAK	L	J	E	R	C	S	M	I	H	B	V	A	G	X	Z	T	Y	D	F	P	W	K	U	N	Q	O
ZAL	L	T	H	S	Q	Y	N	C	M	W	U	A	I	G	Z	V	E	X	D	B	K	P	J	R	F	O
ZAM	T	O	P	Z	I	V	K	R	E	Q	G	W	S	U	B	C	J	H	M	A	N	F	L	Y	X	D
ZAN	L	H	W	V	I	P	J	B	E	G	R	A	U	Q	S	F	N	K	O	Y	M	D	C	Z	T	X
ZAO	K	D	S	B	V	Y	I	P	G	W	A	Z	U	X	Q	H	O	T	C	R	M	E	J	N	F	L
ZAP	V	J	H	R	W	S	K	C	O	B	G	U	P	Z	I	M	T	D	F	Q	L	A	E	Y	X	N
ZAQ	W	Q	R	N	H	S	X	E	U	O	V	Z	T	D	J	Y	B	C	F	M	I	K	A	G	P	L
ZAR	J	Z	X	R	H	N	V	E	U	A	O	W	P	F	K	M	Y	D	T	S	I	G	L	C	Q	B
ZAS	K	C	B	X	O	S	T	Y	P	U	A	R	W	Z	E	I	V	L	F	G	J	Q	M	D	H	N
ZAT	I	G	D	C	H	T	B	E	A	V	O	R	Z	X	K	Q	P	L	Y	F	W	J	U	N	S	M
ZAU	Y	Z	X	M	I	P	V	Q	E	L	T	J	D	U	R	F	H	O	W	K	N	G	S	C	A	B
ZAV	X	J	K	E	D	Q	N	P	R	B	C	Y	O	G	M	H	F	I	V	W	Z	S	T	A	L	U
ZAW	D	S	P	A	Q	V	U	W	J	I	M	T	K	X	Y	C	E	Z	B	L	G	F	H	N	O	R
ZAX	B	A	L	N	V	W	M	K	J	I	H	C	G	D	Y	R	S	P	Q	Z	X	E	F	U	O	T
ZAY	F	L	V	S	J	A	O	U	M	E	R	B	I	Y	G	Q	P	K	D	Z	H	C	X	W	N	T
ZAZ	N	P	I	S	R	Z	O	Y	C	U	V	X	Q	A	G	B	M	E	D	W	J	K	T	L	H	F

Table 5.27   Double scrambler outputs for the Wehrmacht Enigma configuration in Exercise 5 (with reflector **C**, rotor order **I**, **IV**, **V**, and ring settings 19, 22, 24).

	A	B	C	D	E	F	G	H	I	J	K	L	M	N	O	P	Q	R	S	T	U	V	W	X	Y	Z
ZAA	Q	V	F	J	N	C	T	W	K	D	I	P	U	E	Y	L	A	Z	X	G	M	B	H	S	O	R
ZAB	B	A	S	R	K	X	H	G	L	V	E	I	Q	Z	T	Y	M	D	C	O	W	J	U	F	P	N
ZAC	V	C	B	Q	Z	I	L	J	F	H	O	G	N	M	K	W	D	U	Y	X	R	A	P	T	S	E
ZAD	L	J	Z	G	P	N	D	K	Y	B	H	A	X	F	T	E	S	W	Q	O	V	U	R	M	I	C
ZAE	Q	K	H	V	U	R	I	C	G	N	B	S	T	J	X	Y	A	F	L	M	E	D	Z	O	P	W
ZAF	F	P	S	R	M	A	Q	J	U	H	N	T	E	K	W	B	G	D	C	L	I	X	O	V	Z	Y
ZAG	Q	S	T	W	J	H	Z	F	O	E	V	U	X	P	I	N	A	Y	B	C	L	K	D	M	R	G
ZAH	S	Y	U	M	X	I	Q	L	F	N	V	H	D	J	P	O	G	W	A	Z	C	K	R	E	B	T
ZAI	P	O	V	I	N	W	J	K	D	G	H	Q	Z	E	B	A	L	S	R	X	Y	C	F	T	U	M
ZAJ	V	F	J	Q	H	B	S	E	Z	C	U	O	R	X	L	W	D	M	G	Y	K	A	P	N	T	I
ZAK	R	X	H	F	U	D	K	C	S	N	G	P	O	J	M	L	Z	A	I	V	E	T	Y	B	W	Q
ZAL	M	J	L	F	N	D	K	W	Y	B	G	C	A	E	Z	X	U	T	V	R	Q	S	H	P	I	O
ZAM	U	H	Z	K	J	W	O	B	R	E	D	N	Y	L	G	Q	P	I	V	X	A	S	F	T	M	C
ZAN	V	R	G	P	K	X	C	T	O	Y	E	W	Q	Z	I	D	M	B	U	H	S	A	L	F	J	N
ZAO	P	D	N	B	H	V	K	E	M	Y	G	S	I	C	X	A	W	Z	L	U	T	F	Q	O	J	R
ZAP	Z	D	O	B	L	W	U	S	K	Y	I	E	X	V	C	R	T	P	H	Q	G	N	F	M	J	A
ZAQ	D	Y	U	A	Z	O	P	R	L	V	X	I	S	W	F	G	T	H	M	Q	C	J	N	K	B	E
ZAR	D	H	Q	A	L	J	Z	B	O	F	P	E	U	S	I	K	C	V	N	Y	M	R	X	W	T	G
ZAS	Z	F	T	G	U	B	D	X	O	M	W	R	J	S	I	V	Y	L	N	C	E	P	K	H	Q	A
ZAT	P	J	E	L	C	I	R	Y	F	B	W	D	V	Z	T	A	U	G	X	O	Q	M	K	S	H	N
ZAU	Z	M	W	J	L	U	Y	T	V	D	R	E	B	Q	S	X	N	K	O	H	F	I	C	P	G	A
ZAV	P	Q	J	Y	G	U	E	T	N	C	Z	S	V	I	X	A	B	W	L	H	F	M	R	O	D	K
ZAW	B	A	Y	Q	L	V	I	U	G	T	P	E	W	R	X	K	D	N	Z	J	H	F	M	O	C	S
ZAX	G	T	D	C	P	V	A	L	Y	U	M	H	K	X	S	E	W	Z	O	B	J	F	Q	N	I	R
ZAY	I	F	D	C	K	B	S	O	A	T	E	W	X	U	H	Y	V	Z	G	J	N	Q	L	M	P	R
ZAZ	K	I	Y	S	V	Q	U	L	B	M	A	H	J	R	P	O	F	N	D	W	G	E	T	Z	C	X

Table 5.28   Double scrambler outputs for the Wehrmacht Enigma configuration in Exercise 6 (with reflector **B**, rotor order **I**, **II**, **III**, and ring settings 26, 11, 9).

	A	B	C	D	E	F	G	H	I	J	K	L	M	N	O	P	Q	R	S	T	U	V	W	X	Y	Z
ZAA	K	O	N	J	L	I	V	M	F	D	A	E	H	C	B	W	Z	S	R	U	T	G	P	Y	X	Q
ZAB	S	I	V	Q	U	N	Z	P	B	M	X	R	J	F	W	H	D	L	A	Y	E	C	O	K	T	G
ZAC	Q	N	Z	J	Y	S	P	L	X	D	W	H	T	B	V	G	A	U	F	M	R	O	K	I	E	C
ZAD	K	X	R	U	G	H	E	F	S	O	A	Y	N	M	J	V	Z	C	I	W	D	P	T	B	L	Q
ZAE	Z	N	G	V	M	Q	C	W	X	L	S	J	E	B	P	O	F	U	K	Y	R	D	H	I	T	A
ZAF	X	I	D	C	K	T	Z	L	B	S	E	H	P	Q	R	M	N	O	J	F	Y	W	V	A	U	G
ZAG	C	N	A	I	H	M	L	E	D	O	W	G	F	B	J	R	U	P	V	Y	Q	S	K	Z	T	X
ZAH	S	C	B	U	J	Z	N	O	V	E	T	X	Y	G	H	Q	P	W	A	K	D	I	R	L	M	F
ZAI	G	Q	M	O	U	L	A	I	H	V	X	F	C	W	D	Z	B	Y	T	S	E	J	N	K	R	P
ZAJ	C	G	A	E	D	V	B	N	M	K	J	U	I	H	S	Q	P	T	O	R	L	F	Y	Z	W	X
ZAK	G	D	J	B	I	P	A	K	E	C	H	X	N	M	U	F	T	Z	Y	Q	O	W	V	L	S	R
ZAL	X	R	K	I	P	J	W	Q	D	F	C	U	S	V	T	E	H	B	M	O	L	N	G	A	Z	Y
ZAM	R	K	Z	J	F	E	U	S	W	D	B	X	Q	O	N	Y	M	A	H	V	G	T	I	L	P	C
ZAN	D	V	F	A	X	C	M	K	Z	R	H	U	G	Q	T	Y	N	J	W	O	L	B	S	E	P	I
ZAO	L	T	I	N	G	H	E	F	C	U	S	A	V	D	Q	X	O	Z	K	B	J	M	Y	P	W	R
ZAP	C	M	A	L	X	P	O	N	U	Y	T	D	B	H	G	F	V	W	Z	K	I	Q	R	E	J	S
ZAQ	K	R	J	S	Q	T	L	X	Z	C	A	G	V	Y	U	W	E	B	D	F	O	M	P	H	N	I
ZAR	F	S	P	K	L	A	N	Z	J	I	D	E	Q	G	W	C	M	V	B	X	Y	R	O	T	U	H
ZAS	V	R	T	L	P	Z	I	U	G	O	S	D	Q	X	J	E	M	B	K	C	H	A	Y	N	W	F
ZAT	X	C	B	R	P	W	J	T	O	G	Z	U	V	Q	I	E	N	D	Y	H	L	M	F	A	S	K
ZAU	E	Q	I	R	A	T	O	L	C	W	V	H	N	M	G	Y	B	D	U	F	S	K	J	Z	P	X
ZAV	G	L	F	Y	W	C	A	S	N	O	R	B	V	I	J	X	Z	K	H	U	T	M	E	P	D	Q
ZAW	C	Q	A	F	X	D	K	L	R	Z	G	H	T	V	P	O	B	I	Y	M	W	N	U	E	S	J
ZAX	I	D	O	B	L	K	S	T	A	U	F	E	P	Y	C	M	R	Q	G	H	J	Z	X	W	N	V
ZAY	U	Z	H	P	F	E	T	C	L	N	Y	I	V	J	W	D	R	Q	X	G	A	M	O	S	K	B
ZAZ	L	O	D	C	H	I	Z	E	F	W	V	A	Y	X	B	S	U	T	P	R	Q	K	J	N	M	G

Table 5.29 Double scrambler outputs for the Wehrmacht Enigma configuration in Exercise 7 (with reflector **C**, rotor order **V**, **III**, **I**, and ring settings 17, 26, 16).

5. As a continuation of Exercise 8 in Section 5.5, suppose the ciphertext is extended to WPMOI NNMUM GNGEL URGXT OYZUN AWT. Find the window letter position where the middle rotor rotated during encryption, and use this information with the double scrambler outputs in Table 5.20 on page 136 and additional double scrambler outputs in Table 5.27 on page 147 to decrypt the full extended ciphertext.

6.* As a continuation of Exercise 9 in Section 5.5, suppose the ciphertext is extended to EMHSS MBNSX NLWWR HISAO JPPNX YFCOH JF. Find the window letter position where the middle rotor rotated during encryption, and use this information with the double scrambler outputs in Table 5.21 on page 137 and additional double scrambler outputs in Table 5.28 on page 148 to decrypt the full extended ciphertext.

7. As a continuation of Exercise 10 in Section 5.5, suppose the ciphertext is extended to AJJRJ EHMRI OFURZ JOTIB TFWHV GEGSG WMY. Find the window letter position where the middle rotor rotated during encryption, and use this information with the double scrambler outputs in Table 5.22 on page 138 and additional double scrambler outputs in Table 5.29 on page 149 to decrypt the full extended ciphertext.

5.7   Clonking

We have seen how the combination of a crib, menu, bombe with a diagonal board, and checking machine could sometimes enable the cryptanalysts at Bletchley Park to decrypt individual Wehrmacht Enigma ciphertexts. When it worked, the cryptanalysis process resulted in ring settings that gave rotor core starting positions based on the very unlikely assumption that the last window letters before the crib were ZZZ. Recall though that German Enigma operators, while provided with a codebook that indicated which reflector, rotor order, ring settings, and plugboard connections they should use on all messages encrypted on each particular day, chose their own initial window letters. Thus, when the cryptanalysis process worked on an individual Enigma ciphertext, although it would give the reflector, rotor order, and plugboard connections indicated in the codebook for the day the ciphertext was formed, it would almost certainly not give the ring settings indicated in the codebook. Finding the ring settings indicated in the codebook was important, for it might allow the cryptanalysts at Bletchley Park to be able to decrypt most or all of the other Enigma ciphertexts formed on the same day without having to go through the entire cryptanalysis process.

    When the cryptanalysis process worked on an individual Enigma ciphertext, the first step in finding the ring settings indicated in the codebook for the day the ciphertext was formed was to determine the actual initial window letter and ring setting of the rightmost rotor when the ciphertext was formed. These could be found from the knowledge of where a turnover of the middle rotor first occurred, the notch letter of the rightmost rotor, and the rightmost initial window letter that resulted from the assumption that the last window letters before the crib were ZZZ. The notch letters for each of the Enigma rotors **I–V**, which were originally included in Table 4.6 on page 57, are shown again in Table 5.30.

Rotor	Notch letter	Rotor to left rotates when window letter changes (from) → (to)
I	Q	Q → R
II	E	E → F
III	V	V → W
IV	J	J → K
V	Z	Z → A

Table 5.30   Notch letters for Enigma rotors **I–V**.

Example 5.10 Consider the decryption at the end of Example 5.9 on page 140, for which a turnover of the middle rotor first occurred at the 22nd letter in the message, the rightmost rotor was **III** (which was given in Example 5.8 on page 126) with notch letter V, and the rightmost initial window letter that resulted from the assumption that the last window letters before the crib were ZZZ was X (the rightmost letter in ZZX that would rotate to ZZY for the encryption of the first letter). To find the actual initial window letter for the rightmost rotor when the ciphertext was formed, note that since a turnover of the middle rotor first occurred at the 22nd letter in the message, then after 21 turnovers of the rightmost rotor during encryption, the notch letter V would have been the rightmost window letter. Thus, if we went backwards on the circle of letters around the rotor (i.e., backwards in the alphabet, wrapping from the start of the alphabet to the end if necessary) 21 positions from V, the result would be the actual rightmost initial window letter. Going backwards on the circle of letters around the rotor 21 positions from V gives A, which was thus the actual rightmost initial window letter when the ciphertext was formed. To find the actual ring setting of the rightmost rotor when the ciphertext was formed, we can use the fact that for either the rightmost initial window letter X and ring setting 18 (which was given in Example 5.8) that resulted from the assumption that the last window letters before the crib were ZZZ, or the actual rightmost initial window letter A and actual ring setting, the rotor core starting positions would have to be the same. For the rightmost initial window letter X and ring setting 18, since X is the 24th letter in the alphabet, the rotor core starting position would be $24 - 18 = 6$. Thus, because subtracting the rotor core starting position from the number of the initial window letter would give the ring setting, and the actual rightmost initial window letter A is the 1st letter in the alphabet, the actual initial ring setting would be 21, since $1 - 6 = -5$ with $-5 + 26 = 21$.   □

After finding the actual initial window letter and ring setting of the rightmost rotor when an individual Wehrmacht Enigma ciphertext was formed, the cryptanalysts at Bletchley Park could find the actual ring settings of the middle and leftmost rotors by exploiting how they knew Enigma operators transmitted their initial window letters. Enigma operators actually used several methods to transmit their initial window letters, but after May 1940, always used the following process. An operator would start by choosing two sequences of three letters each. After configuring an Enigma with the reflector, rotor order, ring settings, and plugboard connections indicated in the codebook, the operator would turn the rotors so one of these three-letter sequences was showing in the windows. The cryptanalysts at Bletchley Park called this three-letter sequence the *indicator setting*. The operator would then use the machine with this configuration to encrypt the second three-letter sequence. Next the operator would turn the rotors so that the unencrypted second three-letter sequence was showing in the windows, and use the machine with this configuration to encrypt the actual message. Finally, the operator would transmit the encrypted message, followed by the indicator setting and encrypted second three-letter sequence.

Example 5.11 As an example of the method used by Wehrmacht Enigma operators after May 1940 to transmit their initial window letters, suppose an operator chose indicator setting BOX and second three-letter sequence TRA. After configuring an Enigma with the reflector, rotor order, ring settings, and plugboard connections indicated in the codebook, the operator would turn the rotors so BOX was showing in the windows, and use the machine with this configuration to encrypt TRA. Suppose this resulted in FUW. Next the operator would turn the rotors so TRA was showing in the windows, and use the machine with this configuration to encrypt the actual message. Finally, the operator would transmit the encrypted message, followed by BOXFUW.   □

For the cryptanalysts at Bletchley Park, assuming that the cryptanalysis process had worked on an individual Wehrmacht Enigma ciphertext, and that they had found the actual initial window letter and ring setting of the rightmost rotor when the ciphertext was formed, to see how they could find the actual ring settings of the middle and leftmost rotors when the ciphertext was formed by exploiting how they knew the Enigma operator transmitted the initial window letters, consider an Enigma configured with the correct reflector, rotor order, plugboard connections, and actual ring setting of the rightmost rotor, and with the rotors turned so that the indicator setting was showing through the windows. There would only be $26 \cdot 26 = 676$ different possible combinations of ring settings of the middle and leftmost rotors. If the letters in the operator's encrypted second three-letter sequence were decrypted assuming each of these 676 different possible combinations, only a relatively small number should cause the third letter to decrypt to the actual rightmost initial window letter. It would certainly not be unreasonable to try all 676 different possible combinations to see which ones caused the third letter in the operator's encrypted second three-letter sequence to decrypt to the actual rightmost initial window letter, and this is exactly what the cryptanalysts at Bletchley Park did. They actually did not even have to decrypt all of the letters in the operator's encrypted second three-letter sequence for every trial. By rotating the rightmost rotor two positions forward to replicate the rotations that would occur when the first two letters were decrypted, they could initially test a combination by decrypting only the third letter, and then decrypt the first two letters only if the third letter decrypted to the actual rightmost initial window letter.

Example 5.12 As a continuation of Example 5.11, the cryptanalysts at Bletchley Park could, with the rotors turned so BOZ was showing in the windows, try all 676 different possible combinations of ring settings of the middle and leftmost rotors to see which ones caused W to decrypt as A. Then only for those that did, they could turn the rotors back so BOX was showing in the windows and decrypt FU. □

The cryptanalysts at Bletchley Park referred to the process that we just described as *clonking*. Afterwards, from the combinations of ring settings of the middle and leftmost rotors that caused the third letter in the operator's encrypted second three-letter sequence to decrypt to the actual rightmost initial window letter, they could identify the correct actual ring settings of the middle and leftmost rotors by checking to see which ones together with the resulting decrypted letters gave the same rotor core starting positions as the initial window letters and ring settings that had worked in the cryptanalysis process on the individual ciphertext.

Example 5.13 As a continuation of Example 5.10, suppose the last six letters in the intercepted transmission were BOXFUW. That is, suppose the Enigma operator's indicator setting was BOX, and encrypted second three-letter sequence was FUW. In Example 5.10, we found that the actual initial window letter and ring setting of the rightmost rotor when the ciphertext was formed were A and 21, respectively. Clonking in this example would thus involve, with a Wehrmacht Enigma configured with reflector **C** and rotor order **I, V, III** (which were given in Example 5.8 on page 126), the plugboard connections we found in Example 5.8, ring setting 21 of the rightmost rotor, and initial window letters BOX, trying all 676 different possible combinations of ring settings of the middle and leftmost rotors to see which ones caused FUW to decrypt as ..A, and then for each combination that did, finding the resulting rotor core starting positions. For example, the ring settings 1, 13, 21 cause FUW to decrypt as ZHA. Then since ZHA are the alphabet letters in positions 26, 8, 1, the ring settings 1, 13, 21 with these initial window letters result in the rotor core starting positions 25, 21, 6, since $26 - 1 = 25$, $8 - 13 = -5$ with $-5 + 26 = 21$, and $1 - 21 = -20$ with $-20 + 26 = 6$. A complete list of the ring settings of the middle and leftmost rotors that cause FUW to decrypt as ..A along with the resulting decrypted letters and rotor core starting positions are shown in the following table.

Ring settings	Decrypted letters	Rotor offsets
1, 12, 21	BXA	1, 12, 6
1, 13, 21	ZHA	25, 21, 6
2, 10, 21	RPA	16, 6, 6
3, 22, 21	JXA	7, 2, 6
3, 23, 21	ZFA	23, 9, 6
4, 9, 21	DXA	0, 15, 6
4, 19, 21	CMA	25, 20, 6
8, 12, 21	WNA	15, 2, 6
9, 25, 21	JFA	1, 7, 6
10, 15, 21	TRA	10, 3, 6
13, 9, 21	VHA	9, 25, 6
15, 16, 21	MXA	24, 8, 6
18, 22, 21	MIA	21, 13, 6
19, 17, 21	QZA	24, 9, 6
21, 13, 21	GOA	12, 2, 6
21, 26, 21	KTA	16, 20, 6
23, 10, 21	HGA	11, 23, 6
23, 23, 21	DXA	7, 1, 6
24, 15, 21	VQA	24, 2, 6
25, 7, 21	IPA	10, 9, 6

The initial window letters and ring settings that had worked in the cryptanalysis process on the ciphertext in this example were ZZX (which rotated to ZZY for the encryption of the first letter) and 16, 23, 18 (which were given in Example 5.8). Since ZZX are the alphabet letters in positions 26, 26, 24, the ring settings 16, 23, 18 with these initial window letters result in the rotor core starting positions 10, 3, 6, since $26 - 16 = 10$, $26 - 23 = 3$, and $24 - 18 = 6$. Note that in the preceding table, the ring settings 10, 15, 21 and decrypted letters TRA give these same rotor core starting positions.

Thus the actual initial window letters and ring settings when the ciphertext in this example was formed were TRA and 10, 15, 21. As a result, 10, 15, 21 must be the ring settings indicated in the codebook for all Enigma ciphertexts formed on the same day as the ciphertext in this example. □

5.7.1 Exercises

1. Consider a ciphertext formed using a Wehrmacht Enigma with the given rotor in the rightmost position, and suppose a turnover of the middle rotor first occurred during encryption at the given position in the message. Find the actual initial window letter for the rightmost rotor when the ciphertext was formed.

   (a)* Rotor **II**, and a turnover of the middle rotor first occurred at the 19th letter in the message

   (b) Rotor **I**, and a turnover of the middle rotor first occurred at the 21st letter in the message

   (c)* Rotor **III**, and a turnover of the middle rotor first occurred at the 23rd letter in the message

   (d) Rotor **IV**, and a turnover of the middle rotor first occurred at the 21st letter in the message

2. Consider a Wehrmacht Enigma ciphertext on which the cryptanalysis process used by the cryptanalysts at Bletchley Park had worked with the given initial window letter and ring setting of the rightmost rotor, and suppose the given actual initial window letter for the rightmost rotor was used when the ciphertext was formed. Find the actual ring setting of the rightmost rotor when the ciphertext was formed.

   (a)* Initial window letter Z and ring setting 24 of the rightmost rotor in the cryptanalysis process, and actual initial window letter K for the rightmost rotor when the ciphertext was formed

   (b) Initial window letter W and ring setting 6 of the rightmost rotor in the cryptanalysis process, and actual initial window letter B for the rightmost rotor when the ciphertext was formed

   (c)* Initial window letter X and ring setting 17 of the rightmost rotor in the cryptanalysis process, and actual initial window letter W for the rightmost rotor when the ciphertext was formed

   (d) Initial window letter Y and ring setting 4 of the rightmost rotor in the cryptanalysis process, and actual initial window letter S for the rightmost rotor when the ciphertext was formed

3. Describe how the intended recipient of the transmission in Example
   5.11, with the knowledge of the reflector, rotor order, ring settings,
   and plugboard connections indicated in the codebook, could determine the operator's unencrypted second three-letter sequence TRA.

4. As a continuation of Exercise 2 in Section 5.6, complete the following.

   (a) Find the position in the message at which a turnover of the
       middle rotor first occurred during encryption.

   (b) Show that the actual initial window letter for the rightmost rotor
       when the ciphertext was formed was V.

   (c) Show that the actual ring setting of the rightmost rotor when
       the ciphertext was formed was 20.

   (d) For the initial window letters and ring settings that had worked
       in the cryptanalysis process used by the cryptanalysts at Bletch-
       ley Park, find the rotor core starting positions.

   (e) Suppose that with the ciphertext, the Enigma operator had
       transmitted indicator setting BAD and encrypted second three-
       letter sequence JCG. A complete list of the ring settings of the
       middle and leftmost rotors that with ring setting 20 of the right-
       most rotor cause JCG to decrypt as ..V along with the resulting
       decrypted letters and rotor core starting positions are shown in
       the following table.

Ring settings	Decrypted letters	Rotor offsets
1, 22, 20	ROV	17, 19, 2
3, 9, 20	ZAV	23, 18, 2
6, 17, 20	TKV	14, 20, 2
8, 11, 20	ROV	10, 4, 2
8, 26, 20	WLV	15, 12, 2
9, 9, 20	THV	11, 25, 2
11, 11, 20	CEV	18, 20, 2
12, 1, 20	ISV	23, 18, 2
13, 8, 20	XVV	11, 14, 2
13, 11, 20	ZEV	13, 20, 2
18, 5, 20	CWV	11, 18, 2
20, 14, 20	LUV	18, 7, 2
21, 1, 20	UOV	0, 14, 2
24, 20, 20	YAV	1, 7, 2
26, 5, 20	VUV	22, 16, 2

       Find the actual initial window letters and ring settings when the
       ciphertext was formed.

5. As a continuation of Exercise 3 in Section 5.6, complete the following.

   (a) Find the position in the message at which a turnover of the middle rotor first occurred during encryption.

   (b) Show that the actual initial window letter for the rightmost rotor when the ciphertext was formed was S.

   (c) Show that the actual ring setting of the rightmost rotor when the ciphertext was formed was 20.

   (d) For the initial window letters and ring settings that had worked in the cryptanalysis process used by the cryptanalysts at Bletchley Park, find the rotor core starting positions.

   (e) Suppose that with the ciphertext, the Enigma operator had transmitted indicator setting NCS and encrypted second three-letter sequence EGR. A complete list of the ring settings of the middle and leftmost rotors that with ring setting 20 of the rightmost rotor cause EGR to decrypt as ..S along with the resulting decrypted letters and rotor core starting positions are shown in Table 5.31 on page 158. Find the actual initial window letters and ring settings when the ciphertext was formed.

6.* As a continuation of Exercise 4 in Section 5.6, complete the following.

   (a) Find the position in the message at which a turnover of the middle rotor first occurred during encryption.

   (b) Show that the actual initial window letter for the rightmost rotor when the ciphertext was formed was I.

   (c) Show that the actual ring setting of the rightmost rotor when the ciphertext was formed was 3.

   (d) For the initial window letters and ring settings that had worked in the cryptanalysis process used by the cryptanalysts at Bletchley Park, find the rotor core starting positions.

   (e) Suppose that with the ciphertext, the Enigma operator had transmitted indicator setting HAT and encrypted second three-letter sequence NEH. A complete list of the ring settings of the middle and leftmost rotors that with ring setting 3 of the rightmost rotor cause NEH to decrypt as ..I along with the resulting decrypted letters and rotor core starting positions are shown in Table 5.32 on page 159. Find the actual initial window letters and ring settings when the ciphertext was formed.

Ring settings	Decrypted letters	Rotor offsets
2, 7, 20	SAS	17, 20, 25
3, 10, 20	BTS	25, 10, 25
4, 1, 20	WCS	19, 2, 25
4, 13, 20	CMS	25, 0, 25
4, 14, 20	ZHS	22, 20, 25
5, 21, 20	QOS	12, 20, 25
7, 7, 20	NXS	7, 17, 25
7, 14, 20	KXS	4, 10, 25
7, 16, 20	KQS	4, 1, 25
11, 22, 20	UNS	10, 18, 25
11, 25, 20	CTS	18, 21, 25
12, 5, 20	DCS	18, 24, 25
13, 4, 20	VLS	9, 8, 25
15, 8, 20	LMS	23, 5, 25
15, 15, 20	IHS	20, 19, 25
16, 13, 20	MQS	23, 4, 25
16, 15, 20	JBS	20, 13, 25
19, 10, 20	MIS	20, 25, 25
19, 19, 20	UTS	2, 1, 25
19, 23, 20	WPS	4, 19, 25
23, 1, 20	NMS	17, 12, 25
23, 3, 20	JJS	13, 7, 25
24, 3, 20	ZHS	2, 5, 25
25, 12, 20	WFS	24, 20, 25
26, 11, 20	FUS	6, 10, 25
26, 20, 20	MPS	13, 22, 25

Table 5.31   Clonking results for Exercise 5.

Ring settings	Decrypted letters	Rotor offsets
2, 14, 3	TRI	18, 4, 6
4, 7, 3	TCI	16, 22, 6
4, 12, 3	TUI	16, 9, 6
5, 6, 3	TGI	15, 1, 6
6, 13, 3	TXI	14, 11, 6
6, 26, 3	TNI	14, 14, 6
7, 5, 3	TSI	13, 14, 6
7, 12, 3	TMI	13, 1, 6
8, 4, 3	THI	12, 4, 6
8, 6, 3	TNI	12, 8, 6
8, 16, 3	TKI	12, 21, 6
9, 26, 3	TTI	11, 20, 6
10, 3, 3	TQI	10, 14, 6
10, 10, 3	TCI	10, 19, 6
10, 22, 3	TOI	10, 19, 6
10, 23, 3	TBI	10, 5, 6
12, 18, 3	TOI	8, 23, 6
12, 19, 3	TII	8, 16, 6
12, 22, 3	TMI	8, 17, 6
14, 14, 3	TII	6, 21, 6
15, 6, 3	TNI	5, 8, 6
16, 10, 3	TFI	4, 22, 6
17, 2, 3	TYI	3, 23, 6
18, 21, 3	TMI	2, 18, 6
20, 2, 3	TOI	0, 13, 6
20, 10, 3	TMI	0, 3, 6
21, 12, 3	TQI	25, 5, 6
22, 6, 3	TTI	24, 14, 6
22, 18, 3	TRI	24, 0, 6
22, 26, 3	TUI	24, 21, 6
23, 13, 3	TOI	23, 2, 6
24, 26, 3	TYI	22, 25, 6
25, 1, 3	TQI	21, 16, 6
26, 19, 3	TCI	20, 10, 6

Table 5.32   Clonking results for Exercise 6.

7. As a continuation of Exercise 5 in Section 5.6, complete the following.

   (a) Find the position in the message at which a turnover of the middle rotor first occurred during encryption.

   (b) Show that the actual initial window letter for the rightmost rotor when the ciphertext was formed was D.

   (c) Show that the actual ring setting of the rightmost rotor when the ciphertext was formed was 2.

   (d) For the initial window letters and ring settings that had worked in the cryptanalysis process used by the cryptanalysts at Bletchley Park, find the rotor core starting positions.

   (e) Suppose that with the ciphertext, the Enigma operator had transmitted indicator setting MIL and encrypted second three-letter sequence ETE. A complete list of the ring settings of the middle and leftmost rotors that with ring setting 2 of the rightmost rotor cause ETE to decrypt as ..D along with the resulting decrypted letters and rotor core starting positions are shown in Table 5.33 on page 161. Find the actual initial window letters and ring settings when the ciphertext was formed.

8.* As a continuation of Exercise 6 in Section 5.6, complete the following.

   (a) Find the position in the message at which a turnover of the middle rotor first occurred during encryption.

   (b) Show that the actual initial window letter for the rightmost rotor when the ciphertext was formed was Y.

   (c) Show that the actual ring setting of the rightmost rotor when the ciphertext was formed was 8.

   (d) For the initial window letters and ring settings that had worked in the cryptanalysis process used by the cryptanalysts at Bletchley Park, find the rotor core starting positions.

   (e) Suppose that with the ciphertext, the Enigma operator had transmitted indicator setting LAC and encrypted second three-letter sequence HDV. A complete list of the ring settings of the middle and leftmost rotors that with ring setting 8 of the rightmost rotor cause HDV to decrypt as ..Y along with the resulting decrypted letters and rotor core starting positions are shown in Table 5.34 on page 162. Find the actual initial window letters and ring settings when the ciphertext was formed.

Ring settings	Decrypted letters	Rotor offsets
1, 24, 2	TAD	19, 3, 2
4, 21, 2	TAD	16, 6, 2
4, 22, 2	FQD	2, 21, 2
5, 7, 2	YID	20, 2, 2
5, 15, 2	FHD	1, 19, 2
6, 23, 2	MAD	7, 4, 2
7, 22, 2	HND	1, 18, 2
12, 8, 2	DVD	18, 14, 2
13, 20, 2	JOD	23, 21, 2
15, 12, 2	BID	13, 23, 2
15, 19, 2	YND	10, 21, 2
18, 9, 2	CDD	11, 21, 2
19, 22, 2	KKD	18, 15, 2
19, 25, 2	VQD	3, 18, 2
20, 5, 2	MED	19, 0, 2
20, 9, 2	FGD	12, 24, 2
21, 13, 2	NAD	19, 14, 2
22, 10, 2	DAD	8, 17, 2
22, 21, 2	FUD	10, 0, 2
23, 7, 2	NLD	17, 5, 2
23, 15, 2	PYD	19, 10, 2
26, 18, 2	KFD	11, 14, 2

Table 5.33 Clonking results for Exercise 7.

Ring settings	Decrypted letters	Rotor offsets
1, 14, 8	ZRY	25, 4, 17
1, 23, 8	ALY	0, 15, 17
1, 25, 8	OUY	14, 22, 17
3, 16, 8	XFY	21, 16, 17
4, 14, 8	WZY	19, 12, 17
6, 8, 8	KBY	5, 20, 17
6, 23, 8	CNY	23, 17, 17
7, 5, 8	ZPY	19, 11, 17
8, 14, 8	MNY	5, 0, 17
9, 10, 8	MXY	4, 14, 17
10, 17, 8	EHY	21, 17, 17
12, 14, 8	ALY	15, 24, 17
14, 25, 8	ZUY	12, 22, 17
15, 1, 8	ZQY	11, 16, 17
15, 23, 8	TMY	5, 16, 17
15, 24, 8	VGY	7, 9, 17
16, 12, 8	DKY	14, 25, 17
17, 14, 8	PWY	25, 9, 17
18, 2, 8	DPY	12, 14, 17
18, 4, 8	PXY	24, 20, 17
18, 9, 8	UPY	3, 7, 17
19, 15, 8	UYY	2, 10, 17
19, 16, 8	QTY	24, 4, 17
20, 13, 8	RZY	24, 13, 17
24, 20, 8	KXY	13, 4, 17
24, 25, 8	OVY	17, 23, 17
25, 18, 8	MEY	14, 13, 17

Table 5.34   Clonking results for Exercise 8.

Ring settings	Decrypted letters	Rotor offsets
1, 17, 8	MFR	12, 15, 10
2, 5, 8	MVR	11, 17, 10
2, 18, 8	MGR	11, 15, 10
3, 13, 8	MWR	10, 10, 10
4, 1, 8	MAR	9, 0, 10
4, 18, 8	MPR	9, 24, 10
6, 16, 8	MNR	7, 24, 10
7, 7, 8	MFR	6, 25, 10
7, 15, 8	MSR	6, 4, 10
8, 4, 8	MIR	5, 5, 10
8, 20, 8	MAR	5, 7, 10
10, 11, 8	MDR	3, 19, 10
14, 3, 8	MAR	25, 24, 10
15, 21, 8	MGR	24, 12, 10
15, 24, 8	MAR	24, 3, 10
18, 5, 8	MQR	21, 12, 10
19, 16, 8	MLR	20, 22, 10
21, 1, 8	MOR	18, 14, 10
22, 3, 8	MLR	17, 9, 10
22, 11, 8	MVR	17, 11, 10
24, 10, 8	MIR	15, 25, 10
26, 19, 8	MFR	13, 13, 10

Table 5.35   Clonking results for Exercise 9.

9. As a continuation of Exercise 7 in Section 5.6, complete the following.

(a) Find the position in the message at which a turnover of the middle rotor first occurred during encryption.

(b) Show that the actual initial window letter for the rightmost rotor when the ciphertext was formed was R.

(c) Show that the actual ring setting of the rightmost rotor when the ciphertext was formed was 8.

(d) For the initial window letters and ring settings that had worked in the cryptanalysis process used by the cryptanalysts at Bletchley Park, find the rotor core starting positions.

(e) Suppose that with the ciphertext, the Enigma operator had transmitted indicator setting NPS and encrypted second three-letter sequence JBE. A complete list of the ring settings of the middle and leftmost rotors that with ring setting 8 of the rightmost rotor cause JBE to decrypt as ..R along with the resulting decrypted letters and rotor core starting positions are shown in Table 5.35 on page 163. Find the actual initial window letters and ring settings when the ciphertext was formed.

5.8   Final Observations

Recall that for a given menu that resulted from a Wehrmacht Enigma crib/ciphertext alignment, the cryptanalysts at Bletchley Park had to consider 17,576 possible ring settings for each of $2 \cdot 60 = 120$ possible combinations of a reflector and rotor arrangement. It typically took about 20 minutes for a bombe to test all possible ring settings on a single combination of a reflector and rotor arrangement. Thus, if ring settings could be tested on only one combination at a time, it could take longer than 24 hours in total for a single bombe to test all possible ring settings on all possible combinations of a reflector and rotor arrangement. Although, recall that depending on the menu, a single bombe could sometimes test ring settings on three different combinations of a reflector and rotor arrangement at the same time. Multiple bombes could also be used to attack the same message at the same time as well.

A standard German operational practice called the *rule of keys* also served to reduce the total time required to test all possible ring settings on all possible combinations of a reflector and rotor arrangement. The rule of keys said that the codebook indicating which reflector, rotor order, ring settings, and plugboard connections should be used on all messages encrypted on any particular day would never instruct Enigma operators to place a specific rotor in the machine in the same position as it had been placed the previous day. For example, if the rotor order on one day was **I**, **V**, **III**, then on the next day rotor **I** would not be in the leftmost position, rotor **V** would not be in the middle position, and rotor **III** would not be in the rightmost position. This reduced the number of possible rotor arrangements each day from 60 to 32.

We could say much more about the cryptanalysis of the Enigma at Bletchley Park. For instance, we have not discussed in any real detail the problems the cryptanalysts at Bletchley Park faced with finding cribs and matching them correctly within ciphertexts, working with menus that had no loops, or dealing with rotor turnovers within menus. We have also not addressed the urgency the cryptanalysts must have felt to decrypt ciphertexts in a timely manner while the intelligence they contained would still have operational value.

We have also not discussed how much more difficult the problem became when the German Navy began using Enigmas with a fourth rotor in February 1942. Even though the cryptanalysts at Bletchley Park knew that the four-rotor Enigma was coming, it still took them almost a year after it began being used before they were able to consistently decrypt ciphertexts that were formed using it. To underscore the importance of this fact, more than four times the amount of Allied shipping was sunk by German U-boats

---

Joan Clarke: Enigma Cryptanalyst at Bletchley Park

Among the nearly 10,000 people who took part in the cryptologic efforts at Bletchley Park, about 75% were women. However, most of these women performed primarily clerical tasks. Only a very small number served as cryptanalysts.

One woman who served as a cryptanalyst at Bletchley Park was Joan Clarke. Clarke attended Newnham College, Cambridge, where in 1939 she earned a double first degree in mathematics, the highest she could earn at the time, as Cambridge did not begin awarding full degrees to women until 1948. Gordon Welchman found Clarke though, and in 1940 recruited her to Bletchley Park. After Clarke initially did some clerical work, her mathematical abilities led to her being assigned to Hut 8, were she practiced Banburismus, a process developed by Alan Turing for breaking German Kriegsmarine M4 Enigma ciphers. Despite being paid less than her male coworkers, Clarke was recognized as one of the best Banburists in Hut 8. Clarke was also one of Turing's closest friends during this time.

in the North Atlantic during the second half of 1942 when compared to the
same time period from the previous year, and virtually all of the increase
can be attributed to the blackout caused by the fourth Enigma rotor.

5.8.1   Exercises

1.* How long would it take on average for one bombe to test all possible ring settings on all possible combinations of a reflector and rotor
arrangement, assuming 60 possible rotor arrangements and that ring
settings could be tested on only one combination at a time? Give
your answer in hours.

2. How long would it take on average for one bombe to test all possible ring settings on all possible combinations of a reflector and rotor
arrangement, assuming 60 possible rotor arrangements and that ring
settings could be tested on three combinations at a time? Give your
answer in hours.

3.* Assuming the rotor order indicated in the codebook for a particular
day was **I, V, III**, make a list of the 32 possible rotor arrangements
that could be indicated in the codebook for the next day.

4.* How long would it take on average for one bombe to test all possible ring settings on all possible combinations of a reflector and rotor
arrangement, assuming 32 possible rotor arrangements and that ring
settings could be tested on only one combination at a time? Give
your answer in hours.

5. How long would it take on average for one bombe to test all possible ring settings on all possible combinations of a reflector and rotor
arrangement, assuming 32 possible rotor arrangements and that ring
settings could be tested on three combinations at a time? Give your
answer in hours.

6. Find some information about the cryptanalysis of the German Navy's
four-rotor Enigma at Bletchley Park, and write a summary of your
findings.

7. Find some additional information about the career in cryptology of
Joan Clarke, and write a summary of your findings.

8. Find some information about the process of Banburismus, and write
a summary of your findings.

9. Find some information about the *Colossus* machine that was designed
at Bletchley Park for the cryptanalysis of the German Lorenz cipher
machine, and write a summary of your findings.

Chapter 6

Shift and Affine Ciphers

Mathematics and cryptology are connected for several reasons, one being that finding success in breaking ciphers is easier for someone who is a good problem solver, and mathematical training is an excellent way to become a good problem solver. For example, as we mentioned in the previous chapter, several mathematicians played crucial roles in the cryptanalysis of the Enigma cipher machine. This was not because they actually used mathematics in attacking the Enigma, but rather because the mathematical training they had received helped them to become good problem solvers.

Also, since finding success in breaking ciphers is easier for someone who is a good problem solver, then finding success in developing ciphers is easier for someone who is a good problem solver as well, because when developing a cipher, it must be anticipated how the cipher will be attacked. In modern cryptology, understanding mathematics is not just helpful, but essential. This is why the National Security Agency, the agency for cryptology in the United States, is this country's largest employer of mathematicians.

Another reason why mathematics and cryptology are connected is that many ciphers use mathematical operations in their encryption and decryption procedures. For most of the rest of this book, we will consider ciphers for which these procedures can or must be expressed using mathematical operations, including a cipher in this chapter that was described by Julius Caesar in his writings on the Gallic Wars. The encryption and decryption procedures for these ciphers can be expressed using *modular* arithmetic.

6.1 Modular Arithmetic

We first learn about the process of dividing integers (i.e., whole numbers) in grade school. Consider $37 \div 3$, with the *dividend* 37 divided by the *divisor*

3, giving *quotient* 12 and *remainder* 1. The result of this division can be expressed in tableau form as follows.

$$
\begin{array}{r}
12 \\
3 \overline{)\ 37} \\
-36 \\
\hline
1
\end{array}
$$

The result of this division can also be expressed using the following equation.

$$
\frac{\text{Dividend}}{\text{Divisor}} = \text{Quotient} + \frac{\text{Remainder}}{\text{Divisor}}
$$

For $37 \div 3$, this equation is the following.

$$
\frac{37}{3} = 12 + \frac{1}{3}
$$

Multiplying both sides of this equation by the divisor 3 gives the following.

$$
3\left(\frac{37}{3}\right) = 3\left(12 + \frac{1}{3}\right)
$$

$$
37 = 3 \cdot 12 + 3\left(\frac{1}{3}\right)
$$

$$
37 = 3 \cdot 12 + 1
$$

This last equation is a special case of the following theorem, commonly called the *division algorithm*.

Theorem 6.1 *(**The Division Algorithm**) If b and m are integers with m positive, then there is exactly one pair of integers q (the quotient) and r (the remainder) such that $b = mq + r$ and $0 \le r < m$.*

Example 6.1

- For $b = 25$ and $m = 7$, the division algorithm gives $q = 3$ and $r = 4$.

- For $b = 6$ and $m = 5$, the division algorithm gives $q = 1$ and $r = 1$.

- For $b = 5$ and $m = 6$, the division algorithm gives $q = 0$ and $r = 5$.

  □

It is important to note that the remainder r in the division algorithm is specified as nonnegative. Thus, for example, for $b = -25$ and $m = 7$, it would be incorrect to say that the division algorithm gives $q = -3$ and $r = -4$, since the remainder is negative. For $b = -25$ and $m = 7$, the division algorithm gives $q = -4$ and $r = 3$.

Example 6.2

- For $b = -25$ and $m = 7$, the division algorithm gives $q = -4$ and $r = 3$.

- For $b = -6$ and $m = 5$, the division algorithm gives $q = -2$ and $r = 4$.

- For $b = -18$ and $m = 6$, the division algorithm gives $q = -3$ and $r = 0$. □

A number of primary interest to us throughout the rest of this book is the remainder r in the division algorithm. This number is needed so frequently in cryptology that finding it is described with a special term.

Definition 6.2 *For integers b and m with m positive, the remainder r in the division algorithm is the value of b modulo m, written $b \bmod m = r$, or, equivalently, $b = r \bmod m$. In these equations, m is called the modulus.*

Example 6.3

- $25 = 4 \bmod 7$

- $6 = 1 \bmod 5$

- $5 = 5 \bmod 6$

- $-25 = 3 \bmod 7$

- $-6 = 4 \bmod 5$

- $-18 = 0 \bmod 6$ □

For larger b and m, values of $b \bmod m$ can be found using a calculator. Many calculators actually have predefined functions for finding $b \bmod m$, but even calculators that can only do basic arithmetic can be used to find $b \bmod m$.

Example 6.4 To find $1024 \bmod 37$, we begin by using a calculator to compute $1024 \div 37 = 27.68$. The quotient q in the division algorithm is the integer part of this result, namely $q = 27$. Since the product of the modulus $m = 37$ and the quotient $q = 27$ is $mq = 999$, subtracting 999 from 1024 gives the remainder $1024 \bmod 37$. This is summarized in the following expression of the division $1024 \div 37$ in tableau form.

$$
\begin{array}{r}
27 \\
37 \overline{)1024} \\
-999 \\
\hline
25
\end{array}
$$

Since the remainder is $r = 25$, $1024 \bmod 37 = 25$. □

Example 6.5 To find $-3071 \bmod 107$, we begin by using a calculator to compute $-3071 \div 107 = -28.78$. Since this result is negative, in order for the remainder in the division algorithm to be nonnegative, the quotient in the division algorithm will have to be -29 rather than -28. Since the product of the modulus $m = 107$ and the quotient $q = -29$ is $mq = -3103$, subtracting -3103 from -3071 gives the remainder $-3071 \bmod 107$. This is summarized in the following expression of the division $-3071 \div 107$ in tableau form.

$$
\begin{array}{r}
-29 \\
107 \overline{)\ -3071} \\
\underline{-(-3103)} \\
32
\end{array}
$$

Since the remainder is $r = 32$, $-3071 \bmod 107 = 32$.   □

Occasionally, throughout the rest of this book we will need to be able to solve equations involving a modulus. For example, consider solving the following equation for b.

$$b = 4 \bmod 7$$

One solution to this equation is $b = 4$, since $4 = 4 \bmod 7$. However, there are also many other solutions, including $b = 11$, $b = 18$, $b = -3$, and $b = -10$. Note that these other solutions all differ from $b = 4$ by a multiple of 7. In fact, any integer that differs from $b = 4$ by a multiple of 7 will be a solution to this equation. Thus, there are an infinite number of solutions, the integers in the following set.

$$\{\ldots, -17, -10, -3, 4, 11, 18, 25, \ldots\}$$

A more concise way to represent this infinite set of solutions is with the equation $b = 7k + 4$ given in the division algorithm, where k represents any integer.

The set $\{\ldots, -17, -10, -3, 4, 11, 18, 25, \ldots\}$ in the previous paragraph is called a *congruence class* for $m = 7$, and what connects the integers in this congruence class is that all of them as b with $m = 7$ in the division algorithm give the same remainder, $r = 4$. Another example of a congruence class for $m = 7$ is the following set.

$$\{\ldots, -19, -12, -5, 2, 9, 16, 23, \ldots\}$$

What connects the integers in this congruence class is that all of them as b with $m = 7$ in the division algorithm give the same remainder, $r = 2$.

There are exactly seven distinct congruence classes for $m = 7$, since there are exactly seven possible remainders in the division algorithm with $m = 7$. These seven possible remainders are the integers from 0 through

6, which form a set called *the integers modulo 7*, or $\mathbb{Z}_7$ for short.[1] That is, $\mathbb{Z}_7 = \{0, 1, 2, 3, 4, 5, 6\}$. More generally, for any positive integer m, the following is the set of integers modulo m.

$$\mathbb{Z}_m = \{0, 1, 2, 3, \ldots, m - 1\}$$

As we have seen, solutions to equations involving a modulus can be sets called congruence classes. As an example, the solution to the equation $b = 4 \bmod 7$ is the congruence class $\{\ldots, -17, -10, -3, 4, 11, 18, 25, \ldots\}$. For simplicity, when solving equations involving a modulus, it is usually understood that solutions are not entire congruence classes, but rather just the smallest nonnegative integers in these congruence classes. For example, we would say that the solution to the equation $b = 4 \bmod 7$ is just $b = 4$, which is, not coincidentally, the remainder given in the division algorithm with any of the integers in the congruence class $\{\ldots, -17, -10, -3, 4, 11, 18, 25, \ldots\}$ as b and $m = 7$.

The understanding that solutions to equations involving a modulus are just the smallest nonnegative integers in congruence classes allows solutions to be integers rather than sets. Just as importantly, it also allows equations involving a modulus to be manipulated in some of the same sorts of ways that we are accustomed to manipulating equations not involving a modulus, including adding or subtracting the same thing on both sides of an equation. More specifically, if

$$x = y \bmod m,$$

then for any integer k,

$$x + k = (y + k) \bmod m,$$

and

$$x - k = (y - k) \bmod m.$$

Example 6.6 Consider the following equation.

$$x + 7 = 2 \bmod 26$$

We can solve this equation for x as follows.

$$x + 7 - 7 = (2 - 7) \bmod 26$$
$$x = -5 \bmod 26$$
$$x = 21$$

□

---

[1] In mathematics, the set of all integers is often represented with the symbol $\mathbb{Z}$. It is short for *zahlen*, the German word for *numbers*.

Example 6.7 Consider the following equation.

$$x - 18 = 13 \bmod 26$$

We can solve this equation for x as follows.

$$x - 18 + 18 = (13 + 18) \bmod 26$$
$$x = 31 \bmod 26$$
$$x = 5 \qquad\qquad \square$$

Since multiplication by an integer can be thought of as repeated addition or subtraction, we can extend the properties given on page 171 for modular addition and subtraction to include multiplying the same thing on both sides of an equation. More specifically, if

$$x = y \bmod m,$$

then for any integer k,

$$kx = ky \bmod m.$$

Also, although it was not illustrated in either of the previous two examples, when manipulating equations involving a modulus, it is possible (but not necessary) at any time to convert any integer in an equation into its remainder modulo the modulus, whether the integer is multiplied by a variable or not.

Example 6.8 Consider the following equation.

$$35x + 79 = -30 \bmod 26$$

To solve this equation for x, we can begin by converting the integers 35, 79, and -30 into their remainders modulo 26. That is, we can convert this equation into the following.

$$9x + 1 = 22 \bmod 26$$

We can then solve this equation for x as follows.

$$9x + 1 - 1 = (22 - 1) \bmod 26$$
$$9x = 21 \bmod 26$$
$$3 \cdot 9x = (3 \cdot 21) \bmod 26$$
$$27x = 63 \bmod 26$$
$$1x = 11 \bmod 26$$
$$x = 11 \qquad\qquad \square$$

In the second equation in the list of equalities at the bottom of page 172, to change the $9x$ on the left side into x, it would not have made sense to divide both sides of the equation by 9, since the right side of the equation would have become $\frac{21}{9}$ mod 26, and fractions are not defined in modular arithmetic. Modular arithmetic is based on the division algorithm, which only works with integers.

Upon seeing the complete previous example, it is clear that in the second equation in the list of equalities at the bottom of page 172, multiplying both sides of the equation by 3 was a correct thing to do, since it ultimately changed the left side of the equation from $9x$ into x. However, it is one thing to be able to see the complete previous example and understand it, and another to be able to reproduce something similar to it in a different setting. For instance, what if isolating the x term on the left side had yielded $17x$ instead of $9x$? Then what would have been a correct thing by which to multiply both sides of the equation? Or perhaps is it possible that there would have been no correct thing by which to multiply both sides of the equation? To answer these questions, we need to consider multiplicative inverses, and to do this we need to take a short detour through greatest common divisors.

Definition 6.3 *For positive integers a and m, the greatest common divisor $gcd(a, m)$ of a and m is the largest positive integer that divides both a and m evenly (i.e., with remainders zero).*

For two small positive integers, it is usually easy to find the greatest common divisor by guessing and checking. For example, it is easy to find that $gcd(20, 30) = 10$ and $gcd(60, 75) = 15$ by guessing and checking. On the other hand, as the two integers become larger, finding the greatest common divisor by guessing and checking can become more difficult. For example, it is certainly more difficult to find that $gcd(935, 1190) = 85$ by guessing and checking. Later we will present an algorithm for finding the greatest common divisor of two large positive integers, even extremely large, very quickly. For now, guessing and checking will suffice.

One thing related to greatest common divisors that will be important throughout the rest of this book is the idea of a pair of integers being *relatively prime*.

Definition 6.4 *Positive integers a and m are said to be relatively prime if $gcd(a, m) = 1$.*

Example 6.9

- Since $gcd(8, 26) \neq 1$, 8 and 26 are not relatively prime.

- Since $gcd(9, 26) = 1$, 9 and 26 are relatively prime. □

Greatest common divisors are necessary to consider when studying multiplicative inverses with modular arithmetic. We learn about the concept of multiplicative inverses with normal (not modular) arithmetic very early in school when solving simple equations. For example, consider the equation $9x = 21$. To solve this equation for x, if we are not restricted in the type of numbers we can use, and if the arithmetic is normal arithmetic, then we can just divide both sides of the equation by 9. This is equivalent to multiplying both sides by $\frac{1}{9}$, of course, and is a correct thing to do because $\frac{1}{9} \cdot 9 = 1$, which changes the left side of the equation into just x.

Because the result is 1 when 9 is multiplied by $\frac{1}{9}$ with normal multiplication, we call $\frac{1}{9}$ the *multiplicative inverse* of 9 for normal multiplication. More generally, for any nonzero real number a, if we are not restricted in the type of numbers we can consider, then with normal multiplication the multiplicative inverse of a exists and is equal to $\frac{1}{a}$. Recall also that we sometimes use the notation a^{-1} to represent this multiplicative inverse.

Things are different if we are restricted in the type of numbers we can consider. For example, if we are only allowed to use integers, then with normal multiplication the multiplicative inverse of a exists only if $a = \pm 1$ (and these two integers are their own inverses). Things are different further if we are solving equations that use modular arithmetic instead of normal arithmetic. For example, with modulo 26 multiplication, the multiplicative inverse of 9 exists, but it is not $\frac{1}{9}$, since fractions are not defined in modular arithmetic. As we saw in Example 6.8, with modulo 26 arithmetic, the multiplicative inverse of 9 is 3, since $3 \cdot 9 = 1 \bmod 26$.

Actually, with modular arithmetic, multiplicative inverses need not be unique. For example, with modulo 26 arithmetic, 29 is also a multiplicative inverse of 9, since $29 \cdot 9 = 1 \bmod 26$. However, with modulo 26 arithmetic, 3 is the only multiplicative inverse of 9 in $\mathbb{Z}_{26}$. More generally, with modulo m arithmetic, if a has a multiplicative inverse, then it has a *unique* multiplicative inverse in $\mathbb{Z}_m$. This unique multiplicative inverse in $\mathbb{Z}_m$ is usually called *the* multiplicative inverse of a modulo m. For consistency, this multiplicative inverse is also represented using the notation a^{-1}, although it is not represented as $\frac{1}{a}$, since fractions are not defined in modular arithmetic.

Not all nonzero integers have multiplicative inverses with modular arithmetic, though. For example, with modulo 26 arithmetic, although 9 has a multiplicative inverse, 8 and 10 do not. The connection between multiplicative inverses with modular arithmetic and greatest common divisors is summarized in the following theorem.

Theorem 6.5 *If a and m are positive integers, then a has a multiplicative inverse modulo m if and only if a and m are relatively prime. Also, if a has a multiplicative inverse modulo m, then a has a unique multiplicative inverse in $\mathbb{Z}_m$, written $a^{-1} \bmod m$.*

Example 6.10

- Since 8 and 26 are not relative prime, 8 does not have a multiplicative inverse modulo 26.

- Since 9 and 26 are relatively prime, 9 has a multiplicative inverse modulo 26. Because $3 \cdot 9 = 1 \bmod 26$, $9^{-1} = 3 \bmod 26$.

- Since 17 and 26 are relatively prime, 17 has a multiplicative inverse modulo 26. Because $23 \cdot 17 = 1 \bmod 26$, $17^{-1} = 23 \bmod 26$. □

For the next several chapters in this book, almost all our computations will be done with modulo 26 arithmetic, in correspondence with the fact that our messages will be written using our 26-letter alphabet. For convenience, the numbers in $\mathbb{Z}_{26}$ that have multiplicative inverses modulo 26 and their corresponding inverses are shown in Table 6.1.

a	1	3	5	7	9	11	15	17	19	21	23	25
$a^{-1} \bmod m$	1	9	21	15	3	19	7	23	11	5	17	25

Table 6.1   Multiplicative inverses in $\mathbb{Z}_{26}$.

Multiplicative inverses with modular arithmetic are useful not just for solving single equations, but also for solving systems of equations, which we will need to do for cryptanalysis later in this chapter.

Example 6.11 Consider the following system of equations.

$$4a + b = 2 \bmod 26$$
$$19a + b = 23 \bmod 26$$

To solve this system of equations for a and b, we can begin by subtracting the second equation from the first to eliminate the variable b. This gives the following equation.

$$-15a = -21 \bmod 26$$

Since $-15 = 11 \bmod 26$ and $-21 = 5 \bmod 26$, this new equation is equivalent to the following.

$$11a = 5 \bmod 26$$

Now to solve this equation for a, we can multiply both sides by $11^{-1} \bmod 26$. From Table 6.1, we can see that $11^{-1} = 19 \bmod 26$.

$$11^{-1} \cdot 11a = (11^{-1} \cdot 5) \bmod 26$$
$$1a = (19 \cdot 5) \bmod 26$$
$$a = 95 \bmod 26$$
$$a = 17$$

Now to find b, we can substitute $a = 17$ into either of the original equations, and then solve for b. We will use the first original equation.

$$(4 \cdot 17) + b = 2 \bmod 26$$
$$68 + b = 2 \bmod 26$$
$$16 + b = 2 \bmod 26$$
$$16 + b - 16 = (2 - 16) \bmod 26$$
$$b = -14 \bmod 26$$
$$b = 12$$

Thus, the solution to the original equations is $a = 17$ and $b = 12$.   □

6.1.1   Exercises

1. For the following integers b and m, find the quotient q and remainder r given by the division algorithm.

   (a)* $b = 38$, $m = 7$

   (b) $b = -38$, $m = 7$

   (c) $b = 100$, $m = 26$

   (d)* $b = -100$, $m = 26$

   (e) $b = 2047$, $m = 137$

   (f) $b = -2047$, $m = 137$

   (g) $b = 124452$, $m = 10371$

   (h) $b = -124452$, $m = 10371$

2. For the following integers b and m, find $b \bmod m$.

   (a)* $b = 38$, $m = 7$

   (b) $b = -38$, $m = 7$

   (c) $b = 100$, $m = 26$

   (d)* $b = -100$, $m = 26$

   (e) $b = 2047$, $m = 137$

   (f) $b = -2047$, $m = 137$

   (g) $b = 124452$, $m = 10371$

   (h) $b = -124452$, $m = 10371$

3. Modulo 12 arithmetic is sometimes called *clock arithmetic*, since when whole hours are added to or subtracted from the time showing on a twelve-hour clock with no AM/PM designator, the hour part of the resulting time showing on the clock is given by modulo 12 addition or subtraction (with 12 used in place of 0). For example, for a twelve-hour clock with no AM/PM designator currently showing 7 o'clock, the time showing on the clock 68 hours from now will be given by $(7 + 68) \bmod 12 = 3$ o'clock. For this same clock (currently showing 7 o'clock), find the following.

   (a)* The time showing on the clock 50 hours from now

   (b) The time showing on the clock 50 hours ago

   (c) The time showing on the clock 500 hours from now

   (d)* The time showing on the clock 500 hours ago

   (e) The time showing on the clock 5000 hours from now

   (f) The time showing on the clock 5000 hours ago

4.* Find two congruence classes for the modulus $m = 7$ different from the two given in this section.

5.* Find two congruence classes for the modulus $m = 10$.

6. Solve the following equations for x.

   (a)* $x - 8 = 7 \bmod 12$

   (b)* $x + 13 = 2 \bmod 19$

   (c) $x - 45 = 24 \bmod 26$

   (d) $x + 75 = 35 \bmod 26$

7. Solve the following equations for x, if it is possible to do so.

   (a)* $3x - 5 = 6 \bmod 26$

   (b) $17x + 23 = 2 \bmod 26$

   (c) $25x - 19 = 25 \bmod 26$

   (d)* $2x + 3 = 10 \bmod 26$

8. Determine whether the following pairs of integers are relatively prime, and justify your answer.

   (a)* 11, 26

   (b) 13, 26

   (c) 15, 26

9. For the following numbers and sets, determine whether the number has a multiplicative inverse in the set. If it does, find this inverse. If it does not, explain how you know.

   (a)* The number 3 in the set $\mathbb{Z}_{10}$

   (b)* The number 5 in the set $\mathbb{Z}_{10}$

   (c) The number 3 in the set $\mathbb{Z}_{12}$

   (d) The number 5 in the set $\mathbb{Z}_{12}$

10. For the following sets, make a list of the numbers in the set that have a multiplicative inverse in the set.

    (a)* $\mathbb{Z}_{10}$

    (b) $\mathbb{Z}_{12}$

11. For the following sets, make a list of the numbers in the set that have a multiplicative inverse in the set, and these inverses.

    (a)* $\mathbb{Z}_6$

    (b) $\mathbb{Z}_7$

12. Solve the following systems of equations for a and b, if it is possible to do so.

    (a) $17a + b = 18 \bmod 26$
    $6a + b = 14 \bmod 26$

    (b)* $8a + b = 12 \bmod 26$
    $19a + b = 5 \bmod 26$

    (c) $8a + b = 8 \bmod 26$
    $25a + b = 17 \bmod 26$

    (d) $10a + b = 15 \bmod 26$
    $8a + b = 8 \bmod 26$

6.2   Shift Ciphers

For shift ciphers, users agree upon an order for the alphabet letters, like for instance the natural order A, B, C, ..., Z of letters in our alphabet, and then encrypt each plaintext letter by replacing it with the letter some agreed-upon number of positions to the right in the alphabet, wrapping from the end of the alphabet to the start whenever necessary. For example, for a shift cipher with our alphabet letters in the natural order and in which each plaintext letter is replaced with the letter three positions to the right, the

plaintext letter A would be replaced with the letter D, the plaintext letter
B with E, C with F, ..., W with Z, X with A, Y with B, and Z with C. Such a
cipher is called a *shift* cipher because the cipher alphabet can be formed by
listing the alphabet letters in order to represent the plaintext letters, and
then shifting these letters to the left the agreed-upon number of positions,
wrapping from the start of the alphabet to the end when necessary.

Example 6.12 Consider a shift cipher with our alphabet letters in the
natural order and a shift of three positions to the right for encryption.
This yields the following cipher alphabet.

Plain: A B C D E F G H I J K L M N O P Q R S T U V W X Y Z
Cipher: D E F G H I J K L M N O P Q R S T U V W X Y Z A B C

Using this cipher alphabet, the plaintext I CAME, I SAW, I CONQUERED[2]
encrypts to LFDPH LVDZL FRQTX HUHG.   □

The Roman Emperor Julius Caesar described a shift cipher with a shift
of three positions to the right for encryption in his writings on the Gallic
Wars. However, shift ciphers are not just something from the distant past.
Shift ciphers were used by the Russian military as recently as 1915, and
the modern *ROT13* cipher (whose name stands for "rotate 13 positions")
is just a shift cipher with our alphabet letters in the natural order and a
shift of 13 positions to the right for encryption. Caesar's cipher was likely
secure in its day, given that most of his enemies were illiterate or unfamiliar
with his language. On the other hand, the shift ciphers used by the Russian
military were easily broken by the Germans and Austrians. ROT13, despite
not being secure, is still widely used to give a casual disguise to things that
users do not want to just state in the clear, such as puzzle answers, movie
or television spoilers, and potentially offensive statements.

One reason for introducing modular arithmetic in this chapter is because we can use it to represent shift ciphers mathematically. If we have
a plaintext expressed using only the letters in the alphabet A, B, C, ..., Z,
and we convert these letters into numbers using the correspondences $A = 0$,
$B = 1$, $C = 2$, ..., $Z = 25$, then we can apply a shift cipher with a shift of b
positions to the right for encryption by adding b to the plaintext numbers
and doing the arithmetic modulo 26. That is, for each plaintext number x in
the set $\mathbb{Z}_{26} = \{0, 1, 2, 3, \ldots, 25\}$, we can obtain a corresponding ciphertext
number y in $\mathbb{Z}_{26}$ using the following formula.

$$y = (x + b) \bmod 26$$

For example, encryption in Caesar's cipher with our alphabet letters in the
natural order (which is the cipher in Example 6.12) can be done using the

---

[2] Julius Caesar (100–44 BCE), quote.

formula $y = (x + 3) \bmod 26$, and encryption in ROT13 can be done using the formula $y = (x + 13) \bmod 26$. Also, for a shift cipher with encryption done using modular addition, ciphertext numbers can be converted back into letters using the same correspondences A = 0, B = 1, C = 2, ... , Z = 25, yielding a list of ciphertext letters.

Example 6.13 Consider a shift cipher with our alphabet letters in the natural order and a shift of 18 positions to the right for encryption. That is, consider a shift cipher in which ciphertext numbers y are formed from plaintext numbers x using the following formula.

$$y = (x + 18) \bmod 26$$

Using this formula, the plaintext JULIUS encrypts as follows.

$$
\begin{array}{ccccccc}
\text{J} & \to & x = 9 & \to & y = (\ 9 + 18) \bmod 26 = 1 & \to & \text{B} \\
\text{U} & \to & x = 20 & \to & y = (20 + 18) \bmod 26 = 12 & \to & \text{M} \\
\text{L} & \to & x = 11 & \to & y = (11 + 18) \bmod 26 = 3 & \to & \text{D} \\
\text{I} & \to & x = 8 & \to & y = (\ 8 + 18) \bmod 26 = 0 & \to & \text{A} \\
\text{U} & & & & & \to & \text{M} \\
\text{S} & \to & x = 18 & \to & y = (18 + 18) \bmod 26 = 10 & \to & \text{K} \\
\end{array}
$$

Thus, the ciphertext is BMDAMK. (The encryption of the second U is not written out or necessary to be calculated, since it would be identical to the encryption of the first U.) □

To decrypt a ciphertext that was formed using a shift cipher, we must only undo what was done to encrypt the message. That is, a shift cipher with a shift of b positions to the right for encryption would use a shift of b positions to the left for decryption. For a shift cipher with encryption done using modular addition, for each ciphertext number y in $\mathbb{Z}_{26}$, we can obtain the corresponding plaintext number x in $\mathbb{Z}_{26}$ using the following formula.

$$x = (y - b) \bmod 26$$

Often in literature, with an actual number for b, this formula is expressed using addition instead of subtraction. This is not difficult to do, only requiring the $-b$ in the formula to be changed into $-b \bmod 26$. For example, for Caesar's cipher with encryption formula $y = (x+3) \bmod 26$, the decryption formula $x = (y-3) \bmod 26$ is often written $x = (y+23) \bmod 26$. Also, for ROT13 with encryption formula $y = (x + 13) \bmod 26$, the decryption formula $x = (y - 13) \bmod 26$ is often written $x = (y + 13) \bmod 26$. Note that the encryption and decryption calculations for ROT13 are identical. This happens because with a 26-letter alphabet, shifting to the right or left 13 positions always ends at the same place.

Example 6.14 Consider the ciphertext WJNAFY, which was formed using a shift cipher with our alphabet letters in the natural order and encryption formula $y = (x + 18) \bmod 26$. The decryption formula for this cipher is $x = (y - 18) \bmod 26$, or, equivalently, the following.

$$x = (y + 8) \bmod 26$$

Using this formula, the ciphertext WJNAFY decrypts as follows.

$$
\begin{array}{ccccccc}
\text{W} & \rightarrow & y = 22 & \rightarrow & x = (22 + 8) \bmod 26 = & 4 & \rightarrow & \text{E} \\
\text{J} & \rightarrow & y = 9 & \rightarrow & x = (9 + 8) \bmod 26 = 17 & & \rightarrow & \text{R} \\
\text{N} & \rightarrow & y = 13 & \rightarrow & x = (13 + 8) \bmod 26 = 21 & & \rightarrow & \text{V} \\
\text{A} & \rightarrow & y = 0 & \rightarrow & x = (0 + 8) \bmod 26 = & 8 & \rightarrow & \text{I} \\
\text{F} & \rightarrow & y = 5 & \rightarrow & x = (5 + 8) \bmod 26 = 13 & & \rightarrow & \text{N} \\
\text{Y} & \rightarrow & y = 24 & \rightarrow & x = (24 + 8) \bmod 26 = & 6 & \rightarrow & \text{G}
\end{array}
$$

Thus, the plaintext is ERVING. □

We should also note that with our alphabet, it would generally be understood that for shift ciphers, whether encryption is represented using modular addition like we did in Examples 6.13 and 6.14, or expressed using words like we did in Example 6.12, the number of positions shifted for encryption is in $\mathbb{Z}_{26}$, since this range of integers gives every possible distinct shift cipher.

6.2.1 Exercises

1. Consider Caesar's cipher with our alphabet letters in the natural order.

   (a)* Use this cipher to encrypt ET TU, BRUTE.[3]

   (b) Use this cipher to encrypt THEN FALL, CAESAR.[4]

   (c) Decrypt HASHU LHQFH LVWKH WHDFK HURID OOWKL QJV,[5] which was formed using this cipher.

2. (a)* Use ROT13 to encrypt BOB SACAMANO.

   (b) Use ROT13 to encrypt CORKY RAMIREZ.

   (c) Decrypt SENAX YVAQR YNABE BZNAB JFXV, which was formed using ROT13.

---

[3] William Shakespeare (1564–1616), from *Julius Caesar*.
[4] William Shakespeare, from *Julius Caesar*.
[5] Julius Caesar, quote.

3. Consider a shift cipher with our alphabet letters in the natural order and a shift of six positions to the right for encryption.

   (a) Give an encryption formula for this cipher that has the form $y = (x + b) \bmod 26$ for some b in $\mathbb{Z}_{26}$.

   (b)* Use this cipher to encrypt THE PROBLEM WITH HAVING.

   (c) Use this cipher to encrypt EVERYTHING YOU WANT IS THAT.

   (d)* Give a decryption formula for this cipher that has the form $x = (y + d) \bmod 26$ for some d in $\mathbb{Z}_{26}$.

   (e) Decrypt TUUTK QTUCY CNGZZ UMKZL UXEUA XHOXZ NJGE, which was formed using this cipher.

4. Consider a shift cipher with our alphabet letters in the natural order and encryption formula $y = (x + 16) \bmod 26$.

   (a)* Use this cipher to encrypt THE TROUBLE WITH DOING.

   (b) Use this cipher to encrypt SOMETHING RIGHT THE FIRST TIME IS THAT.

   (c) Give a decryption formula for this cipher that has the form $x = (y + d) \bmod 26$ for some d in $\mathbb{Z}_{26}$.

   (d) Decrypt DERET OQFFH USYQJ UIXEM TYVVY SKBJY JMQI,[6] which was formed using this cipher.

5. Consider a shift cipher with our alphabet letters in the natural order and encryption formula $y = (x + 20) \bmod 26$.

   (a)* Use this cipher to encrypt THE DOWNSIDE OF BEING.

   (b) Use this cipher to encrypt BETTER THAN EVERYONE ELSE IS THAT.

   (c) Give a decryption formula for this cipher that has the form $x = (y + d) \bmod 26$ for some d in $\mathbb{Z}_{26}$.

   (d) Decrypt JYIJF YNYHX NIUMM OGYSI OLYJL YNYHN CIOM,[7] which was formed using this cipher.

6. Create a shift cipher and use it to encrypt a plaintext of your choice with at least 20 letters.

7. Although no shift ciphers are secure, explain why a shift cipher in which ciphertext numbers y are formed from plaintext numbers x using the formula $y = (x + b) \bmod 26$ with $b = 0$ is especially insecure.

---

[6] Walt West (1917–1984), quote.
[7] Despair, Inc.

8. Find some information about Julius Caesar's use of encryption during the Gallic Wars, and write a summary of your findings.

9. Find some information about some actual uses of ROT13 in modern culture, and write a summary of your findings.

10. Find a copy of the spoof academic paper *On the 2ROT13 Encryption Algorithm*, and write a summary of the description of 2ROT13 and opinions about 2ROT13 and other ciphers given in this article.

11. Find some information about the ROT47 cipher and ASCII alphabet, and write a summary of your findings.

6.3 Cryptanalysis of Shift Ciphers

Shift ciphers are no harder to break than substitution ciphers, of course, because they are substitution ciphers. In fact, shift ciphers are much easier to break than substitution ciphers in which the correspondences between plaintext and ciphertext letters are assigned randomly or via a keyword, since with a shift cipher, if the correspondence between one plaintext letter and one ciphertext letter is known, the rest of the correspondences follow. This makes a brute force attack effective against a shift cipher, and often allows a short ciphertext formed using a shift cipher to be cryptanalyzed much more easily than if it had been formed using a non-shift substitution cipher.

For a message written using our alphabet letters and encrypted with a shift cipher, the ciphertext could be the result of a maximum of only 25 distinct shifts (assuming that a shift of zero positions is not used). A brute force attack could be done by simply trying to decrypt the ciphertext assuming each of these 25 possible encryption shifts one at a time, and stopping when the correct plaintext is revealed. It would almost certainly be known immediately when the correct plaintext was revealed, since of the results of the various attempts at decryption, it is almost certain that only the letters in the correct plaintext would make sense when strung together. In addition, it may be possible to save a significant amount of time in cryptanalysis by trying to decrypt just a small portion of the ciphertext, and then decrypting the full ciphertext only after the correct shift is determined.

Example 6.15 Consider the ciphertext HVSDF CPZSA KWHVG CQWOZ WGAWG HVOHS JSBHI OZZMM CIFIB CIHCT CHVSF DSCDZ SGACB SM, which was formed using a shift cipher with our alphabet letters in the natural order. For each number b of possible positions shifted to the right for encryption that could have produced this ciphertext, the following shows the result of trying to decrypt the first 10 letters in the ciphertext, starting with $b = 1$, and stopping when plaintext letters that make sense when strung together are obtained.

$$
\begin{aligned}
b = 1: &\quad \text{GURCEBOYRZ} \\
b = 2: &\quad \text{FTQBDANXQY} \\
b = 3: &\quad \text{ESPACZMWPX} \\
b = 4: &\quad \text{DROZBYLVOW} \\
b = 5: &\quad \text{CQNYAXKUNV} \\
b = 6: &\quad \text{BPMXZWJTMU} \\
b = 7: &\quad \text{AOLWYVISLT} \\
b = 8: &\quad \text{ZNKVXUHRKS} \\
b = 9: &\quad \text{YMJUWTGQJR} \\
b = 10: &\quad \text{XLITVSFPIQ} \\
b = 11: &\quad \text{WKHSUREOHP} \\
b = 12: &\quad \text{VJGRTQDNGO} \\
b = 13: &\quad \text{UIFQSPCMFN} \\
b = 14: &\quad \text{THEPROBLEM}
\end{aligned}
$$

Thus, the number of positions shifted to the right for encryption was almost certainly $b = 14$. Trying to decrypt the rest of the ciphertext for this value of b yields the full plaintext: THE PROBLEM WITH SOCIALISM IS THAT EVENTUALLY YOU RUN OUT OF OTHER PEOPLE'S MONEY.[8]   □

To further save time when breaking a shift cipher, frequency analysis may be used to identify some likely correspondences between plaintext and ciphertext letters, and then a portion of the ciphertext decrypted assuming the encryption shifts that result from these likely correspondences first. Recall that in ordinary English, the letters that naturally occur the most often are, in order, E, T, A, O, I, N, and S. So for a plaintext written in ordinary English and encrypted using a shift cipher, it is reasonable to suppose that the letters that occur in the ciphertext with the highest frequency will correspond to E, T, A, O, I, N, or S in the plaintext. Trying the decrypt a portion of the ciphertext assuming the encryption shifts that result from these correspondences first should limit the number of shifts that must be checked.

Example 6.16 Consider the ciphertext HVSDF CPZSA KWHVG CQWOZ WGAWG HVOHS JSBHI OZZMM CIFIB CIHCT CHVSF DSCDZ SGACB SM in Example 6.15. The letters that occur in this ciphertext with the highest frequency are C and S, occurring eight times each. Assuming first that the ciphertext letter C corresponds to E in the plaintext, then with E = 4 for x and C = 2 for y,

---

[8]Margaret Thatcher (1925–2013), quote.

the shift cipher encryption formula $y = (x+b) \bmod 26$ is $2 = (4+b) \bmod 26$. The solution to this equation is $b = 24$; however, trying to decrypt the first 10 letters in the ciphertext for this value of b yields JXUFHERBUC, which is clearly not part of the plaintext. Assuming next that the ciphertext letter S corresponds to E in the plaintext, then with E = 4 for x and S = 18 for y, the shift cipher encryption formula is $18 = (4+b) \bmod 26$. The solution to this equation is $b = 14$, and trying to decrypt the first 10 letters in the ciphertext for this value of b yields THEPROBLEM. Thus, the correct value of b is almost certainly $b = 14$. Trying to decrypt the rest of the ciphertext for this value of b would then yield the full plaintext given in Example 6.15. An important thing to note about this is that by using frequency analysis, we only had to check two shifts, as opposed to 14, which we had to check in Example 6.15. □

Although shift ciphers are not secure, they are still useful, for they do at least prevent someone from, upon glancing at a particular ciphertext, being able to immediately read the corresponding plaintext. This could allow, for example, data to be stored electronically in a way so as to thwart an automated word scanner searching for words. Also, as we noted for ROT13 in Section 6.2, shift ciphers can be used to give a casual disguise to things that users do not want to just state in the clear. Most importantly, shift ciphers have been included as parts of larger ciphers, such as Vigenère ciphers, which we will consider in Chapter 7, and the modern Advanced Encryption Standard, which we will consider in Chapter 11, and which was secure enough that it was selected by the National Institute of Standards and Technology in 2001 to serve as a Federal Information Processing Standard.

6.3.1 Exercises

1. The following ciphertexts were formed using shift ciphers with our alphabet letters in the natural order. For each, use a brute force attack to cryptanalyze the ciphertext.

   (a) ESPOZ RDCLY BFTNV WJ

   (b) YKHFF XFIAB LMHYE TZLMT YY

   (c) QXYNR BJPXX MCQRW PVJHK NCQNK NBCXO CQRWP BJWMW XPXXM
       CQRWP NENAM RNB[9]

2. The following ciphertexts were formed using shift ciphers with our alphabet letters in the natural order. For each, use frequency analysis to identify some likely correspondences between plaintext and

---

[9]Stephen King, quote.

ciphertext letters, and then use this information to cryptanalyze the ciphertext.

(a) XKBVB CLOBV BLKIV BKAPR MJXHF KDQEB TELIB TLOIA YIFKA[10]

(b) KYVFE CPKZD VKFVR KUZVK WFFUZ JNYZC VPFLI VNRZK ZEXWF IKYVJ KVRBK FTFFB[11]

(c) AMLDG BCLAC AMKCQ DPMKF MSPQY LBBYW QYLBU CCIQY LBWCY PQMDA MLQRY LRUMP IYLBB CBGAY RGML[12]

3. Cryptanalyze the following ciphertexts, which were formed using shift ciphers with our alphabet letters in the natural order.

(a) OAOBG UCHHC YBCKV WGZWA WHOHW CBG[13]

(b) AYMWK KAVGF LKGEM UZEAF VTWAF YGDVS KAEAF VTWAF YXSLS FVGDV[14]

(c) QCHYC MWIHM NUHNJ LIIZN BUNAI XFIPY MOMUH XQUHN MNIMY YOMBU JJS[15]

(d) JXUYD LUDJY EDEVR QIAUJ RQBBM QIDEJ QDQSS YTUDJ YJMQI TULUB EFUTJ ECUUJ QDUUT JXEIU REOII YCFBO MEKBT DEJFB QOTHE FJXUX QDTAU HSXYU V[16]

4. Recall from Exercise 5 in Section 2.2 the concept of *superencryption*.

(a)* Use shift ciphers with our alphabet letters in the natural order and shifts of four and six positions to the right for encryption (in that order) to superencrypt PEOPLE WHO SAY NEEDLESS TO SAY.

(b) Decrypt KVWYC DSXFK BSKLV ICKIS DKXIG KI, which was superencrypted using shift ciphers with our alphabet letters in the natural order and shifts of six and four positions to the right for encryption (in that order).

(c) Does superencryption by two shift ciphers yield more security than encryption by one shift cipher? In other words, if a plaintext P is encrypted using a shift cipher, yielding M, and then M is encrypted using another shift cipher, yielding C, would C be harder in general to cryptanalyze than M? Explain your answer completely, and be as specific as possible.

[10] Mahatma Gandhi (1869–1948), quote.
[11] Julia Child (1912–2004), quote.
[12] Roger Staubach, quote.
[13] Harry Callahan, quote.
[14] Benjamin Franklin (1706–1790), quote.
[15] Benjamin Franklin, quote.
[16] James Naismith (1861–1939), quote.

6.4 Affine Ciphers

As we noted in Section 6.2, one reason for introducing modular arithmetic in this chapter is because we can use it to represent shift ciphers mathematically. However, shift ciphers are easy to understand without using modular arithmetic, which might make you wonder why we would want to use modular arithmetic to represent them. To see why, recall that no shift ciphers are secure, regardless of how they are represented. We did mention a couple of reasons in Section 6.2 why they are still useful to consider, though, most importantly because they can be used as parts of larger more secure ciphers. Shift ciphers can also be generalized into other ciphers that have more security, some even a legitimate amount of security, and some of these ciphers are much easier to understand if modular arithmetic is used to represent the shift ciphers from which they are generalized. More specifically, by using modular arithmetic to represent shift ciphers, we can more easily understand how security can be increased by using mathematical operations that are more sophisticated than just modular addition. In this section, we will consider what happens when we use not just modular addition in the encryption step for a cipher, but modular multiplication as well, yielding an *affine* cipher.

Recall that if we have a plaintext expressed using only the letters in the alphabet A, B, C, ... , Z, and we convert these letters into numbers using the correspondences $A = 0$, $B = 1$, $C = 2$, ... , $Z = 25$, then we can apply a shift cipher with a shift of b positions to the right for encryption with the formula $y = (x + b) \bmod 26$ for some b in $\mathbb{Z}_{26}$. For an affine cipher, the encryption step is done with the following formula, for some a and b in $\mathbb{Z}_{26}$ with $\gcd(a, 26) = 1$.

$$y = (ax + b) \bmod 26$$

Example 6.17 Consider an affine cipher with our alphabet letters in the natural order and the following encryption formula.

$$y = (5x + 4) \bmod 26$$

Using this formula, the plaintext RADFORD encrypts as follows.

R	$\to$	$x = 17$	$\to$	$y = (5 \cdot 17 + 4) \bmod 26 = 11$	$\to$	L
A	$\to$	$x = 0$	$\to$	$y = (5 \cdot 0 + 4) \bmod 26 = 4$	$\to$	E
D	$\to$	$x = 3$	$\to$	$y = (5 \cdot 3 + 4) \bmod 26 = 19$	$\to$	T
F	$\to$	$x = 5$	$\to$	$y = (5 \cdot 5 + 4) \bmod 26 = 3$	$\to$	D
O	$\to$	$x = 14$	$\to$	$y = (5 \cdot 14 + 4) \bmod 26 = 22$	$\to$	W
R					$\to$	L
D					$\to$	T

Thus, the ciphertext is LETDWLT.   □

The following example shows why the requirement that $\gcd(a, 26) = 1$ in an affine cipher is essential.

Example 6.18 Consider an affine cipher with our alphabet letters in the natural order and the following encryption formula.

$$y = (2x + 1) \bmod 26$$

Using this formula, the plaintext letters A and N would encrypt as follows.

$$\text{A} \quad \rightarrow \quad x = 0 \quad \rightarrow \quad y = (2 \cdot 0 + 1) \bmod 26 = 1 \quad \rightarrow \quad \text{B}$$
$$\text{N} \quad \rightarrow \quad x = 13 \quad \rightarrow \quad y = (2 \cdot 13 + 1) \bmod 26 = 1 \quad \rightarrow \quad \text{B}$$

This is a problem, of course, because if a plaintext containing the letters A and N were encrypted using this cipher, the intended recipient of the message would not know whether each letter B in the ciphertext was supposed to be decrypted as A or N. Similar results occur for this cipher with other pairs of letters as well. □

The root of the problem in Example 6.18 is that $\gcd(2, 26) \neq 1$, which violates the requirement for an affine cipher that $\gcd(a, 26) = 1$. This requirement guarantees that $a^{-1} \bmod 26$ will exist, which not only assures that different plaintext letters will never encrypt to the same ciphertext letter, but is also necessary for decryption. For an affine cipher, we can obtain a formula giving the plaintext number x that corresponds to each ciphertext number y by solving $y = (ax + b) \bmod 26$ for x:

$$y = (ax + b) \bmod 26, \text{ or, equivalently,}$$
$$ax + b = y \bmod 26$$
$$ax + b - b = (y - b) \bmod 26$$
$$ax = (y - b) \bmod 26$$
$$a^{-1}ax = a^{-1}(y - b) \bmod 26, \text{ or, finally,}$$
$$x = a^{-1}(y - b) \bmod 26.$$

As was the case with the analogous decryption formula for shift ciphers, often in literature with an actual value of b, this formula is expressed using addition instead of subtraction.

Example 6.19 Consider the ciphertext FSLISRSE, which was formed using an affine cipher with our alphabet letters in the natural order and encryption formula $y = (5x + 4) \bmod 26$. The decryption formula for this cipher is $x = 5^{-1}(y - 4) \bmod 26$, or, equivalently, since $5^{-1} = 21 \bmod 26$ (from Table 6.1 on page 175) and $-4 = 22 \bmod 26$, the following.

$$x = 21(y + 22) \bmod 26$$

Using this formula, the ciphertext FSLISRSE decrypts as follows.

$$
\begin{array}{ccccccc}
\text{F} & \rightarrow & y = 5 & \rightarrow & x = 21(\ 5 + 22) \bmod 26 = 21 & \rightarrow & \text{V} \\
\text{S} & \rightarrow & y = 18 & \rightarrow & x = 21(18 + 22) \bmod 26 = \ 8 & \rightarrow & \text{I} \\
\text{L} & \rightarrow & y = 11 & \rightarrow & x = 21(11 + 22) \bmod 26 = 17 & \rightarrow & \text{R} \\
\text{I} & \rightarrow & y = 8 & \rightarrow & x = 21(\ 8 + 22) \bmod 26 = \ 6 & \rightarrow & \text{G} \\
\text{S} & & & & & \rightarrow & \text{I} \\
\text{R} & \rightarrow & y = 17 & \rightarrow & x = 21(17 + 22) \bmod 26 = 13 & \rightarrow & \text{N} \\
\text{S} & & & & & \rightarrow & \text{I} \\
\text{E} & \rightarrow & y = 4 & \rightarrow & x = 21(\ 4 + 22) \bmod 26 = \ 0 & \rightarrow & \text{A}
\end{array}
$$

Thus, the plaintext is VIRGINIA. □

6.4.1 Exercises

1. Consider an affine cipher with our alphabet letters in the natural order and encryption formula $y = (7x + 18) \bmod 26$.

   (a)* Use this cipher to encrypt WOLFPACK.

   (b) Use this cipher to encrypt HIGHLANDERS.

   (c)* Give a decryption formula for this cipher that has the form $x = c(y + d) \bmod 26$ for some c and d in $\mathbb{Z}_{26}$.

   (d) Decrypt YMCFV SWFUU HO, which was formed using this cipher.

2. Consider an affine cipher with our alphabet letters in the natural order and encryption formula $y = (19x + 6) \bmod 26$.

   (a)* Use this cipher to encrypt FIVE TOWNS.

   (b) Use this cipher to encrypt VIKING QUEST.

   (c) Give a decryption formula for this cipher that has the form $x = c(y + d) \bmod 26$ for some c and d in $\mathbb{Z}_{26}$.

   (d) Decrypt AEHRM KEFHG SE, which was formed using this cipher.

3. Consider an affine cipher with our alphabet letters in the natural order and encryption formula $y = (17x + 4) \bmod 26$.

   (a)* Use this cipher to encrypt IT'S LIKE A SAUNA IN HERE.[17]

   (b) Use this cipher to encrypt I'M GOING THROUGH THIS STUFF LIKE WATER.[18]

---

[17]Cosmo Kramer, quote.
[18]Cosmo Kramer, quote.

(c) Give a decryption formula for this cipher that has the form $x = c(y + d)$ mod 26 for some c and d in $\mathbb{Z}_{26}$.

(d) Decrypt PTUYU ZHUPN UJYEH UAESK RCAUP TKHYP W,[19] which was formed using this cipher.

4. Consider an affine cipher with our alphabet letters in the natural order and encryption formula $y = (15x + 22)$ mod 26.

    (a)* Use this cipher to encrypt APPALACHIAN IS THE SECOND-LARGEST ASU, BEHIND ARIZONA STATE.

    (b) Use this cipher to encrypt RADFORD IS THE FOURTH-LARGEST RU, BEHIND REGIS, ROWAN, AND RUTGERS.

    (c) Give a decryption formula for this cipher that has the form $x = c(y + d)$ mod 26 for some c and d in $\mathbb{Z}_{26}$.

    (d) Decrypt JAGVW VEMGJ YVYJF SVXEF WRIEG VJAGK LKVWF GYVXE GUWFF EGV, which was formed using this cipher.

5. Create an affine cipher and use it to encrypt a plaintext of your choice with at least 15 letters.

6. Although no affine ciphers are secure, explain why an affine cipher in which ciphertext numbers y are formed from plaintext numbers x using the formula $y = (ax + b)$ mod 26 with $a = 1$ and $b = 0$ is especially insecure.

7. Explain why all shift ciphers are affine ciphers.

8. The ancient Hebrew Atbash cipher was a substitution cipher in which the cipher alphabet was formed by reversing the Hebrew alphabet letters in their natural order. This same cipher with our alphabet letters in the natural order would encrypt the plaintext letter A as the ciphertext letter Z, the plaintext letter B as Y, C as X, ... , Y as B, and Z as A. Show that this cipher with our alphabet letters in the natural order is also an affine cipher with encryption formula $y = (25x + 25)$ mod 26.

6.5   Cryptanalysis of Affine Ciphers

As with shift ciphers, affine ciphers are no harder to break than substitution ciphers, because they are substitution ciphers. However, affine ciphers do at least not share the property with shift ciphers that if the correspondence

---

[19] Cosmo Kramer, quote.

between one plaintext letter and one ciphertext letter is known, the rest of the correspondences follow.

A brute force attack is still effective against an affine cipher, though, although unlike with a shift cipher, a brute force attack against an affine cipher is not necessarily easier than an attack by frequency analysis alone. Recall that for a message written using our alphabet letters and encrypted with a shift cipher, the ciphertext could be the result of a maximum of only 25 distinct shifts (assuming a shift of zero positions is not used). These encryption shifts correspond to the 25 possible nonzero keys for a shift cipher, the 25 possible nonzero values of b in the shift cipher encryption formula $y = (x + b) \bmod 26$. Affine ciphers have two keys, the multiplicative key a and the additive key b in the affine cipher encryption formula $y = (ax + b) \bmod 26$. Even with our small 26-letter alphabet, there are many more possible pairs of keys for an affine cipher than the number of possible keys for a shift cipher. For an affine cipher, there are 12 possible values of a (the values of a in Table 6.1 on page 175) and 26 possible values of b, yielding $(12 \cdot 26) - 1 = 311$ possible pairs of keys (assuming the pair $a = 1$ and $b = 0$ is not used). A brute force attack could be done by simply trying to decrypt the ciphertext assuming each of these 311 possible pairs of keys, stopping when the correct plaintext is revealed. In addition, like with a shift cipher, it may be possible to save a significant amount of time by trying to decrypt just a small portion of the ciphertext, and then decrypting the full ciphertext only after the correct keys are determined.

Also as with shift ciphers, to further save time when breaking an affine cipher, frequency analysis may be used to identify some likely correspondences between plaintext and ciphertext letters, and then a portion of the ciphertext decrypted assuming the keys that result from these likely correspondences first. Because affine ciphers have two keys, this would require matching two ciphertext letters with corresponding plaintext letters, and then, if it is possible, solving a system of two equations for the pair of keys.

Example 6.20 Consider the ciphertext XBCKQ QLXBC DMLMZ LXBCO KRE, which was formed using an affine cipher with our alphabet letters in the natural order. Suppose it is somehow known that the ciphertext letters C and X correspond to E and T, respectively, in the plaintext. With E = 4 for x and C = 2 for y, the affine cipher encryption formula $y = (ax + b) \bmod 26$ is $2 = (a \cdot 4 + b) \bmod 26$. Also, with T = 19 for x and X = 23 for y, the affine cipher encryption formula is $23 = (a \cdot 19 + b) \bmod 26$. Thus, the values of a and b in the affine cipher encryption formula satisfy the following system of equations.

$$4a + b = 2 \bmod 26$$
$$19a + b = 23 \bmod 26$$

We solved this system of equations in Example 6.11 on page 175, finding that $a = 17$ and $b = 12$. Thus, the encryption formula for the cipher is $y = (17x+12) \bmod 26$. The decryption formula is $x = 17^{-1}(y-12) \bmod 26$, or, equivalently, since $17^{-1} = 23 \bmod 26$ (from Table 6.1 on page 175) and $-12 = 14 \bmod 26$, $x = 23(y + 14) \bmod 26$. Using this formula, the ciphertext XBCKQ QLXBC DMLMZ LXBCO KRE decrypts as THE GOOD, THE BAD, AND THE UGLY.   □

When trying to determine affine cipher keys a and b by solving a system of equations like in Example 6.20, the solutions for a and b to the system may not be unique. For example, consider the following system of equations, which would result from a ciphertext formed using an affine cipher with our alphabet letters in the natural order, if it were somehow known that the ciphertext letters S and Q correspond to K and G, respectively, in the plaintext.

$$10a + b = 18 \bmod 26$$
$$6a + b = 16 \bmod 26$$

To solve this system for a and b, we can begin by subtracting the second equation from the first to eliminate the variable b. This gives the equation $4a = 2 \bmod 26$, but since $\gcd(4, 26) \neq 1$, then $4^{-1} \bmod 26$ does not exist. This does not mean that we cannot still recover the values of a and b, though. If we try by trial and error to find values of a that satisfy $4a = 2 \bmod 26$, we will find that there are two, $a = 7$ and $a = 20$. However, $a = 20$ does not satisfy the affine cipher requirement that $\gcd(a, 26) = 1$. Thus, the multiplicative key for the cipher must be $a = 7$. Using this value of a in either of the original equations will yield the additive key $b = 0$.

6.5.1   Exercises

1. The following ciphertexts were formed using affine ciphers with our alphabet letters in the natural order. For each, the correspondences between two plaintext and ciphertext letters are given. Use this information to determine the encryption and decryption formulas for the cipher, and cryptanalyze the ciphertext.

    (a)* The ciphertext is LNUWN CZCZY CWWQM HI, and the ciphertext letters C and N correspond to I and H, respectively, in the plaintext.

    (b) The ciphertext is COGCZ JSSNO FYGCZ, and the ciphertext letters S and Z correspond to E and T, respectively, in the plaintext.

    (c)* The ciphertext is YLNNY ELQXP HSNSY N, and the ciphertext letters L and N correspond to N and D, respectively, in the plaintext.

   (d) The ciphertext is TSDRG DOFES RGBDF MXMEX, and the ciphertext letters S and G correspond to U and O, respectively, in the
   plaintext.

2. The following ciphertexts were formed using affine ciphers with our
   alphabet letters in the natural order. For each, use frequency analysis to identify some likely correspondences between plaintext and
   ciphertext letters, and then use this information to cryptanalyze the
   ciphertext.

   (a)* LPDVQ LCYZR VADVG YVE

   (b)* GYJHL GSPVI HBJFV PSJB

   (c)* PJOTB QHCOI GVSMH QIGZA GWZPY

   (d)* WAIXA VXZGO WPATX YLXXZ OLKIO LXZGQ ATHRZ YBGNG GLYCC
        AWPHO IZGRN UPGAP HGQZA ZYBGS GPXAL XTUOL KQZGL XZGTG
        IGGWG RXANG LAZAP GYXYH H[20]

   (e) WFORF BQNYD BKRXY DYNEB KNEYF OJNVE RLNFQ LYNBK RQUJN
       VEPIN WRWFO RFQYI RLEDQ UBYDQ UCVBY PDYXI FQLPI DYLNR
       BNQFQ TJNKF QFNQJ NVERP DBYFQ LJNVE WFOR[21]

   (f) NPUCX RAJUT UALJC JKPRC ALLJU XYUXL JCSIP BRLIC DTBKU
       XULAX ILAIM CLJUX YOIWB CKPXU XAGJI IBVWL UNOIW JKHCX
       LBCKP XCRLJ CMCKX UXYIN NPUCX RAJUT OIWPC KBBOJ KHCXL
       BCKPX CRKXO LJUXY[22]

3. Recall from Exercise 5 in Section 2.2 the concept of *superencryption*.

   . (a)* Use affine ciphers with our alphabet letters in the natural
        order and encryption formulas $\dot{y} = (11x + 22) \bmod 26$ and
        $y = (3x + 12) \bmod 26$ (in that order) to superencrypt MAGNUM
        FORCE.

   (b) Decrypt EYOLI EHSNM A, which was superencrypted using affine
       ciphers with our alphabet letters in the natural order and encryp-
       tion formulas $y = (3x + 12) \bmod 26$ and $y = (11x + 22) \bmod 26$
       (in that order).

   (c) Does superencryption by two affine ciphers yield more security
       than encryption by one affine cipher? In other words, if a plain-
       text P is encrypted using an affine cipher, yielding M, and then
       M is encrypted using another affine cipher, yielding C, would
       C be harder in general to cryptanalyze than M? Explain your
       answer completely, and be as specific as possible.

---

[20] Dale Carnegie (1888–1955), quote.
[21] Jackie Robinson (1919–1972), quote.
[22] Muhammad Ali (1942–2016), quote.

4.* For a ciphertext formed using an affine cipher with our alphabet letters in the natural order, suppose it takes three minutes to try to
decrypt the ciphertext assuming a pair of keys for the cipher. How
long would it take on average (i.e., trying half of the possible pairs
of keys) to break the cipher using a brute force attack? Give your
answer in hours.

5. The ASCII alphabet is a complete alphabet for the English language,
and contains 95 characters, including both capital and lowercase letters, the digits 0–9, punctuation, a blank space, and some others.

   (a)* For an affine cipher defined for an alphabet with 95 characters
   (i.e., with encryption formula $y = (ax + b)$ mod 95, for some
   a and b in $\mathbb{Z}_{95} = \{0, 1, 2, 3, \ldots, 94\}$ with $\gcd(a, 95) = 1$), how
   many possible pairs of keys are there?

   (b) For a ciphertext that was formed using an affine cipher defined
   for an alphabet with 95 characters (i.e., with encryption formula
   $y = (ax+b)$ mod 95, for some a and b in $\mathbb{Z}_{95} = \{0, 1, 2, 3, \ldots, 94\}$
   with $\gcd(a, 95) = 1$), suppose it takes three minutes to try to
   decrypt the ciphertext assuming a pair of keys for the cipher.
   How long would it take on average (i.e., trying half of the possible
   pairs of keys) to break the cipher using a brute force attack? Give
   your answer in hours.

Chapter 7

Alberti and Vigenère Ciphers

A shift, affine, or any other type of substitution cipher is said to be *monoalphabetic* because a single cipher alphabet is used throughout the entire encryption process. The fact that a single cipher alphabet is used throughout the entire encryption process is what generally makes a substitution cipher easy to break, by frequency analysis, since the distribution of letter frequencies in plaintexts is preserved into ciphertexts. One way to increase security is to change the cipher alphabet one or more times while encrypting a message. Such a cipher, in which more than one cipher alphabet is used, is said to be *polyalphabetic.*

Example 7.1 Consider the following pair of cipher alphabets.

Plain:	A B C D E F G H I J K L M N O P Q R S T U V W X Y Z
Cipher 1:	E F G H I J K L M N O P Q R S T U V W X T Z A B C D
Cipher 2:	H I J K L M N O P Q R S T U V W X Y Z A B C D E F G

Using a polyalphabetic cipher in which the cipher alphabet **Cipher 1** is used to encrypt letters in odd-numbered positions (i.e., every other letter beginning with the first letter), and the cipher alphabet **Cipher 2** is used to encrypt letters in even-numbered positions (i.e., every other letter beginning with the second letter), the plaintext HELLO THERE HENRY encrypts to LLPSS ALLVL LLRYC. □

Note in Example 7.1 that the plaintext letters L and R both encrypt to two different ciphertext letters, and the ciphertext letters L and S both are encrypted from two different plaintext letters. None of these things could have happened if the cipher in the example were monoalphabetic.

For a ciphertext formed using a polyalphabetic cipher, since identical ciphertext letters will not necessarily correspond to identical plaintext letters, breaking a polyalphabetic cipher is generally more difficult than breaking a monoalphabetic cipher. The changing cipher alphabets in polyalphabetic ciphers typically have the effect of causing letter frequencies in ciphertexts to be more evenly distributed than in ciphertexts formed using monoalphabetic ciphers. However, the polyalphabetic cipher in Example 7.1 is still not very secure. For this cipher, for someone trying to break the cipher who somehow knew that plaintext letters in odd-numbered positions were all encrypted using a single cipher alphabet, and plaintext letters in even-numbered positions were all encrypted using a different single cipher alphabet, frequency analysis could still be used to break the two "halves" of the cipher relatively easily. Methods for creating more secure polyalphabetic ciphers do exist, though. We will consider three such methods in this chapter.

7.1   Alberti Ciphers

Leon Battista Alberti was an Italian man who lived during the Renaissance, and, like many other influential people of the time, excelled in numerous areas of study. He was an accomplished artist and musician, as well as a prolific author who wrote poetry, music, and books on such diverse topics as law, architecture, and cryptology. In *The Codebreakers* [13], David Kahn refers to Alberti as the "Father of Western Cryptology," not only because Alberti's writings on cryptology were substantial and the earliest in the Western world, but also because they were progressive enough that Alberti's original idea of a polyalphabetic cipher was not put into large-scale practice for hundreds of years.

The type of polyalphabetic cipher created by Alberti was based on the use of a cipher wheel, the design of which is shown in Figure 7.1 on page 197. Alberti's cipher wheel consisted of two disks, an outer stationary disk that contained 24 possible plaintext characters, and an inner rotatable disk that contained 24 possible ciphertext characters. The plaintext characters on Alberti's outer disk were 20 of the 26 capital letters in our alphabet in order, followed by the digits 1 through 4 in order. The six letters in our alphabet not included on Alberti's outer disk were J, U, and W, which were not part of Alberti's alphabet, and H, K, and Y, which Alberti deemed unnecessary. The ciphertext characters on Alberti's inner disk were 23 of the 26 lowercase letters in our alphabet in a scrambled order, and the character &, which was probably used in place of the digraph et. The three letters in our alphabet not included on Alberti's inner disk were j, u, and w.

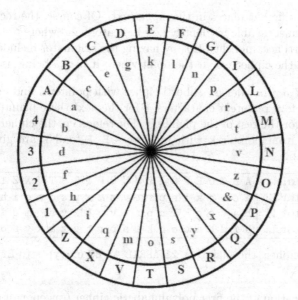

Figure 7.1   Design of Alberti's cipher wheel.

To encrypt a message using an Alberti cipher, the originator and intended recipient first agree upon a *pointer* character on the inner disk. In his own description of the wheel, Alberti used k as the pointer. The originator then chooses an *indicator* character on the outer disk with which to align the pointer. For example, in Figure 7.1, if the pointer is k, then the indicator is E. The originator then forms the ciphertext by writing the indicator followed in order by the characters on the inner disk that correspond to the plaintext characters on the outer disk. To decrypt this ciphertext, the recipient first identifies the indicator from the first character in the ciphertext, aligns the agreed-upon pointer on his or her inner disk with this indicator, and reads in order the plaintext characters on the outer disk that correspond to the remaining ciphertext characters on the inner disk.

Example 7.2 Using an Alberti cipher with pointer k and indicator E, the plaintext 21 APRIL 43 BC encrypts to Efha& yprbd ce.   □

Since a single indicator, and thus a single cipher alphabet, was used through the entire encryption in Example 7.2, the cipher in this example is monoalphabetic, without really any more security than Caesar's cipher that had been created at least 15 centuries earlier. Alberti's truly progressive idea was to rotate the inner disk to different places during the encryption of a message, so that the indicator, and thus the cipher alphabet, could be different at different places in the ciphertext. Of course, the recipient of a ciphertext must be able to identify new indicators and where to begin using them. Alberti accomplished this by having the originator include each new indicator in the ciphertext at the point where it began being used.

Example 7.3 Consider an Alberti cipher with pointer k and indicators E for the first four plaintext characters, F for the next three plaintext characters, and L for the rest of the plaintext. For reference, the cipher alphabets that result from pointer k with indicators E, F, and L are given in the following table.

Outer disk:	A B C D E F G I L M N O P Q R S T V X Z 1 2 3 4
Indicator E:	a c e g k l n p r t v z & x y s o m q i h f d b
Indicator F:	b a c e g k l n p r t v z & x y s o m q i h f d
Indicator L:	h f d b a c e g k l n p r t v z & x y s o m q i

Using this cipher, the plaintext 21 APRIL 43 BC encrypts to `Efha& FxnpL iqfd`.   □

Alberti's creation of the first polyalphabetic cipher was enough to earn his position in the annals of cryptology. However, the polyalphabetic cipher was not Alberti's only progressive idea about cryptology. Alberti included the digits 1 through 4 on the outer disk of his wheel not just so that these digits could be used in plaintexts, but also so that common words in plaintexts could be replaced by these digits in two-, three-, and four-digit blocks. This allowed users to create codes through which up to 336 common words could be replaced by blocks of these digits. This feature of Alberti ciphers is an example of a *nomenclator*, and increases the security of the system.

Example 7.4 Consider an Alberti cipher with pointer k and indicators E for the first 11 plaintext characters, Z for the next 10 plaintext characters, and P for the rest of the plaintext. For reference, the cipher alphabets that result from pointer k with indicators E, Z, and P are in the following table.

Outer disk:	A B C D E F G I L M N O P Q R S T V X Z 1 2 3 4
Indicator E:	a c e g k l n p r t v z & x y s o m q i h f d b
Indicator Z:	t v z & x y s o m q i h f d b a c e g k l n p r
Indicator P:	o m q i h f d b a c e g k l n p r t v z & x y s

For convenience, we will also use the following double letter substitutions for the six missing letters on Alberti's outer disk: FF for H, II for J, QQ for K, VV for U, XX for W, and ZZ for Y. We will also use the following nomenclator substitutions: 1332 for IF, and 41 for MEN. Now consider the plaintext MEN CAN DO ALL THINGS IF THEY WILL.[1] Applying the double

---

[1] Leon Battista Alberti (1404–1472), quote.

letter and nomenclator substitutions to this plaintext yields 41 CAN DO ALL TFFINGS 1332 TFFEZZ XXILL, which then encrypts to Ebhea vgzar roZyy oisal ppnPr ffhzz vvbaa.   □

As with his idea of a polyalphabetic cipher, Alberti's original idea of encrypting nomenclator substitutions was so progressive that it was not put into large-scale practice for hundreds of years. Because of this, and the fact that his writings on cryptology were not published until almost a hundred years after his death, Alberti is a recognized historical figure mostly for his noncryptologic talents rather than his cryptologic ingenuity. However, this may also be attributable to the level to which his numerous other talents were developed and acknowledged during his lifetime.

7.1.1 Exercises

1. Consider an Alberti cipher with pointer k, indicators G for the first 10 plaintext characters and M for the rest of the plaintext, double letter substitutions FF for H, II for J, QQ for K, VV for U, XX for W, and ZZ for Y for the six missing letters on Alberti's outer disk, and no nomenclator substitutions. Use this cipher to encrypt the following plaintexts.

   (a)* DE PICTURA

   (b) PHILODOXUS

   (c)* DESCRIPTIO URBIS ROMAE

   (d) HYPNEROTOMACHIA POLIPHILI

2. Consider an Alberti cipher with pointer k, indicators G for the first 10 plaintext characters, M for the next five plaintext characters, and N for the rest of the plaintext, double letter substitutions FF for H, II for J, QQ for K, VV for U, XX for W, and ZZ for Y for the six missing letters on Alberti's outer disk, and nomenclator substitutions 4324 for IS, and 213 for WE. Use this cipher to encrypt the following plaintexts.

   (a)* CLIMATE IS WHAT WE EXPECT.

   (b) WEATHER IS WHAT WE GET.[2]

   (c) MY FIRST WISH IS TO SEE THIS PLAGUE OF MANKIND, WAR, BANISHED FROM THE EARTH.[3]

   (d) ONE OF THE MOST POWERFUL INSTRUMENTS OF OUR RISING PROSPERITY IS THAT WE ARE AT ALL TIMES READY FOR WAR.[4]

---

[2]Mark Twain (1835–1910), quote.
[3]George Washington (1732–1799), quote.
[4]George Washington, quote.

3. Consider an Alberti cipher with pointer k, double letter substitutions FF for H, II for J, QQ for K, VV for U, XX for W, and ZZ for Y for the six missing letters on Alberti's outer disk, and nomenclator substitutions 12 for AND, 1423 for CAN, 234 for GOVERNMENT, 1223 for NEWSPAPERS, and 21 for THE. Decrypt the following ciphertexts, which were formed using this cipher.

   (a) Gzzlr crexx lxiqn drkss dke

   (b) Googg laggi qMdbi asqom aabit

   (c) Gqiiq bnMel dsqoN otdd[5]

   (d) Giqpd roogg t&edc xrtyg gMelc izigg evhbz zbtbd &&fiz bdzaa ilNsy gqk&& bblrd qftkl vbbck aizzv ysso[6]

   (e) Gooe& elyne gyytp eytce aMedb xxaab zaabt xxbva an&&g daai& biNso m&&cv bblPt tr&xx ygn&x xyvvb rffgt troxy sbpff gttai egrff hpbro rhocg cherr gknhf hnx&a orrhn[7]

4. Create an Alberti cipher with two indicators, double letter substitutions for the six missing letters on Alberti's outer disk, and no nomenclator substitutions. Use this cipher to encrypt a plaintext of your choice with at least 10 characters that allows both indicators and at least one double letter substitution to be used.

5. Create an Alberti cipher with three indicators, double letter substitutions for the six missing letters on Alberti's outer disk, and two nomenclator substitutions. Use this cipher to encrypt a plaintext of your choice with at least 20 characters that allows all three indicators, at least two double letter substitutions, and both nomenclator substitutions to be used.

6. Did the German Enigma machines described in Chapter 4 produce ciphers that were monoalphabetic or polyalphabetic? Explain your answer completely.

7. Find an English translation of Alberti's own description of his type of cipher, and write a summary of this description.

8. Find some information about a type of polyalphabetic cipher created by German author Johannes Trithemius, including how it worked, and write a summary of your findings.

---

[5]Mark Twain, quote.
[6]Thomas Jefferson (1743–1826), quote.
[7]Thomas Jefferson, quote.

7.2 Vigenère Ciphers

A second type of polyalphabetic cipher that we will present in this chapter
is attributed to Frenchman Blaise de Vigenère, who holds his own lofty position in the annals of cryptology. Born to a common family about fifty years
after Alberti's death, Vigenère was also talented in many areas. While not
the equal of Alberti, Vigenère wrote on such topics as astronomy, alchemy,
and religion. Vigenère also wrote on the subject of cryptology, and in these
writings described his own type of polyalphabetic cipher. Centuries later,
scholars, probably mistakenly, attached Vigenère's name to a different type
of polyalphabetic cipher, one that in *The Codebreakers* [13] David Kahn
refers to as "probably the most famous cipher system of all time." In this
section, we will present both of these types of ciphers, the one that Vigenère
described and the "most famous" one to which his name is attached.

7.2.1 Vigenère Autokey Ciphers

The type of polyalphabetic cipher that Vigenère described in his own writings uses a rectangular array of letters referred to as the *Vigenère square*,
shown in Table 7.1.

	(Plaintext Letter)																									
	A	B	C	D	E	F	G	H	I	J	K	L	M	N	O	P	Q	R	S	T	U	V	W	X	Y	Z
A	A	B	C	D	E	F	G	H	I	J	K	L	M	N	O	P	Q	R	S	T	U	V	W	X	Y	Z
B	B	C	D	E	F	G	H	I	J	K	L	M	N	O	P	Q	R	S	T	U	V	W	X	Y	Z	A
C	C	D	E	F	G	H	I	J	K	L	M	N	O	P	Q	R	S	T	U	V	W	X	Y	Z	A	B
D	D	E	F	G	H	I	J	K	L	M	N	O	P	Q	R	S	T	U	V	W	X	Y	Z	A	B	C
E	E	F	G	H	I	J	K	L	M	N	O	P	Q	R	S	T	U	V	W	X	Y	Z	A	B	C	D
F	F	G	H	I	J	K	L	M	N	O	P	Q	R	S	T	U	V	W	X	Y	Z	A	B	C	D	E
G	G	H	I	J	K	L	M	N	O	P	Q	R	S	T	U	V	W	X	Y	Z	A	B	C	D	E	F
H	H	I	J	K	L	M	N	O	P	Q	R	S	T	U	V	W	X	Y	Z	A	B	C	D	E	F	G
I	I	J	K	L	M	N	O	P	Q	R	S	T	U	V	W	X	Y	Z	A	B	C	D	E	F	G	H
J	J	K	L	M	N	O	P	Q	R	S	T	U	V	W	X	Y	Z	A	B	C	D	E	F	G	H	I
K	K	L	M	N	O	P	Q	R	S	T	U	V	W	X	Y	Z	A	B	C	D	E	F	G	H	I	J
L	L	M	N	O	P	Q	R	S	T	U	V	W	X	Y	Z	A	B	C	D	E	F	G	H	I	J	K
M	M	N	O	P	Q	R	S	T	U	V	W	X	Y	Z	A	B	C	D	E	F	G	H	I	J	K	L
N	N	O	P	Q	R	S	T	U	V	W	X	Y	Z	A	B	C	D	E	F	G	H	I	J	K	L	M
O	O	P	Q	R	S	T	U	V	W	X	Y	Z	A	B	C	D	E	F	G	H	I	J	K	L	M	N
P	P	Q	R	S	T	U	V	W	X	Y	Z	A	B	C	D	E	F	G	H	I	J	K	L	M	N	O
Q	Q	R	S	T	U	V	W	X	Y	Z	A	B	C	D	E	F	G	H	I	J	K	L	M	N	O	P
R	R	S	T	U	V	W	X	Y	Z	A	B	C	D	E	F	G	H	I	J	K	L	M	N	O	P	Q
S	S	T	U	V	W	X	Y	Z	A	B	C	D	E	F	G	H	I	J	K	L	M	N	O	P	Q	R
T	T	U	V	W	X	Y	Z	A	B	C	D	E	F	G	H	I	J	K	L	M	N	O	P	Q	R	S
U	U	V	W	X	Y	Z	A	B	C	D	E	F	G	H	I	J	K	L	M	N	O	P	Q	R	S	T
V	V	W	X	Y	Z	A	B	C	D	E	F	G	H	I	J	K	L	M	N	O	P	Q	R	S	T	U
W	W	X	Y	Z	A	B	C	D	E	F	G	H	I	J	K	L	M	N	O	P	Q	R	S	T	U	V
X	X	Y	Z	A	B	C	D	E	F	G	H	I	J	K	L	M	N	O	P	Q	R	S	T	U	V	W
Y	Y	Z	A	B	C	D	E	F	G	H	I	J	K	L	M	N	O	P	Q	R	S	T	U	V	W	X
Z	Z	A	B	C	D	E	F	G	H	I	J	K	L	M	N	O	P	Q	R	S	T	U	V	W	X	Y

Table 7.1 The Vigenère square.

A version of the Vigenère square (with letters appropriate for the language and time) actually first appeared in the writings on cryptology by German author Johannes Trithemius. The square should thus be named for Trithemius; however, it is always named in honor of Vigenère.

The top row of the Vigenère square consists of the letters A through Z, in order from left to right, representing plaintext letters. The leftmost column of the square also consists of the letters A through Z, in order from top to bottom, representing key letters. With these plaintext letters viewed as column labels and key letters viewed as row labels, the letters in the inner part of the square represent ciphertext letters that correspond to pairs of plaintext and key letters, with the ciphertext letter that corresponds to a particular pair of plaintext and key letters being the letter in the inner part of the square where the column labeled with the plaintext letter intersects the row labeled with the key letter.

We will refer to the type of cipher that Vigenère described in his own writings as a Vigenère *autokey* cipher. This type of cipher requires the originator and intended recipient of a message to agree upon only a single key letter, called the *priming* key. Subsequent key letters are then the plaintext letters, with the first plaintext letter being the second key letter, the second plaintext letter being the third key letter, and so on, continuing through the next-to-last plaintext letter, which is the last key letter.

Example 7.5 Consider a Vigenère autokey cipher with priming key P. Using this cipher, the plaintext A LOVING HEART IS THE TRUEST WISDOM[8] encrypts as follows. (For example, the third ciphertext letter Z is the letter in the inner part of the Vigenère square where the column labeled with the third plaintext letter O intersects the row labeled with the third key letter L.)

Plain: A L O V I N G H E A R T I S T H E T R U E S T W I S D O M
Key: P A L O V I N G H E A R T I S T H E T R U E S T W I S D O
Cipher: P L Z J D V T N L E R K B A L A L X K L Y W L P E A V R A

Thus, the ciphertext is PLZJD VTNLE RKBAL ALXKL YWLPE AVRA. □

To decrypt a ciphertext that was formed using a Vigenère autokey cipher, for a particular pair of ciphertext and key letters, we go to the row of the Vigenère square labeled with the key letter, and then find the ciphertext letter in this row. The label on the column in which this ciphertext letter appears is the corresponding plaintext letter. When the decryption of a message is begun, the full list of key letters will not be known, since each key letter after the first is the previous plaintext letter. Thus, the key letters, like the plaintext letters, must be determined one at a time.

---

[8]Charles Dickens (1812–1870), quote.

Example 7.6 Consider the ciphertext TFFRR DDVJU TFCUK, which was formed using a Vigenère autokey cipher with priming key P. To decrypt this ciphertext, we begin with the first pair of ciphertext and key letters, T and P, respectively. We go to the row of the Vigenère square labeled with the key letter P, and then find the ciphertext letter T in this row. The label on the column in which this ciphertext letter appears is E, which is thus the first plaintext letter.

> **Plain:** E
> **Key:** P
> **Cipher:** T F F R R D D V J U T F C U K

Since the first plaintext letter is E, E is also the second key letter. So next we go to the row of the Vigenère square labeled with E, and then find the second ciphertext letter F in this row. The label on the column in which this ciphertext letter appears is B, which is thus the second plaintext letter.

> **Plain:** E B
> **Key:** P E
> **Cipher:** T F F R R D D V J U T F C U K

Continuing in this manner yields the following full decryption.

> **Plain:** E B E N E Z E R S C R O O G E
> **Key:** P E B E N E Z E R S C R O O G
> **Cipher:** T F F R R D D V J U T F C U K

Thus, the plaintext is EBENEZER SCROOGE. □

Vigenère autokey ciphers are polyalphabetic because the cipher alphabet changes throughout the encryption process. As with all polyalphabetic ciphers, the changing cipher alphabets typically cause letter frequencies in ciphertexts to be more evenly distributed than in ciphertexts formed using monoalphabetic ciphers, which generally make Vigenère autokey ciphers harder to break than monoalphabetic ciphers. On the other hand, since, with a Vigenère autokey cipher, the plaintext letters dictate the key letters, some key letters will be used more often than others. This could potentially lower the difficulty in breaking the cipher when compared to other types of polyalphabetic ciphers, especially when attacking longer ciphertexts.

Scholars have differing opinions as to the strength of Vigenère autokey ciphers. For example, in *Invitation to Cryptology* [1], Thomas Barr describes Vigenère autokey ciphers as "not much more secure than a Caesar-type cipher," while in *The Codebreakers* [13], David Kahn describes Vigenère keyword ciphers, which we will consider in the rest of this section, and which are significantly more difficult to break than a Caesar-type cipher, as a "degrading" of Vigenère autokey ciphers.

7.2.2   Vigenère Keyword Ciphers

The type of cipher most frequently attributed to Vigenère is not a Vigenère autokey cipher. In fact, a description of Vigenère autokey ciphers is often not included in books on cryptology, even books on cryptologic history. The type of cipher most frequently attributed to Vigenère is one that Vigenère himself likely did not even create. Even so, we will refer to this type of cipher as a *Vigenère keyword* cipher, since it is always labeled with Vigenère's name in literature, and because Vigenère keyword ciphers operate identically to Vigenère autokey ciphers, with the exception that the key letters are dictated by the letters in one or more keywords as opposed to the plaintext.

Vigenère keyword ciphers require the originator and intended recipient of a message to agree upon one or more keywords. Encryption is then done using the Vigenère square (Table 7.1) just as in Vigenère autokey ciphers, but with the key letters determined by repeating the letters in the keyword(s) as many times as necessary until the total number of key letters matches the total number of plaintext letters.

Example 7.7 Consider a Vigenère keyword cipher with keyword TRIXIE. Using this cipher, the plaintext HAVING A PET CAN MAKE YOU HAPPY encrypts as follows. (For example, the first ciphertext letter A is the letter in the inner part of the Vigenère square where the column labeled with the first plaintext letter H intersects the row labeled with the first key letter T.)

  Plain: H A V I N G A P E T C A N M A K E Y O U H A P P Y
  Key: T R I X I E T R I X I E T R I X I E T R I X I E T
  Cipher: A R D F V K T G M Q K E G D I H M C H L P X X T R

Thus, the ciphertext is ARDFV KTGMQ KEGDI HMCHL PXXTR.   □

Example 7.8 Consider a Vigenère keyword cipher with keyword SOPHIE. Using this cipher, the ciphertext VCVZI VWSHW MGAOA SGKGC SWMXK decrypts as follows. (For example, to find the first plaintext letter, we go to the row of the Vigenère square labeled with the first key letter S, and then find the first ciphertext letter V in this row. The label on the column in which this ciphertext letter appears is D, which is thus the first plaintext letter.)

  Plain: D O G S A R E E S P E C I A L L Y G O O D P E T S
  Key: S O P H I E S O P H I E S O P H I E S O P H I E S
  Cipher: V C V Z I V W S H W M G A O A S G K G C S W M X K

Thus, the plaintext is DOGS ARE ESPECIALLY GOOD PETS.   □

Note that unlike with Vigenère autokey ciphers, when the decryption of a ciphertext that was formed using a Vigenère keyword cipher is begun, the full list of key letters will be known.

Scholars have differing opinions as to which type of cipher, Vigenère autokey ciphers or Vigenère keyword ciphers, is better. Recall that, in *The Codebreakers* [13], David Kahn describes Vigenère keyword ciphers as a "degrading" of Vigenère autokey ciphers. Kahn goes on to clarify that Vigenère keyword ciphers are "far more susceptible" to being broken than Vigenère autokey ciphers. On the other hand, in *Invitation to Cryptology* [1], Thomas Barr states that Vigenère keyword ciphers are "much more secure" than Vigenère autokey ciphers.

We agree with Barr that Vigenère keyword ciphers are stronger, for the sole reason that a brute force attack against a Vigenère autokey cipher would require testing a maximum of only 26 possible priming keys, whereas a Vigenère keyword cipher could have any number of different keywords of various lengths. The reality though in comparison of the two is that whichever is better, Vigenère keyword ciphers are presented and discussed in literature much more frequently than Vigenère autokey ciphers. This may be because Vigenère keyword ciphers were used in practice much more frequently, or because effective (and enthralling) cryptanalytic attacks against Vigenère keyword ciphers have been developed and formalized.

7.2.3 Exercises

1. Consider a Vigenère autokey cipher with priming key P.

   (a)* Use this cipher to encrypt GEROLAMO CARDANO.

   (b) Use this cipher to encrypt BLAISE DE VIGENÈRE.

   (c) Decrypt VOWJV NOBTM BALTB FPWLS G, which was formed using this cipher.

2. Consider a Vigenère autokey cipher with priming key S.

   (a)* Use this cipher to encrypt WISDOM BEGINS IN WONDER.[9]

   (b) Use this cipher to encrypt PHILOSOPHY BEGINS IN WONDER.[10]

   (c) Decrypt ABBAM HOFGQ AUVTL TFPMI HTSQR GHINM XVDMX FUE,[11] which was formed using this cipher.

3. Create a Vigenère autokey cipher and use it to encrypt a plaintext of your choice with at least 20 letters.

4. Consider a Vigenère keyword cipher with keyword EMILY.

   (a)* Use this cipher to encrypt JOHN THWAITES.

---

[9]Socrates (469–399 BCE), quote.
[10]Plato (428–347 BCE), quote.
[11]Aristotle (384–322 BCE), quote.

(b) Use this cipher to encrypt CHARLES BABBAGE.

(c) Decrypt XTMGG WUWYM JEQY, which was formed using this cipher.

5. Consider a Vigenère keyword cipher with keyword SOUTH.

(a)* Use this cipher to encrypt MANCHESTER BLUFF.

(b) Use this cipher to encrypt COMPLETE VICTORY.

(c) Decrypt UCGXY WHLBI MHCHU, which was formed using this cipher.

6. Consider a Vigenère keyword cipher with keyword VIGENÈRE.

(a)* Use this cipher to encrypt ALL THE THINGS IN THE WORLD CON-
STITUTE A CIPHER.[12]

(b) Use this cipher to encrypt ALL NATURE IS MERELY A CIPHER AND
A SECRET WRITING.[13]

(c) Decrypt OPKZR VPHZM JWCVF NZKZW JSIHN IIXVS EWVVJ HRQVE
IWXSS QRRFQ THJLR XVZKX UIPJJ ZZLRQ FWOXG VGFLX VKOTU
II,[14] which was formed using this cipher.

7. Create a Vigenère keyword cipher and use it to encrypt a plaintext
of your choice with at least 20 letters.

8. A second type of autokey cipher that Vigenère described in his own
writings is identical to the type of autokey cipher that we considered
in this section, but uses the ciphertext letters instead of the plaintext
letters to dictate the key letters. Is this second type of autokey cipher
in general more secure than, less secure than, or equally secure as the
type of autokey cipher that we considered in this section? Explain
your answer completely.

9. Find an English translation of Vigenère's own description of the type
of autokey cipher that we considered in this section, and write a
summary of this description.

10. Find some information about a type of autokey cipher created by
Italian mathematician Gerolamo Cardano, including how it worked,
and write a summary of your findings.

11. Find some information about a type of autokey cipher created by
Italian cryptologist Giovan Battista Bellaso, including how it worked,
and write a summary of your findings.

---

[12]Blaise de Vigenère (1523–1596), quote.
[13]Blaise de Vigenère, quote.
[14]Blaise de Vigenère, quote.

12. Find some information about the use of Vigenère keyword ciphers by the Confederacy during the American Civil War, and write a summary of your findings.

13. Find some information about two variants of Vigenère keyword ciphers created by British naval officer Sir Francis Beaufort, including how they worked, and write a summary of your findings.

7.3 Probability

Our goal for the rest of this chapter is to understand one of the most ingenious feats in cryptologic history—cryptanalysis of Vigenère keyword ciphers. To do this, we first need to briefly review some basics about probability.

Probability begins with the idea of a *sample space*, which is the set of all possible outcomes for an experiment. For example, for the experiment of rolling a single die and observing the number showing on the top face of the die after the roll, the sample space is $\{1, 2, 3, 4, 5, 6\}$. An *event* is a subset of a sample space. For example, for rolling a single die, some events are rolling (i.e., observing) a 5, rolling an even number, and rolling a number less than 5; these events are the subsets $\{5\}$, $\{2, 4, 6\}$, and $\{1, 2, 3, 4\}$, respectively, of the sample space.

Definition 7.1 *For an event A for an experiment, the probability $P(A)$ is a real number from 0 to 1 that gives the chance that if the experiment were performed, the outcome would be in A.*

Often all outcomes in the sample space for an experiment are equally likely to occur. For example, when considering rolling a single die, it is usually assumed that the die is fair, meaning that all of the numbers in the sample space $\{1, 2, 3, 4, 5, 6\}$ are equally likely to be rolled. As you may already know or be able to deduce, for rolling a single fair die, the probability of rolling a 5 is $\frac{1}{6}$, or, using the notation in Definition 7.1, $P(\text{rolling a 5}) = \frac{1}{6}$. Similarly, $P(\text{rolling an even number}) = \frac{3}{6} = \frac{1}{2}$, and $P(\text{rolling a number less than 5}) = \frac{4}{6} = \frac{2}{3}$. These probabilities are special cases of the following theorem.

Theorem 7.2 *If all outcomes in the sample space for an experiment are equally likely to occur, then the probability of an event A is given by the following formula.*

$$P(A) = \frac{\text{number of outcomes in } A}{\text{number of outcomes in the sample space}}$$

Example 7.9 Consider the experiment of drawing a single card from a deck of 52 standard playing cards that is shuffled, meaning that all 52 cards are equally likely to be drawn. Theorem 7.2 gives the following probabilities.

$$P(\text{drawing the ace of spades}) = \frac{1}{52} \approx 0.019$$

$$P(\text{drawing an ace}) = \frac{4}{52} \approx 0.077$$

$$P(\text{drawing a face card}^{15}) = \frac{12}{52} \approx 0.231$$

$$P(\text{drawing a spade}) = \frac{13}{52} = 0.25$$

$\square$

The probability of an event can, of course, be exactly zero or one. For example, for rolling a single fair die, $P(\text{rolling a 7}) = \frac{0}{6} = 0$, and $P(\text{rolling a number less than 7}) = \frac{6}{6} = 1$. Also, the sum of the probabilities of all outcomes in a sample space must be exactly one. For example, for rolling a single fair die, the probability of rolling each of the six individual numbers in the sample space $\{1, 2, 3, 4, 5, 6\}$ is $\frac{1}{6}$. The sum of these six probabilities is $\frac{1}{6} + \frac{1}{6} + \frac{1}{6} + \frac{1}{6} + \frac{1}{6} + \frac{1}{6} = 6 \cdot \left(\frac{1}{6}\right) = 1$.

For a pair of events for an experiment, if it is impossible for the outcome to be in both events simultaneously, then the events are said to be *mutually exclusive*. For example, for rolling a single die, it is impossible to roll both a 5 and an even number simultaneously, and thus the events of rolling a 5 and rolling an even number are mutually exclusive. On the other hand, it is possible to roll both an even number and a number less than 5 simultaneously, and thus the events of rolling an even number and rolling a number less than 5 are not mutually exclusive.

Mutually exclusive events have the handy property that the probability of one event or the other occurring can be found by adding the probabilities of either event occurring. For example, for rolling a single fair die, since the events of rolling a 5 and rolling an even number are mutually exclusive, we have the following.

$P(\text{rolling a 5 or an even number})$

$$= P(\text{rolling a 5}) + P(\text{rolling an even number})$$

$$= \frac{1}{6} + \frac{3}{6} = \frac{4}{6} = \frac{2}{3}$$

This probability is a special case of the following theorem.

---

[15]In a deck of cards, the *face cards* are the jacks, queens, and kings.

Theorem 7.3 *If A and B are events that are mutually exclusive, then* $P(A \text{ or } B) = P(A) + P(B)$.

Example 7.10 Consider drawing a single card from a shuffled deck of 52 standard playing cards. Since the events of drawing an ace and drawing a face card are mutually exclusive, as are the events of drawing a spade and drawing a heart, Theorem 7.3 gives the following probabilities.

P(drawing an ace or a face card)

$$= P(\text{drawing an ace}) + P(\text{drawing a face card})$$

$$= \frac{4}{52} + \frac{12}{52} = \frac{16}{52} \approx 0.308$$

P(drawing a spade or a heart)

$$= P(\text{drawing a spade}) + P(\text{drawing a heart})$$

$$= \frac{13}{52} + \frac{13}{52} = \frac{26}{52} = 0.5$$

□

Theorem 7.3 actually works for any number of mutually exclusive events. For example, if A, B, and C are mutually exclusive events (meaning that it is impossible for the outcome to be in any pair of the events simultaneously), then $P(A \text{ or } B \text{ or } C) = P(A) + P(B) + P(C)$.

Example 7.11 For drawing a single card from a shuffled deck of 52 standard playing cards, the event of drawing a face card is equivalent to the mutually exclusive events of drawing a jack, drawing a queen, and drawing a king. Theorem 7.3 gives the following probability, which is consistent with one given in Example 7.9.

P(drawing a face card)

$$= P(\text{drawing a jack or a queen or a king})$$

$$= P(\text{drawing a jack}) + P(\text{drawing a queen}) + P(\text{drawing a king})$$

$$= \frac{4}{52} + \frac{4}{52} + \frac{4}{52} = 3 \cdot \left(\frac{4}{52}\right) = \frac{12}{52} \approx 0.231$$

□

Consider now a pair of events for an experiment for which it is possible for the outcome to be in both events simultaneously. For example, for rolling a single fair die, suppose we somehow know the roll will be an odd number, and we wish to find the probability that the roll will be a 5. Since we know the roll will be odd, the result must be 1, 3, or 5, and thus the probability of rolling a 5 is $\frac{1}{3}$. This is an example of *conditional* probability.

Definition 7.4 *For a pair of events A and B for an experiment, the conditional probability P(B|A), called the "probability of B given A," is the probability that the outcome would be in B if it is known that the outcome will be in A.*

Example 7.12 Consider drawing a pair of cards from a shuffled deck of 52 standard playing cards without replacement (i.e., without replacing the first card in the deck before drawing the second), and suppose we wish to find the probability that the second card drawn will be an ace given that the first was an ace. If the first card drawn was an ace, then when the second is drawn there will be three aces and 51 total cards in the deck. Thus, the conditional probability that the second card drawn will be an ace given that the first was an ace is $\frac{3}{51} \approx 0.059$, or, using the notation in Definition 7.4, $P(\text{second card is an ace}|\text{first card is an ace}) = \frac{3}{51} \approx 0.059$. Similarly, $P(\text{second card is a face card}|\text{first card is a face card}) = \frac{11}{51} \approx 0.216$, and $P(\text{second card is a heart}|\text{first card is a spade}) = \frac{13}{51} \approx 0.255$. $\square$

Sometimes for a pair of events for an experiment for which it is possible for the outcome to be in both events simultaneously, knowing that an outcome is in the first event has no effect on its chances of being in the second event. For example, consider rolling a single fair die twice, and suppose we wish to find the probability that the second roll will be a 5 given that the first was a 5. In this case, the first roll being a 5 has no effect on the result of the second, and so $P(\text{second roll is a 5}|\text{first roll is a 5}) = \frac{1}{6}$, the same as the probability that the second roll would be a 5 given no information about the first. We say that these two events of rolling a 5 on the first roll and rolling a 5 on the second are *independent*, since the result of each has no effect on the result of the other.

Example 7.13 Consider drawing a pair of cards from a shuffled deck of 52 standard playing cards with replacement (i.e., replacing the first card in the deck and reshuffling before drawing the second). For this experiment, each pair of events we considered in Example 7.12 is independent. For example, suppose we wish to find the probability that the second card drawn will be an ace given that the first was an ace. Since the first card will be replaced and the deck reshuffled before the second is drawn, $P(\text{second card is an ace}|\text{first card is an ace}) = \frac{4}{52} \approx 0.077$. Similarly, $P(\text{second card is a face card}|\text{first card is a face card}) = \frac{12}{52} \approx 0.231$, and $P(\text{second card is a heart}|\text{first card is a spade}) = \frac{13}{52} = 0.25$. $\square$

For a pair of events for an experiment, to find the probability that the outcome will be in both events, we use the *multiplication principle of probability*. This principle states that for a pair of events A and B for an experiment, the probability that the outcome will be in both A and B is given by the following formula.

$$P(A \text{ and } B) = P(A) \cdot P(B|A)$$

Example 7.14 Consider drawing a pair of cards from a shuffled deck of 52 standard playing cards without replacement. For the probability that both cards will be aces, the multiplication principle of probability gives the following.

$P(\text{both cards are aces})$

$= \ P(\text{first card is an ace}) \cdot P(\text{second card is an ace}|\text{first card is an ace})$

$= \ \dfrac{4}{52} \cdot \dfrac{3}{51} \ = \ \dfrac{12}{2652} \ \approx \ 0.005$

Similarly, the multiplication principle of probability gives the following.

$P(\text{both cards are face cards}) \ = \ \dfrac{12}{52} \cdot \dfrac{11}{51} \ = \ \dfrac{132}{2652} \ \approx \ 0.050$

$P(\text{first card is a spade and second card is a heart})$

$$= \ \dfrac{13}{52} \cdot \dfrac{13}{51} \ = \ \dfrac{169}{2652} \ \approx \ 0.064$$

$\square$

Example 7.15 Consider drawing a pair of cards from a shuffled deck of 52 standard playing cards without replacement, and suppose we wish to find the probability that the two cards drawn will be a spade and a heart. In Example 7.14, we used the multiplication principle of probability to find the probability that the first card drawn will be a spade and the second a heart. However, if the two cards drawn are a spade and a heart, it could also be the case that the first card drawn is a heart and the second a spade. This gives two mutually exclusive events (a spade followed by a heart, and a heart followed by a spade), for which Theorem 7.3 gives the following probability.

$P(\text{a spade and a heart})$

$= \ P(\text{spade followed by heart}) + P(\text{heart followed by spade})$

$= \ \dfrac{13}{52} \cdot \dfrac{13}{51} + \dfrac{13}{52} \cdot \dfrac{13}{51} \ = \ \dfrac{169}{2652} + \dfrac{169}{2652} \ = \ \dfrac{338}{2652} \ \approx \ 0.127$

Next, suppose we wish to find the probability that the two cards drawn will both be face cards of the same rank (i.e., both are jacks, both are queens, or both are kings). Since the events that both cards are jacks, both are queens, and both are kings are mutually exclusive, Theorem 7.3 gives the following probability.

P(face cards of the same rank)

$$= \quad P(\text{both are jacks}) + P(\text{both are queens}) + P(\text{both are kings})$$

$$= \quad \frac{4}{52} \cdot \frac{3}{51} + \frac{4}{52} \cdot \frac{3}{51} + \frac{4}{52} \cdot \frac{3}{51}$$

$$= \quad 3\left(\frac{4}{52} \cdot \frac{3}{51}\right) \quad = \quad 3\left(\frac{12}{2652}\right) \quad = \quad \frac{36}{2652} \approx 0.014$$

Similarly, suppose we wish to find the probability that the two cards drawn will both be any cards of the same rank (i.e., both cards are aces, both cards are twos, ... , both cards are queens, or both cards are kings). Since the events that both cards are aces, both cards are twos, ... , both cards are queens, and both cards are kings are mutually exclusive, Theorem 7.3 gives the following probability.

P(any cards of the same rank)

$$= \quad P(\text{both are aces}) + P(\text{both are twos}) + \cdots + P(\text{both are kings})$$

$$= \quad \frac{4}{52} \cdot \frac{3}{51} + \frac{4}{52} \cdot \frac{3}{51} + \cdots + \frac{4}{52} \cdot \frac{3}{51}$$

$$= \quad 13\left(\frac{4}{52} \cdot \frac{3}{51}\right) \quad = \quad 13\left(\frac{12}{2652}\right) \quad = \quad \frac{156}{2652} \approx 0.059$$

$\square$

Finally, independent events have the handy property that the probability of both events occurring can be found by multiplying the probabilities of either event occurring. The reason for this is because if the events A and B are independent, then $P(B|A) = P(B)$, and the multiplication principle of probability becomes the following.

$$P(A \text{ and } B) = P(A) \cdot P(B)$$

For example, for rolling a single fair die twice, $P(\text{first roll is a 5}) = \frac{1}{6}$, and $P(\text{second roll is a 5}) = \frac{1}{6}$. Since the events of rolling a 5 on the first roll and rolling a 5 on the second are independent, this formula gives $P(\text{both rolls are 5s}) = \frac{1}{6} \cdot \frac{1}{6} = \frac{1}{36}$. Similarly, for rolling a single fair die twice, $P(\text{first roll is even and second roll is less than 5}) = \frac{1}{2} \cdot \frac{2}{3} = \frac{2}{6} = \frac{1}{3}$.

7.3.1 Exercises

1. For rolling a single fair die, find the probability of the following.

    (a)* Not rolling a 6

    (b) Not rolling an odd number

    (c)* Rolling an odd number or a number less than 6

    (d) Rolling an even number or a number less than 6

    (e)* Rolling an odd number given that the number rolled will be greater than 1

    (f) Rolling an even number given that the number rolled will be greater than 1

2. For rolling a single fair die twice, find the probability of the following.

    (a)* Rolling *boxcars* (i.e., both rolls are 6s).

    (b) Rolling *snake eyes* (i.e., both rolls are 1s).

    (c)* The first roll is a 6, and the second is a 1.

    (d) The results of the two rolls are a 6 and a 1 (in either order).

    (e)* Rolling *doubles* (i.e., both rolls are the same number).

    (f) Not rolling doubles.

3. Consider the experiment of rolling a single fair die twice and observing the sum of the results of the two rolls.

    (a)* Determine whether 14 is in the sample space for this experiment, and justify your answer.

    (b) Find the sample space for this experiment.

    (c)* Find the probability that the sum will be 11.

    (d) Find the probability that the sum will be at least 11.

    (e)* Find the probability that the sum will be 11 given that the first roll is a 6.

    (f) Find the probability that the sum will be at least 11 given that the first roll is a 6.

4. For drawing a single card from a shuffled deck of 52 standard playing cards, find the probability of the following.

    (a)* Not drawing an ace

    (b) Not drawing a spade

    (c)* Drawing a king or a queen

    (d) Drawing an ace or a face card

    (e)* Drawing a spade given that the card drawn will not be a heart

    (f) Drawing a face card given that the card drawn will not be an ace

5. For drawing a pair of cards from a shuffled deck of 52 standard playing cards without replacement, find the probability of the following.

    (a)* The first card drawn is a king, and the second is a queen.

    (b) The two cards drawn are a king and a queen (in either order).

    (c)* The second card drawn will be a face card given that the first is a king.

    (d) The second card drawn will not be a face card given that the first is a king.

    (e)* Drawing a pair (i.e., both cards are of the same rank).

    (f) Not drawing a pair.

6.* Repeat Exercise 5 for drawing a pair of cards from a shuffled deck of 52 standard playing cards with replacement.

For Exercises 7–9, the card game pinochle is played with a deck of 48 cards that includes only nines, tens, jacks, queens, kings, and aces, each in all four suits, and with two of every type of card in each suit. For example, the 48 cards include two nines of clubs, two jacks of diamonds, and two aces of spades, but no threes, sixes, or eights.

7.* Repeat Exercise 4 for drawing a single card from a shuffled deck of 48 pinochle playing cards.

8.* Repeat Exercise 5 for drawing a pair of cards from a shuffled deck of 48 pinochle playing cards without replacement.

9.* Repeat Exercise 5 for drawing a pair of cards from a shuffled deck of 48 pinochle playing cards with replacement.

10. From a group of 100 college students, of which 10 are freshman, 25 are sophomores, 30 are juniors, 20 are seniors, and 15 are graduate students, suppose a random drawing is to be held to select the recipient of a scholarship. Find the probability of the following.

    (a)* The scholarship will not be awarded to a freshman.

(b) The scholarship will not be awarded to a freshman or a sophomore.

(c)* The scholarship will be awarded to a senior or a graduate student.

(d) The scholarship will be awarded to a junior, a senior, or a graduate student.

(e)* The scholarship will be awarded to a senior given that it will not be awarded to a freshman.

(f) The scholarship will not be awarded to a senior given that it will not be awarded to a freshman.

11. From a group of 100 college students, of which 10 are freshman, 25 are sophomores, 30 are juniors, 20 are seniors, and 15 are graduate students, suppose a random drawing is to be held to select the recipients of a pair of scholarships. If the recipient of the second scholarship cannot be the same as the recipient of the first (i.e., the recipient of the first scholarship will be removed from the group before the second recipient is selected), find the probability of the following.

(a)* The first scholarship will be awarded to a senior, and the second to a freshman.

(b) The two scholarships will be awarded to a senior and a freshman (in either order).

(c)* Neither scholarship will be awarded to a senior.

(d) The two scholarships will be awarded to students at the same class level.

(e)* The two scholarships will be awarded to different students.

(f) The two scholarships will be awarded to the same student.

12.* Repeat Exercise 11 assuming the recipient of the second scholarship can be the same as the recipient of the first (i.e., the recipient of the first scholarship will not be removed from the group before the second recipient is selected).

7.4 The Friedman Test

In order to break a Vigenère keyword cipher, the keyword for the cipher must be determined. A first step in determining the keyword is to estimate its length, for which several methods exist. We will consider three such methods, the first being the Friedman test.

The Friedman test was developed in the 1920s by William Friedman, the Dean of American Cryptology. In *The Codebreakers* [13], David Kahn calls Friedman the "greatest of all cryptologists." Originally trained as a geneticist, Friedman was drawn to cryptology while courting his wife, Elizebeth, herself an accomplished cryptologist and pioneer of American cryptology. William Friedman quickly became a lead researcher for a succession of U.S. cryptologic agencies, including the Army's Signals Intelligence Service, the Armed Forces Security Agency, and the National Security Agency.

Friedman authored a number of seminal papers and books on cryptology, and coined the term *cryptanalysis* in his book *Elements of Cryptanalysis* [8]. In one of Friedman's earliest papers, *The Index of Coincidence and Its Applications in Cryptography* [9], which he wrote while Director of the Department of Codes and Ciphers at Riverbank Laboratories, he introduced the *index of coincidence*, an important tool in cryptanalysis. In *The Codebreakers* [13], David Kahn describes this paper as "the most important single publication in cryptology," and although the index of coincidence is not Friedman's most technically advanced original idea, Friedman himself called it his single greatest creation. In this section, we will present how Friedman's index of coincidence can help determine whether a given ciphertext was formed using a Vigenère keyword cipher, and if so be used to estimate the length of the keyword for the cipher.

7.4.1   The Index of Coincidence

In Section 2.2, we considered the frequencies with which the 26 letters in our alphabet occur in ordinary English. For convenience, these frequencies are shown again in Table 7.2 on page 217.

In decimal form, these frequencies can be considered as probabilities. For example, the first numeric entry in Table 7.2 indicates that if a single letter is selected at random from a very large sample of ordinary English text, the probability that the letter will be an A should be 0.0817. If the sample of text is large enough, then if a second letter is selected at random, the frequencies in Table 7.2 should still be the probabilities for which letter will be selected. For example, if a second letter is selected at random, the probability that the letter will be an A should still be 0.0817.

Example 7.16 Consider selecting a pair of letters at random from a very large sample of ordinary English text. The probability that both letters will be As should be as follows.

P(both letters are As)

$$= \quad P(\text{first letter is an A}) \cdot P(\text{second letter is an A})$$

$$= \quad 0.0817 \cdot 0.0817 \quad = \quad (0.0817)^2 \quad \approx \quad 0.0076$$

Letter	Frequency	Letter	Frequency
A	8.17%	N	6.75%
B	1.49%	O	7.51%
C	2.78%	P	1.93%
D	4.25%	Q	0.10%
E	12.70%	R	5.99%
F	2.23%	S	6.33%
G	2.02%	T	9.06%
H	6.09%	U	2.76%
I	6.97%	V	0.98%
J	0.15%	W	2.36%
K	0.77%	X	0.15%
L	4.03%	Y	1.97%
M	2.41%	Z	0.07%

Table 7.2   Letter frequencies in ordinary English.

Similarly, the probability that both letters will be Bs should be given by $(0.0149)^2 \approx 0.0002$. The probability that the letters will be any matching pair should be as follows.

P(any matching pair of letters)

$$= \quad P(\text{both are As}) + P(\text{both are Bs}) + \cdots + P(\text{both are Zs})$$

$$= \quad (0.0817)^2 + (0.0149)^2 + \cdots + (0.0007)^2 \quad \approx \quad 0.0655$$

$\square$

If the letter frequencies in a particular ciphertext are highly varied, then the cipher that produced the ciphertext is more likely to be monoalphabetic than polyalphabetic. On the other hand, if the letter frequencies are more evenly distributed, then the cipher is more likely to be polyalphabetic. The *index of coincidence* is a number that measures variation in letter frequencies, which for a ciphertext indicates whether the cipher that produced the ciphertext is more likely to be monoalphabetic or polyalphabetic. More specifically, the index of coincidence for a piece of text is the probability that a pair of letters selected at random from the text will match. In Example 7.16, we found that the index of coincidence for the ordinary English language is approximately 0.0655.

Now consider a mythical language in which the frequencies with which the 26 letters in our alphabet occur are all identical. From a very large sample of text written in this language, for any first letter selected at random, the probability that a second letter selected at random will match the first should be $\frac{1}{26} \approx 0.0385$. Thus, the index of coincidence for this mythical language is approximately 0.0385.

Monoalphabetic ciphers preserve letter frequencies from plaintexts into ciphertexts. As a result of this, the index of coincidence for a large ciphertext formed using a monoalphabetic cipher should be closer to 0.0655 than 0.0385. Polyalphabetic ciphers typically cause letter frequencies in ciphertexts to be more evenly distributed. Thus, the index of coincidence for a large ciphertext formed using a polyalphabetic cipher should be closer to 0.0385.

In order to find the index of coincidence for a given ciphertext, let n_1, n_2, ... , n_{26} be the frequencies with which the letters A, B, ... , Z occur in the ciphertext, respectively, and let $n = n_1 + n_2 + \cdots + n_{26}$ be the total number of letters in the entire ciphertext. If two letters are selected at random without replacement from the ciphertext, then the probability that both letters will be As, Bs, ... , Ys, or Zs will be given by the following formulas.

$$P(\text{both letters are As}) \quad = \quad \frac{n_1}{n} \cdot \frac{n_1 - 1}{n - 1} \quad = \quad \frac{n_1(n_1 - 1)}{n(n - 1)}$$

$$P(\text{both letters are Bs}) \quad = \quad \frac{n_2}{n} \cdot \frac{n_2 - 1}{n - 1} \quad = \quad \frac{n_2(n_2 - 1)}{n(n - 1)}$$

$$\vdots$$

$$P(\text{both letters are Zs}) \quad = \quad \frac{n_{26}}{n} \cdot \frac{n_{26} - 1}{n - 1} \quad = \quad \frac{n_{26}(n_{26} - 1)}{n(n - 1)}$$

The probability that the letters will be any matching pair, which is the index of coincidence for the ciphertext, should be as follows.

$P(\text{any matching pair of letters})$

$$= \quad P(\text{both are As}) + P(\text{both are Bs}) + \cdots + P(\text{both are Zs})$$

$$= \quad \frac{n_1(n_1 - 1)}{n(n - 1)} + \frac{n_2(n_2 - 1)}{n(n - 1)} + \cdots + \frac{n_{26}(n_{26} - 1)}{n(n - 1)}$$

$$= \quad \frac{1}{n(n - 1)} \left(n_1(n_1 - 1) + n_2(n_2 - 1) + \cdots + n_{26}(n_{26} - 1) \right)$$

In summary, for a particular ciphertext, the index of coincidence I, which gives the probability that a pair of letters selected at random from the ciphertext will match, is given by the following formula.

$$I = \frac{1}{n(n - 1)} \left(n_1(n_1 - 1) + n_2(n_2 - 1) + \cdots + n_{26}(n_{26} - 1) \right)$$

In this formula, n is the total number of letters in the entire ciphertext, and $n_1, n_2, \ldots, n_{26}$ are the number of As, Bs, $\ldots$, Zs in the ciphertext, respectively.

Example 7.17 Consider the ciphertext UZRZE GNJEN VLISE XRHLY PYEGT ESBJH JCSBP TGDYF XXBHE EIFTC CHVRK PNHWX PCTUQ TGDJH TBIPR FEMJC NHVTC FSAII IFNRE GSALH XHWZW RZXGT TVWGD HTEYX ISAGQ TCJPR SIAPT UMGZA LHXHH SOHPW CZLBR ZTCBR GHCDI QIKTO AAEFT OPYEG TENRA IALNR XLPCE PYKGP NGPRQ PIAKW XDCBZ XGPDN RWXEI FZXGJ LVOXA JTUEM BLNLQ HGPWV PEQPI AXATY ENVYJ EUEI, which was formed using a Vigenère keyword cipher. There are 269 letters in this ciphertext, and the frequency with which each particular letter occurs is shown in the following table.

Letter:	A	B	C	D	E	F	G	H	I	J	K	L	M
Count:	13	8	12	6	20	7	16	16	15	9	4	10	3
Letter:	N	O	P	Q	R	S	T	U	V	W	X	Y	Z
Count:	11	4	18	6	13	8	18	5	7	8	15	8	9

For the index of coincidence for this ciphertext, we have the following.

$$I = \frac{1}{269 \cdot 268} \left((13 \cdot 12) + (8 \cdot 7) + (12 \cdot 11) + \cdots + (9 \cdot 8) \right) \approx 0.0430$$

Since this index of coincidence is closer to 0.0385 than 0.0655, the cipher that produced this ciphertext is more likely to be polyalphabetic than monoalphabetic. In contrast, consider the ciphertext OJJOT OWDUO BIZOZ EQBUH ERIUZ GUITS WOZEN UBFSS BEUBZ DEKSQ BZOUB SQIOR EOSVY EIZER BBSRZ DWORS TUBOB SRZDW ORSTU BOIZO ZEQBU HERIU ZGUIT SWOZE NUBRO TEUMD UBZDE EOIZE RBJSR ZUSBS VZDEB SRZDW ORSTU BOJUE NKSBZ REMUS BRONV SRNQB UHERI UZGUI TSWOZ ENUBR ONVSR NUBZD EKSQB ZOUBS QIORE OSVIS QZDYE IZERB HURMU BUO, which was formed using an affine cipher. There are 258 letters in this ciphertext, and the frequency with which each particular letter occurs is shown in the following table.

Letter:	A	B	C	D	E	F	G	H	I	J	K	L	M
Count:	0	28	0	10	24	1	3	4	14	4	3	0	3
Letter:	N	O	P	Q	R	S	T	U	V	W	X	Y	Z
Count:	8	26	0	8	22	25	8	28	5	7	0	2	25

For the index of coincidence for this ciphertext, we have the following.

$$I = \frac{1}{258 \cdot 257} \left((0 \cdot (-1)) + (28 \cdot 27) + \cdots + (25 \cdot 24) \right) \approx 0.0742$$

Since this index of coincidence is closer to 0.0655 than 0.0385, the cipher that produced this ciphertext is more likely to be monoalphabetic than polyalphabetic. $\qquad\square$

7.4.2   Estimating the Keyword Length

For a ciphertext formed using a Vigenère keyword cipher, the index of coincidence can also be used to estimate the length of the keyword for the cipher. This estimate k can be found using the following formula.

$$k \approx \frac{0.0270n}{0.0655 - I + n(I - 0.0385)}$$

In this formula, n is the total number of letters in the ciphertext, and I is the index of coincidence for the ciphertext. An explanation for why this formula is true will be given after the following example.

Example 7.18 Consider the ciphertext UZRZE GNJEN VLISE XRHLY PYEGT ESBJH JCSBP TGDYF XXBHE EIFTC CHVRK PNHWX PCTUQ TGDJH TBIPR FEMJC NHVTC FSAII IFNRE GSALH XHWZW RZXGT TVWGD HTEYX ISAGQ TCJPR SIAPT UMGZA LHXHH SOHPW CZLBR ZTCBR GHCDI QIKTO AAEFT OPYEG TENRA IALNR XLPCE PYKGP NGPRQ PIAKW XDCBZ XGPDN RWXEI FZXGJ LVOXA JTUEM BLNLQ HGPWV PEQPI AXATY ENVYJ EUEI, which was formed using a Vigenère keyword cipher. There are 269 letters in this ciphertext, and in Example 7.17 we found that the index of coincidence for the ciphertext is approximately 0.0430. For an estimate for the length of the keyword for the cipher, we have the following.

$$k \approx \frac{0.0270 \cdot 269}{0.0655 - 0.0430 + 269(0.0430 - 0.0385)} \approx 5.8905.$$

Since this estimate is close to the integer 6, the most likely length of the keyword is six letters.   □

To see why this formula for estimating the length of the keyword for a Vigenère keyword cipher is true, consider a ciphertext of length n formed using a Vigenère keyword cipher with a keyword of length k. Suppose this ciphertext is separated into k groups of length $\frac{n}{k}$ each, in which the letters in each group were all encrypted using the same keyword letter. Then the letters in each group should be distributed with approximately the same frequencies as the letters in ordinary English. Consequently, if two letters were selected at random from any particular group, the probability that the letters match should be approximately 0.0655. On the other hand, if two letters were selected at random from different groups, the probability that the letters match should be approximately 0.0385.

Consider now the operation of choosing two letters from different groups. Since there are k groups of length $\frac{n}{k}$ each, the formula in Theorem 4.6 on page 71 for counting combinations gives the following.

Number of ways to choose two ciphertext

$$\text{letters from different groups} = C(k,2) \cdot \frac{n}{k} \cdot \frac{n}{k}$$

$$= \frac{k(k-1)}{2} \cdot \frac{n}{k} \cdot \frac{n}{k}$$

$$= \frac{n^2(k-1)}{2k}$$

Also, since the letters must be chosen from different groups, the portion of these combinations for which the letters will match should be approximately $0.0385 \cdot \frac{n^2(k-1)}{2k}$.

Similarly, for the operation of choosing two letters from the same group, the formula in Theorem 4.6 for counting combinations gives the following.

Number of ways to choose two ciphertext

$$\text{letters from the same group} = C\left(\frac{n}{k},2\right) \cdot k$$

$$= \frac{\frac{n}{k}\left(\frac{n}{k}-1\right)}{2} \cdot k$$

$$= \frac{1}{2} \cdot n \cdot \left(\frac{n}{k}-1\right)$$

$$= \frac{n(n-k)}{2k}$$

Also, since the letters must be chosen from the same group, the portion of these combinations for which the letters will match should be approximately $0.0655 \cdot \frac{n(n-k)}{2k}$.

Further, since the entire ciphertext is of length n, the formula in Theorem 4.6 for counting combinations gives the following.

$$\text{Number of ways to choose two ciphertext letters} = C(n,2)$$

$$= \frac{n(n-1)}{2}$$

Using all of these results and Theorem 7.2, we can express the probability that two letters chosen at random from the ciphertext will match, that is, the index of coincidence I for the ciphertext, as follows.

$$I \approx \frac{0.0385 \cdot \frac{n^2(k-1)}{2k} + 0.0655 \cdot \frac{n(n-k)}{2k}}{\frac{n(n-1)}{2}}$$

$$\approx \frac{0.0385n(k-1) + 0.0655(n-k)}{k(n-1)}$$

Solving for k in terms of n and I in this last approximation equation is left as an exercise (Exercise 6 at the end of this section), and gives the desired formula.

7.4.3 Exercises

1. Consider selecting a pair of letters at random from a very large sample of ordinary English text.

   (a)* Find the probability that the letters will be a matching pair of vowels.

   (b) Find the probability that the letters will be a matching pair of consonants.

2. Consider a ciphertext for which the frequency with which each letter occurs is shown in the following table.

Letter:	A	B	C	D	E	F	G	H	I	J	K	L	M
Count:	2	14	9	8	8	12	7	5	19	14	8	4	9
Letter:	N	O	P	Q	R	S	T	U	V	W	X	Y	Z
Count:	8	12	10	3	14	15	10	7	13	2	10	15	4

   (a) Find the index of coincidence for the ciphertext, and use it to determine whether the cipher that produced the ciphertext is more likely to be monoalphabetic or polyalphabetic.

   (b) Assuming the cipher that produced the ciphertext is a Vigenère keyword cipher, estimate the length of the keyword for the cipher.

3.* Consider a ciphertext for which the frequency with which each letter occurs is shown in the following table.

Letter:	A	B	C	D	E	F	G	H	I	J	K	L	M
Count:	8	31	17	19	0	4	5	1	7	4	0	6	14
Letter:	N	O	P	Q	R	S	T	U	V	W	X	Y	Z
Count:	10	22	5	3	6	0	18	3	19	11	2	8	19

   (a) Find the index of coincidence for the ciphertext, and use it to determine whether the cipher that produced the ciphertext is more likely to be monoalphabetic or polyalphabetic.

   (b) Assuming the cipher that produced the ciphertext is a Vigenère keyword cipher, estimate the length of the keyword for the cipher.

4. For the following lengths n and indexes of coincidence I for a ciphertext formed using a Vigenère keyword cipher, estimate the length of the keyword for the cipher.

   (a)* $n = 362$, $I = 0.0481$

(b) $n = 523$, $I = 0.0423$

(c) $n = 3568$, $I = 0.0425$

(d) $n = 5284$, $I = 0.0458$

5. For the following indexes of coincidence I for a ciphertext formed using a Vigenère keyword cipher, find an estimate for the length of the keyword for the cipher.

(a)* 0.0385

(b) 0.0655

6. Show that the formula in this section for estimating the keyword length for a Vigenère keyword cipher is correct by solving the following approximation equation for k in terms of n and I.

$$I \approx \frac{0.0385n(k-1) + 0.0655(n-k)}{k(n-1)}$$

7. The index of coincidence and keyword length formulas given in this section are for texts written in the ordinary English language. Find a source that gives the index of coincidence and keyword length formulas for another language, and write a summary of your findings.

8. Find some information about William Friedman's career in cryptology, include his background, accomplishments, and importance in its history, and write a summary of your findings.

9. Find some information about a "cryptographic system" for which William Friedman applied for a patent on July 25, 1933, and write a summary of your findings.

10. Find some information about Elizebeth Friedman's career in cryptology, include her background, accomplishments, and importance in its history, and write a summary of your findings.

11. Elizebeth Friedman played a role in the apprehension of Velvalee Dickinson, an American convicted of spying for Japan against the United States during World War II. Find some information about Dickinson's espionage operation and how it was discovered, and write a summary of your findings.

12. Find some information about William and Elizebeth Friedman's book *The Shakespearean Ciphers Examined*, including the hypotheses and conclusions they put forth in it, and write a summary of your findings.

7.5   The Kasiski Test

For a ciphertext formed using a Vigenère keyword cipher, the Friedman test gives an estimate for the length of the keyword for the cipher. While likely to be accurate, this estimate is not guaranteed to be so. Fortunately, other tests exist that may be accurate when the Friedman test is not. The Kasiski test is one such test.

Friedrich Kasiski lived during the nineteenth century, and was a career officer in the Prussian military. Kasiski's work in cryptology took place after he had resigned from official military status, and like for others we have seen, was not recognized for its importance during his lifetime, perhaps even by Kasiski himself. The Kasiski test appeared in a short book written by Kasiski and published in Germany in 1863. Although named for Kasiski, the test had actually been discovered independently almost a decade earlier by British mathematician and inventor Charles Babbage, who is best known for originating the idea of the electronic programmable computer.

The Kasiski test relies on the occasional coincidental alignment of letter groups in the plaintext with letters in the keyword. In a ciphertext formed using a Vigenère keyword cipher, if a group of letters occurs repeatedly, it is possible (though not certain) that the distance between the start of these occurrences will be a multiple of the length of the keyword. To demonstrate this, consider the plaintext THEY SAY THE BEST MOVIE IS THE GODFATHER encrypted using a Vigenère keyword cipher with the keyword COPPOLA.

Plain: T H E Y S A Y T H E B E S T M O V I E I S T H E ...
Key: C O P P O L A C O P P O L A C O P P O L A C O P ...
Cipher: <u>V V T</u> N G L Y <u>V V T</u> Q S D T O C K X S T S <u>V V T</u> ...

In this ciphertext, the trigraph VVT occurs three times, with the second occurrence beginning seven letters after the start of the first, and the third beginning 14 letters after the start of the second. Note that both of these separations are multiples of the length of the keyword for the cipher. While we could obviously concoct a small example to illustrate anything, the fact is that often in ciphertexts formed using a Vigenère keyword cipher, repeated groups of letters occur, with the separations between the beginnings of these groups being multiples of the length of the keyword.

In Definition 6.3 on page 173, we defined the greatest common divisor of a pair of positive integers. This definition can be extended for any collection of integers. For example, $\gcd(10, 15, 25, 40) = 5$, since 5 is the largest positive integer that divides 10, 15, 25, and 40 evenly. The Kasiski test states that for a ciphertext formed using a Vigenère keyword cipher, a common divisor, with the greatest common divisor being the most likely, of the separations between some of the beginnings of repeated groups of letters in the ciphertext stands a good chance of being equal to or a multiple of the length of the keyword for the cipher.

Example 7.19 Consider the ciphertext UZRZE GNJEN VLISE XRHLY PYEGT ESBJH JCSBP TGDYF XXBHE EIFTC CHVRK PNHWX PCTUQ TGDJH TBIPR FEMJC NHVTC FSAII IFNRE GSALH XHWZW RZXGT TVWGD HTEYX ISAGQ TCJPR SIAPT UMGZA LHXHH SOHPW CZLBR ZTCBR GHCDI QIKTO AAEFT OPYEG TENRA IALNR XLPCE PYKGP NGPRQ PIAKW XDCBZ XGPDN RWXEI FZXGJ LVOXA JTUEM BLNLQ HGPWV PEQPI AXATY ENVYJ EUEI, which was formed using a Vigenère keyword cipher. Some of the repeated groups of letters in this ciphertext along with the separations between these groups are shown in the following table.

Letters	Occurrences	Separations
ZXG	3	108, 12
TGD	2	30
QPIA	2	48
ALHXH	2	42
PYEGTE	2	156

Since $\gcd(12, 30, 42, 48, 108, 156) = 6$, the Kasiski test gives that the length of the keyword for the cipher is likely to be 6, with 2 and 3 being other potential possibilities. This is consistent with the estimate 5.8905 for the length of the keyword that we found using the Friedman test with this same ciphertext in Example 7.18 on page 220. □

7.5.1 Exercises

1.* In a ciphertext formed using a Vigenère keyword cipher, some of the repeated groups of letters, along with the separations between these groups and the greatest common divisor of these separations, are shown in the following table.

Letters	Occurrences	Separations
AEY	4	12, 24, 36
PEZ	3	18, 36
TSRK	2	42
GTEM	2	48
QPIAP	2	24

With only the information in this table, use the Kasiski test to find the most likely length of the keyword for the cipher.

2. In a ciphertext formed using a Vigenère keyword cipher, some of the repeated groups of letters, along with the separations between these

groups and the greatest common divisor of these separations, are shown in the following table.

Letters	Occurrences	Separations
PFY	5	16, 32, 48, 64
AQZ	4	24, 48, 72
RSVP	3	32, 48
NEET	2	24
MIWX	2	48

With only the information in this table, use the Kasiski test to find the most likely length of the keyword for the cipher.

3. For the following ciphertexts, which were formed using Vigenère keyword ciphers, identify the longest repeated group of letters. Then, with only this information, use the Kasiski test to find the most likely lengths of the keywords for the ciphers.

   (a)* LVVMG MZLLV VUSRU XKHFY LVVFS ZC

   (b) TYZXL FMFMU TYZXL MHWMF WZBAA TMKBA TYZXL

   (c)* UCIJM KWICD JGYHP HMKAC CKSJW ICDXG WICDQ SU

   (d) IOVCZ MAKZS CMCKF FOMAK SULAT LUIKG TYKAA RAFEO LJTLU

4. Find some information about Friedrich Kasiski and his efforts to break Vigenère keyword ciphers, and write a summary of your findings.

5. Find some information about Charles Babbage and his efforts to break Vigenère keyword ciphers, and write a summary of your findings.

7.6   Cryptanalysis of Vigenère Keyword Ciphers

For a ciphertext formed using a Vigenère keyword cipher, the Friedman and Kasiski tests both end at the same place, with an estimate for the length of the keyword. Unfortunately, neither test is guaranteed to be accurate, nor does either provide a method for determining what the actual keyword is, despite the fact that knowing the keyword is necessary in order to break the cipher. However, knowing the length of the keyword is a notable first step in breaking the cipher. To see why, we need to examine how Vigenère keyword ciphers work a little more closely.

Consider again the following encryption from Example 7.7 on page 204, done using a Vigenère keyword cipher.

Plain: H̲ A V I N G A̲ P E T C A N̲ M A K E Y O̲ U H A P P Y̲
Key: T̲ R I X I E T̲ R I X I E T̲ R I X I E T̲ R I X I E T̲
Cipher: A̲ R D F V K T̲ G M Q K E G̲ D I H M C H̲ L P X X T̲ R

Note that in this encryption, every sixth plaintext letter is encrypted with the same keyword letter. For example, each underlined plaintext letter above is encrypted with the same keyword letter T. In addition, due to the shifting pattern of the Vigenère square (Table 7.1 on page 201), each keyword letter in a Vigenère keyword cipher dictates a shift cipher to be applied to any plaintext letter designated to be encrypted with it. For example, for the correspondences A = 0, B = 1, C = 2, ... , Z = 25, since T = 19, then any plaintext letter encrypted with the keyword letter T is equivalently encrypted using a shift cipher with our alphabet letters in the natural order and a shift of 19 positions to the right for encryption. Thus, a Vigenère keyword cipher is completely equivalent to a collection of shift ciphers, one for each of the letters in the keyword, and the keyword letters themselves can be determined by simply breaking this collection of shift ciphers.

To break these shift ciphers, we begin by separating a ciphertext into groups in which the letters have all been encrypted with a common keyword letter, and thus a common shift cipher. For instance, in the example above, if we knew that the length of the keyword was six, then we would know that each of the underlined ciphertext letters A, T, G, H, and R resulted from a common shift cipher. Similarly, we would know that each of the ciphertext letters R, G, D, and L resulted from a common shift cipher. These groups of letters are called *cosets*. A full collection of cosets for this ciphertext is ATGHR, RGDL, DMIP, FQHX, VKMX, and KECT.[16]

Each of these cosets contains ciphertext letters that all result from a common shift cipher. Breaking these six "smaller" shift ciphers can be more complicated than breaking a "full" shift cipher, though, since even when decrypted correctly, the plaintext letters that correspond to the letters in a coset should not form sensible English when strung together. Fortunately, several processes for breaking these shift ciphers have been developed and refined. We will consider one such process in this section.

Before considering this process, we will first see another method for finding the length of the keyword for a Vigenère keyword cipher. We have already done this using the Friedman and Kasiski tests, but wish to consider a third method that relies on technology that was not available to Friedman or Kasiski when they created their tests. The idea for this third method

---

[16] Although the letters in the cosets DMIP and VKMX were coincidentally formed using the same ciphertext letter, when trying to break the cipher knowing only that the length of the keyword is six, we would not know this. Exploiting repeated keyword letters has not historically been part of breaking Vigenère keyword ciphers.

actually originated with Charles Babbage in the mid 1800s, but was more recently refined by Andrew Simoson and Thomas Barr.

7.6.1   Finding the Keyword Length Using Signatures

In this section, we will demonstrate a graphical method that involves plotting frequencies to find the length of the keyword for a Vigenère keyword cipher. To begin, consider again the frequencies with which the 26 letters in our alphabet occur in ordinary English, which we showed previously in Table 7.2 on page 217, and show again in decimal form in Table 7.3.

Letter	Frequency	Letter	Frequency
A	0.0817	N	0.0675
B	0.0149	O	0.0751
C	0.0278	P	0.0193
D	0.0425	Q	0.0010
E	0.1270	R	0.0599
F	0.0223	S	0.0633
G	0.0202	T	0.0906
H	0.0609	U	0.0276
I	0.0697	V	0.0098
J	0.0015	W	0.0236
K	0.0077	X	0.0015
L	0.0403	Y	0.0197
M	0.0241	Z	0.0007

Table 7.3   Letter frequencies in ordinary English.

The *signature* of ordinary English is a graph of these frequencies, plotted in order from smallest to largest, with each pair of consecutive points connected by a straight line. The signature of ordinary English is shown in Figure 7.2 on page 229. This graph starts on the left even with the smallest frequency 0.0007 in Table 7.3, and ends on the right even with the largest 0.1270.

Similarly, the signature of a sample of text (e.g., a plaintext or a ciphertext) is a graph of the frequencies with which the letters occur in the sample, again plotted from smallest to largest and connected by straight lines. Two facts are important to note when considering the signature of a sample. First, in a sample of reasonable length, it is likely that some letters will not actually occur. The frequencies of these letters would of course then be 0. Second, since the sum of the frequencies of all letters, whether for the full English language or just a sample of text, must be 1, to compensate for the existence of letters with frequency 0 in a sample,

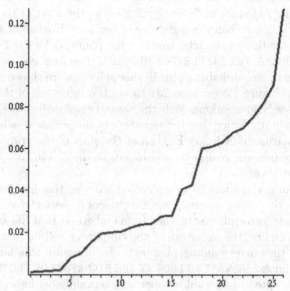

Figure 7.2   Signature of ordinary English.

Letter	Frequency	Letter	Frequency
A	0.1277	N	0.0426
B	0	O	0.1064
C	0	P	0.0213
D	0	Q	0
E	0.1489	R	0.0213
F	0.0426	S	0.0851
G	0.0426	T	0.1277
H	0.0426	U	0.0213
I	0.0213	V	0
J	0	W	0.0426
K	0.0213	X	0
L	0.0426	Y	0
M	0.0426	Z	0

Table 7.4   Letter frequencies in a sample plaintext.

it is likely that some letters in a sample will have higher frequencies than in ordinary English. Thus, the signature of a sample should in general be lower than the signature of ordinary English at the start of the graph (on the left), and higher than the signature of ordinary English at the end.

To illustrate these two facts, consider the plaintext WE WANT TO LOOK AT THE SIGNATURE OF THE SAMPLE OF A MESSAGE. The frequencies with which the 26 letters in our alphabet occur in this plaintext are shown in Table 7.4 on page 229. Figure 7.3 on page 231 shows the signature of this plaintext (the thinner segments) along with the signature of ordinary English (the thicker segments). Notice that the signature of the plaintext is indeed lower than the signature of ordinary English at the start of the graph due to the frequencies of 0 in the plaintext, and higher at the end in compensation of these frequencies of 0.

Another important fact to note when considering the signature of a sample of text is that since monoalphabetic ciphers preserve letter frequencies from plaintexts into ciphertexts, then for a ciphertext that has been formed using a shift cipher, the signature of the ciphertext will be identical to the signature of the corresponding plaintext. To illustrate this fact, consider again the plaintext WE WANT TO LOOK AT THE SIGNATURE OF THE SAMPLE OF A MESSAGE, encrypted using a shift cipher with our alphabet letters in the natural order and a shift of 19 positions to the right for encryption, for which the resulting ciphertext is PXPTG MMHEH HDTMM AXLBZ GTMNK XHYMA XLTFI EXHYT FXLLT ZX. The frequencies with which the 26 letters in our alphabet occur in this ciphertext are shown in Table 7.5 on page 231. Note that Tables 7.4 and 7.5 contain identical frequencies, but that these frequencies are associated with different letters. Thus, the signature of this ciphertext would be identical to the signature of the corresponding plaintext. That is, Figure 7.3 on page 231, which we noted previously shows the signatures of the plaintext and ordinary English, equivalently shows the signatures of the ciphertext and ordinary English.

We can now state how signatures can be used to find the length of the keyword for a Vigenère keyword cipher. Recall that each keyword letter dictates a shift cipher to be applied to any plaintext letter designated to be encrypted with it. Recall also that the ciphertext letters encrypted with a common shift cipher form a coset. If a ciphertext formed using a Vigenère keyword cipher is separated into the correct number of cosets, then the signatures of these cosets should for the most part exhibit the behavior we noted previously of starting lower and ending higher when compared with the signature of ordinary English. On the other hand, if the ciphertext is separated into an incorrect number of cosets, then the signatures of these cosets should for the most part fail to exhibit this behavior. Looking for this behavior can help identify the likely correct number of cosets, which would then be the likely length of the keyword.

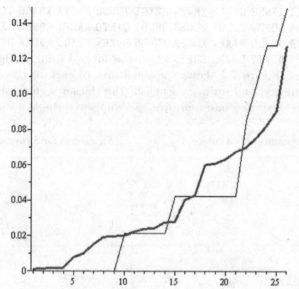

Figure 7.3   Signatures of ordinary English and a sample.

Letter	Frequency	Letter	Frequency
A	0.0426	N	0.0213
B	0.0213	O	0
C	0	P	0.0426
D	0.0213	Q	0
E	0.0426	R	0
F	0.0426	S	0
G	0.0426	T	0.1277
H	0.1064	U	0
I	0.0213	V	0
J	0	W	0
K	0.0213	X	0.1489
L	0.0851	Y	0.0426
M	0.1277	Z	0.0426

Table 7.5   Letter frequencies in a sample ciphertext.

Example 7.20 Consider the ciphertext UZRZE GNJEN VLISE XRHLY PYEGT ESBJH JCSBP TGDYF XXBHE EIFTC CHVRK PNHWX PCTUQ TGDJH TBIPR FEMJC NHVTC FSAII IFNRE GSALH XHWZW RZXGT TVWGD HTEYX ISAGQ TCJPR SIAPT UMGZA LHXHH SOHPW CZLBR ZTCBR GHCDI QIKTO AAEFT OPYEG TENRA IALNR XLPCE PYKGP NGPRQ PIAKW XDCBZ XGPDN RWXEI FZXGJ LVOXA JTUEM BLNLQ HGPWV PEQPI AXATY ENVYJ EUEI, which was formed using a Vigenère key-word cipher. Figure 7.4 shows the signatures of each of the cosets (the thinner segments) and ordinary English (the thicker segments) when this ciphertext is separated into four, five, six, and seven cosets.

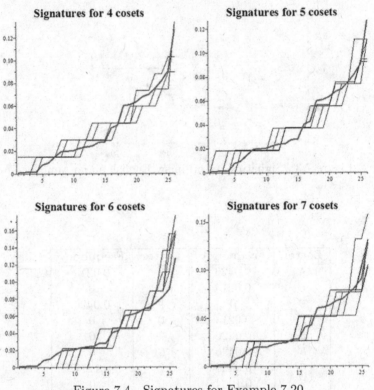

Figure 7.4   Signatures for Example 7.20.

The likely length of the keyword for the cipher is the number of cosets for which the signatures of the cosets best exhibit the behavior of starting lower and ending higher than the signature of ordinary English. This behavior is best exhibited for six cosets, indicating a likely keyword length of six. This is consistent with the estimate for the length of the keyword that we found using the Friedman and Kasiski tests with this same ciphertext in Examples 7.18 on page 220 and 7.19 on page 225, respectively.   □

7.6.2 Finding the Keyword Letters Using Scrawls

We have now seen three methods, the Friedman test, the Kasiski test, and signature plotting, for estimating the length of the keyword for a Vigenère keyword cipher. Once the number of letters in the keyword is known, these letters must still be determined. To do this, consider again the frequencies with which the 26 letters in our alphabet occur in ordinary English, which are shown in Table 7.3. The *scrawl* of ordinary English is like the signature, except the frequencies are plotted in alphabetical order instead of increasing order. The scrawl of ordinary English is shown in Figure 7.5. The graph starts on the left even with the first frequency 0.0817 in Table 7.3, and ends on the right even with the last frequency 0.0007.

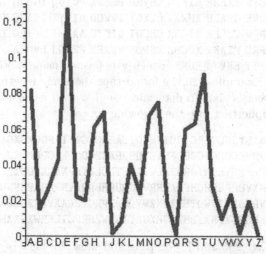

Figure 7.5   Scrawl of ordinary English.

Similarly, the scrawl of a sample of text is like the signature, but with the frequencies plotted in alphabetical order. For a plaintext or a ciphertext formed using a monoalphabetic cipher, even for a relatively small sample of text, the scrawl of the sample should have a similar appearance to the scrawl of ordinary English, with roughly the same peaks and dips. However, for a coset in a ciphertext formed using a Vigenère keyword cipher, recall that the letters all result from a common shift cipher. So while the scrawl of a coset should have roughly the same peaks and dips as the scrawl of ordinary English, these peaks and dips should all be shifted to the right some number of positions, wrapping from the right edge of the graph to the left when necessary, with the number of positions shifted corresponding to the keyword letter that dictated the shift.

More precisely, with the correspondences A = 0, B = 1, . . . , Z = 25, for a particular keyword letter, all ciphertext letters encrypted with the keyword letter will be shifted to the right the corresponding number of positions. For the coset that contains these ciphertext letters, if the scrawl were shifted to the *left* the same number of positions, wrapping from the left edge of the graph to the right when necessary, the scrawl should be as closely aligned as possible with the scrawl of ordinary English. Thus, shifting the scrawl of a coset, looking for where it aligns as closely as possible with the scrawl of ordinary English, can help identify the likely keyword letter that produced the shift. Doing this for each coset should allow the entire keyword to be determined one letter at a time, and the cipher to finally be broken.

Example 7.21 Consider the ciphertext UZRZE GNJEN VLISE XRHLY PYEGT ESBJH JCSBP TGDYF XXBHE EIFTC CHVRK PNHWX PCTUQ TGDJH TBIPR FEMJC NHVTC FSAII IFNRE GSALH XHWZW RZXGT TVWGD HTEYX ISAGQ TCJPR SIAPT UMGZA LHXHH SOHPW CZLBR ZTCBR GHCDI QIKTO AAEFT OPYEG TENRA IALNR XLPCE PYKGP NGPRQ PIAKW XDCBZ XGPDN RWXEI FZXGJ LVOXA JTUEM BLNLQ HGPWV PEQPI AXATY ENVYJ EUEI, which was formed using a Vigenère key-word cipher. In Example 7.20, we found that the likely length of the key-word for the cipher is six. To determine the keyword letters, we begin by separating the ciphertext into the following six cosets.

Coset 1: UNILTJGBTKPGIJCISWGDICAZHCTCTTTAPGQXGXGABGQTJ
Coset 2: ZJSYECDHCPCDPCFFAZTHSJPASZCDOOELCPPDPEJJLPPYE
Coset 3: REEPSSYECNTJRNSNLWTTAPTLOLBIAPNNENICDILTNWIEU
Coset 4: ZNXYBBFEHHUHFHARHRVEGRUHHBRQAYRRPGABNFVULVANE
Coset 5: EVREJPXIVWQTEVIEXZWYQSMXPRGIEEAXYPKZRZOEQPXVI
Coset 6: GLHGHTXFRXTBMTIGHXGXTIGHWZHKFGILKRWXWXXMHEAY

Figures 7.6 on page 235 and 7.7 on page 236 show, for each of these cosets individually, the scrawls of the coset (the thinner segments) and ordinary English (the thicker segments) for two shifts, one being the correct shift, and the other shift chosen randomly. As we can see from these graphs, the shifts to the left for which the scrawls of the cosets more closely align with the scrawls of ordinary English are, for the keyword letters in order, 15, 11, 0, 13, 4, and 19 positions. With the correspondences A = 0, B = 1, C = 2, . . . , Z = 25, these shift values give the keyword letters P, L, A, N, E, and T. With the keyword PLANET, the entire ciphertext decrypts as follows: FOR MANY YEARS, THE KNOWN PLANETS OF OUR SOLAR SYSTEM WERE MERCURY, VENUS, EARTH, MARS, JUPITER, SATURN, URANUS, NEPTUNE, AND PLUTO. HOWEVER, IT IS NOW TRUE THAT MANY PEOPLE THINK PLUTO SHOULD NO LONGER BE CONSIDERED A NAMED PLANET. NEW PLANETS ARE CURRENTLY BEING DISCOVERED, AND IT IS VERY LIKELY THAT MANY MORE WILL BE IN THE NEAR FUTURE.   □

Keyword letter 1

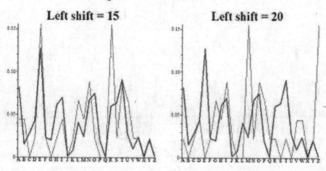

Keyword letter 2

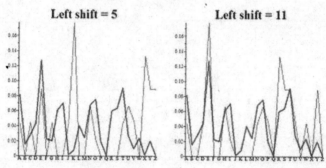

Keyword letter 3

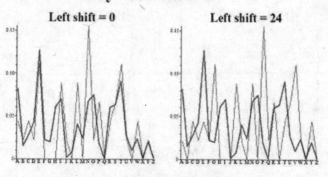

Figure 7.6   Scrawls for Example 7.21.

Keyword letter 4

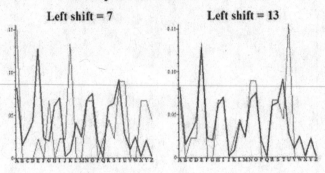

Keyword letter 5

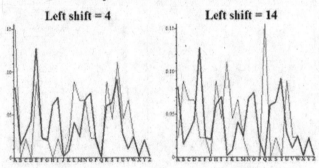

Keyword letter 6

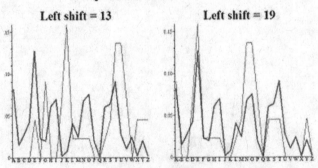

Figure 7.7   Scrawls for Example 7.21.

Choosing the shifts that lead to the correct keyword letters in Example 7.21 was made much easier by the fact that only two alignments had to be considered for each letter, as opposed to the 26 that would have to be considered for each letter in actual practice. This method for breaking Vigenère keyword ciphers is obviously reliant upon the production of several carefully constructed graphs. While it would be possible to draw these graphs by hand, it might not be practical to do so. Fortunately, in modern society, technology exists that facilitates the production of these graphs. However, even without the use of technology, constructing these graphs is still possible, since signatures and scrawls are just plotted points connected by straight lines.

7.6.3   Exercises

1.* Consider the ciphertext LPOFE MYFSO KVKQT GWVJR VNUEK BAQWV ZFLWS BJMLH DTEHF LKEKB GAVPU JKQMA SBAEW AGAVA IESER FUITO WCYRV BUQWB KEEQT RLPKX WGCBJ LRRFO ZUSVJ XWGCB JLAFW LOALP KIAOK AWZKP AXNRJ, which was formed using a Vigenère keyword cipher. Figure 7.8 on page 238 shows the signatures of each of the cosets (the thinner segments) and the signature of ordinary English (the thicker segments) when this ciphertext is separated into two, three, four, and five cosets.

   (a) Using only these signatures, determine the most likely length of the keyword for the cipher.

   (b) Determine the cosets in this ciphertext that result from the likely keyword length you found in part (a).

2. Consider the ciphertext AJWWV NVROO WALIR LVRKT RABTO SUDRA GXBWG BHPES GNFXI XKRVM DFCGK SMWHV EISLO FLIAV JRJWI LMCJS SHSEA XYOCH DHNGH OWWOO SYUSM W,[17] which was formed using a Vigenère keyword cipher. Figure 7.9 on page 239 shows the signatures of each of the cosets (the thinner segments) and ordinary English (the thicker segments) when this ciphertext is separated into three, four, five, and six cosets.

   (a) Using only these signatures, determine the most likely length of the keyword for the cipher.

   (b) Determine the cosets in this ciphertext that result from the likely keyword length you found in part (a).

---

[17] Anne Bradstreet (1612–1672), quote.

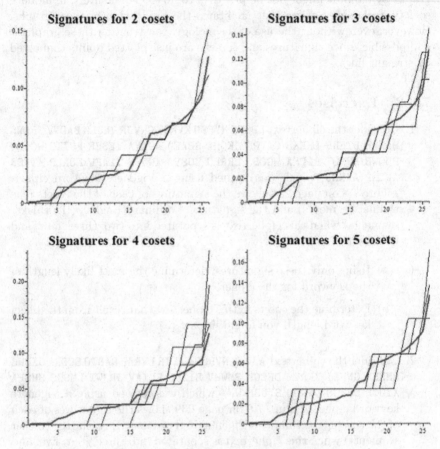

Figure 7.8   Signatures for Exercise 1.

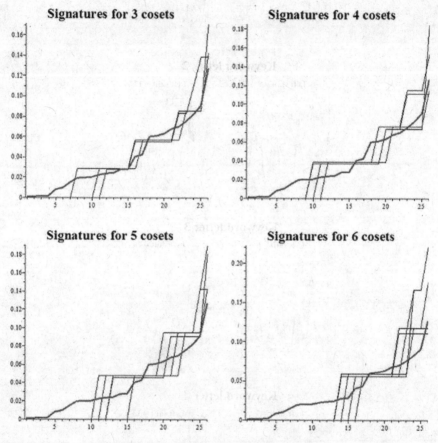

Figure 7.9   Signatures for Exercise 2.

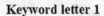

Keyword letter 1

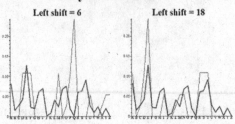

Keyword letter 2

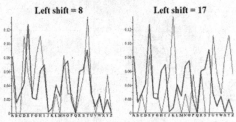

Keyword letter 3

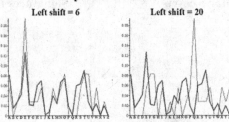

Keyword letter 4

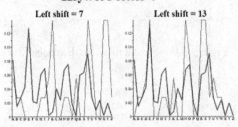

Figure 7.10   Scrawls for Exercise 3.

Keyword letter 1

Left shift = 5   ### Left shift = 18

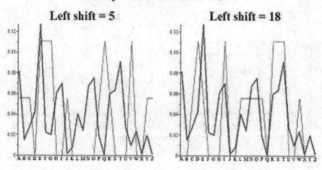

Keyword letter 2

Left shift = 4   ### Left shift = 24

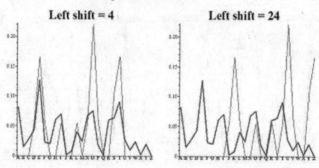

Keyword letter 3

Left shift = 0   ### Left shift = 11

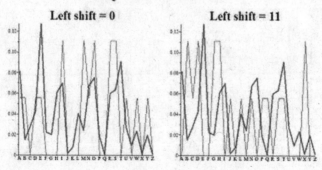

Figure 7.11   Scrawls for Exercise 4.

Keyword letter 4

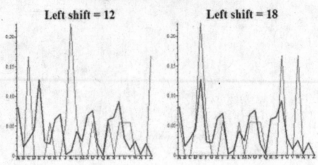

Keyword letter 5

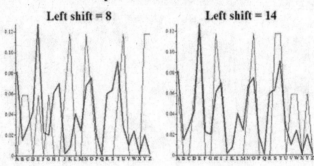

Keyword letter 6

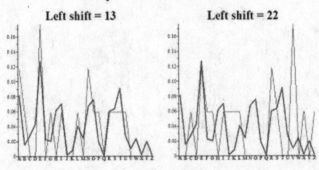

Figure 7.12   Scrawls for Exercise 4.

3. Consider the ciphertext LPOFE MYFSO KVKQT GWVJR VNUEK BAQWV ZFLWS BJMLH DTEHF LKEKB GAVPU JKQMA SBAEW AGAVA IESER FUITO WCYRV BUQWB KEEQT RLPKX WGCBJ LRRFO ZUSVJ XWGCB JLAFW LOALP KIAOK AWZKP AXNRJ (the same ciphertext as Exercise 1), which was formed using a Vigenère keyword cipher. Suppose that the length of the keyword for the cipher is known to be four, and that this ciphertext is separated into four cosets. Figure 7.10 on page 240 shows, for each of these cosets individually, the scrawls of the coset (the thinner segments) and ordinary English (the thicker segments) for two shifts, one being the correct shift, and the other chosen randomly.

   (a)* Using only these scrawls, determine the keyword for the cipher.

   (b) Decrypt the ciphertext using the keyword you found in part (a).

4. Consider the ciphertext AJWWV NVROO WALIR LVRKT RABTO SUDRA GXBWG BHPES GNFXI XKRVM DFCGK SMWHV EISLO FLIAV JRJWI LMCJS SHSEA XYOCH DHNGH OWWOO SYUSM W (the same ciphertext as Exercise 2), which was formed using a Vigenère keyword cipher. Suppose that the length of the keyword for the cipher is known to be six, and that this ciphertext is separated into six cosets. Figures 7.11 on page 241 and 7.12 on page 242 show, for each of these cosets individually, the scrawls of the coset (the thinner segments) and ordinary English (the thicker segments) for two shifts, one being the correct shift, and the other chosen randomly.

   (a) Using only these scrawls, determine the keyword for the cipher.

   (b) Decrypt the ciphertext using the keyword you found in part (a).

5. Consider the following pair of ciphertexts, one formed using a monoalphabetic cipher, and the other using a Vigenère keyword cipher.

   Ciphertext A: EAYQR MYAGE EAGFT QDZOU HUXIM DSQZQ DMXEU ZOXGP
   QDANQ DFQXQ QEFAZ QIMXX VMOWE AZVMY QEXAZ SEFDQ
   QFMZP VAEQB TVATZ EFAZ

   Ciphertext B: UWHMQ CUJCD PWMBS GZIKT XQGEL TOZVP TIGAT PKGCO
   GCGGD UMNAR TIIBH KTGQL OBNPP TUVVR GWMOP OKXTP
   NTVVL PLBMZ TOZUP CLZ

   (a)* Figure 7.13 on page 244 shows, for each of these ciphertexts separately, the signatures of the ciphertext (the thinner segments) and ordinary English (the thicker segments), with ciphertext A in the graph on the left in Figure 7.13, and ciphertext B in the graph on the right. Using only these signatures, determine which ciphertext was formed using a monoalphabetic cipher.

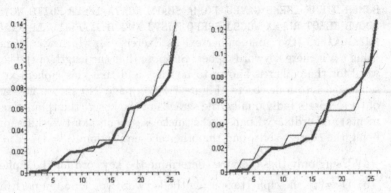

Figure 7.13   Signatures for Exercise 5a.

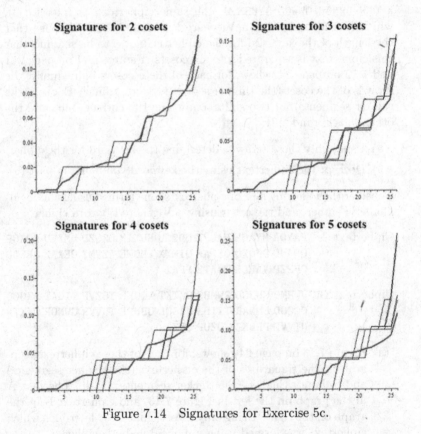

Figure 7.14   Signatures for Exercise 5c.

Keyword letter 1

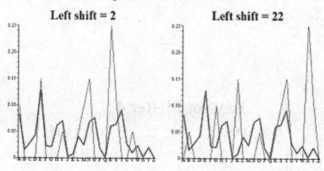

Keyword letter 2

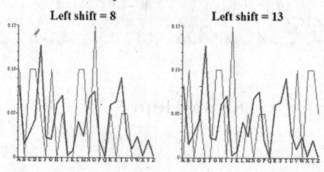

Keyword letter 3

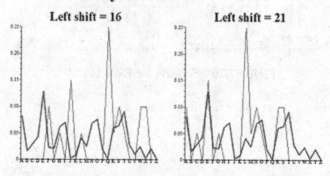

Figure 7.15 Scrawls for Exercise 5d.

Keyword letter 4

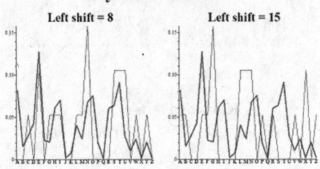

Keyword letter 5

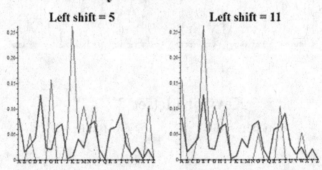

Figure 7.16   Scrawls for Exercise 5d.

(b) Decrypt the ciphertext that was formed using a monoalphabetic cipher.

(c)* For the ciphertext that was formed using a Vigenère keyword cipher, Figure 7.14 on page 244 shows the signatures of each of the cosets (the thinner segments) and ordinary English (the thicker segments) when the ciphertext is separated into two, three, four, and five cosets. Using only these signatures, determine the most likely length of the keyword for the cipher. Then, determine the cosets that result from this likely keyword length.

(d) For the ciphertext that was formed using a Vigenère keyword cipher, suppose that the most likely length of the keyword for the cipher is known to be five letters, and that the ciphertext is separated into five cosets. Figures 7.15 on page 245 and 7.16 on page 246 show, for each of these cosets individually, the scrawls of the coset (the thinner segments) and ordinary English (the thicker segments) for two shifts, one being the correct shift, and the other chosen randomly. Using only these scrawls, determine the keyword for the cipher. Then, decrypt the ciphertext.

6. Cryptanalyze the following ciphertexts, which were formed using Vigenère keyword ciphers.

(a) PAPCP SRSIC RKILT GYFXG ETWAI JIUPG RLTGH ACMOQ RWXYT
    JIEDF NVEAC ZUUEJ TLOHA WHEET RFDCT JGSGZ LKRSC ZRVLU
    PCONM FPDTC XWJYI XIJHT TAMKA ZCCXW STNTE DTTGJ MFISE
    GEKIP RPTGG EIQRG UEHGR GGEHE EJDWI PEHXP DOSFI CEIMG
    CCAFJ GGOUP MNTCS KXQXD LQGSI PDKRJ POFQV VXYTJ IEDFN
    VEACZ UUEJT LOHWG JEHYI KIPRP ZAGRI PMS

(b)* FWPSI DTSAG SDODY MGYUI PXJRS GLPXH HQSEC THRJV ACNMC
    YSLTH MCIRK TSAAY DBPWN JSZSO JUGLU ELYBV WEUXG YEXNO
    OTYFX YAEXI HRURT LBGXY IDRYN RJTZY BREME CFART VRDTH
    VJPOF LIGSR FFSYE ECYZZ LRFVT EJLBJ WIYYB RGRSV JNGLR
    NOTCA KKHPJ OYSXY[18]

(c) EVKEF XYELZ XTZUK MGKLE WUIPR EMNWN VIGKL CWHKY KPEPW
    LJSGE EMWXK LHDES TYRPX SYRAY UKHCH AFXAI PVPSO IIXAR
    LUYYJ YFFJM GHVCG VERTY UJHIH VALXM GZEBW UCIXE GRQJK
    IWKLE DITEM ZSNGZ KLXKV ESMLV XRRDA NJFXE IFSWK SKJMN
    LBIIX USCMG VRMJS NDSFR XFJTZ YJIWF GUEYE XLYES WPVVU
    VINVY TVRGX EVNYI XEGRQ JKIWU SCMGV RMFYT DCEMG XXHWJ
    IIVZW EDITE MZSNG ZKLXK VESML VXYES FYMIK SIEFX VGKPT
    TWXVZ XEXHG OXLFR RYHYF TEVMN UFLHB EKSGG VJKFQ TZYEW
    TYEVW NIMXU

---

[18] Jerry Seinfeld, quote.

(d) BOOHE VFTOD MMLRU NLIIC ZYGAM AFTFH FZVWH PAULG SUCQB
SSEFJ PQBFJ HYREI FBREI GTCOF IGOSR GENSH EIPGI EEFXS
CITZF WUCJT JSAFL IIGCS NURFH WGSLW SPPBS AJUMC WILIS
XOFHM NRNHP BFLQU STSII HFFIW TAHZW XUOYE KBVER MOLBU
NFVQB PVMOF QEZFV EHMMO HTEXJ VSEHX WCEIL DSNDM LASPO
XUINS YGKFQ DSOHF ZYKAS QNHPY NOWHM UALVB RDELB YHKEA
HLTSY GKFQD SOHTH IECSB OMOHS ZJLPZ HEJEI PZWBA SPNGW
IIRLM LSZNA OITPI GAVQS WDSNO ATZAM DWEIC PMZOH QEFUA
EYXRV BQAFE GOYWB ZHQDG GRIYI MLSZH MOHRP HTURE ENFRP
ZYGKG AFYPP DLRWA KQLNF LUYHK LRMNV FMGSX RLWSH LPJST
POLFM LKPNE HIEZC NTSJR EOMGZ HXOMJ WIYIQ JVMNY FXODE
OLHDA FTTOC XTAWA NSOHV LPNLR MTLIM REIXU HTOMT ENOHH
SZMRK ULELF HCSUS KFGUC IEFDM CCFHI YMKVB BOLTA IELBY
CZCGW IRDXA LJMUD UMSCS NNVXY DJREO ABAVE TGOIA YHMOS
HEKTI LDVXZ HANKP PIOWM VBQAF EERPG HCSDE VXMTS SMOSD
SHBTE CRNTP QRGOI DPWVY WNEKU LEPBT JHXOU BPIEC HMHTE
NBYLE WHAVM TFPHI QJBJI XTQXM LWFXO OPIFG MNOMG NWF

7. Recall from Exercise 5 in Section 2.2 the concept of *superencryption*.

(a)* Use Vigenère keyword ciphers with keywords BEN and MATT (in that order) to superencrypt GOOD WILL HUNTING.

(b) Decrypt TSUXM VFIGV KZVRM, which was superencrypted using Vigenère keyword ciphers with keywords MATT and BEN (in that order).

(c) Does superencryption by two Vigenère keyword ciphers yield more security than encryption by one Vigenère keyword cipher? In other words, if a plaintext P is encrypted using a Vigenère keyword cipher, yielding M, and then M is encrypted using another Vigenère keyword cipher, yielding C, would C be harder in general to cryptanalyze than M? Explain your answer completely, and be as specific as possible.

8. In *Invitation to Cryptology* [1], Thomas Barr outlines a way to quantify the method presented in this section for using signatures and scrawls to break Vigenère keyword ciphers. Find a copy of this book, and write a summary of Barr's quantified way for performing this method. Include at least one example in your summary.

Chapter 8

Hill Ciphers

Recall that substitution ciphers, in which characters are encrypted one at a time using a single cipher alphabet, are susceptible to attack by frequency analysis. However, recall also that substitution ciphers would be less susceptible to attack if plaintext characters were encrypted in pairs (i.e., digraphs) rather than one at a time. This is the idea that forms the basis for Playfair ciphers, which we considered in Chapter 2. In this chapter, we will consider *Hill* ciphers, which can loosely be thought of as doing for mathematical substitution ciphers what Playfair ciphers do for nonmathematical substitution ciphers. Specifically, Hill ciphers are designed to allow for mathematical encryption of characters in groups of more than one at a time. As we will see, while Playfair ciphers are designed to allow for encryption of characters only in pairs, Hill ciphers are designed to allow for encryption of characters in groups of any size.

Hill ciphers are examples of *block* ciphers. Block ciphers include all ciphers in which encryption of characters occurs in groups, or blocks, of more than one at a time. Technically, Playfair ciphers are block ciphers, although the term *block cipher* is usually reserved for more sophisticated types of ciphers, such as the Advanced Encryption Standard, and other ciphers that use mathematics for encryption, as Hill ciphers do. Hill ciphers specifically use mathematical objects called *matrices* (singular, *matrix*) and mathematical operations involving them.

8.1 Matrices

Matrices have a rich assortment of applications in many areas, for example, computer graphics and game theory, as well as in mathematical areas such as linear regression and stochastic processes. Our interest in matrices lies of course in the fact that they can be used in cryptology. The purpose of this section is to review some basic facts about matrices and their operations, so that we can use them later in Hill ciphers.

8.1.1   Definition and Basic Terminology

A *matrix* is a rectangular array of numbers, usually surrounded by square brackets to separate it from other items on a page. The *size* of a matrix refers to the number of rows and columns in the array, and is expressed as (number of rows) × (number of columns). A matrix is typically denoted using an italicized capital letter if it contains more than one row and more than one column or is of an unspecified size, and an unitalicized boldface lowercase letter if it contains only one row or column. The following are examples of matrices A, B, $\mathbf{c}$, and $\mathbf{d}$, of sizes 2×4, 3×2, 1×4, and 3×1, respectively.

$$A = \begin{bmatrix} -5 & 10 & 0 & 1 \\ 3 & 2 & -5 & 2 \end{bmatrix} \qquad B = \begin{bmatrix} 2 & 1 \\ -4 & 0 \\ 5 & 2 \end{bmatrix}$$

$$\mathbf{c} = \begin{bmatrix} 3 & -1 & 8 & 1 \end{bmatrix} \qquad \mathbf{d} = \begin{bmatrix} 0 \\ 1 \\ 4 \end{bmatrix}$$

We will refer to a matrix that contains a single row as a *row matrix*, and a matrix that contains a single column as a *column matrix*.[1] Thus, the matrix $\mathbf{c}$ defined above is a row matrix, and $\mathbf{d}$ is a column matrix.

Individual numbers, or *entries*, in a matrix are usually denoted using the same letter as denoting the entire matrix, but in an italicized lowercase version, with subscripts $_{ij}$ to indicate the row number i (starting with row 1 at the top) and column number j (starting with column 1 on the left) of the position of a particular entry in the matrix. For example, for the matrices A, B, $\mathbf{c}$, and $\mathbf{d}$ defined above, $a_{12} = 10$, $b_{32} = 2$, $c_{11} = 3$, and $d_{31} = 4$. In general, the entries in a matrix A of size $m \times n$ are denoted as follows.

$$A = \begin{bmatrix} a_{11} & a_{12} & \cdots & a_{1n} \\ a_{21} & a_{22} & \cdots & a_{2n} \\ \vdots & \vdots & & \vdots \\ a_{m1} & a_{m2} & \cdots & a_{mn} \end{bmatrix}$$

A matrix that contains the same number of rows as columns is said to be *square*. That is, a matrix of size $m \times n$ is square if $m = n$. Also, a square

---

[1] A matrix that contains a single row or column is sometimes called a row or column *vector*. For simplicity, we will use the terms row and column *matrix* to refer to these types of matrices.

matrix of size $n \times n$ is said to have *order* n. The following are square matrices of orders 2 and 3, respectively.

$$\begin{bmatrix} 8 & 1 \\ 19 & 14 \end{bmatrix} \qquad \begin{bmatrix} 11 & 6 & 8 \\ 0 & 3 & 14 \\ 24 & 0 & 9 \end{bmatrix}$$

8.1.2   Matrix Operations

A matrix A can be added to another matrix B if and only if A and B have the same size, with corresponding entries in A and B added to form the entries in the sum $A + B$. Similarly, B can be subtracted from A if and only if A and B have the same size, with corresponding entries subtracted to form the entries in the difference $A - B$.

Example 8.1 Consider the following matrices A, B, and C.

$$A = \begin{bmatrix} 2 & 1 & 1 \\ -1 & -1 & 4 \end{bmatrix} \qquad B = \begin{bmatrix} 2 & -3 & 4 \\ -3 & 1 & -2 \end{bmatrix} \qquad C = \begin{bmatrix} 2 & 3 \\ 1 & 0 \end{bmatrix}$$

Since A and B have the same size, $A + B$ exists and can be found as follows.

$$\begin{aligned} A + B &= \begin{bmatrix} 2 & 1 & 1 \\ -1 & -1 & 4 \end{bmatrix} + \begin{bmatrix} 2 & -3 & 4 \\ -3 & 1 & -2 \end{bmatrix} \\ &= \begin{bmatrix} 2+2 & 1+(-3) & 1+4 \\ -1+(-3) & -1+1 & 4+(-2) \end{bmatrix} \\ &= \begin{bmatrix} 4 & -2 & 5 \\ -4 & 0 & 2 \end{bmatrix} \end{aligned}$$

Similarly, $A - B$ exists and can be found as follows.

$$\begin{aligned} A - B &= \begin{bmatrix} 2 & 1 & 1 \\ -1 & -1 & 4 \end{bmatrix} - \begin{bmatrix} 2 & -3 & 4 \\ -3 & 1 & -2 \end{bmatrix} \\ &= \begin{bmatrix} 2-2 & 1-(-3) & 1-4 \\ -1-(-3) & -1-1 & 4-(-2) \end{bmatrix} \\ &= \begin{bmatrix} 0 & 4 & -3 \\ 2 & -2 & 6 \end{bmatrix} \end{aligned}$$

However, since A and C do not have the same size, $A + C$ and $A - C$ do not exist.   □

A matrix A can be multiplied by a number n, with all entries in A multiplied by n to form the entries in the product nA. This operation is called *scalar multiplication*, with n referred to as a *scalar*.

Example 8.2 For the matrix C in Example 8.1, the scalar multiple $4C$ can be found as follows.

$$4C = 4 \begin{bmatrix} 2 & 3 \\ 1 & 0 \end{bmatrix} = \begin{bmatrix} 4 \cdot 2 & 4 \cdot 3 \\ 4 \cdot 1 & 4 \cdot 0 \end{bmatrix} = \begin{bmatrix} 8 & 12 \\ 4 & 0 \end{bmatrix}$$

Similarly, for the matrices A and B in Example 8.1, the matrix $5A - 4B$ can be found as follows.

$$\begin{aligned} 5A - 4B &= 5 \begin{bmatrix} 2 & 1 & 1 \\ -1 & -1 & 4 \end{bmatrix} - 4 \begin{bmatrix} 2 & -3 & 4 \\ -3 & 1 & -2 \end{bmatrix} \\ &= \begin{bmatrix} 10 & 5 & 5 \\ -5 & -5 & 20 \end{bmatrix} - \begin{bmatrix} 8 & -12 & 16 \\ -12 & 4 & -8 \end{bmatrix} \\ &= \begin{bmatrix} 2 & 17 & -11 \\ 7 & -9 & 28 \end{bmatrix} \end{aligned}$$

However, since the matrices A and C in Example 8.1 do not have the same size, $5A - 4C$ does not exist (although both $5A$ and $4C$ do exist). □

A matrix can sometimes be multiplied by another matrix, but the rules for this operation, called *matrix multiplication*, are very different from the simple rules for matrix addition, matrix subtraction, and scalar multiplication. Based on how these other types of operations are done, it would seem natural for matrix multiplication to be done by simply multiplying corresponding entries in a pair of matrices. However, matrix multiplication is not done this way.

In order to understand how matrix multiplication is done, we must first understand how to multiply a row matrix times a column matrix. Consider the following $1 \times n$ row matrix $\mathbf{a}$ and $n \times 1$ column matrix $\mathbf{b}$.

$$\mathbf{a} = \begin{bmatrix} a_{11} & a_{12} & \cdots & a_{1n} \end{bmatrix} \qquad \mathbf{b} = \begin{bmatrix} b_{11} \\ b_{21} \\ \vdots \\ b_{n1} \end{bmatrix}$$

For this pair of matrices $\mathbf{a}$ and $\mathbf{b}$, the product $\mathbf{ab}$ exists, and is a number. To find this number, we multiply each pair of corresponding entries in $\mathbf{a}$ and $\mathbf{b}$, reading from left-to-right across $\mathbf{a}$ and top-to-bottom down $\mathbf{b}$, and add the results together. The number obtained from this process is the product $\mathbf{ab}$. That is, the product $\mathbf{ab}$ is the following number.

$$\mathbf{ab} = a_{11}b_{11} + a_{12}b_{21} + \cdots + a_{1n}b_{n1}$$

This product $\mathbf{ab}$ exists if and only if $\mathbf{a}$ and $\mathbf{b}$ contain the same number of entries. A row matrix of size $1 \times n$ can only be multiplied times a column matrix of size $m \times 1$ if $n = m$.

Example 8.3 The result of the row matrix $\mathbf{a} = \begin{bmatrix} 2 & 1 & 3 \end{bmatrix}$ times the column matrix $\mathbf{b} = \begin{bmatrix} 2 \\ -4 \\ 5 \end{bmatrix}$ is the following.

$$\mathbf{ab} = \begin{bmatrix} 2 & 1 & 3 \end{bmatrix} \begin{bmatrix} 2 \\ -4 \\ 5 \end{bmatrix} = 2 \cdot 2 + 1 \cdot (-4) + 3 \cdot 5 = 15$$

However, for the row matrix $\mathbf{d} = \begin{bmatrix} 6 & 7 \end{bmatrix}$, since $\mathbf{d}$ and $\mathbf{b}$ do not contain the same number of entries, $\mathbf{db}$ does not exist. ☐

Matrix multiplication in general involves multiple multiplications of rows times columns. Specifically, if A is an $m \times n$ matrix and B is an $n \times p$ matrix, then the product $C = AB$ exists and has size $m \times p$, and each entry c_{ij} is formed by multiplying the ith row of A times the jth column of B.

$$c_{ij} = \begin{bmatrix} i\text{th row of } A \end{bmatrix} \begin{bmatrix} j\text{th} \\ \text{column} \\ \text{of} \\ B \end{bmatrix}$$

It is important to note that for a pair of matrices A and B, the product AB exists if and only if the number of columns in A (the second of its size parameters $m \times n$) equals the number of rows in B (the first of its size parameters $n \times p$). It is sometimes helpful for remembering this and identifying the size of a product to write the sizes of the matrices next to each other, as in the following diagram.

$$\begin{array}{cc} A & \cdot & B \\ m \times n & & n \times p \end{array}$$

The product AB exists if and only if the two inner size parameters n in this diagram are equal. Also, if AB exists, then its size will be given by the outer size parameters $m \times p$.

Example 8.4 Consider the following matrices $\mathbf{a}$ and B.

$$\mathbf{a} = \begin{bmatrix} 2 & 1 & 3 \end{bmatrix} \qquad B = \begin{bmatrix} 2 & 5 & 4 \\ -4 & 10 & 2 \\ 5 & 4 & 0 \end{bmatrix}$$

To see if the product $\mathbf{a}B$ exists, we form the following diagram.

$$\underset{1 \times 3}{\mathbf{a}} \quad \cdot \quad \underset{3 \times 3}{B}$$

Since the inner size parameters in this diagram are equal, the product $\mathbf{a}B$ exists, and its size is given by the outer size parameters 1×3. For the product $\mathbf{c} = \mathbf{a}B$, the entries are found as follows.

$$c_{11} \;=\; \begin{bmatrix} 2 & 1 & 3 \end{bmatrix} \begin{bmatrix} 2 \\ -4 \\ 5 \end{bmatrix} \;=\; 2 \cdot 2 + 1 \cdot (-4) + 3 \cdot 5 \;=\; 15$$

$$c_{12} \;=\; \begin{bmatrix} 2 & 1 & 3 \end{bmatrix} \begin{bmatrix} 5 \\ 10 \\ 4 \end{bmatrix} \;=\; 2 \cdot 5 + 1 \cdot 10 + 3 \cdot 4 \;=\; 32$$

$$c_{13} \;=\; \begin{bmatrix} 2 & 1 & 3 \end{bmatrix} \begin{bmatrix} 4 \\ 2 \\ 0 \end{bmatrix} \;=\; 2 \cdot 4 + 1 \cdot 2 + 3 \cdot 0 \;=\; 10$$

Thus, $\mathbf{c} = \mathbf{a}B = \begin{bmatrix} 15 & 32 & 10 \end{bmatrix}$. □

As you may have observed, for a pair of matrices A and B, the operation of matrix multiplication allows for the possibility that AB will exist but BA will not. For example, for the matrices $\mathbf{a}$ and B for which we calculated $\mathbf{a}B$ in Example 8.4, consider the following diagram.

$$\underset{3 \times 3}{B} \quad \cdot \quad \underset{1 \times 3}{\mathbf{a}}$$

Since the inner size parameters in this diagram are not equal, the product $B\mathbf{a}$ does not exist.

As you may have also observed, the operation of matrix multiplication allows for the possibility that both AB and BA will exist, but not be equal to each other. For example, for the row matrix $\mathbf{a}$ and column matrix $\mathbf{b}$ in Example 8.3, consider the following diagram.

$$\underset{3 \times 1}{\mathbf{b}} \quad \cdot \quad \underset{1 \times 3}{\mathbf{a}}$$

Since the inner size parameters in this diagram are equal, the product $\mathbf{ba}$ exists, and its size is given by the outer size parameters 3×3. However, in Example 8.3, we found that the product $\mathbf{ab}$ was the single number 15, or, equivalently, the 1×1 matrix $[15]$. Thus, although both $\mathbf{ab}$ and $\mathbf{ba}$ exist, they can certainly not be equal to each other, since $\mathbf{ab}$ is of size 1×1 and $\mathbf{ba}$ is of size 3×3.

Example 8.5 Consider the following matrices A and B.

$$A = \begin{bmatrix} 3 & 15 \\ 2 & 4 \\ 9 & 20 \end{bmatrix} \qquad B = \begin{bmatrix} 0 & 4 & 14 \\ 2 & 1 & 6 \end{bmatrix}$$

To see if the product AB exists, we form the following diagram.

$$\begin{array}{ccc} A & \cdot & B \\ {\scriptstyle 3\times 2} & & {\scriptstyle 2\times 3} \end{array}$$

Since the inner size parameters in this diagram are equal, the product AB exists, and its size is given by the outer size parameters 3×3. The product AB can be found as follows.

$$AB = \begin{bmatrix} 3 & 15 \\ 2 & 4 \\ 9 & 20 \end{bmatrix} \begin{bmatrix} 0 & 4 & 14 \\ 2 & 1 & 6 \end{bmatrix}$$

$$= \begin{bmatrix} 3\cdot 0 + 15\cdot 2 & 3\cdot 4 + 15\cdot 1 & 3\cdot 14 + 15\cdot 6 \\ 2\cdot 0 + 4\cdot 2 & 2\cdot 4 + 4\cdot 1 & 2\cdot 14 + 4\cdot 6 \\ 9\cdot 0 + 20\cdot 2 & 9\cdot 4 + 20\cdot 1 & 9\cdot 14 + 20\cdot 6 \end{bmatrix}$$

$$= \begin{bmatrix} 30 & 27 & 132 \\ 8 & 12 & 52 \\ 40 & 56 & 246 \end{bmatrix}$$

Next, to see if the product BA exists, we form the following diagram.

$$\begin{array}{ccc} B & \cdot & A \\ {\scriptstyle 2\times 3} & & {\scriptstyle 3\times 2} \end{array}$$

Since the inner size parameters in this diagram are equal, the product BA exists, and its size is given by the outer size parameters 2×2. The product BA can be found as follows.

$$BA = \begin{bmatrix} 0 & 4 & 14 \\ 2 & 1 & 6 \end{bmatrix} \begin{bmatrix} 3 & 15 \\ 2 & 4 \\ 9 & 20 \end{bmatrix}$$

$$= \begin{bmatrix} 0\cdot 3 + 4\cdot 2 + 14\cdot 9 & 0\cdot 15 + 4\cdot 4 + 14\cdot 20 \\ 2\cdot 3 + 1\cdot 2 + 6\cdot 9 & 2\cdot 15 + 1\cdot 4 + 6\cdot 20 \end{bmatrix}$$

$$= \begin{bmatrix} 134 & 296 \\ 62 & 154 \end{bmatrix}$$

Note in this example that although AB and BA both exist, they are not equal to each other. □

8.1.3   Identity and Inverse Matrices

An *identity* matrix (or, more accurately, a *multiplicative* identity) is a square matrix I containing ones diagonally from the upper left corner to the lower right corner, and zeros elsewhere.

$$I = \begin{bmatrix} 1 & 0 & \cdots & 0 & 0 \\ 0 & 1 & \cdots & 0 & 0 \\ \vdots & \vdots & \ddots & \vdots & \vdots \\ 0 & 0 & \cdots & 1 & 0 \\ 0 & 0 & \cdots & 0 & 1 \end{bmatrix}$$

The reason a matrix of this form is called an *identity* is because, as can easily be verified, for a given matrix A and appropriately sized choice(s) for I, it will be true that $AI = A$ and $IA = A$.

Since an identity matrix must be square, a single parameter indicating its number of rows or columns can be used to represent its size. This parameter is sometimes written as a subscript on the name I. For example, the following are the identity matrices I_2, I_3, and I_4, of sizes 2×2, 3×3, and 4×4, respectively.

$$I_2 = \begin{bmatrix} 1 & 0 \\ 0 & 1 \end{bmatrix} \quad I_3 = \begin{bmatrix} 1 & 0 & 0 \\ 0 & 1 & 0 \\ 0 & 0 & 1 \end{bmatrix} \quad I_4 = \begin{bmatrix} 1 & 0 & 0 & 0 \\ 0 & 1 & 0 & 0 \\ 0 & 0 & 1 & 0 \\ 0 & 0 & 0 & 1 \end{bmatrix}$$

The subscript indicating the size of an identity matrix is typically only included if the size must be specified for some reason. Otherwise, the identity is usually just expressed as I.

For a given matrix A, the size of I in the products AI and IA depends on the size of A and the side of A on which the identity is written. For instance, for a matrix A of size $m \times n$, in the product AI the identity would be I_n, but in the product IA it would be I_m. Only for a square matrix A would a particular identity be able to be used on both sides of A.

Example 8.6 For the 3×2 matrix A in Example 8.5, in the product AI the identity would be I_2.

$$AI = \begin{bmatrix} 3 & 15 \\ 2 & 4 \\ 9 & 20 \end{bmatrix} \begin{bmatrix} 1 & 0 \\ 0 & 1 \end{bmatrix} = \begin{bmatrix} 3 & 15 \\ 2 & 4 \\ 9 & 20 \end{bmatrix} = A$$

For this same matrix A, in the product IA the identity would be I_3.

$$IA = \begin{bmatrix} 1 & 0 & 0 \\ 0 & 1 & 0 \\ 0 & 0 & 1 \end{bmatrix} \begin{bmatrix} 3 & 15 \\ 2 & 4 \\ 9 & 20 \end{bmatrix} = \begin{bmatrix} 3 & 15 \\ 2 & 4 \\ 9 & 20 \end{bmatrix} = A$$

$\square$

With any square matrix A, we associate a number called the *determinant*, denoted $\det(A)$. Often the determinant of a matrix is very easy to find. For a matrix of size 1×1, the determinant is the lone entry in the matrix. For example, for the matrix $[12]$, the determinant is 12.

For a matrix of size 2×2, the determinant is also very easy to find. Specifically, for a 2×2 matrix, the determinant is found using the following formula.

$$\det \left(\begin{bmatrix} a_{11} & a_{12} \\ a_{21} & a_{22} \end{bmatrix} \right) = a_{11}a_{22} - a_{12}a_{21}$$

Example 8.7 For $A = \begin{bmatrix} 8 & 1 \\ 19 & 14 \end{bmatrix}$, $\det(A) = 8 \cdot 14 - 1 \cdot 19 = 93$.

$\square$

In Exercise 19 at the end of this section, we will show a method for finding the determinant of a 3×3 matrix. We will not show how to find the determinant of a matrix larger than 3×3 though, as finding determinants becomes difficult very quickly as the sizes of the matrices under consideration grow larger.

For a given square matrix A, the *inverse* (or, more accurately, the *multiplicative* inverse) of A is a matrix B with the property that $AB = I$ and $BA = I$.

Example 8.8 Consider the following matrices A and B.

$$A = \begin{bmatrix} 8 & 1 \\ 19 & 14 \end{bmatrix} \qquad B = \begin{bmatrix} 14/93 & -1/93 \\ -19/93 & 8/93 \end{bmatrix}$$

For these matrices, the following gives the product AB.

$$AB = \begin{bmatrix} 8 & 1 \\ 19 & 14 \end{bmatrix} \begin{bmatrix} 14/93 & -1/93 \\ -19/93 & 8/93 \end{bmatrix}$$

$$= \begin{bmatrix} 112/93 - 19/93 & -8/93 + 8/93 \\ 266/93 - 266/93 & -19/93 + 112/93 \end{bmatrix}$$

$$= \begin{bmatrix} 1 & 0 \\ 0 & 1 \end{bmatrix}$$

A similar calculation shows $BA = I$. Thus, B is the inverse of A. $\square$

The following are several important and relevant facts about matrices and their inverses.

- For a pair of matrices A and B, if $AB = I$, then it would have to be true that $BA = I$. Thus, to verify that B is the inverse of A, it is only necessary to check that $AB = I$.

- The inverse of a square matrix A will be unique, meaning that if B is a matrix such that $AB = I$, then B is the *only* matrix such that $AB = I$. Thus, it is accurate to use the word "the" when discussing the inverse of a matrix (as we have done).

- Since the inverse of a matrix A is unique to A, it is expressed using the notation A^{-1} (read "A-inverse"). For example, for the matrices A and B in Example 8.8, $B = A^{-1}$.

- Matrices and inverses come in pairs—if A^{-1} is the inverse of A, then A is the inverse of A^{-1}. For example, for the matrices A and B in Example 8.8 for which $B = A^{-1}$, it is also true that $A = B^{-1}$.

One additional important fact about matrices and their inverses bears mentioning—not every matrix has an inverse. For instance, for the matrix $C = \begin{bmatrix} 1 & 2 \\ 2 & 4 \end{bmatrix}$, there is no matrix C^{-1} such that $CC^{-1} = I$. A matrix that has an inverse is said to be *invertible*. For example, the matrix A in Example 8.8 is invertible, while the matrix $C = \begin{bmatrix} 1 & 2 \\ 2 & 4 \end{bmatrix}$ is not.

Given a matrix A, how can we figure out whether A^{-1} exists? Also, when A^{-1} exists, how can we find it? It turns out that for matrices of size 2×2 these questions are both easy to answer. Consider the following 2×2 matrix A.

$$A = \begin{bmatrix} a_{11} & a_{12} \\ a_{21} & a_{22} \end{bmatrix}$$

Then A^{-1} will exist if and only if $\det(A) \neq 0$, and when A^{-1} exists, it will be given by the following formula.

$$A^{-1} = (1/\det(A)) \begin{bmatrix} a_{22} & -a_{12} \\ -a_{21} & a_{11} \end{bmatrix}$$

That is, for a 2×2 matrix A, when A^{-1} exists we can find it by swapping the entries in the upper left and lower right corners of A, changing the sign of the other two entries in A, and multiplying the resulting matrix by the scalar $(1/\det(A))$.

Example 8.9 Consider the 2×2 matrix $A = \begin{bmatrix} 8 & 1 \\ 19 & 14 \end{bmatrix}$. Note first that $\det(A) = 8 \cdot 14 - 1 \cdot 19 = 93$. Since $\det(A) \neq 0$, A^{-1} exists, and we can find A^{-1} as follows.

$$A^{-1} = (1/93) \begin{bmatrix} 14 & -1 \\ -19 & 8 \end{bmatrix} = \begin{bmatrix} 14/93 & -1/93 \\ -19/93 & 8/93 \end{bmatrix}$$

□

Example 8.10 Consider the 2×2 matrix $C = \begin{bmatrix} 1 & 2 \\ 2 & 4 \end{bmatrix}$. Note that since $\det(C) = 1 \cdot 4 - 2 \cdot 2 = 0$, C^{-1} does not exist.   □

8.1.4 Matrices with Modular Arithmetic

Hill ciphers combine matrix operations with modular arithmetic. For a given matrix A, the matrix $A \bmod m$ is the result of converting each entry in A into its remainder modulo m. For example, for $A = \begin{bmatrix} 98 & -7 \\ -133 & 56 \end{bmatrix}$, we can find $A \bmod 26$ as follows.

$$\begin{bmatrix} 98 & -7 \\ -133 & 56 \end{bmatrix} \bmod 26 = \begin{bmatrix} 98 \bmod 26 & -7 \bmod 26 \\ -133 \bmod 26 & 56 \bmod 26 \end{bmatrix}$$

$$= \begin{bmatrix} 20 & 19 \\ 23 & 4 \end{bmatrix}$$

The operations of matrix addition, matrix subtraction, scalar multiplication, and matrix multiplication can all be done with modular arithmetic modulo m by finding the sum, difference, or product as usual, and then converting each entry in the resulting matrix into its remainder modulo m.

Example 8.11 Consider the following matrices A and B.

$$A = \begin{bmatrix} 2 & 1 & 1 \\ -1 & -1 & 4 \end{bmatrix} \qquad B = \begin{bmatrix} 2 & -3 & 4 \\ -3 & 1 & -2 \end{bmatrix}$$

In Example 8.2 on page 252, we found that $5A - 4B = \begin{bmatrix} 2 & 17 & -11 \\ 7 & -9 & 28 \end{bmatrix}$. Thus, the following gives $(5A - 4B) \bmod 15$.

$$(5A - 4B) \bmod 15 = \begin{bmatrix} 2 & 17 & -11 \\ 7 & -9 & 28 \end{bmatrix} \bmod 15$$

$$= \begin{bmatrix} 2 & 2 & 4 \\ 7 & 6 & 13 \end{bmatrix}$$

□

Example 8.12 Consider the following matrices A and B.

$$A = \begin{bmatrix} 3 & 15 \\ 2 & 4 \\ 9 & 20 \end{bmatrix} \qquad B = \begin{bmatrix} 0 & 4 & 14 \\ 2 & 1 & 6 \end{bmatrix}$$

In Example 8.5, we found that $AB = \begin{bmatrix} 30 & 27 & 132 \\ 8 & 12 & 52 \\ 40 & 56 & 246 \end{bmatrix}$. Thus, the

following gives $AB \bmod 26$.

$$AB \bmod 26 = \begin{bmatrix} 30 & 27 & 132 \\ 8 & 12 & 52 \\ 40 & 56 & 246 \end{bmatrix} \bmod 26$$

$$= \begin{bmatrix} 4 & 1 & 2 \\ 8 & 12 & 0 \\ 14 & 4 & 12 \end{bmatrix}$$

$\square$

The inverse of a matrix can exist for modular arithmetic modulo m as well. For a square matrix A with entries in $\mathbb{Z}_m = \{0, 1, 2, 3, \ldots, m - 1\}$, the inverse of A modulo m, if it exists, is a matrix B with entries in $\mathbb{Z}_m$ and with the property that $AB \bmod m$ is I. This inverse, when it exists, is denoted $A^{-1} \bmod m$.

Example 8.13 Consider the following matrices A and B.

$$A = \begin{bmatrix} 11 & 6 & 8 \\ 0 & 3 & 14 \\ 24 & 0 & 9 \end{bmatrix} \qquad B = \begin{bmatrix} 5 & 16 & 14 \\ 16 & 3 & 10 \\ 4 & 18 & 9 \end{bmatrix}$$

To see whether $B = A^{-1} \bmod 26$, we compute the following.

$$AB \bmod 26 = \begin{bmatrix} 11 & 6 & 8 \\ 0 & 3 & 14 \\ 24 & 0 & 9 \end{bmatrix} \begin{bmatrix} 5 & 16 & 14 \\ 16 & 3 & 10 \\ 4 & 18 & 9 \end{bmatrix} \bmod 26$$

$$= \begin{bmatrix} 183 & 338 & 286 \\ 104 & 261 & 156 \\ 156 & 546 & 417 \end{bmatrix} \bmod 26$$

$$= \begin{bmatrix} 1 & 0 & 0 \\ 0 & 1 & 0 \\ 0 & 0 & 1 \end{bmatrix}$$

Thus, $B = A^{-1} \bmod 26$.

$\square$

The formula for A^{-1} given on page 258 can be modified slightly to give the inverse of a 2×2 matrix for modular arithmetic modulo m. The matrix part of the formula is easy, since each entry in the matrix must just be converted into its remainder modulo m. The scalar part of the formula requires a little more thought, since fractions are not defined in modular arithmetic. We can account for this in the same way as when we introduced modular arithmetic in Section 6.1, by considering $(1/\det(A))$ as the multiplicative inverse of $\det(A)$ modulo m, which we will continue to represent like we did in Section 6.1 with the notation $(\det(A))^{-1} \bmod m$. By Theorem 6.5 on page 174, we know that $(\det(A))^{-1} \bmod m$ exists if and only if $\det(A)$ and m are relatively prime, that is, if $\gcd(\det(A), m) = 1$.

With Hill ciphers, all of our calculations will be done modulo 26. For convenience, each value of $\det(A) \bmod 26$ for which $(\det(A))^{-1} \bmod 26$ exists is shown along with $(\det(A))^{-1} \bmod 26$ in Table 8.1.

$\det(A)$	1	3	5	7	9	11	15	17	19	21	23	25
$(\det(A))^{-1}$	1	9	21	15	3	19	7	23	11	5	17	25

Table 8.1   Corresponding values of $\det(A)$ and $(\det(A))^{-1} \bmod 26$.

To clarify, for a particular given matrix $A = \begin{bmatrix} a_{11} & a_{12} \\ a_{21} & a_{22} \end{bmatrix}$ with entries in $\mathbb{Z}_{26} = \{0, 1, 2, 3, \ldots, 25\}$, $A^{-1} \bmod 26$ exists if and only if $\det(A)$ and 26 are relatively prime, that is, if $\gcd(\det(A), 26) = 1$. The values of $\det(A) \bmod 26$ for which $\gcd(\det(A), 26) = 1$ are listed in the first row in Table 8.1. Also, when $A^{-1} \bmod 26$ exists, it will be given by the following formula.

$$A^{-1} \bmod 26 = (\det(A))^{-1} \begin{bmatrix} a_{22} & -a_{12} \\ -a_{21} & a_{11} \end{bmatrix} \bmod 26$$

For this formula, the values of $(\det(A))^{-1} \bmod 26$ are listed in the second row in Table 8.1, in correspondence with the values of $\det(A) \bmod 26$ listed in the first row.

Example 8.14 Consider the matrix $A = \begin{bmatrix} 8 & 1 \\ 19 & 14 \end{bmatrix}$. Note first that we can find $\det(A) \bmod 26$ as follows.

$$\det(A) = 8 \cdot 14 - 1 \cdot 19 = 93 = 15 \bmod 26$$

Since $\gcd(15, 26) = 1$, then we know that $A^{-1} \bmod 26$ exists. Also, since $\det(A) = 15 \bmod 26$, Table 8.1 gives $(\det(A))^{-1} = 7 \bmod 26$, and we can find A^{-1} mod 26 as follows.

$$A^{-1} \bmod 26 \; = \; 15^{-1} \begin{bmatrix} 14 & -1 \\ -19 & 8 \end{bmatrix} \bmod 26$$

$$= \; 7 \begin{bmatrix} 14 & -1 \\ -19 & 8 \end{bmatrix} \bmod 26$$

$$= \; \begin{bmatrix} 98 & -7 \\ -133 & 56 \end{bmatrix} \bmod 26$$

$$= \; \begin{bmatrix} 20 & 19 \\ 23 & 4 \end{bmatrix}$$

□

Example 8.15 Consider the matrix $A = \begin{bmatrix} 8 & 1 \\ 4 & 6 \end{bmatrix}$. Note that we can find det(A) mod 26 as follows.

$$\det(A) = 8 \cdot 6 - 1 \cdot 4 = 44 = 18 \bmod 26$$

Since $\gcd(18, 26) \neq 1$, A^{-1} mod 26 does not exist.   □

8.1.5   Exercises

1. Consider the following matrices A and B.

$$A = \begin{bmatrix} 5 & 10 \\ -6 & 7 \\ 1 & 0 \end{bmatrix} \qquad B = \begin{bmatrix} 2 & 0 \\ 4 & -9 \\ 3 & 8 \end{bmatrix}$$

Find the following.

(a)* $A + B$

(b) $A - B$

(c)* $3A$

(d) $4B$

(e) $3A + 4B$

2.* Repeat Exercise 1, but for each part, find the result modulo 10.

3. For the following matrices $\mathbf{a}$ and B, find $\mathbf{a}B$.

(a)* $\mathbf{a} = \begin{bmatrix} 1 & 3 \end{bmatrix}$, $B = \begin{bmatrix} 1 & 3 \\ 4 & 2 \end{bmatrix}$

(b) $\mathbf{a} = \begin{bmatrix} 4 & 2 \end{bmatrix}$, $B = \begin{bmatrix} 10 & 4 \\ 7 & 6 \end{bmatrix}$

(c)* $\mathbf{a} = \begin{bmatrix} 4 & 5 & 0 \end{bmatrix}$, $B = \begin{bmatrix} 2 & 3 & 5 \\ 6 & 2 & 4 \\ 0 & 1 & 5 \end{bmatrix}$

(d) $\mathbf{a} = \begin{bmatrix} -2 & 2 & -8 \end{bmatrix}$, $B = \begin{bmatrix} 4 & -3 & 6 \\ 0 & 1 & 2 \\ 7 & 4 & -3 \end{bmatrix}$

(e) $\mathbf{a} = \begin{bmatrix} 8 & 10 & 2 & 7 \end{bmatrix}$, $B = \begin{bmatrix} 13 & 2 & 15 & 17 \\ 19 & 17 & 0 & 4 \\ 19 & 2 & 18 & 11 \\ 14 & 14 & 1 & 8 \end{bmatrix}$

4.* Repeat Exercise 3, but for each part, find $\mathbf{a}B$ mod 26.

5. For the following matrices A and B, find AB and BA.

(a)* $A = \begin{bmatrix} 1 & 2 \\ 3 & 4 \end{bmatrix}$, $B = \begin{bmatrix} 5 & 6 \\ 7 & 8 \end{bmatrix}$

(b) $A = \begin{bmatrix} 13 & 10 \\ 5 & -4 \end{bmatrix}$, $B = \begin{bmatrix} 10 & 4 \\ 7 & 6 \end{bmatrix}$

(c)* $A = \begin{bmatrix} 4 & 5 & 0 \\ -10 & 0 & -2 \\ 8 & -8 & 1 \end{bmatrix}$, $B = \begin{bmatrix} 2 & 3 & 5 \\ 6 & 2 & 4 \\ 0 & 1 & 5 \end{bmatrix}$

(d) $A = \begin{bmatrix} 2 & 2 & 1 \\ 0 & 5 & 2 \\ 5 & 3 & 7 \end{bmatrix}$, $B = \begin{bmatrix} 4 & 2 & 6 \\ 0 & 1 & 2 \\ 7 & 4 & 2 \end{bmatrix}$

6.* Repeat Exercise 5, but for each part, find AB mod 26 and BA mod 26.

7. For the following matrices A and B, find, if they exist, AB and BA.

(a)* $A = \begin{bmatrix} 1 & 2 \\ 3 & 4 \end{bmatrix}$, $B = \begin{bmatrix} 5 & 6 & 7 \\ 7 & 8 & 9 \end{bmatrix}$

(b) $A = \begin{bmatrix} 13 & 10 & 0 \\ -5 & 4 & -1 \end{bmatrix}$, $B = \begin{bmatrix} 0 & 4 \\ 7 & 6 \\ 2 & 4 \end{bmatrix}$

(c)* $A = \begin{bmatrix} 2 & 1 \\ 0 & 5 \end{bmatrix}$, $B = \begin{bmatrix} 4 & 2 \\ 0 & 10 \\ 9 & 7 \end{bmatrix}$

(d) $A = \begin{bmatrix} 12 & 0 & -4 & -19 \\ 4 & 12 & 10 & -3 \end{bmatrix}$, $B = \begin{bmatrix} 13 & 2 \\ -15 & -10 \\ -12 & -9 \\ 0 & 2 \end{bmatrix}$

8.* Repeat Exercise 7, but for each part, find, if they exist, $AB \bmod 26$ and $BA \bmod 26$.

9.* For the following matrices A and b, find, if they exist, Ab and bA.

$$A = \begin{bmatrix} 4 & 5 & 0 \\ -10 & 0 & -2 \\ 8 & -8 & 1 \end{bmatrix}, b = \begin{bmatrix} 2 & 3 & 5 \end{bmatrix}$$

10.* Repeat Exercise 9, but find, if they exist, $Ab \bmod 26$ and $bA \bmod 26$.

11.* For the following matrices a and b, find, if they exist, ab and ba.

$$a = \begin{bmatrix} 12 & 0 & -4 & -19 & 2 \end{bmatrix}, b = \begin{bmatrix} 13 \\ -17 \\ -12 \\ 10 \\ 19 \end{bmatrix}$$

12.* Repeat Exercise 11, but find, if they exist, $ab \bmod 26$ and $ba \bmod 26$.

13. Find the determinant of the following matrices.

(a)* $\begin{bmatrix} 15 & 4 \\ 17 & 3 \end{bmatrix}$

(b) $\begin{bmatrix} 13 & 7 \\ 8 & 21 \end{bmatrix}$

(c)* $\begin{bmatrix} 6 & 7 \\ 8 & 15 \end{bmatrix}$

(d) $\begin{bmatrix} 2 & 11 \\ 1 & 21 \end{bmatrix}$

14.* Repeat Exercise 13, but for each part, find the determinant modulo 26.

15. Find the inverse, if it exists, of the following matrices.

(a)* $\begin{bmatrix} 15 & 4 \\ 17 & 3 \end{bmatrix}$

(b) $\begin{bmatrix} 13 & 7 \\ 8 & 21 \end{bmatrix}$

(c)* $\begin{bmatrix} 6 & 7 \\ 8 & 15 \end{bmatrix}$

(d) $\begin{bmatrix} 2 & 11 \\ 1 & 21 \end{bmatrix}$

16.* Repeat Exercise 15, but for each part, find the inverse, if it exists, modulo 26.

17. For the following matrices A and B, determine if $B = A^{-1} \bmod 26$.

(a)* $A = \begin{bmatrix} 1 & 2 & 1 \\ 3 & 1 & 0 \\ 0 & 2 & 1 \end{bmatrix}$, $B = \begin{bmatrix} 1 & 0 & 25 \\ 23 & 1 & 3 \\ 6 & 24 & 21 \end{bmatrix}$

(b) $A = \begin{bmatrix} 3 & 1 & 2 \\ 17 & 8 & 0 \\ 21 & 10 & 3 \end{bmatrix}$, $B = \begin{bmatrix} 2 & 9 & 16 \\ 25 & 7 & 18 \\ 24 & 9 & 19 \end{bmatrix}$

(c) $A = \begin{bmatrix} 5 & 7 & 8 \\ 10 & 1 & 0 \\ 1 & 2 & 1 \end{bmatrix}$, $B = \begin{bmatrix} 3 & 1 & 2 \\ 23 & 16 & 6 \\ 5 & 17 & 13 \end{bmatrix}$

18. For the following matrices A and B, assuming $\det(A) \neq 0$, verify that B is the inverse of A.

$$A = \begin{bmatrix} a_{11} & a_{12} \\ a_{21} & a_{22} \end{bmatrix} \qquad B = (1/\det(A)) \begin{bmatrix} a_{22} & -a_{12} \\ -a_{21} & a_{11} \end{bmatrix}$$

Exercises 19 and 20 use an alternate notation for the determinant of a matrix. This alternate notation is to replace the square brackets surrounding the matrix with vertical lines, indicating that the expression is the determinant of the matrix, rather than the matrix itself. For example, the determinant of a 2×2 matrix can be expressed as $\begin{vmatrix} a & b \\ c & d \end{vmatrix} = ad - bc$.

19. For a matrix $A = \begin{bmatrix} a & b & c \\ d & e & f \\ g & h & i \end{bmatrix}$, the following is a formula for the determinant of A.

$$\det(A) = a \begin{vmatrix} e & f \\ h & i \end{vmatrix} - b \begin{vmatrix} d & f \\ g & i \end{vmatrix} + c \begin{vmatrix} d & e \\ g & h \end{vmatrix}$$

Use this formula to find the determinant modulo 26 of the following matrices.

(a)* $\begin{bmatrix} 3 & 1 & 2 \\ 17 & 8 & 0 \\ 21 & 10 & 3 \end{bmatrix}$

(b) $\begin{bmatrix} 1 & 2 & 1 \\ 3 & 1 & 0 \\ 0 & 2 & 1 \end{bmatrix}$

(c) $\begin{bmatrix} 5 & 7 & 8 \\ 10 & 1 & 0 \\ 1 & 2 & 1 \end{bmatrix}$

20.* For a matrix $A = \begin{bmatrix} a & b & c \\ d & e & f \\ g & h & i \end{bmatrix}$ with entries in $\mathbb{Z}_{26}$, $A^{-1} \bmod 26$ exists if and only if $\gcd(\det(A), 26) = 1$, and when $A^{-1} \bmod 26$ exists, it will be given by the following expression.

$$(\det(A))^{-1} \begin{bmatrix} \begin{vmatrix} e & f \\ h & i \end{vmatrix} & -\begin{vmatrix} b & c \\ h & i \end{vmatrix} & \begin{vmatrix} b & c \\ e & f \end{vmatrix} \\ -\begin{vmatrix} d & f \\ g & i \end{vmatrix} & \begin{vmatrix} a & c \\ g & i \end{vmatrix} & -\begin{vmatrix} a & c \\ d & f \end{vmatrix} \\ \begin{vmatrix} d & e \\ g & h \end{vmatrix} & -\begin{vmatrix} a & b \\ g & h \end{vmatrix} & \begin{vmatrix} a & b \\ d & e \end{vmatrix} \end{bmatrix} \bmod 26$$

Use this formula to find the inverse modulo 26, if it exists, of the matrices in Exercise 19.

8.2　Hill Ciphers

One way mathematics and cryptology are connected is that, in many types of ciphers, the encryption and decryption procedures can or must be expressed using mathematical operations. However, the art of cryptology existed long before its connections with mathematics were observed. William Friedman was one of the first to connect mathematics and cryptology with his index of coincidence, but while Friedman used mathematics in breaking Vigenère keyword ciphers, the encryption and decryption procedures in these ciphers do not have to be expressed using mathematical operations. The first type of cipher that actually used mathematics in its encryption and decryption procedures appeared in the 1929 article *Cryptography in an Algebraic Alphabet* [10], written by a little-known mathematician from Hunter College named Lester Hill.

Prior to publishing this new type of cipher, Hill had applied for a patent for a technique he designed for using mathematics in error checking of numbers transmitted via telegraph. Hill subsequently published his error checking technique in the journal *Telegraph and Telephone Age*, perhaps in the hope that it would be adopted by the industry and lead to a financial windfall for himself. While this did not happen, Hill was not discouraged from trying again with a new cryptographic apparatus for which he was actually awarded a patent. Shortly after the 1929 article in which Hill described the first type of cipher that used mathematics in its encryption and decryption procedures, Hill lectured on this type of cipher before the American Mathematical Society. In this lecture, Hill also presented his cryptographic apparatus, and may have even given precedence to the apparatus over the type of cipher, as his lecture was later published under the title *Concerning Certain Linear Transformation Apparatus of Cryptography* [11]. Hill once again failed to find fortune through his ingenuity, but Hill's new type of cipher was and continues to be recognized for its brilliance, as is Hill himself alongside Friedman among a select group of cryptologists acknowledged for moving cryptology from being considered mainly an art to being considered equally as a science.

With Hill ciphers, we will again consider plaintext messages expressed using only the letters in the alphabet A, B, C, ... , Z, and convert these

Jack Levine: Prolific Researcher of Hill Ciphers

Jack Levine was a faculty member at North Carolina State University for 60 years. Much of Levine's research was devoted to improving and cryptanalyzing Hill ciphers. In fact, approximately three years before the 1929 article in which Lester Hill first described Hill ciphers, Levine, a teenager at the time, published an article in a detective magazine in which a very similar method for encrypting messages was used. Levine, who coined the term *algebraic cryptography* for a method that uses an algebraic system to encrypt messages, served as a cryptanalyst for the U.S. Army Security Agency during World War II, and received the Legion of Merit for his classified work. During his academic career, Levine published over 100 papers in scholarly journals, more than 40 of which were devoted to cryptology, with the last published in 1993 when Levine was 85 years of age. Levine also advised a dozen doctoral students at NC State, most of whose research was related to cryptology.

letters into numbers using the correspondences $A = 0$, $B = 1$, $C = 2$, ... ,
$Z = 25$. Recall that encryption in an affine cipher is done with a calculation
of the form $y = (ax + b) \bmod 26$, or, equivalently, $y = (xa + b) \bmod 26$, for
some a and b in $\mathbb{Z}_{26}$, where a is chosen so that $a^{-1} \bmod 26$ exists. Consider
now just the affine ciphers for which $b = 0$, so encryption is done with a
calculation of the form $y = xa \bmod 26$. Hill ciphers are a generalization of
these affine ciphers, where the plaintext x is a row matrix $\mathbf{x}$ containing a list
of plaintext numbers, and a is a square matrix A with the same number of
rows (or columns) as the number of entries in $\mathbf{x}$. That is, for a Hill cipher,
we begin by grouping the plaintext numbers in order into row matrices[2] of
some fixed size, say $1 \times n$, and then choosing an encryption matrix A of
size $n \times n$ with entries in $\mathbb{Z}_{26}$, where A is chosen so that $A^{-1} \bmod 26$ exists.
Then, for each plaintext row matrix $\mathbf{x}$, we form a corresponding ciphertext
row matrix $\mathbf{y}$ with a calculation of the form $\mathbf{y} = \mathbf{x}A \bmod 26$. The entries in
the ciphertext row matrices $\mathbf{y}$, when strung together in order and converted
back into letters, form the ciphertext.

More specifically, and for example, let A be a 2×2 matrix for which
$A^{-1} \bmod 26$ exists, and suppose we wish to encrypt a plaintext using a Hill
cipher with encryption matrix A. We begin by converting the list of letters
in the plaintext into a list of numbers, and then grouping these numbers in
order into row matrices of size 1×2. Note that the last plaintext row matrix
will be filled only if the length of the plaintext is a multiple of 2 (or, more
generally, a multiple of the number of rows in A). If this is not the case,
then the plaintext should be padded at the end with a letter (or letters,
possibly, if A were larger than 2×2) so that the length of the plaintext is
a multiple of 2. Assuming this, suppose the plaintext is of length n, with
plaintext numbers $x_1, x_2, x_3, \ldots, x_n$ in order. Then the corresponding list
of ciphertext numbers $y_1, y_2, y_3, \ldots, y_n$ is found by forming the following
matrix products.

$$\begin{bmatrix} y_1 & y_2 \end{bmatrix} = \begin{bmatrix} x_1 & x_2 \end{bmatrix} A \bmod 26$$
$$\begin{bmatrix} y_3 & y_4 \end{bmatrix} = \begin{bmatrix} x_3 & x_4 \end{bmatrix} A \bmod 26$$
$$\begin{bmatrix} y_5 & y_6 \end{bmatrix} = \begin{bmatrix} x_5 & x_6 \end{bmatrix} A \bmod 26$$
$$\vdots$$
$$\begin{bmatrix} y_{n-1} & y_n \end{bmatrix} = \begin{bmatrix} x_{n-1} & x_n \end{bmatrix} A \bmod 26$$

The ciphertext numbers $y_1, y_2, y_3, \ldots, y_n$ are then converted back into
letters to give the final ciphertext.

---

[2]With Hill ciphers, it is possible to use column matrices here instead of row matrices.
However, since using column matrices does not increase (or decrease) the security of
Hill ciphers, for consistency we will only use row matrices.

Example 8.16 Consider a Hill cipher with the following encryption matrix A.

$$A = \begin{bmatrix} 8 & 1 \\ 19 & 14 \end{bmatrix}$$

To use this matrix in encrypting the plaintext BE HERE AT SEVEN, since this plaintext contains 13 letters, we begin by padding an A at the end of the plaintext so that its length will be a multiple of 2. Next, we use the correspondences A = 0, B = 1, C = 2, ..., Z = 25 to convert the plaintext from a list of letters into a list of numbers.

B	E	H	E	R	E	A	T	S	E	V	E	N	A
1	4	7	4	17	4	0	19	18	4	21	4	13	0

To encrypt the plaintext, we form the following matrix products.

$$\begin{bmatrix} 1 & 4 \end{bmatrix} A = \begin{bmatrix} 1 & 4 \end{bmatrix} \begin{bmatrix} 8 & 1 \\ 19 & 14 \end{bmatrix} = \begin{bmatrix} 6 & 5 \end{bmatrix} \bmod 26$$

$$\begin{bmatrix} 7 & 4 \end{bmatrix} A = \begin{bmatrix} 7 & 4 \end{bmatrix} \begin{bmatrix} 8 & 1 \\ 19 & 14 \end{bmatrix} = \begin{bmatrix} 2 & 11 \end{bmatrix} \bmod 26$$

$$\begin{bmatrix} 17 & 4 \end{bmatrix} A = \begin{bmatrix} 17 & 4 \end{bmatrix} \begin{bmatrix} 8 & 1 \\ 19 & 14 \end{bmatrix} = \begin{bmatrix} 4 & 21 \end{bmatrix} \bmod 26$$

$$\begin{bmatrix} 0 & 19 \end{bmatrix} A = \begin{bmatrix} 0 & 19 \end{bmatrix} \begin{bmatrix} 8 & 1 \\ 19 & 14 \end{bmatrix} = \begin{bmatrix} 23 & 6 \end{bmatrix} \bmod 26$$

$$\begin{bmatrix} 18 & 4 \end{bmatrix} A = \begin{bmatrix} 18 & 4 \end{bmatrix} \begin{bmatrix} 8 & 1 \\ 19 & 14 \end{bmatrix} = \begin{bmatrix} 12 & 22 \end{bmatrix} \bmod 26$$

$$\begin{bmatrix} 21 & 4 \end{bmatrix} A = \begin{bmatrix} 21 & 4 \end{bmatrix} \begin{bmatrix} 8 & 1 \\ 19 & 14 \end{bmatrix} = \begin{bmatrix} 10 & 25 \end{bmatrix} \bmod 26$$

$$\begin{bmatrix} 13 & 0 \end{bmatrix} A = \begin{bmatrix} 13 & 0 \end{bmatrix} \begin{bmatrix} 8 & 1 \\ 19 & 14 \end{bmatrix} = \begin{bmatrix} 0 & 13 \end{bmatrix} \bmod 26$$

Converting the ciphertext numbers back into letters yields the following.

6	5	2	11	4	21	23	6	12	22	10	25	0	13
G	F	C	L	E	V	X	G	M	W	K	Z	A	N

Thus, the ciphertext is GFCLE VXGMW KZAN.   □

Recall that the encryption matrix A in a Hill cipher must be chosen so that $A^{-1} \bmod 26$ exists. This is necessary for decryption. For a ciphertext formed using a Hill cipher with encryption calculation $\mathbf{y} = \mathbf{x}A \bmod 26$, the corresponding decryption calculation is $\mathbf{x} = \mathbf{y}A^{-1} \bmod 26$.

More specifically, and for example, let A be a 2×2 matrix for which $A^{-1} \bmod 26$ exists, and suppose we wish to decrypt a ciphertext that was formed using a Hill cipher with encryption matrix A. We begin by converting the list of letters in the ciphertext into a list of numbers, and then grouping these numbers in order into row matrices of size 1×2. Suppose the ciphertext is of length n (which would already be a multiple of 2, or, more generally, a multiple of the number of rows in A). For the list of ciphertext numbers $y_1, y_2, y_3, \ldots, y_n$, the corresponding list of plaintext numbers x_1, $x_2, x_3, \ldots, x_n$ is found by forming the following matrix products.

$$\begin{bmatrix} x_1 & x_2 \end{bmatrix} = \begin{bmatrix} y_1 & y_2 \end{bmatrix} A^{-1} \bmod 26$$
$$\begin{bmatrix} x_3 & x_4 \end{bmatrix} = \begin{bmatrix} y_3 & y_4 \end{bmatrix} A^{-1} \bmod 26$$
$$\vdots$$
$$\begin{bmatrix} x_{n-1} & x_n \end{bmatrix} = \begin{bmatrix} y_{n-1} & y_n \end{bmatrix} A^{-1} \bmod 26$$

The plaintext numbers $x_1, x_2, x_3, \ldots, x_n$ are then converted back into letters to give the final plaintext.

Example 8.17 Consider the ciphertext OENGD ZHCXG GE, which was formed using a Hill cipher with the encryption matrix A in Example 8.16. For this matrix A, we found in Example 8.14 on page 261 that $A^{-1} \bmod 26$ is the following.

$$A^{-1} \bmod 26 = \begin{bmatrix} 20 & 19 \\ 23 & 4 \end{bmatrix}$$

Converting the ciphertext letters into numbers yields the following.

O	E	N	G	D	Z	H	C	X	G	G	E
14	4	13	6	3	25	7	2	23	6	6	4

To decrypt the ciphertext, we form the following matrix products.

$$\begin{bmatrix} 14 & 4 \end{bmatrix} A^{-1} = \begin{bmatrix} 14 & 4 \end{bmatrix} \begin{bmatrix} 20 & 19 \\ 23 & 4 \end{bmatrix} = \begin{bmatrix} 8 & 22 \end{bmatrix} \bmod 26$$

$$\begin{bmatrix} 13 & 6 \end{bmatrix} A^{-1} = \begin{bmatrix} 13 & 6 \end{bmatrix} \begin{bmatrix} 20 & 19 \\ 23 & 4 \end{bmatrix} = \begin{bmatrix} 8 & 11 \end{bmatrix} \bmod 26$$

$$\begin{bmatrix} 3 & 25 \end{bmatrix} A^{-1} = \begin{bmatrix} 3 & 25 \end{bmatrix} \begin{bmatrix} 20 & 19 \\ 23 & 4 \end{bmatrix} = \begin{bmatrix} 11 & 1 \end{bmatrix} \bmod 26$$

$$\begin{bmatrix} 7 & 2 \end{bmatrix} A^{-1} = \begin{bmatrix} 7 & 2 \end{bmatrix} \begin{bmatrix} 20 & 19 \\ 23 & 4 \end{bmatrix} = \begin{bmatrix} 4 & 11 \end{bmatrix} \bmod 26$$

$$\begin{bmatrix} 23 & 6 \end{bmatrix} A^{-1} = \begin{bmatrix} 23 & 6 \end{bmatrix} \begin{bmatrix} 20 & 19 \\ 23 & 4 \end{bmatrix} = \begin{bmatrix} 0 & 19 \end{bmatrix} \bmod 26$$

$$\begin{bmatrix} 6 & 4 \end{bmatrix} A^{-1} = \begin{bmatrix} 6 & 4 \end{bmatrix} \begin{bmatrix} 20 & 19 \\ 23 & 4 \end{bmatrix} = \begin{bmatrix} 4 & 0 \end{bmatrix} \bmod 26$$

Converting the plaintext numbers back into letters yields the following.

$$\begin{array}{cccccccccccc} 8 & 22 & 8 & 11 & 11 & 1 & 4 & 11 & 0 & 19 & 4 & 0 \\ I & W & I & L & L & B & E & L & A & T & E & A \end{array}$$

Thus, the plaintext is I WILL BE LATE.   □

Each of the previous two examples used a Hill cipher in which the encryption matrix A was of size 2×2. It is not required, of course, that A be of size 2×2, but rather just that A be square (and that A^{-1} mod 26 exist).

Example 8.18 Consider a Hill cipher with the following encryption matrix A.

$$A = \begin{bmatrix} 11 & 6 & 8 \\ 0 & 3 & 14 \\ 24 & 0 & 9 \end{bmatrix}$$

To use this matrix in encrypting the plaintext BE ON TIME AT TEN, since this plaintext contains 13 letters, we begin by padding two As at the end of the plaintext so that its length will be a multiple of 3. Next, we convert the plaintext from a list of letters into a list of numbers.

$$\begin{array}{cccccccccccccc} B & E & O & N & T & I & M & E & A & T & T & E & N & A & A \\ 1 & 4 & 14 & 13 & 19 & 8 & 12 & 4 & 0 & 19 & 19 & 4 & 13 & 0 & 0 \end{array}$$

To encrypt the plaintext, we form the following matrix products.

$$\begin{bmatrix} 1 & 4 & 14 \end{bmatrix} \begin{bmatrix} 11 & 6 & 8 \\ 0 & 3 & 14 \\ 24 & 0 & 9 \end{bmatrix} = \begin{bmatrix} 9 & 18 & 8 \end{bmatrix} \bmod 26$$

$$\begin{bmatrix} 13 & 19 & 8 \end{bmatrix} \begin{bmatrix} 11 & 6 & 8 \\ 0 & 3 & 14 \\ 24 & 0 & 9 \end{bmatrix} = \begin{bmatrix} 23 & 5 & 0 \end{bmatrix} \bmod 26$$

$$\begin{bmatrix} 12 & 4 & 0 \end{bmatrix} \begin{bmatrix} 11 & 6 & 8 \\ 0 & 3 & 14 \\ 24 & 0 & 9 \end{bmatrix} = \begin{bmatrix} 2 & 6 & 22 \end{bmatrix} \bmod 26$$

$$\begin{bmatrix} 19 & 19 & 4 \end{bmatrix} \begin{bmatrix} 11 & 6 & 8 \\ 0 & 3 & 14 \\ 24 & 0 & 9 \end{bmatrix} = \begin{bmatrix} 19 & 15 & 12 \end{bmatrix} \bmod 26$$

$$\begin{bmatrix} 13 & 0 & 0 \end{bmatrix} \begin{bmatrix} 11 & 6 & 8 \\ 0 & 3 & 14 \\ 24 & 0 & 9 \end{bmatrix} = \begin{bmatrix} 13 & 0 & 0 \end{bmatrix} \bmod 26$$

Converting the ciphertext numbers back into letters yields the following.

9	18	8	23	5	0	2	6	22	19	15	12	13	0	0
J	S	I	X	F	A	C	G	W	T	P	M	N	A	A

Thus, the ciphertext is JSIXF ACGWT PMNAA.   □

Example 8.19 Consider the ciphertext TFNGK ZGLVK AECUS, which was formed using a Hill cipher with the encryption matrix A in Example 8.18. For this matrix A, we showed in Example 8.13 on page 260 that A^{-1} mod 26 is the following.

$$A^{-1} \bmod 26 = \begin{bmatrix} 5 & 16 & 14 \\ 16 & 3 & 10 \\ 4 & 18 & 9 \end{bmatrix}$$

Converting the ciphertext letters into numbers yields the following.

T	F	N	G	K	Z	G	L	V	K	A	E	C	U	S
19	5	13	6	10	25	6	11	21	10	0	4	2	20	18

Thus, to decrypt the ciphertext, we form the following matrix products.

$$\begin{bmatrix} 19 & 5 & 13 \end{bmatrix} \begin{bmatrix} 5 & 16 & 14 \\ 16 & 3 & 10 \\ 4 & 18 & 9 \end{bmatrix} = \begin{bmatrix} 19 & 7 & 17 \end{bmatrix} \bmod 26$$

$$\begin{bmatrix} 6 & 10 & 25 \end{bmatrix} \begin{bmatrix} 5 & 16 & 14 \\ 16 & 3 & 10 \\ 4 & 18 & 9 \end{bmatrix} = \begin{bmatrix} 4 & 4 & 19 \end{bmatrix} \bmod 26$$

$$\begin{bmatrix} 6 & 11 & 21 \end{bmatrix} \begin{bmatrix} 5 & 16 & 14 \\ 16 & 3 & 10 \\ 4 & 18 & 9 \end{bmatrix} = \begin{bmatrix} 4 & 13 & 19 \end{bmatrix} \bmod 26$$

$$\begin{bmatrix} 10 & 0 & 4 \end{bmatrix} \begin{bmatrix} 5 & 16 & 14 \\ 16 & 3 & 10 \\ 4 & 18 & 9 \end{bmatrix} = \begin{bmatrix} 14 & 24 & 20 \end{bmatrix} \bmod 26$$

$$\begin{bmatrix} 2 & 20 & 18 \end{bmatrix} \begin{bmatrix} 5 & 16 & 14 \\ 16 & 3 & 10 \\ 4 & 18 & 9 \end{bmatrix} = \begin{bmatrix} 12 & 0 & 0 \end{bmatrix} \bmod 26$$

Converting the plaintext numbers back into letters yields the following.

19	7	17	4	4	19	4	13	19	14	24	20	12	0	0
T	H	R	E	E	T	E	N	T	O	Y	U	M	A	A

Thus, the plaintext is THREE TEN TO YUMA.   □

The fact that Hill ciphers encrypt plaintext numbers in groups rather than one at a time can in a sense be viewed as a negative feature since a transcription error in a single ciphertext letter usually leads to more than one error in the decrypted message. (See Exercises 10 and 11 at the end of this section.) However, this negative aspect is far outweighed by the additional security gained by encrypting plaintext numbers in groups. In Section 8.3, we will consider the additional security that this brings.

8.2.1 Exercises

1. Consider a Hill cipher with the encryption matrix A in Example 8.16.

   (a)* Use this cipher to encrypt WHEN YOU PLAY, PLAY HARD.

   (b) Use this cipher to encrypt WHEN YOU WORK.

   (c) Decrypt ERXTR NOYXG BYKL,[3] which was formed using this cipher. (Recall that A^{-1} mod 26 is given in Example 8.17.)

2. Consider a Hill cipher with the following encryption matrix A.

$$A = \begin{bmatrix} 15 & 4 \\ 17 & 3 \end{bmatrix}$$

   (a)* Use this cipher to encrypt MITCHELL.

   (b) Use this cipher to encrypt WHITNEY.

   (c)* Find A^{-1} mod 26.

   (d) Decrypt CMBPV ISMDT ZEWS, which was formed using this cipher.

3. Consider a Hill cipher with the following encryption matrix A.

$$A = \begin{bmatrix} 2 & 11 \\ 1 & 21 \end{bmatrix}$$

   (a)* Use this cipher to encrypt COLD MOUNTAIN.

   (b) Use this cipher to encrypt GONE WITH THE WIND.

   (c) Find A^{-1} mod 26.

   (d) Decrypt TSZLL DCLMD WAMTW CIFQU, which was formed using this cipher.

4. Consider a Hill cipher with the encryption matrix A in Example 8.18.

   (a)* Use this cipher to encrypt NOBODY CARES HOW MUCH YOU KNOW.

---

[3]Theodore Roosevelt (1858–1919), quote.

(b) Use this cipher to encrypt UNTIL THEY KNOW.

(c) Decrypt HGIYC EXKYM WGFKK,[4] which was formed using this cipher. (Recall that A^{-1} mod 26 is given in Example 8.19.)

5. Consider a Hill cipher with the following encryption matrix A.

$$A = \begin{bmatrix} 3 & 1 & 2 \\ 17 & 8 & 0 \\ 21 & 10 & 3 \end{bmatrix}$$

(a)* Use this cipher to encrypt ELBRUS.

(b) Use this cipher to encrypt KILIMANJARO.

(c) Show that A^{-1} mod 26 $= \begin{bmatrix} 2 & 9 & 16 \\ 25 & 7 & 18 \\ 24 & 9 & 19 \end{bmatrix}$.

(d) Decrypt SMGEX MUPPP AN, which was formed using this cipher.

6. Consider a Hill cipher with the following encryption matrix A.

$$A = \begin{bmatrix} 5 & 7 & 8 \\ 10 & 1 & 0 \\ 1 & 2 & 1 \end{bmatrix}$$

(This is the matrix in Exercise 19c in Section 8.1.)

(a)* Use this cipher to encrypt MUIRFIELD.

(b) Use this cipher to encrypt CARNOUSTIE.

(c) Use the formulas given in Exercises 19 and 20 in Section 8.1 to find A^{-1} mod 26.

(d) Decrypt NSABL LOQKJ PK, which was formed using this cipher.

7. Create a Hill cipher with an encryption matrix A of size 2×2, and use it to encrypt a plaintext of your choice with at least 12 letters. Also, show how you know that A^{-1} mod 26 exists.

8. Create a Hill cipher with an encryption matrix A of size 3×3, and use it to encrypt a plaintext of your choice with at least 12 letters. Also, use the formula given in Exercise 19 in Section 8.1 to show how you know that A^{-1} mod 26 exists.

9. Explain why a Hill cipher with encryption matrix $A = \begin{bmatrix} 1 & 0 \\ 0 & 1 \end{bmatrix}$ is especially insecure.

---

[4]Theodore Roosevelt, quote.

10.* Suppose the third ciphertext letter in Example 8.17 is incorrectly transcribed as M instead of N, so the received ciphertext is OEMGD ZHCXG GE. Decrypt this ciphertext using A^{-1} mod 26 from Example 8.17. How many errors appear in the result? Why does this occur?

11. Suppose the third ciphertext letter in Example 8.19 is incorrectly transcribed as M instead of N, so the received ciphertext is TFMGK ZGLVK AECUS. Decrypt this ciphertext using A^{-1} mod 26 from Example 8.19. How many errors appear in the result? Why does this occur?

12. Find a copy of the 1929 article in which Lester Hill first described Hill ciphers, and write a summary of how Hill ciphers are described in this article as compared to how they are presented in this book.

13. Find some information about the cryptographic apparatus for which Lester Hill was awarded a patent, and write a summary of your findings.

14. Recall that approximately three years before the 1929 article in which Lester Hill first described Hill ciphers, a paper in which a very similar type of cipher was used appeared in a detective magazine, written by the young mathematician Jack Levine.

    (a) Find a copy of this paper, and write a summary of how Levine's cipher compares to Hill ciphers.

    (b) Find copies of some of Levine's later writings on Hill ciphers and cryptology in general, and write a summary of your findings.

    (c) Find some additional information about Levine's career in cryptology, include his background, accomplishments, and importance in its history, and write a summary of your findings.

8.3 Cryptanalysis of Hill Ciphers

Recall that in Hill ciphers, the fact that plaintext numbers are encrypted in groups rather than one at a time can be viewed as a negative feature since a transcription error in a single ciphertext letter usually leads to more than one error in the decrypted message. However, this negative aspect is far outweighed by the additional security gained by Hill ciphers not being substitution ciphers, which also occurs because plaintext numbers are encrypted in groups rather than one at a time.

Put more plainly, note that while in the substitution ciphers that we generalized to form Hill ciphers (affine ciphers with encryption formula

$y = xa \bmod 26$) there are only 26 possible plaintext numbers, in a Hill cipher with a 2×2 encryption matrix there are $26^2 = 676$ possible plaintext row matrices.[5] Similarly, in a Hill cipher with a 3×3 encryption matrix, there are $26^3 = 17{,}576$ possible plaintext row matrices, and in a Hill cipher with a 4×4 encryption matrix, there are $26^4 = 456{,}976$ possible plaintext row matrices. Since the encryption matrix A for a Hill cipher can be of any size (provided $A^{-1} \bmod 26$ exists), and larger encryption matrices allow for more possible plaintext row matrices, the security of Hill ciphers grows as the size of the encryption matrix increases. More precisely, in a Hill cipher with an $n \times n$ encryption matrix, the number of possible plaintext row matrices is 26^n, a number that grows, and very quickly, as n increases.

From another perspective, note that in an affine cipher with encryption formula $y = xa \bmod 26$, there are only 12 possible values of a (the values of a in Table 6.1 on page 175), yielding 11 possible keys for the cipher (assuming $a = 1$ is not used). A brute force attack on the cipher could be done by simply trying to decrypt the ciphertext assuming each of these 11 possible keys, stopping when the correct plaintext is revealed. However, a brute force attack on a Hill cipher with a 2×2 key (encryption) matrix would require trying to decrypt the ciphertext assuming a much larger number of possible keys. The number of possible 2×2 matrices with entries in $\mathbb{Z}_{26}$ is $26^4 = 456{,}976$, and while many of these matrices would not have an inverse modulo 26 and thus not be a valid key matrix for a Hill cipher, it may not be known whether a particular matrix has an inverse modulo 26 until the matrix is at least formed. Thus, a brute force attack on a Hill cipher with a 2×2 key matrix would require some level of testing with up to a maximum of almost 456,976 matrices. Similarly, a brute force attack on a Hill cipher with a 3×3 key matrix would require some level of testing with up to a maximum of almost $26^9 = 5{,}429{,}503{,}678{,}976$ matrices. So even for relatively small key matrices, Hill ciphers are much more resistant to a brute force attack than substitution ciphers. More importantly, Hill ciphers can be constructed with any desired level of security by simply using a key matrix that is sufficiently large.

Hill ciphers do have one notable vulnerability, though. It is not unreasonable to suppose that someone in possession of a ciphertext formed using a Hill cipher and trying to break the cipher might know or be able to correctly guess a small crib (that is, a small known part of the plaintext). For example, it may be known where or from whom the message originated, and correctly guessed that the first several letters in the plaintext are a time or location stamp or that the last few letters in the plaintext are the originator's name. As it turns out, it is sometimes possible to break

---

[5]This follows from the multiplication principle, since both of the entries in a plaintext row matrix could be any of the 26 numbers in $\mathbb{Z}_{26}$.

a Hill cipher relatively easily if a small crib is known. More specifically, for a ciphertext formed using a Hill cipher with an $n \times n$ key matrix, it is sometimes possible to break the cipher relatively easily if a crib of length n^2 letters is known.

Example 8.20 Consider the ciphertext HJGID OZKEJ LPYIO TAIRB XXDTU WRQYF HAGEL FPKPS TF, which was formed using a Hill cipher with a 2×2 key matrix A. Suppose it is somehow known that the first three words in the plaintext are I BEG TO. As a result, it is known that the key matrix A encrypts the first and second plaintext letters IB to the ciphertext letters HJ, the third and fourth plaintext letters EG to GI, and the fifth and sixth plaintext letters TO to DO. We will use these plaintext/ciphertext digraph pairs in groups of 2 (corresponding to the fact that the key matrix is of size 2×2) to try to find the key matrix for the cipher. For example, consider the plaintext/ciphertext digraphs IB/HJ and EG/GI. With these letters converted into numeric form, the key matrix A must satisfy the following matrix equations.

$$\begin{bmatrix} 8 & 1 \end{bmatrix} A = \begin{bmatrix} 7 & 9 \end{bmatrix} \bmod 26$$
$$\begin{bmatrix} 4 & 6 \end{bmatrix} A = \begin{bmatrix} 6 & 8 \end{bmatrix} \bmod 26$$

Equivalently, these two matrix equations can be expressed as the following single matrix equation.

$$\begin{bmatrix} 8 & 1 \\ 4 & 6 \end{bmatrix} A = \begin{bmatrix} 7 & 9 \\ 6 & 8 \end{bmatrix} \bmod 26$$

Thus, to find the key matrix for the cipher, we must only solve this equation for A, which we could do by multiplying both sides of the equation on the left by, if it exists, $\begin{bmatrix} 8 & 1 \\ 4 & 6 \end{bmatrix}^{-1} \bmod 26$. However, we showed in Example 8.15 on page 262 that this inverse modulo 26 does not exist. So, next we will consider the plaintext/ciphertext digraphs IB/HJ and TO/DO. With these letters converted into numeric form, the key matrix A must satisfy the following matrix equations.

$$\begin{bmatrix} 8 & 1 \end{bmatrix} A = \begin{bmatrix} 7 & 9 \end{bmatrix} \bmod 26$$
$$\begin{bmatrix} 19 & 14 \end{bmatrix} A = \begin{bmatrix} 3 & 14 \end{bmatrix} \bmod 26$$

Equivalently, these two matrix equations can be expressed as the following single matrix equation.

$$\begin{bmatrix} 8 & 1 \\ 19 & 14 \end{bmatrix} A = \begin{bmatrix} 7 & 9 \\ 3 & 14 \end{bmatrix} \bmod 26$$

We found in Example 8.14 that $\begin{bmatrix} 8 & 1 \\ 19 & 14 \end{bmatrix}^{-1} \bmod 26 = \begin{bmatrix} 20 & 19 \\ 23 & 4 \end{bmatrix}$. The key matrix A can then be determined as follows.

$$\begin{bmatrix} 8 & 1 \\ 19 & 14 \end{bmatrix}^{-1} \begin{bmatrix} 8 & 1 \\ 19 & 14 \end{bmatrix} A = \begin{bmatrix} 8 & 1 \\ 19 & 14 \end{bmatrix}^{-1} \begin{bmatrix} 7 & 9 \\ 3 & 14 \end{bmatrix} \bmod 26$$

$$IA = \begin{bmatrix} 20 & 19 \\ 23 & 4 \end{bmatrix} \begin{bmatrix} 7 & 9 \\ 3 & 14 \end{bmatrix} \bmod 26$$

$$A = \begin{bmatrix} 197 & 446 \\ 173 & 263 \end{bmatrix} \bmod 26$$

$$A = \begin{bmatrix} 15 & 4 \\ 17 & 3 \end{bmatrix}$$

For this key matrix, note that $\det(A) = 15 \cdot 3 - 4 \cdot 17 = -23 = 3 \bmod 26$, and thus we can use the formula on page 261 to find $A^{-1} \bmod 26$ in the following way.

$$A^{-1} \bmod 26 = 3^{-1} \begin{bmatrix} 3 & -4 \\ -17 & 15 \end{bmatrix} \bmod 26$$

$$= 9 \begin{bmatrix} 3 & -4 \\ -17 & 15 \end{bmatrix} \bmod 26$$

$$= \begin{bmatrix} 27 & -36 \\ -153 & 135 \end{bmatrix} \bmod 26$$

$$= \begin{bmatrix} 1 & 16 \\ 3 & 5 \end{bmatrix}$$

Using $A^{-1} \bmod 26$ to decrypt the rest of the ciphertext yields the full plaintext: I BEG TO DIFFER WITH YOUR OPINION ON THIS MATTER SIR.   □

Curiously, Hill ciphers are only known to have actually been used in one capacity, by the U.S. government to encrypt three-letter groups of radio call signals [13]. While it could be the case that Hill ciphers were never widely used because of their vulnerability that we showed in the last example, a more likely reason is that when Hill ciphers came along in the early twentieth century, the lack of technology available at the time made them very tedious to use. If this is indeed the case, then it would be just one of numerous instances in history in which a person's mathematical ingenuity was ahead of society's ability to put his or her ideas into practice. Unfortunately for Lester Hill, by the time technology advanced to the point of making Hill ciphers practical for implementation, more advanced systems that were easier to use and even more secure had been created.

8.3.1 Exercises

1.* Consider the ciphertext ZQRVY USKNA, which was formed using a Hill cipher with a 2×2 key matrix, and suppose it is somehow known that the first four letters in the plaintext are HALL. Find the key matrix, and cryptanalyze the ciphertext.

2. Consider the ciphertext FLBIP URCRG AO, which was formed using a Hill cipher with a 2×2 key matrix, and suppose it is somehow known that the first four letters in the plaintext are NCST. Find the key matrix, and cryptanalyze the ciphertext.

3.* Consider the ciphertext ETGYX OIMOI NGQMV EJGPM NNNNZ CLOIG, which was formed using a Hill cipher with a 2×2 key matrix, and suppose it is somehow known that the first two words in the plaintext are THE ALAMO. Find the key matrix, and cryptanalyze the ciphertext.

4. Consider the ciphertext FBVNO MBMWA SCNWH UPTCI SIELQ EVY, which was formed using a Hill cipher with a 2×2 key matrix, and suppose it is somehow known that the first two words in the plaintext are I BELIEVE. Find the key matrix, and cryptanalyze the ciphertext.

5.* Consider the ciphertext OAXEL QLSMT KTCOQ, which was formed using a Hill cipher with a 3×3 key matrix, and suppose it is somehow known that the first nine letters in the plaintext are GONAVYBEA. Find the key matrix, and cryptanalyze the ciphertext.

6.* Consider the ciphertext ALYUE FKMER OZIRU HHHBM GQSRU PKWZD WCGMP YLJAE RMBWS, which was formed using a Hill cipher with a 3×3 key matrix, and suppose it is somehow known that the first three words in the plaintext are THE ATTACK WILL. Find the key matrix, and cryptanalyze the ciphertext.

7. Recall from Exercise 5 in Section 2.2 the concept of *superencryption*.

   (a)* Use Hill ciphers with the key matrices A in Exercises 2 and 3 in Section 8.2 (in that order) to superencrypt MIDSHIPMEN.

   (b) Decrypt QOPDX RXPKV, which was superencrypted using Hill ciphers with the key matrices A in Exercises 3 and 2 in Section 8.2 (in that order).

(c) Does superencryption by two Hill ciphers yield more security than encryption by one Hill cipher? In other words, if a plaintext P is encrypted using a Hill cipher, yielding M, and then M is encrypted using another Hill cipher, yielding C, would C be harder in general to cryptanalyze than M? Explain your answer completely, and be as specific as possible.

8.* For a ciphertext formed using a Hill cipher with a 2×2 key matrix, as part of a brute force attack against the cipher suppose it takes on average five minutes to test a potential key matrix (whether the matrix is invertible modulo 26 or not). How long would it take on average (i.e., testing half of the potential key matrices) to break the cipher using a brute force attack? Give your answer in days.

9. Repeat Exercise 8 for a ciphertext formed using a Hill cipher with a 3×3 key matrix. Give your answer in years.

10. We viewed Hill ciphers with the encryption formula $y = xA \bmod 26$ as a generalization of affine ciphers that have the encryption formula $y = (xa + b) \bmod 26$ with $b = 0$. Hill ciphers can also include the analogue of a nonzero affine additive key b. Such Hill ciphers would have the encryption formula $y = (xA + b) \bmod 26$, where b is a fixed row matrix of the same size as the plaintext row matrices x and with entries in $\mathbb{Z}_{26}$.

   (a)* Consider a Hill cipher with the key matrices A in Exercise 2 in Section 8.2 and $b = \begin{bmatrix} 1 & 2 \end{bmatrix}$ and the encryption formula $y = (xA + b) \bmod 26$. Use this cipher to encrypt ANNAPOLIS.

   (b) Use the cipher in part (a) to encrypt WEST POINT.

   (c) For a Hill cipher with encryption formula $y = (xA + b) \bmod 26$, show that the decryption formula is $x = (y - b)A^{-1} \bmod 26$.

   (d) Use the decryption formula given in part (c) to decrypt JAOKW SYEGP CQMUL W, which was formed using the cipher in part (a).

   (e) Is a Hill cipher with encryption formula $y = (xA + b) \bmod 26$ more secure against a brute force attack than one with encryption formula $y = xA \bmod 26$? If so, how much more secure? Explain your answer completely, and be as specific as possible.

11. The stronger Hill cipher encryption formula $y = (xA + b) \bmod 26$ described in Exercise 10 can be made even more secure by allowing the row matrix b to change during the encryption process. To make it easier to explain how this might be done, we will denote the sequence of plaintext row matrices to be encrypted as $(x_1, x_2, \ldots)$, with

corresponding ciphertext row matrices $(\mathbf{y}_1, \mathbf{y}_2, \ldots)$. That is, we will denote the first plaintext row matrix as $\mathbf{x}_1$, with corresponding ciphertext row matrix $\mathbf{y}_1$, the second plaintext row matrix as $\mathbf{x}_2$, with corresponding ciphertext row matrix $\mathbf{y}_2$, and so on. This makes it possible for us to express the stronger Hill cipher encryption formula with changing row matrices $\mathbf{b}$ as $\mathbf{y}_i = (\mathbf{x}_i A + \mathbf{b}_i) \bmod 26$, emphasizing that the row matrices in the sequence $(\mathbf{b}_1, \mathbf{b}_2, \ldots)$ can change as the plaintext row matrices $(\mathbf{x}_1, \mathbf{x}_2, \ldots)$ and corresponding ciphertext row matrices $(\mathbf{y}_1, \mathbf{y}_2, \ldots)$ change during the encryption process. To reduce the difficulty in keeping a record of the $\mathbf{b}_i$, they can be chosen to depend uniquely on the $\mathbf{x}_i$ or the $\mathbf{y}_i$. For example, the following are two possible methods for choosing the $\mathbf{b}_i$.

- $\mathbf{b}_i = \mathbf{x}_{i-1} B \bmod 26$, where B is a fixed matrix of the same size as A and entries in $\mathbb{Z}_{26}$, and $\mathbf{x}_0$ is defined separately.

- $\mathbf{b}_i = \mathbf{y}_{i-1} B \bmod 26$, where B is a fixed matrix of the same size as A and entries in $\mathbb{Z}_{26}$, and $\mathbf{y}_0$ is defined separately.

(a)* Consider a Hill cipher with the following key matrices A, B, and $\mathbf{x}_0$ and the encryption formula $\mathbf{y}_i = (\mathbf{x}_i A + \mathbf{x}_{i-1} B) \bmod 26$.

$$A = \begin{bmatrix} 2 & 5 \\ 1 & 4 \end{bmatrix} \qquad B = \begin{bmatrix} 1 & 0 \\ 1 & 1 \end{bmatrix} \qquad \mathbf{x}_0 = \begin{bmatrix} 1 & 2 \end{bmatrix}$$

Use this cipher to encrypt ASAP AS.

(b) Use the cipher in part (a) to encrypt POSSIBLE.[6]

(c) For a Hill cipher with $\mathbf{y}_i = (\mathbf{x}_i A + \mathbf{x}_{i-1} B) \bmod 26$ for encryption, show that decryption is done by $\mathbf{x}_i = (\mathbf{y}_i - \mathbf{x}_{i-1} B) A^{-1} \bmod 26$.

(d) Use the decryption formula given in part (c) to decrypt PWCJK LVL, which was formed using the cipher in part (a).

(e) For a Hill cipher of the type described in this exercise with the $\mathbf{b}_i$ depending on the $\mathbf{x}_i$, why does it make more sense to use the previous plaintext row matrix $\mathbf{x}_{i-1}$ when forming $\mathbf{b}_i$ than the current plaintext row matrix $\mathbf{x}_i$? That is, why does it make more sense to use $\mathbf{y}_i = (\mathbf{x}_i A + \mathbf{x}_{i-1} B) \bmod 26$ for encryption rather than $\mathbf{y}_i = (\mathbf{x}_i A + \mathbf{x}_i B) \bmod 26$?

(f)* Consider a Hill cipher with the key matrices A and B in part (a) and $\mathbf{y}_0 = \begin{bmatrix} 1 & 2 \end{bmatrix}$, and $\mathbf{y}_i = (\mathbf{x}_i A + \mathbf{y}_{i-1} B) \bmod 26$ for encryption. Use this cipher to encrypt YEPPERS.

(g) Use the cipher in part (f) to encrypt YESH.

---

[6] Michael Scott, quote

(h) For a Hill cipher with $\mathbf{y}_i = (\mathbf{x}_i A + \mathbf{y}_{i-1}B)$ mod 26 for encryption, show that decryption is done by $\mathbf{x}_i = (\mathbf{y}_i - \mathbf{y}_{i-1}B)A^{-1}$ mod 26.

(i) Use the decryption formula given in part (h) to decrypt DTZSL W, which was formed using the cipher in part (f).

(j) For a Hill cipher of the type described in this exercise with the $\mathbf{b}_i$ depending on the $\mathbf{y}_i$, why does it make more sense to use the previous ciphertext row matrix $\mathbf{y}_{i-1}$ when forming $\mathbf{b}_i$ than the current ciphertext row matrix $\mathbf{y}_i$? That is, why does it make more sense to use $\mathbf{y}_i = (\mathbf{x}_i A + \mathbf{y}_{i-1}B)$ mod 26 for encryption rather than $\mathbf{y}_i = (\mathbf{x}_i A + \mathbf{y}_i B)$ mod 26?

Chapter 9

RSA Ciphers

Advances in cryptology have followed many notable changes in direction over the centuries, for example, the use of polyalphabetic over monoalphabetic ciphers begun by Alberti in the 1400s, and the use of mathematics in cryptology begun by Friedman and Hill in the early 1900s. The rapid progress in technology that our society achieved over the last several decades, specifically in computing, has had a dramatic effect on our abilities in cryptology. This effect is felt not just on classical ciphers though. While classical ciphers, even those considered progressive in their time, are indeed often easy to implement and analyze through the use of technology, recent progress in technology has also given rise to an entirely new kind of cipher, and yet another change in direction for the subject of cryptology itself. This change began in 1976, when Stanford University graduate student Whitfield Diffie and his faculty mentor Martin Hellman published a paper, aptly titled *New Directions in Cryptography* [6], in which they described the idea of a *public-key* cipher.

9.1 Introduction to Public-Key Ciphers

When assessing the security of a cipher, it is generally assumed that the encryption keys are the only thing about the cipher kept secret between the communicating parties. For example, for a mathematical cipher in which plaintext letters must be converted into numbers before being encrypted, we would not consider the correspondences between letters and numbers as part of the security of the cipher. We would also not consider keeping secret the type of mathematical operation used in the cipher as part of its security. For all we care, the world could know that the cipher was, for instance, a Hill cipher with a 5×5 key matrix. The security of the cipher, we would assume, comes only from keeping the key matrix itself secret from outsiders.

Similarly, for a nonmathematical cipher such as a Vigenère keyword cipher, we would not consider keeping secret the fact that the cipher was a Vigenère keyword cipher as part of its security. The security of the cipher we would assume comes only from keeping the keyword itself secret. This is because for a Hill cipher or a Vigenère keyword cipher or any other type of cipher, it should be possible to easily change the encryption keys without affecting the basic operation of the cipher.

On the other hand, it only makes sense that the encryption keys for a cipher must be kept secret in order to prevent an outsider from acting as the intended recipient of a message, determining the corresponding decryption keys, and decrypting the message. However, remarkably, as Diffie and Hellman first explained to the world in their 1976 paper, the encryption keys for a cipher need not always be kept secret. Such ciphers in which the encryption keys need not be kept secret are called *public-key* ciphers, since the encryption keys can be known publicly without ruining their security. The (tremendous) benefit to this is that a public-key cipher can be used by two parties who have no way to secretly agree upon encryption keys. Before public-key ciphers, two parties wishing to communicate secretly over an insecure communication line had to somehow be able to secretly agree

Pioneers of Public-Key Cryptography

| James Ellis | Clifford Cocks | Malcolm Williamson |

Pioneering work in cryptology does not always lead to public recognition, as seen in the work of James Ellis, Clifford Cocks, and Malcolm Williamson in a classified environment at the British intelligence agency GCHQ. It was at GCHQ that Ellis first developed the idea of public-key ciphers in 1970, and Cocks RSA ciphers in 1973. These discoveries, along with Williamson's own pioneering work, were not revealed publicly until Cocks gave a presentation about them in 1997, less than a month after Ellis's death. Ellis, Cocks, and Williamson also missed the fortunes their discoveries would have brought in a non-classified environment. In 1996, the RSA Data Security corporation created by Rivest, Shamir, and Adleman to market RSA ciphers sold for $400 million. Ellis, Cocks, and Williamson were good-humored about this. Cocks has stated: "You don't get into this business for public recognition."

upon encryption keys. By using a public-key cipher, the two parties would be able to openly agree upon encryption keys over the insecure communication line without having to consider whether these keys were intercepted by an outsider.

While Diffie and Hellman described the idea of public-key ciphers in their 1976 paper, they did not actually provide a specific type of public-key cipher. The first specific type of public-key cipher was revealed to the world two years later by Ron Rivest, Adi Shamir, and Len Adleman, a trio of researchers at MIT, in their paper *A Method for Obtaining Digital Signatures and Public-Key Cryptosystems* [20]. In this chapter, we will consider the details of this type of cipher, called *RSA* ciphers in honor of the people who first published them.

Even ignoring the fact that RSA ciphers are public-key ciphers, which alone makes them as much worth studying as any other type of cipher we have considered in this book, RSA ciphers are every bit as fascinating as the other types of ciphers we have considered, perhaps even more so, because of their reliance upon several very famous, old, and relatively simple mathematical facts. Also, unlike many of the other types of ciphers we considered in this book, RSA ciphers very quickly became widely used worldwide. You have most likely even used an RSA cipher yourself, probably unknowingly, if you have ever used an ATM or purchased something with a credit card over the Internet. In fact, RSA ciphers became widely used worldwide so quickly that the RSA Data Security corporation formed by Rivest, Shamir, and Adleman in 1982 to market their type of cipher was sold a mere 14 years later for $400 million. However, despite this irrefutable evidence as to the practical importance of RSA ciphers, they are not difficult to understand, as we will see in this chapter.

9.1.1   Exercises

1. Find some information about the careers in cryptology of Whitfield Diffie and Martin Hellman, including their backgrounds, accomplishments, and importance in its history, and write a summary of your findings.

2. Find a copy of the paper [6] in which Whitfield Diffie and Martin Hellman first publicly described the idea of public-key ciphers, and as best you can write a summary of how public-key ciphers are described in this paper.

3. Find some information about the careers in cryptology of Ron Rivest, Adi Shamir, and Len Adleman, including their backgrounds, accomplishments, and importance in its history, and write a summary of your findings.

4. Find some information about the RSA Data Security corporation, including its history, whom it typically serves, and the products and services that it provides, and write a summary of your findings.

5. Find some information about the careers in cryptology of James Ellis, Clifford Cocks, and Malcolm Williamson, and write a summary of your findings.

9.2   Introduction to RSA Ciphers

From a mathematical perspective RSA ciphers are not fundamentally different from the other types of mathematical ciphers we have considered in this book. RSA ciphers just use modular arithmetic with a different type of mathematical operation. We have already considered ciphers that use modular arithmetic with addition (shift ciphers), multiplication (affine ciphers), and matrix multiplication (Hill ciphers); RSA ciphers use modular arithmetic with the mathematical operation of raising to powers, or *exponentiation*. RSA ciphers also use prime numbers. A positive integer p is said to be *prime* if the only positive integers that divide p evenly are 1 and p. For reference, the prime numbers less than 100 are 2, 3, 5, 7, 11, 13, 17, 19, 23, 29, 31, 37, 41, 43, 47, 53, 59, 61, 67, 71, 73, 79, 83, 89, and 97.

With RSA ciphers, we will again convert plaintext messages into numeric form. Suppose the originator of a numeric plaintext message wishes to send the message to an intended recipient over an insecure communication line, and wants to use a cipher to disguise the message to protect it from outsiders who may observe it along the way. The basic steps in using an RSA cipher to do this can be summarized as follows.

1. The intended recipient of the message initiates the process by choosing prime numbers p and q that are not equal to each other, and forming $m = p \cdot q$ and $f = (p - 1) \cdot (q - 1)$. The intended recipient then chooses an integer e between 1 and f with $\gcd(e, f) = 1$, and sends the values of e and m to the originator of the message over the insecure communication line.

2. Suppose the numeric plaintext message is expressed as one or more positive integers x less than m. Then, for each plaintext integer x, the originator of the message encrypts x by forming the following quantity y.

$$y = x^e \bmod m$$

The originator then sends the resulting ciphertext integer(s) y to the intended recipient over the insecure communication line.

3. To decrypt the ciphertext integer(s) y, the recipient must first find the multiplicative inverse $d = e^{-1} \bmod f$. That is, the recipient must first find an integer d between 1 and f with $e \cdot d = 1 \bmod f$. Then, for each ciphertext integer y, the recipient decrypts y by forming the following quantity, which results in the plaintext integer x from which y was formed.

$$x = y^d \bmod m$$

One notable difference between RSA ciphers and the other types of ciphers we have considered in this book is the form of ciphertexts. Because the modulus m in an RSA cipher does not have to match the number of alphabet characters, and indeed is typically much larger than the number of alphabet characters, it is almost always impossible to convert numbers y that result from the encryption calculation in step 2 above back into characters. As such, RSA ciphertexts are always just left in numeric form. This is reflected in our summary of the basic steps in using an RSA cipher. Ciphertext integers y that result from step 2 are sent at the end of step 2 and received at the beginning of step 3 as integers.

Example 9.1 Suppose you wish to send the secret message B.B. KING to a colleague over an insecure communication line using an RSA cipher with the correspondences A = 0, B = 1, C = 2, ... , Z = 25, under which the plaintext converts into the list of integers 1, 1, 10, 8, 13, 6. Your colleague initiates the process by choosing primes $p = 3$ and $q = 11$, and forming $m = p \cdot q = 33$ and $f = (p - 1) \cdot (q - 1) = 20$. Your colleague then chooses $e = 7$, and sends the values of e and m to you over the insecure communication line. With the values $e = 7$ and $m = 33$ received from your colleague, you encrypt the plaintext integers as follows.

$$
\begin{aligned}
\text{B} &\rightarrow x = 1 \rightarrow y = 1^7 \bmod 33 = 1 \bmod 33 = 1 \\
\text{B} &\rightarrow x = 1 \rightarrow y = 1^7 \bmod 33 = 1 \bmod 33 = 1 \\
\text{K} &\rightarrow x = 10 \rightarrow y = 10^7 \bmod 33 = 10000000 \bmod 33 = 10 \\
\text{I} &\rightarrow x = 8 \rightarrow y = 8^7 \bmod 33 = 2097152 \bmod 33 = 2 \\
\text{N} &\rightarrow x = 13 \rightarrow y = 13^7 \bmod 33 = 62748517 \bmod 33 = 7 \\
\text{G} &\rightarrow x = 6 \rightarrow y = 6^7 \bmod 33 = 279936 \bmod 33 = 30
\end{aligned}
$$

You then send the ciphertext integers 1, 1, 10, 2, 7, 30 to your colleague over the insecure communication line. To decrypt the message, your colleague must first find the value of $d = e^{-1} \bmod f = 7^{-1} \bmod 20$. Since $7 \cdot 3 = 21 = 1 \bmod 20$, this value is $d = 3$. Your colleague can then decrypt the message as follows.

$$
\begin{aligned}
y = 1 &\rightarrow x = 1^3 \bmod 33 = 1 \bmod 33 = 1 \rightarrow \text{B} \\
y = 1 &\rightarrow x = 1^3 \bmod 33 = 1 \bmod 33 = 1 \rightarrow \text{B}
\end{aligned}
$$

$$y = 10 \;\rightarrow\; x = 10^3 \bmod 33 = 1000 \bmod 33 = 10 \;\rightarrow\; \texttt{K}$$
$$y = 2 \;\rightarrow\; x = 2^3 \bmod 33 = 8 \bmod 33 = 8 \;\rightarrow\; \texttt{I}$$
$$y = 7 \;\rightarrow\; x = 7^3 \bmod 33 = 343 \bmod 33 = 13 \;\rightarrow\; \texttt{N}$$
$$y = 30 \;\rightarrow\; x = 30^3 \bmod 33 = 27000 \bmod 33 = 6 \;\rightarrow\; \texttt{G} \qquad \square$$

Several points about the process summarized in general before Example 9.1 and illustrated in this example bear mentioning.

- The work in using an RSA cipher is done primarily by the intended recipient of a message, not the originator. This is unlike all other types of ciphers we have considered so far in this book, and is because RSA ciphers are public-key ciphers. For all public-key ciphers, the work in using the cipher is done primarily by the intended recipient of a message. The reason for this is that with a public-key cipher, while anyone can know the encryption keys, no one but the intended recipient of a message, excluding even the originator of the message, needs to know the decryption keys.

- It cannot be assumed that the encryption keys e and m are kept secret between the originator and intended recipient of a message, because they are sent over an insecure communication line. This is precisely why RSA ciphers are called *public-key*, since the encryption keys must be assumed to be public knowledge. Remarkably, this fact does not in general prevent RSA ciphers from being secure, the reason for which we will see in Section 9.7.

- The encryption exponent e must be chosen so that $\gcd(e, f) = 1$ because this guarantees the existence of a corresponding decryption exponent, the reason for which we will see in Section 9.3. Since $\gcd(e, f) = 1$, the multiplicative inverse $e^{-1} \bmod f$ will exist, and this multiplicative inverse will be the corresponding decryption exponent d.

- For a ciphertext integer y, it is by no means obvious that the calculation $y^d \bmod m$ will result in the plaintext integer x from which y was formed, especially considering that e and d are multiplicative inverses of each other modulo f, but encryption and decryption are done modulo m. We will explain why this happens in Section 9.6.

Finally, the RSA cipher in Example 9.1 is grossly insecure, for more than one reason. First, it is a substitution cipher, and thus if it were used to encrypt a longer plaintext, it would be susceptible to attack by frequency analysis. Further, with such small encryption keys e and m, if these values were public knowledge, as they are assumed to be in RSA ciphers, the cipher could easily be broken without using frequency analysis. Much remains for us to consider about RSA ciphers, enough that by the end of this chapter it will be apparent why they continue to be widely used worldwide.

9.2.1 Exercises

1. Consider an RSA cipher with primes $p = 3$ and $q = 11$, encryption exponent $e = 7$, and correspondences A = 0, B = 1, C = 2, ... , Z = 25.

   (a)* Use this cipher to encrypt BOONE.

   (b) Decrypt 27, 2, 29, 10, 16, 7, 6, which was formed using this cipher.

2. Consider an RSA cipher with primes $p = 5$ and $q = 7$, encryption exponent $e = 5$, and correspondences A = 0, B = 1, C = 2, ... , Z = 25.

   (a) Use this cipher to encrypt RON.

   (b) Decrypt 12, 9, 0, 6, 0, 13, which was formed using this cipher.

3. Consider an RSA cipher with encryption keys $e = 27$ and $m = 55$, and correspondences A = 0, B = 1, C = 2, ... , Z = 25.

   (a) Verify that the decryption exponent for this cipher is $d = 3$.

   (b) Decrypt 17, 28, 0, 20, 0, 23, which was formed using this cipher.

4. Consider an RSA cipher with encryption keys $e = 7$ and $m = 39$, and correspondences A = 0, B = 1, C = 2, ... , Z = 25. Decrypt 19, 4, 2, 2, 1, 14, 15, which was formed using this cipher.

9.3 The Euclidean Algorithm

For an affine cipher with encryption calculation $y = (ax + b)$ mod 26, the multiplicative key a must be chosen so that $\gcd(a, 26) = 1$. This is to guarantee that the multiplicative inverse a^{-1} mod 26 will exist. For an affine cipher, verifying that $\gcd(a, 26) = 1$ and finding a^{-1} mod 26 are usually done by trial and error (as we did in Chapter 6), which works effectively because the modulus is so small. For an RSA cipher with primes p and q and $f = (p-1) \cdot (q-1)$, the encryption exponent e must be chosen so that $\gcd(e, f) = 1$, and e^{-1} mod f must be determined. Since the modulus in an RSA cipher is typically much larger than 26, it is often impractical to verify that $\gcd(e, f) = 1$ and find e^{-1} mod f by trial and error. Luckily for us, someone a very long time ago created a relatively simple tool that can be used to find greatest common divisors and multiplicative inverses when the modulus is large, even extremely large, as is required for RSA ciphers in practice. This tool is the *Euclidean algorithm*, and it is a lot less scary than it sounds.

The fact that the Euclidean algorithm is a useful tool in something as modern as RSA ciphers is itself remarkable. The Euclidean algorithm was described around 300 BC by the Greek mathematician Euclid in one of the most famous and influential textbooks every written, Euclid's *Elements*, which is still in print to this day in an edition numbered well over one thousand. Despite its age, the Euclidean algorithm is still the de facto method for finding greatest common divisors and multiplicative inverses for modular arithmetic when the modulus is large. This is because it is very fast, even for extremely large numbers like those required for RSA ciphers in practice, and it is also very easy to encode on a computer.

Let a and b be a pair of positive integers with $a > b$, and suppose we wish to find $\gcd(a, b)$. If b divides into a evenly, then $\gcd(a, b) = b$. If b does not divide into a evenly, then the Euclidean algorithm uses repeated division to find $\gcd(a, b)$. For the first division, we divide b into a, obtaining quotient q_1 and remainder r_1, which fit with a and b into the following equation.

$$a = q_1 b + r_1$$

For the second division, we divide r_1 into b, obtaining quotient q_2 and remainder r_2, which fit with b and r_1 into the following equation.

$$b = q_2 r_1 + r_2$$

For the third division, we divide r_2 into r_1, obtaining quotient q_3 and remainder r_3, which fit with r_1 and r_2 into the following equation.

$$r_1 = q_3 r_2 + r_3$$

We then continue this process of repeated division, with each division resulting in a similar equation in which the four subscripts are each larger by one than the corresponding subscripts in the equation resulting from the previous division. The following is the general form of the equation resulting from the nth division.

$$r_{n-2} = q_n r_{n-1} + r_n$$

Because in each division, the remainder will be nonnegative and less than the divisor, the sequence of remainders $r_1, r_2, r_3, \ldots$ must be nonnegative and decreasing. This means that this sequence of remainders must eventually reach zero. Assuming the last nonzero remainder occurs in the nth division, the following is the general form of the full resulting sequence of equations.

$$
\begin{aligned}
a &= q_1 b + r_1 & \text{(with } r_1 \text{ nonzero)} \\
b &= q_2 r_1 + r_2 & \text{(with } r_2 \text{ nonzero)} \\
r_1 &= q_3 r_2 + r_3 & \text{(with } r_3 \text{ nonzero)} \\
r_2 &= q_4 r_3 + r_4 & \text{(with } r_4 \text{ nonzero)} \\
&\quad\vdots \\
r_{n-3} &= q_{n-1} r_{n-2} + r_{n-1} & \text{(with } r_{n-1} \text{ nonzero)} \\
r_{n-2} &= q_n r_{n-1} + r_n & \text{(with } r_n \text{ nonzero)} \\
r_{n-1} &= q_{n+1} r_n + 0
\end{aligned}
$$

It is here that the initial part of the algorithm ends, yielding $\gcd(a, b) = r_n$. That is, $\gcd(a, b)$ is the last nonzero remainder r_n in this list of equations.

Example 9.2 Suppose we wish to find $\gcd(2299, 627)$. The following are the equations that result from the divisions in the initial part of the Euclidean algorithm.

$$
\begin{aligned}
2299 &= 3 \cdot 627 + 418 \\
627 &= 1 \cdot 418 + 209 \\
418 &= 2 \cdot 209 + 0
\end{aligned}
$$

Thus, $\gcd(2299, 627) = 209$. □

Example 9.3 Suppose we wish to find $\gcd(160, 17)$. The following are the equations that result from the divisions in the initial part of the Euclidean algorithm.

$$
\begin{aligned}
160 &= 9 \cdot 17 + 7 \\
17 &= 2 \cdot 7 + 3 \\
7 &= 2 \cdot 3 + 1 \\
3 &= 3 \cdot 1 + 0
\end{aligned}
$$

Thus, $\gcd(160, 17) = 1$. □

Example 9.4 Suppose we wish to find $\gcd(52598, 2541)$. The following are the equations that result from the divisions in the initial part of the Euclidean algorithm.

$$
\begin{aligned}
52598 &= 20 \cdot 2541 + 1778 \\
2541 &= 1 \cdot 1778 + 763 \\
1778 &= 2 \cdot 763 + 252 \\
763 &= 3 \cdot 252 + 7 \\
252 &= 36 \cdot 7 + 0
\end{aligned}
$$

Thus, $\gcd(52598, 2541) = 7$. □

For a pair of positive integers a and b, it is always possible to express $\gcd(a, b)$ as the sum of integer multiplies of a and b. That is, the following theorem is true.

Theorem 9.1 *For any pair of positive integers a and b, integers s and t exist such that $\gcd(a, b) = sa + tb$.*

The second and final part of the Euclidean algorithm uses the sequence of equations resulting from the initial part of the algorithm to determine integers s and t such that $\gcd(a, b) = sa + tb$. To do this, we express each remainder, in order starting with the largest, in terms of a and b by substituting for a and b directly or using the expressions for the remainders obtained in the previous steps. The process ends when the last nonzero remainder, which is $\gcd(a, b)$, is expressed in terms of a and b.

Example 9.5 Suppose we wish to determine integers s and t such that $\gcd(a, b) = sa + tb$ for $a = 2299$ and $b = 627$. We begin with the following sequence of equations from Example 9.2.

$$
\begin{aligned}
2299 &= 3 \cdot 627 + 418 \\
627 &= 1 \cdot 418 + 209 \\
418 &= 2 \cdot 209 + 0
\end{aligned}
$$

Expressing the first remainder in this sequence in terms of a and b yields the following.

$$418 = 2299 - 3 \cdot 627 = a - 3b$$

Next, since $418 = a - 3b$, we can express the second remainder in terms of a and b as follows.

$$209 = 627 - 1 \cdot 418 = b - 1(a - 3b) = -a + 4b$$

Since $\gcd(a, b) = 209$, this last equation $209 = -a + 4b$ completes the Euclidean algorithm: $\gcd(a, b) = sa + tb$ with $s = -1$ and $t = 4$. $\square$

Example 9.6 Suppose we wish to determine integers s and t such that $\gcd(a, b) = sa + tb$ for $a = 52598$ and $b = 2541$. We begin with the following sequence of equations from Example 9.4.

$$
\begin{aligned}
52598 &= 20 \cdot 2541 + 1778 \\
2541 &= 1 \cdot 1778 + 763 \\
1778 &= 2 \cdot 763 + 252 \\
763 &= 3 \cdot 252 + 7 \\
252 &= 36 \cdot 7 + 0
\end{aligned}
$$

Expressing the remainders in this sequence in terms of a and b yields the following.

$$1778 = 52598 - 20 \cdot 2541 = a - 20b$$
$$763 = 2541 - 1 \cdot 1778 \quad = b - 1(a - 20b) \qquad\qquad = -a + 21b$$
$$252 = 1778 - 2 \cdot 763 \quad = (a - 20b) - 2(-a + 21b) \quad = 3a - 62b$$
$$7 = 763 - 3 \cdot 252 \quad = (-a + 21b) - 3(3a - 62b) = -10a + 207b$$

Since $\gcd(a, b) = 7$, this last equation $7 = -10a + 207b$ completes the Euclidean algorithm: $\gcd(a, b) = sa + tb$ with $s = -10$ and $t = 207$. $\quad\square$

Recall the reason for our interest in the Euclidean algorithm. For an RSA cipher with primes p and q and $f = (p-1) \cdot (q-1)$, the encryption exponent e must be chosen so that $\gcd(e, f) = 1$, and $e^{-1} \bmod f$ must be determined. It should be clear how the Euclidean algorithm could help with verifying $\gcd(e, f) = 1$, but it is less clear how it could help with determining $e^{-1} \bmod f$. The basis for this is found in Theorem 9.1. If $\gcd(e, f) = 1$, then Theorem 9.1 states that integers s and t will exist such that the following equation is true.

$$sf + te = 1$$

Solving this equation for te gives the following.

$$te = 1 - sf$$

Reducing this equation modulo f then gives the following.

$$
\begin{aligned}
te &= (1 - sf) \bmod f \\
&= (1 - 0) \bmod f \\
&= 1 \bmod f
\end{aligned}
$$

Thus, assuming $\gcd(e, f) = 1$, which we could verify using the initial part of the Euclidean algorithm, to determine $e^{-1} \bmod f$, we can use the final part of the Euclidean algorithm to find integers s and t such that $sf + te = 1$. The value of $t \bmod f$ will be $e^{-1} \bmod f$.

Example 9.7 Suppose we wish to determine, if it exists, $17^{-1} \bmod 160$. In Example 9.3, we used the initial part of the Euclidean algorithm to show that $\gcd(160, 17) = 1$. Thus, $17^{-1} \bmod 160$ exists. To find $17^{-1} \bmod 160$, we begin with the following sequence of equations from Example 9.3.

$$
\begin{aligned}
160 &= 9 \cdot 17 + 7 \\
17 &= 2 \cdot 7 + 3 \\
7 &= 2 \cdot 3 + 1 \\
3 &= 3 \cdot 1 + 0
\end{aligned}
$$

With $f = 160$ and $e = 17$, expressing the remainders in this sequence in terms of f and e yields the following.

$$
\begin{aligned}
7 &= 160 - 9 \cdot 17 &&= f - 9e \\
3 &= 17 - 2 \cdot 7 &&= e - 2(f - 9e) &&= -2f + 19e \\
1 &= 7 - 2 \cdot 3 &&= (f - 9e) - 2(-2f + 19e) &&= 5f - 47e
\end{aligned}
$$

This last equation gives $5f - 47e = 1$, or, equivalently, $5 \cdot 160 - 47 \cdot 17 = 1$. Thus, $-47 \cdot 17 = 1 \bmod 160$, and so $17^{-1} \bmod 160 = -47 \bmod 160 = 113$, the desired result. We can easily verify that this is correct, by computing $113 \cdot 17 = 1921 = 1 \bmod 160$. $\square$

Example 9.8 Suppose we wish to determine, if it exists, the value of $683^{-1} \bmod 1007424$. To find whether $\gcd(1007424, 683) = 1$, we use the initial part of the Euclidean algorithm as follows.

$$
\begin{aligned}
1007424 &= 1474 \cdot 683 + 682 \\
683 &= 1 \cdot 682 + 1 \\
682 &= 682 \cdot 1 + 0
\end{aligned}
$$

Thus, $\gcd(1007424, 683) = 1$, and $683^{-1} \bmod 1007424$ does exist. To find $683^{-1} \bmod 1007424$, we express the remainders in this sequence of equations in terms of $f = 1007424$ and $e = 683$ as follows.

$$
\begin{aligned}
682 &= 1007424 - 1474 \cdot 683 &&= f - 1474e \\
1 &= 683 - 1 \cdot 682 &&= e - 1(f - 1474e) &&= -f + 1475e
\end{aligned}
$$

This last equation gives $-f + 1475e = 1$, or $-1 \cdot 1007424 + 1475 \cdot 683 = 1$. Thus, $1475 \cdot 683 = 1 \bmod 1007424$, and so $683^{-1} \bmod 1007424 = 1475$. To check this, we compute $1475 \cdot 683 = 1007425 = 1 \bmod 1007424$. $\square$

Example 9.9 Suppose we wish to determine, if it exists, $87^{-1} \bmod 537$. To find whether $\gcd(537, 87) = 1$, we use the initial part of the Euclidean algorithm as follows.

$$
\begin{aligned}
537 &= 6 \cdot 87 + 15 \\
87 &= 5 \cdot 15 + 12 \\
15 &= 1 \cdot 12 + 3 \\
12 &= 4 \cdot 3 + 0
\end{aligned}
$$

Thus, $\gcd(537, 87) \neq 1$, and so $87^{-1} \bmod 537$ does not exist. $\square$

9.3.1   Exercises

1. For the following integers a and b, use the initial part of the Euclidean algorithm to find $\gcd(a, b)$.

   (a)* $a = 540$, $b = 360$

(b) $a = 5280$, $b = 127$

(c)* $a = 5432$, $b = 1480$

(d) $a = 10380$, $b = 8880$

(e)* $a = 12345$, $b = 4321$

(f) $a = 3854682$, $b = 1095939$

2.* Repeat Exercise 1, but for each part use the final part of the Euclidean algorithm to determine integers s and t such that $\gcd(a, b) = sa + tb$.

3. Use the Euclidean algorithm to find the multiplicative inverse, if it exists.

(a)* $7^{-1} \bmod 57$

(b) $17^{-1} \bmod 47$

(c)* $27^{-1} \bmod 153$

(d) $101^{-1} \bmod 1023$

(e)* $1542^{-1} \bmod 5017$

(f) $1479^{-1} \bmod 1359$

4. For the following primes p and q and the potential encryption exponent e for an RSA cipher, use the Euclidean algorithm to find, if it exists, the corresponding decryption exponent d.

(a)* $p = 7$, $q = 11$, $e = 13$

(b) $p = 11$, $q = 17$, $e = 27$

(c)* $p = 61$, $q = 89$, $e = 127$

(d) $p = 23$, $q = 37$, $e = 33$

(e)* $p = 151$, $q = 173$, $e = 731$

(f) $p = 1009$, $q = 1523$, $e = 565$

5. For the following primes p and q, use the Euclidean algorithm to determine a valid encryption exponent e and the corresponding decryption exponent d for an RSA cipher.

(a) $p = 7$, $q = 19$

(b) $p = 11$, $q = 53$

(c) $p = 31$, $q = 307$

(d) $p = 257$, $q = 4001$

6. Repeat Exercise 5, but for each part determine values of e and d different from your answers to Exercise 5.

7. Recall, as we noted after the list of equations at the top of page 291 that result from the repeated divisions in the initial part of the Euclidean algorithm, $\gcd(a, b)$ is the last nonzero remainder r_n in this list. To see why this is true, consider the following. From the first equation in this list, explain how you know that any divisor of both a and b must also be a divisor of r_1. Then from the second equation in this list, explain how you know that any divisor of both b and r_1 must also be a divisor of r_2. Continuing in this manner until the next-to-last equation in this list, explain how you know that any divisor of both r_{n-2} and r_{n-1} must also be a divisor of r_n. Finally, explain why all of this means that $\gcd(a, b)$ will be r_n.

9.4   Modular Exponentiation

As we saw in Example 9.1 on page 287, encryption and decryption in RSA ciphers requires the mathematical operation of exponentiation with modular arithmetic. To perform this operation with small numbers, we can complete the exponentiation, and then reduce the result modulo the modulus. We did exactly this in Example 9.1. To compute $8^7 \bmod 33$, which was required in the encryption in Example 9.1, we first found $8^7 = 2097152$, and then reduced this result modulo 33 to find that $8^7 = 2097152 = 2 \bmod 33$. However, this process does not work well with larger numbers. For example, consider $6646^{683} \bmod 1010189$. If we tried to perform this operation by first finding 6646^{683}, we may run into a problem, since $6646^{683} \approx 6.44 \times 10^{2610}$, a number much too large to be stored precisely on standard calculators.

Further, in ignoring the problem of lost precision leading to incorrect results, another problem may arise. RSA ciphers that are used in industry generally require extremely large numbers, with perhaps a hundred or more digits in each. For a value of e this large, evaluating $x^e \bmod m$ by first finding x^e would require $e - 1$ multiplications, far too many to be done in a reasonable amount of time, certainly on a standard calculator, and even on a computer capable of millions of operations per second. In this section, we will present a technique for modular exponentiation that avoids both of these potential problems. This technique is sometimes referred to as *successive squaring* or *binary exponentiation*. It, like so many other things in some way connected to RSA ciphers, is very old, appearing in a book published in India around 200 BC.

To present this technique, suppose we wish to compute $x^e \bmod m$. The binary exponentiation process begins with a list of the values of 2^k for $k = 0, 1, 2, 3, \ldots$, continuing through the largest integer power of 2 smaller than e. Next, for each of these integer powers of 2, the value of $x^{2^k} \bmod m$ is determined, for which each value after the first can be obtained by squaring the previous value and reducing modulo m. The exponent e is then written as the sum of some of these integer powers of 2, with the powers used written in order from largest to smallest. Finally, using the expression of e as the sum of integer powers of 2, $x^e \bmod m$ is written as a product of factors, each of which is one of the $x^{2^k} \bmod m$ values. The product of these factors can then be formed modulo m to achieve the final result, all without ever needing to store a number larger than m, and generally with many fewer than $e - 1$ multiplications.

Example 9.10 Suppose we wish to use the technique of binary exponentiation to compute the value of $11^{47} \bmod 73$. We begin with a list of the following powers of 2, starting with 2^0, and going through the largest integer power of 2 smaller than 47.

$$
\begin{aligned}
2^0 &= 1 \\
2^1 &= 2 \\
2^2 &= 4 \\
2^3 &= 8 \\
2^4 &= 16 \\
2^5 &= 32
\end{aligned}
$$

Next, for each of these values of 2^k, we determine $11^{2^k} \bmod 73$, with each value after the first in the following list obtained by squaring the previous value and reducing modulo 73.

$$
\begin{aligned}
11^1 & & & & &= 11 \bmod 73 \\
11^2 & & & & &= 48 \bmod 73 \\
11^4 &= (11^2)^2 &=& 48^2 &=& 41 \bmod 73 \\
11^8 &= (11^4)^2 &=& 41^2 &=& 2 \bmod 73 \\
11^{16} &= (11^8)^2 &=& 2^2 &=& 4 \bmod 73 \\
11^{32} &= (11^{16})^2 &=& 4^2 &=& 16 \bmod 73
\end{aligned}
$$

Next, we write 47 as the sum of integer powers of 2, with the powers used written in order from largest to smallest. We can do this by starting with the sum considered as the single number 47, and then repeatedly removing the largest possible integer power of 2 from the smallest term in the sum, until the entire sum consists of integer powers of 2.

$$
\begin{aligned}
47 &= 32 + 15 \\
&= 32 + 8 + 7 \\
&= 32 + 8 + 4 + 3 \\
&= 32 + 8 + 4 + 2 + 1
\end{aligned}
$$

Finally, we use this expression of 47 as the sum of integer powers of 2 to find $11^{47} \bmod 73$ as follows.

$$
\begin{aligned}
11^{47} &= 11^{32+8+4+2+1} \bmod 73 \\
&= 11^{32} \cdot 11^8 \cdot 11^4 \cdot 11^2 \cdot 11^1 \bmod 73 \\
&= 16 \cdot 2 \cdot 41 \cdot 48 \cdot 11 \bmod 73 \\
&= 32 \cdot 41 \cdot 48 \cdot 11 \bmod 73 && \text{(since } 16 \cdot 2 = 32) \\
&= 71 \cdot 48 \cdot 11 \bmod 73 && \text{(since } 32 \cdot 41 = 71 \bmod 73) \\
&= 50 \cdot 11 \bmod 73 && \text{(since } 71 \cdot 48 = 50 \bmod 73) \\
&= 39 \bmod 73 && \text{(since } 50 \cdot 11 = 39 \bmod 73)
\end{aligned}
$$

Thus, $11^{47} = 39 \bmod 73$. $\square$

Referring to the two potential problems we mentioned at the beginning of this section, note that in Example 9.10 we were able to find $11^{47} \bmod 73$ with many fewer than 46 total multiplications. The number of multiplications required to complete this example was actually only nine, the five to determine the values of $11^{2^k} \bmod 73$ for $k = 1, 2, 3, 4, 5$, and the four to form $16 \cdot 2 \cdot 41 \cdot 48 \cdot 11 \bmod 73$. Also, note that in this example we were never required to store (i.e., write down) a number larger than 73.

Example 9.11 Suppose we wish to use the technique of binary exponentiation to compute the value of $2^{160} \bmod 161$. We begin with a list of the following powers of 2.

$$
\begin{aligned}
2^0 &= 1 \\
2^1 &= 2 \\
2^2 &= 4 \\
2^3 &= 8 \\
2^4 &= 16 \\
2^5 &= 32 \\
2^6 &= 64 \\
2^7 &= 128
\end{aligned}
$$

Next, for each of these values of 2^k, we determine $2^{2^k} \bmod 161$.

$$
\begin{aligned}
2^1 &&&&&&= 2 \bmod 161 \\
2^2 &&&&&&= 4 \bmod 161 \\
2^4 &= (2^2)^2 &&= 4^2 &&= 16 \bmod 161 \\
2^8 &= (2^4)^2 &&= 16^2 &&= 95 \bmod 161 \\
2^{16} &= (2^8)^2 &&= 95^2 &&= 9 \bmod 161 \\
2^{32} &= (2^{16})^2 &&= 9^2 &&= 81 \bmod 161
\end{aligned}
$$

$$2^{64} = (2^{32})^2 = 81^2 = 121 \bmod 161$$
$$2^{128} = (2^{64})^2 = 121^2 = 151 \bmod 161$$

Next, we write 160 as the sum of integer powers of 2.

$$160 = 128 + 32$$

Finally, we use this expression of 160 as the sum of integer powers of 2 to find $2^{160} \bmod 161$ as follows.

$$
\begin{aligned}
2^{160} &= 2^{128+32} \bmod 161 \\
&= 2^{128} \cdot 2^{32} \bmod 161 \\
&= 151 \cdot 81 \bmod 161 \\
&= 156 \bmod 161
\end{aligned}
$$

Thus, $2^{160} = 156 \bmod 161$. $\qquad\square$

Note that in Example 9.11, we were able to find $2^{160} \bmod 161$ with only eight multiplications, the seven to determine the values of $2^{2^k} \bmod 161$ for $k = 1, 2, 3, 4, 5, 6, 7$, and the one to form $151 \cdot 81 \bmod 161$.

Example 9.12 Suppose we wish to use the technique of binary exponentiation to compute the value of $103^{683} \bmod 1010189$. We begin with a list of the following powers of 2.

$$
\begin{aligned}
2^0 &= 1 \\
2^1 &= 2 \\
2^2 &= 4 \\
2^3 &= 8 \\
2^4 &= 16 \\
2^5 &= 32 \\
2^6 &= 64 \\
2^7 &= 128 \\
2^8 &= 256 \\
2^9 &= 512
\end{aligned}
$$

Next, for each of these values of 2^k, we determine $103^{2^k} \bmod 1010189$.

$$
\begin{aligned}
103^1 &&&= 103 \bmod 1010189 \\
103^2 &&&= 10609 \bmod 1010189 \\
103^4 &= (103^2)^2 &= 10609^2 &= 419902 \bmod 1010189 \\
103^8 &= (103^4)^2 &= 419902^2 &= 311733 \bmod 1010189 \\
103^{16} &= (103^8)^2 &= 311733^2 &= 312056 \bmod 1010189
\end{aligned}
$$

$$103^{32} = (103^{16})^2 = 312056^2 = 768292 \bmod 1010189$$
$$103^{64} = (103^{32})^2 = 768292^2 = 981162 \bmod 1010189$$
$$103^{128} = (103^{64})^2 = 981162^2 = 69103 \bmod 1010189$$
$$103^{256} = (103^{128})^2 = 69103^2 = 61206 \bmod 1010189$$
$$103^{512} = (103^{256})^2 = 61206^2 = 393624 \bmod 1010189$$

Next, we write 683 as the sum of integer powers of 2.

$$
\begin{aligned}
683 &= 512 + 171 \\
&= 512 + 128 + 43 \\
&= 512 + 128 + 32 + 11 \\
&= 512 + 128 + 32 + 8 + 3 \\
&= 512 + 128 + 32 + 8 + 2 + 1
\end{aligned}
$$

Finally, we find $103^{683} \bmod 1010189$ as follows.

$$
\begin{aligned}
103^{683} &= 103^{512+128+32+8+2+1} \bmod 1010189 \\
&= 103^{512} \cdot 103^{128} \cdot 103^{32} \cdot 103^{8} \cdot 103^{2} \cdot 103^{1} \bmod 1010189 \\
&= 393624 \cdot 69103 \cdot 768292 \cdot 311733 \cdot 10609 \cdot 103 \bmod 1010189 \\
&= 250258 \cdot 768292 \cdot 311733 \cdot 10609 \cdot 103 \bmod 1010189 \\
&= 936777 \cdot 311733 \cdot 10609 \cdot 103 \bmod 1010189 \\
&= 888799 \cdot 10609 \cdot 103 \bmod 1010189 \\
&= 164465 \cdot 103 \bmod 1010189 \\
&= 776871 \bmod 1010189
\end{aligned}
$$

Thus, $103^{683} = 776871 \bmod 1010189$. $\square$

Note that in Example 9.12, we were able to find $103^{683} \bmod 1010189$ with a total of only 14 multiplications, the nine to determine the values of $103^{2^k} \bmod 1010189$ for $k = 1, 2, 3, 4, 5, 6, 7, 8, 9$, and the five to form $393624 \cdot 69103 \cdot 768292 \cdot 311733 \cdot 10609 \cdot 103 \bmod 1010189$.

In general, binary exponentiation saves a larger percentage of multiplications for larger exponents. For instance, in Example 9.10, binary exponentiation required only $9/46 = 19.57\%$ of the multiplications that would have been required to find $11^{47} \bmod 73$ by fully finding 11^{47}. In Example 9.11, this percentage was only $8/159 = 5.03\%$, and in Example 9.12, it was only $14/682 = 2.05\%$. This means that as the exponents under consideration grow larger and binary exponentiation becomes more necessary, it also becomes more efficient.

9.4.1 Exercises

1. Use binary exponentiation to compute the following.

   (a)* 4042^{67} mod 9727

   (b) 32^{113} mod 267

   (c)* 14^{173} mod 851

   (d) 20101^{301} mod 50101

   (e)* 7^{983} mod 14123

   (f) 776871^{1475} mod 1010189

2.* Determine the exact number of multiplications necessary to complete each of the binary exponentiations in Exercise 1.

3. For a positive integer k, b^{-k} mod m exists if and only if b^{-1} mod m exists. If b^{-1} mod m exists, then we can find b^{-k} mod m by computing $(b^{-1}$ mod $m)^k$ mod m. Use these facts to compute the following, if they exist.

   (a)* 3564^{-67} mod 9727

   (b) 242^{-113} mod 267

   (c)* 592^{-173} mod 851

   (d) 16535^{-301} mod 50101

   (e)* 10088^{-983} mod 14123

   (f) 70146^{-1475} mod 1010189

9.5   ASCII

Modern cryptographic methods such as RSA are almost exclusively automated on computers. The most common way for text to be stored on computers is through the *American Standard Code for Information Interchange*, or *ASCII* for short. ASCII is simply a list of correspondences between characters and numbers, like the correspondences A = 0, B = 1, C = 2, ... , Z = 25 we have used in the last several chapters of this book. However, characters in ASCII are not restricted to only capital letters. ASCII includes correspondences for all printable characters on a modern keyboard, including both capital and lowercase letters, the digits 0–9, punctuation, a blank space, and some others. These printable characters, 95 in all, correspond in ASCII to the numbers 32–126. The numbers 0–31 were reserved in ASCII for control characters used to format text and space. Although most of these control characters are now obsolete, the numbers

32–126 originally used to represent the printable characters have been pre-served. A full list of the ASCII correspondences for the printable characters on a modern keyboard is shown in Table 9.1.

Char	Num	Char	Num	Char	Num	Char	Num
(space)	32	8	56	P	80	h	104
!	33	9	57	Q	81	i	105
"	34	:	58	R	82	j	106
#	35	;	59	S	83	k	107
$	36	<	60	T	84	l	108
%	37	=	61	U	85	m	109
&	38	>	62	V	86	n	110
,	39	?	63	W	87	o	111
(	40	@	64	X	88	p	112
)	41	A	65	Y	89	q	113
*	42	B	66	Z	90	r	114
+	43	C	67	[	91	s	115
,	44	D	68	\	92	t	116
-	45	E	69	]	93	u	117
.	46	F	70	^	94	v	118
/	47	G	71	_	95	w	119
0	48	H	72	`	96	x	120
1	49	I	73	a	97	y	121
2	50	J	74	b	98	z	122
3	51	K	75	c	99	{	123
4	52	L	76	d	100	|	124
5	53	M	77	e	101	}	125
6	54	N	78	f	102	~	126
7	55	O	79	g	103		

Table 9.1   ASCII correspondences between characters and numbers.

Throughout the rest of this book, we will use the ASCII correspondences between characters and numbers. In addition to being consistent with modern cryptography, this will provide us with much more flexibility in the messages we can consider, since we will be able to consider messages written using all of the printable characters on a modern keyboard.

9.5.1   Exercise

1. Find some information about the ASCII code, including its history, the reasons for its development, and its current uses, and write a summary of your findings.

9.6 RSA Ciphers

For the rest of this book, we will only consider plaintext messages expressed using ASCII characters. Since the corresponding ASCII numbers are larger than 26, this will require us to do calculations in which the modulus is larger than 26. We have actually already considered an example of the basic steps in using an RSA cipher, in Example 9.1 on page 287, with a modulus of 33. However, if we used the ASCII correspondences to convert a plaintext into numeric form, 33 would not be an acceptable modulus either. The modulus in an RSA cipher is typically much larger than 26 or 33, and this is why in the last few sections we considered calculations in which the modulus is much larger than 26 or 33.

Recall also that we noted after Example 9.1 that the RSA cipher in that example is not secure because it is a substitution cipher. In fact, any RSA cipher in which characters are all encrypted separately (as they are in Example 9.1) would be a substitution cipher, regardless of the size of the exponent and modulus. It was still useful to consider the basic steps in using an RSA cipher in Example 9.1, though. This example clearly demonstrated the RSA encryption and decryption procedures, which can be followed without the resulting cipher being a substitution cipher if consecutive plaintext numbers are grouped into blocks before being encrypted. The modulus for such a cipher would have to be larger than the plaintext blocks, but this is again why in the last few sections we began considering calculations in which the modulus is large. Specifically, for prime numbers p and q, and $m = p \cdot q$, the basic steps in using an RSA cipher listed on page 286 can be followed exactly as presented, even if the plaintext integers x result from grouping ASCII numbers into larger blocks, provided the blocks are all less than m. This is how we will implement most of our RSA ciphers in the rest of this chapter, with the plaintext ASCII numbers grouped in order into blocks, using for each block the maximum possible collection of numbers such that the block remains less than m.

Example 9.13 Suppose you wish to send the secret message B.B. King to a colleague over an insecure communication line using an RSA cipher with the ASCII correspondences, under which the plaintext converts into the list of integers 66, 46, 66, 46, 32, 75, 105, 110, 103. Your colleague initiates the process by choosing primes $p = 433$ and $q = 2333$, and forming $m = p \cdot q = 1010189$ and $f = (p - 1) \cdot (q - 1) = 1007424$. Your colleague then chooses $e = 683$ (for which $\gcd(e, f) = 1$, as your colleague could easily verify using the Euclidean algorithm), and sends the values of e and m to you over the insecure communication line. With the value $m = 1010189$ received from your colleague, you can group the first three plaintext numbers into the single block 664666, which remains less than m.

Similarly, you can group the next three plaintext numbers into the block 463275, and the next two into the block 105110, with only the last plaintext number 103 left to form the last block. That is, the full list of plaintext blocks to be encrypted can actually be 664666, 463275, 105110, 103. Then, with the value $e = 683$ received from your colleague, you encrypt these plaintext blocks as follows.

$$664666^{683} = 501396 \bmod 1010189$$
$$463275^{683} = 24780 \bmod 1010189$$
$$105110^{683} = 889984 \bmod 1010189$$
$$103^{683} = 776871 \bmod 1010189 \quad \text{(see Example 9.12)}$$

You then send the ciphertext blocks 501396, 24780, 889984, 776871 to your colleague. To decrypt the message, your colleague must first find the value of $d = e^{-1} \bmod f = 683^{-1} \bmod 1007424$, which we found using the Euclidean algorithm in Example 9.8 on page 294 to be $d = 1475$. Your colleague can then decrypt the message as follows.

$$501396^{1475} = 664666 \bmod 1010189$$
$$24780^{1475} = 463275 \bmod 1010189$$
$$889984^{1475} = 105110 \bmod 1010189$$
$$776871^{1475} = 103 \bmod 1010189$$

Your colleague can then split these plaintext blocks into numbers that correspond to a single ASCII character each, and convert these numbers back into the original plaintext characters. $\qquad\square$

Note that in Example 9.13, it was not possible to group the last three plaintext numbers into a single block, since this block would have been 105110103, which is greater than $m = 1010189$. The reason this is a problem is because if we encrypted the plaintext block 105110103 by forming $c = 105110103^{683} \bmod 1010189$, it would not be possible to recover this plaintext block by forming $c^{1475} \bmod 1010189$, since the result of this calculation would be an integer less than 1010189. Also, note that in Example 9.13, although you have no way of secretly informing your colleague how many plaintext numbers you grouped together to form each block you encrypted, your colleague would still be guaranteed to be able to uniquely convert the decrypted blocks back into the original plaintext characters. This is due to the range of numbers used in the ASCII code, specifically the fact that all three-digit numbers in the code begin with the digit 1, while no two-digit numbers begin with 1.

Although grouping plaintext numbers into blocks prevents an RSA cipher from being broken via single-letter frequency analysis, if the blocks are small it may still be possible for the cipher to be broken via digraph or trigraph frequency analysis. Even this could be prevented, though, if the primes p and q used to initiate the cipher were large enough such that $m = p \cdot q$ would allow for plaintext numbers to be grouped into blocks corresponding to a large number of characters each. In fact, since there are infinitely many prime numbers (as Euclid also demonstrated in his book *Elements* around 300 BC), it would actually always be possible to choose p and q large enough such that any entire plaintext could be encrypted with a single calculation.

Example 9.14 Suppose again that you wish to send the secret message B.B. King to a colleague over an insecure communication line using an RSA cipher with the ASCII correspondences, under which the plaintext converts into the list of integers 66, 46, 66, 46, 32, 75, 105, 110, 103. Your colleague initiates the process by choosing primes $p = 21458121277$ and $q = 59728127341$, and forming $m = p \cdot q = 1281653400131277534457$ and $f = (p-1) \cdot (q-1) = 1281653400050091285840$. Your colleague then chooses $e = 2312412281$ (for which $\gcd(e, f) = 1$, as your colleague could easily verify using the Euclidean algorithm), and sends the values of e and m to you over the insecure communication line. With the value of m received from your colleague, you can group the plaintext numbers into the single block 664666463275105110103, which is less than m. Then, with the value of e received from your colleague, you encrypt this plaintext block as follows.

$$664666463275105110103^e = 189250999109759995893 \bmod m$$

You then send the ciphertext block 189250999109759995893 to your colleague. To decrypt the message, your colleague must first find the value of $d = e^{-1} \bmod f$, which can be found using the Euclidean algorithm to be $d = 453711873428844852761$. Your colleague can then decrypt the message as follows.

$$189250999109759995893^d = 664666463275105110103 \bmod m$$

Your colleague can then split this plaintext block into numbers that correspond to a single ASCII character each, and convert these numbers back into the original plaintext characters. □

Granted, some of the numbers in Example 9.14 may be too large for the calculations to be done on some standard calculators. However, the calculations themselves do not require too many operations for any standard calculator. The Euclidean algorithm typically requires a very small number of calculations, even beginning with extremely large numbers. Also, for example, the decryption calculation in Example 9.14 could be done using binary exponentiation with only 95 multiplications.

Finally, recall as we noted in Section 9.2, for an RSA ciphertext integer y, it is by no means obvious that the calculation $y^d \bmod m$ will result in the plaintext integer x from which y was formed. More specifically, for an RSA cipher with primes p and q that are not equal to each other, values of $m = p \cdot q$ and $f = (p-1) \cdot (q-1)$, encryption exponent e and corresponding decryption exponent $d = e^{-1} \bmod f$, and plaintext integer x with corresponding ciphertext integer $y = x^e \bmod m$, it is certainly not obvious why $x = y^d \bmod m$. The reason why this happens is because of a very famous theorem in mathematics called *Fermat's Little Theorem*, named for Pierre de Fermat, a French lawyer and amateur mathematician who first identified it in 1640.

Theorem 9.2 *(Fermat's Little Theorem) If p is a prime number and r is an integer with $gcd(r,p) = 1$, then $r^{p-1} = 1 \bmod p$.*

Consider an RSA decryption calculation $y^d \bmod m$. Since $y = x^e \bmod m$, this decryption calculation is equivalent to $(x^e)^d \bmod m$, or $x^{ed} \bmod m$. Further, since $d = e^{-1} \bmod f$, then $ed = 1 \bmod f$, from which it follows that $ed = 1 + kf$ for some integer k. This gives the following.

$$x^{ed} = x^{1+kf} = x \cdot x^{kf} = x \cdot x^{k(p-1)(q-1)}$$

If $gcd(x,p) = 1$, then by Fermat's Little Theorem we have the following.

$$x^{ed} \bmod p = x \cdot x^{k(p-1)(q-1)} \bmod p = x \cdot (x^{p-1})^{k(q-1)} \bmod p = x \cdot 1 = x$$

Also, since p is prime, if $gcd(x,p) \neq 1$, then x must be a multiple of p, and so $x^{ed} \bmod p = x$, since both sides of this equation are 0 modulo p. This would all work identically with q in place of p, of course. That is, for any integer x, it must be the case that both $x^{ed} \bmod p = x$ and $x^{ed} \bmod q = x$. This is equivalent to saying that for any integer x, it must be the case that $x^{ed} - x$ is both a multiple of p and a multiple of q. Since p and q are primes that are not equal to each other, this means $x^{ed} - x$ must be a multiple of $m = p \cdot q$. It follows that $x^{ed} \bmod m = x$, which is equivalent to $y^d \bmod m = x$. This is what we were after all along, the decryption calculation that makes RSA ciphers work.

Even before RSA ciphers became known, the fact that $x^{ed} \bmod p = x$ for prime p and integers e and d with $ed = 1 \bmod f$ was being used to create ciphers. In such ciphers, plaintext integers x were encrypted by being raised to the power e and reduced modulo p. Resulting ciphertext integers $x^e \bmod p$ could then be decrypted by being raised to the power d and reduced modulo p, since $(x^e)^d \bmod p = x^{ed} \bmod p = x$. Such ciphers are called *exponentiation* ciphers. Note that in exponentiation ciphers, e and p must be kept secret between the originator and intended recipient of a message, for otherwise outsiders could easily find d using the Euclidean algorithm. Thus, exponentiation ciphers are not public-key. This makes them, while operationally similar to RSA, useful only under entirely different circumstances. Changing the modulus to be the product of two primes, as in RSA, was a major advancement, indeed leading to a full-scale revolution in modern cryptology, because it allowed the resulting ciphers to be public-key.

9.6.1   Exercises

1. Consider an RSA cipher with primes $p = 17$ and $q = 23$, and the ASCII correspondences.

   (a)* Use this cipher with encryption exponent $e = 5$ to encrypt Mr. T, with the plaintext numbers all encrypted separately.

   (b) Use this cipher with encryption exponent $e = 5$ to encrypt Dr. J, with the plaintext numbers all encrypted separately.

   (c) For this cipher with encryption exponent $e = 235$, verify that the decryption exponent is $d = 3$.

   (d) Decrypt 68, 91, 23, 213, 69, 288, 41, 124, which was formed using this cipher with encryption exponent $e = 235$ (for which the decryption exponent is given in part (c)), and in which the plaintext numbers were all encrypted separately.

   (e)* Use this cipher with encryption exponent $e = 47$ to encrypt Boon, with the plaintext numbers all encrypted separately.

   (f) Use this cipher with encryption exponent $e = 47$ to encrypt D-Day, with the plaintext numbers all encrypted separately.

   (g)* For this cipher with encryption exponent $e = 47$, find the decryption exponent.

   (h) Decrypt 218, 300, 300, 16, 367, which was formed using this cipher with encryption exponent $e = 47$ (for which the decryption exponent is the answer to part (g)), and in which the plaintext numbers were all encrypted separately.

2. Consider an RSA cipher with primes $p = 569$ and $q = 227$, and the ASCII correspondences.

   (a)* Use this cipher with encryption exponent $e = 5$ to encrypt AC/DC, with the plaintext numbers grouped into the largest possible blocks before being encrypted.

(b) Use this cipher with encryption exponent $e = 5$ to encrypt B-52s, with the plaintext numbers grouped into the largest possible blocks before being encrypted.

(c) For this cipher with encryption exponent $e = 85579$, verify that the decryption exponent is $d = 3$.

(d) Decrypt 63914, 28751, 16503, 125855, 118031, which was formed using this cipher with encryption exponent $e = 85579$ (for which the decryption exponent is given in part (c)), and in which the plaintext numbers were grouped into the largest possible blocks before being encrypted.

(e)* Use this cipher with encryption exponent $e = 161$ to encrypt WKRP, with the plaintext numbers grouped into the largest possible blocks before being encrypted.

(f) Use this cipher with encryption exponent $e = 161$ to encrypt KACL, with the plaintext numbers grouped into the largest possible blocks before being encrypted.

(g)* For this cipher with encryption exponent $e = 161$, find the decryption exponent.

(h) Decrypt 65410, 49459, which was formed using this cipher with encryption exponent $e = 161$ (for which the decryption exponent is the answer to part (g)), and in which the plaintext numbers were grouped into the largest possible blocks before being encrypted.

3. Consider an RSA cipher with primes $p = 6653$ and $q = 19457$, and the ASCII correspondences.

   (a)* Use this cipher with encryption exponent $e = 5$ to encrypt SF 49ers, with the plaintext numbers grouped into the largest possible blocks before being encrypted.

   (b) Use this cipher with encryption exponent $e = 5$ to encrypt PHI 76ers, with the plaintext numbers grouped into the largest possible blocks before being encrypted.

   (c) For this cipher with encryption exponent $e = 86280875$, verify that the decryption exponent is $d = 3$.

   (d) Decrypt 82592672, 47345524, 128803984, 55541520, which was formed using this cipher with encryption exponent $e = 86280875$ (for which the decryption exponent is given in part (c)), and in which the plaintext numbers were grouped into the largest possible blocks before being encrypted.

(e)* Use this cipher with encryption exponent $e = 683$ to encrypt CIA, with the plaintext numbers grouped into a single block before being encrypted.

(f) Use this cipher with encryption exponent $e = 683$ to encrypt NSA, with the plaintext numbers grouped into a single block before being encrypted.

(g) For this cipher with encryption exponent $e = 683$, find the decryption exponent.

(h) Decrypt 76946937, which was formed using this cipher with encryption exponent $e = 683$ (for which the decryption exponent is the answer to part (g)), and in which the plaintext numbers were grouped into a single block before being encrypted.

4. Recall, as we noted in this section, there are infinitely many prime numbers. To see why this is true, consider the following. Suppose that there were in fact only finitely many prime numbers, with p_1, p_2, p_3, ... , p_k being a complete list. Explain how you know that none of the numbers in this list divides the number $(p_1 \cdot p_2 \cdot p_3 \cdots p_k) + 1$. Then explain why this means that there cannot actually be only finitely many prime numbers.

5. To see why Fermat's Little Theorem is true, consider the following. Suppose p is a prime number, let r be an integer with $1 \leq r \leq p - 1$, and consider the list of numbers r, $2r$, $3r$, ... , $(p - 1)r$, all reduced modulo p. Explain how you know that none of the numbers in this list could equal 0; that is, explain why $kr \neq 0 \bmod p$ for any $1 \leq k \leq p-1$. Then explain how you know that there could be no duplicate numbers in this list; that is, explain why if $1 \leq k \leq p - 1$, $1 \leq j \leq p - 1$, and $k \neq j$, then $kr \neq jr \bmod p$. From these two facts it follows that the list of numbers r, $2r$, $3r$, ... , $(p - 1)r$, all reduced modulo p, must be at most a reordering of the numbers 1, 2, 3, ... , $p - 1$. As such, $r \cdot 2r \cdot 3r \cdots (p-1)r = 1 \cdot 2 \cdot 3 \cdots (p - 1) \bmod p$, or, equivalently, $r^{p-1} \cdot (p - 1)! = (p - 1)! \bmod p$. Explain how you know then that $r^{p-1} = 1 \bmod p$. Then explain why Fermat's Little Theorem follows from this.

9.7 Cryptanalysis of RSA Ciphers

Recall that RSA ciphers came to be during a change in direction for the subject of cryptology, from classical ciphers such as Hill and Vigenère, to public-key ciphers such as RSA. With classical ciphers such as Hill and Vigenère, both the originator and intended recipient of a message must know the encryption keys, but outsiders must not, since this knowledge alone leads to the determination of the decryption keys. With a public-key cipher, outsiders can know the encryption keys, since this knowledge alone does not lead to the determination of the decryption keys. Recall the benefit to this—two parties wishing to communicate secretly over an insecure communication line can openly agree upon encryption keys without having to consider whether these keys were intercepted by an outsider. On the other hand, with a public-key cipher, outsiders must still not be able to determine the decryption keys. Typically, with a public-key cipher, no one but the intended recipient of a message, excluding even the originator of the message, can determine the decryption keys. This explains why the work in using an RSA cipher is done primarily by the intended recipient, not the originator. In this section, we will elaborate on these facts in light of what we now know about how RSA ciphers work.

As we have noted, when assessing the security of a cipher, it is generally assumed that the encryption keys are the only thing about the cipher kept secret between the communicating parties. This means, for example, we would assume an outsider who intercepts a ciphertext formed using an RSA cipher would know that each ciphertext block was formed as $x^e \bmod m$ for some plaintext block x and encryption keys e and m, with $m = p \cdot q$ for some prime numbers p and q, and $\gcd(e, f) = 1$ for $f = (p-1) \cdot (q-1)$. The fact that RSA ciphers are public-key means we would also assume the outsider knows the actual values of e and m used in the encryption calculations. For example, we would assume an outsider who intercepts the ciphertext blocks in Example 9.13 on page 303 would know that each ciphertext block y was formed as $y = x^{683} \bmod 1010189$ for some plaintext block x. This obviously adversely affects the security of the cipher. In fact, it implies that the cipher in Example 9.13 is not mathematically secure. This is because an outsider who intercepts the ciphertext blocks y and knows the values of e and m could use the following steps to break the cipher.

1. Factor $m = p \cdot q$ to find p and q.

2. Form $f = (p-1) \cdot (q-1)$.

3. Find $d = e^{-1} \bmod f$.

4. Recover the plaintext blocks x by forming $x = y^d \bmod m$.

These general steps are not unique to breaking the cipher in Example 9.13, of course. Any RSA cipher could be broken using the same steps 1–4. In addition, even with extremely large ciphertext blocks y and values of e and m, steps 2–4 could all be done relatively easily. After p and q are determined, step 2 would be trivial. Steps 3 and 4 could always be done very quickly using the Euclidean algorithm and binary exponentiation, respectively, both of which are remarkably fast even with extremely large numbers. This leaves step 1 to stand alone against cryptanalysts.

Fortunately (for cryptographers), step 1 is very difficult to complete in general, thereby giving RSA ciphers their high level of security. Actually, if either p or q were very small, step 1 would be easy to complete by trial divisions (i.e., trial and error). Even if m were formed using the primes in Example 9.13 ($p = 433$ and $q = 2333$), step 1 could be completed by trial divisions in no more than a few minutes. However, as p and q grow larger, the time it would take to factor $m = p \cdot q$ by trial divisions increases very quickly. Techniques for factoring integers that are faster than trial divisions do exist, but to factor a number that is the product of two very large primes, even the fastest known factoring techniques would essentially take forever.[1] More specifically, if m were the product of two primes both hundreds of digits in length, the fastest known factoring techniques would in general take millions of years to factor m, even when programmed on a computer capable of millions of operations per second.

It is the general difficulty of factoring m that gives RSA ciphers their high level of security. Knowing m and the encryption exponent e in an RSA cipher is not enough to determine the decryption exponent. To find the decryption exponent, an outsider would have to factor m to find p and q, something that in general would be essentially impossible to do if p and q were both very large. On the other hand, the difficulty of factoring m would not pose a problem for the intended recipient of a ciphertext formed using an RSA cipher, since the intended recipient would have begun the entire process by choosing p and q.[2]

In addition to factoring m, there are other methods an outsider can use to try to break an RSA cipher. Exercise 3 at the end of this section details one such method. However, the success of this and other methods relies primarily on implementation or human error on the part of users of an RSA cipher. The fact is, when implemented correctly, RSA ciphers have proved thus far to be impregnable.

Public-key ciphers have obvious advantages over non-public-key ciphers. However, there are disadvantages as well. One is that public-key ciphers are in general slower than non-public-key ciphers, which can be especially important for users who have a lot of information to transmit confidentially. Also, public-key ciphers can be exploited if *message authentication* (e.g., the assurance to the intended recipient of a message that it actually came from the originator claiming to have sent it) is compromised. We will consider this important aspect of public-key ciphers in Chapter 12.

---

[1] We will briefly discuss factoring techniques that do not require trial divisions in Section 9.9.

[2] We will briefly discuss the problem of finding very large prime numbers in Section 9.8.

9.7.1   Exercises

1. Decrypt the following ciphertexts, which were formed using RSA ciphers with the given values of m and e.

   (a) $m = 381$; $e = 101$;
   ciphertext = 295, 147, 222, 177, 23, 214, 170, 180

   (b) $m = 415$; $e = 193$;
   ciphertext = 42, 241, 241, 163, 296, 319

   (c) $m = 553$; $e = 7$;
   ciphertext = 80, 98, 404, 221, 321

   (d) $m = 649$; $e = 453$;
   ciphertext = 440, 80, 530, 7, 386

2. Recall from Exercise 5 in Section 2.2 the concept of *superencryption*.

   (a)* Use RSA ciphers with the common modulus $m = 141$ and encryption exponents $e_1 = 3$ and $e_2 = 37$ (in that order) to superencrypt Greg.

   (b) Decrypt 122, 24, 78, 136, which was superencrypted using RSA ciphers with the common modulus $m = 141$ and encryption exponents $e_2 = 37$ and $e_1 = 3$ (in that order).

   (c) Does superencryption by two RSA ciphers with a common modulus m yield more security than encryption by one RSA cipher? In other words, if a plaintext P is encrypted using an RSA cipher with modulus m, yielding M, and then M is encrypted using another RSA cipher with the same modulus m (but not necessarily the same encryption exponent), yielding C, would C be harder in general to cryptanalyze than M? Explain your answer completely, and be as specific as possible.

3. One method through which an RSA cipher can be broken without a cryptanalyst having to factor the modulus m is the result of what is called a *common modulus protocol failure*. This cryptanalysis method relies on human error on the part of users of RSA ciphers, specifically if the originator of a message uses RSA ciphers to encrypt the same plaintext for two intended recipients who happen to share a common modulus m. Suppose the originator of a message uses RSA ciphers to encrypt the same plaintext x for two intended recipients, one having modulus m (with $m > x$) and encryption exponent e_1, and the other having the same modulus but a different encryption exponent e_2 that satisfies $\gcd(e_1, e_2) = 1$. If the originator forms ciphertexts

$y_1 = x^{e_1} \bmod m$ and $y_2 = x^{e_2} \bmod m$, and sends each ciphertext to its respective intended recipient, then an outsider who intercepts both ciphertexts can find the plaintext x by first using the Euclidean algorithm to find integers s and t for which $se_1 + te_2 = 1$, and then forming $x = y_1^s \cdot y_2^t \bmod m$.

(a) Consider a pair of intended recipients of a ciphertext formed using an RSA cipher, one having modulus $m = 74663$ and encryption exponent $e_1 = 41$, and the other having the same modulus but encryption exponent $e_2 = 71$. Find integers s and t for which $se_1 + te_2 = 1$.

(b) Suppose the originator of a message uses RSA ciphers to encrypt the same plaintext for the two intended recipients described in part (a), with resulting ciphertexts 21939 and 50711, respectively. Without factoring m, find the plaintext. (Recall from Exercise 3 in Section 9.4 the concept of $b^{-k} \bmod m$ for a positive integer k.)

(c) Show why this cryptanalysis method works. That is, show why if $y_1 = x^{e_1} \bmod m$ and $y_2 = x^{e_2} \bmod m$, and $se_1 + te_2 = 1$, then $x = y_1^s \cdot y_2^t \bmod m$.

4. Find some information about one or more other cryptanalytic attacks against RSA ciphers besides factoring m and the method detailed in Exercise 3, and write a summary of your findings.

5. Find some information about how the size of the primes p and q believed to be necessary for RSA ciphers to be secure has increased over the years, including current standards or opinions, and write a summary of your findings.

6. Find some information about one or more real-life uses of RSA ciphers, and write a summary of your findings.

7. Find a copy of the paper [20] in which Ron Rivest, Adi Shamir, and Len Adleman first publicly described RSA ciphers, and as best you are able, write a summary of how RSA ciphers are described in this paper.

8. Find a copy of the paper in which Clifford Cocks first described RSA ciphers (which has been declassified), and as best you are able, write a summary comparing how RSA ciphers are described in this paper with how they are described in the paper [20] in which Ron Rivest, Adi Shamir, and Len Adleman first publicly described them.

9.8 Primality Testing

Recall that to break an RSA cipher, the only real difficulty an outsider should face is in factoring $m = p \cdot q$ to find p and q. However, recall also that if p and q were both hundreds of digits in length, the fastest known factoring techniques would in general take millions of years to factor m, even when programmed on a computer capable of millions of operations per second. So an RSA cipher can be secure from outsiders, even very technologically savvy ones, if the primes p and q for the cipher are both extremely large. This leads to a couple of questions. First, just how large can prime numbers be? Second, assuming extremely large prime numbers exist, how can we find them?

The first of these questions is easy to answer. As we noted in Section 9.6, there are infinitely many prime numbers, and as such there is no limit to how large they can be. The second question is harder to answer. Motivated at least in part by the development of public-key ciphers like RSA, much research has been done over the past few decades on the problem of finding extremely large prime numbers, in an area of mathematics called *primality testing*.

A primality test indicates whether a given positive integer is likely or certain to be prime. The Fundamental Theorem of Arithmetic guarantees that every nonprime integer greater than 1 can be expressed as a product of primes, with the primes in this product being unique. As a result of this, the most obvious primality test is to just do actual division to see if a given positive integer can be expressed as a product of smaller prime factors. For a given positive integer n, we could complete this process systematically by checking to see if any prime number from 2 through the largest prime less than $\sqrt{n}$ divides n evenly.[3] We will call this the primality test of *trial divisions*.

Example 9.15 To use trial divisions to determine whether 839 is prime, we begin by computing $\sqrt{839} \approx 28.97$. The only primes less than 28.97 are 2, 3, 5, 7, 11, 13, 17, 19, and 23. Since each of these primes fails to divide 839 evenly, 839 is prime. □

Example 9.16 To use trial divisions to determine whether 1073 is prime, we begin by computing $\sqrt{1073} \approx 32.76$. The only primes less than 32.76 are 2, 3, 5, 7, 11, 13, 17, 19, 23, 29, and 31. Since 29 divides 1073 evenly, 1073 is not prime. □

One problem with using trial divisions to determine whether a given positive integer n is prime is that if n were extremely large, there would be a

---

[3] We would only have to check through the largest prime less than $\sqrt{n}$, since if n had a smaller prime factor greater than $\sqrt{n}$, it would also have one less than $\sqrt{n}$.

tremendously large number of primes less than $\sqrt{n}$ that could need to be tested. For a positive integer s, the number of primes less than or equal to s is denoted as $\pi(s)$, called the *prime counting function*. For example, referring to the list of primes in Example 9.15, we see that $\pi(23) = 9$. A known fact regarding the prime counting function is that if s is large, the following formula gives an approximation for $\pi(s)$.

$$\pi(s) \approx \frac{s}{\ln s}$$

For example, suppose we wish to determine whether 84308508309887 is prime. To do this using trial divisions, since $\sqrt{84308508309887} \approx 9181966$, this formula for $\pi(s)$ indicates that we could need to test as many as $\pi(9181966) \approx \frac{9181966}{\ln(9181966)} \approx 572701$ primes. Thus, trial divisions could be a very inefficient method for determining whether 84308508309887 is prime. Even worse, for a given number hundreds of digits in length, it could take millions of years to determine whether the number is prime by trial divisions, even if the process were programmed on a computer capable of millions of divisions per second.

Contrary to what the name *primality test* suggests, most primality tests only allow, with absolute certainty, the conclusion that a given positive integer is *not* prime. The conclusions that can usually be drawn from a primality test are that a given positive integer either "fails" the test and is definitely not prime, or "passes" the test and is likely to be prime, with likelihood depending on the test itself. One relatively simple and efficient primality test is based on our old friend Fermat's Little Theorem, the result that made RSA ciphers work in the first place. For reference, we include Fermat's Little Theorem again here.

Theorem 9.2 *(Fermat's Little Theorem) If p is a prime number and r is an integer with $gcd(r, p) = 1$, then $r^{p-1} = 1 \ mod \ p$.*

As a result of Fermat's Little Theorem, for a given positive integer n, if $r^{n-1} \neq 1 \ mod \ n$ for any integer r from 1 to $n - 1$, then we can conclude that n is definitely not prime. The relatively simple and efficient primality test based on Fermat's Little Theorem is just to calculate $r^{n-1} \ mod \ n$ for some integers r from 1 to $n - 1$. If it ever occurs that $r^{n-1} \neq 1 \ mod \ n$ for any tested value of r, then we can conclude that n is definitely not prime. If $r^{n-1} = 1 \ mod \ n$ for all tested values of r, then we can conclude that n is likely to be prime, with likelihood increasing as we test more values of r.

Example 9.17 In Example 9.11, we showed that $2^{160} = 156 \ mod \ 161$. Thus, $2^{160} \neq 1 \ mod \ 161$, and we can conclude by Fermat's Little Theorem that 161 is definitely not prime. $\square$

Of course, referring to Example 9.17, it would be easier to conclude that 161 is not prime by trial divisions, since one of the first divisions would show that 7 divides 161 evenly. However, recall that the method of trial divisions is inefficient if the number in question is large. The primality test based on Fermat's Little Theorem and illustrated in Example 9.17, on the other hand, is not inefficient, even if the number in question is large. For example, from the single result $2^{1234566} = 899557 \bmod 1234567$, which we could find quickly using binary exponentiation, we can conclude that 1234567 is definitely not prime.

While the primality test based on Fermat's Little Theorem and illustrated in Example 9.17 is very easy to perform, and allows with certainty the conclusion that a given positive integer is not prime, it unfortunately sometimes identifies integers as likely to be prime even when they are not. More specifically, there are values of n and r for which $r^{n-1} = 1 \bmod n$, even though $\gcd(r, n) = 1$ and n is not prime. In such cases, n is called *pseudoprime* to the base r. For example, $2^{340} = 1 \bmod 341$, even though $\gcd(2, 341) = 1$ and 341 is not prime, and so 341 is pseudoprime to the base 2. However, since $\gcd(3, 341) = 1$ and $3^{340} \neq 1 \bmod 341$, then 341 is not pseudoprime to the base 3.

Pseudoprimes are scarce relative to the primes. There are only 245 pseudoprimes to the base 2 less than one million, while there are 78498 primes less than one million. Also, most pseudoprimes to the base 2 are not pseudoprime to many other bases. There do exist positive integers n, though, that are pseudoprime to every base r from 1 to $n - 1$ with $\gcd(r, n) = 1$, but which are not prime. Such numbers are called *Carmichael* numbers, for American mathematician Robert Carmichael. There are 2163 Carmichael numbers less than 25 billion, the smallest of which is 561.

There are many other primality tests besides trial divisions and the test based on Fermat's Little Theorem and illustrated in Example 9.17. For example, a stronger test based on Fermat's Little Theorem misidentifies only a very small number of nonprimes called *strong* pseudoprimes. There is only one strong pseudoprime to the bases 2, 3, 5, and 7 less than 25 billion, and no strong pseudoprime analogue to Carmichael numbers.

As of this writing, the largest known prime number is $2^{77232917} - 1$, a number of length 23249425 digits. This prime was discovered on December 26, 2017 by a volunteer project called the *Great Internet Mersenne Prime Search*, or *GIMPS* for short, on a computer volunteered by Jonathan Pace, an electrical engineer from Germantown, Tennessee. To put the size of this number into perspective, if printed in a standard 12-point font it would stretch for more than 36 miles. This number is an example of a *Mersenne* prime because it is of the form $2^p - 1$, where p is itself prime. Mersenne primes are one of the types researchers look for in their never-ending search for large prime numbers.

9.8.1 Exercises

1. Use trial divisions to determine whether the following integers are prime.

   (a)* 131

   (b) 323

   (c)* 667

   (d) 947

   (e)* 1559

   (f) 2557

   (g)* 5849

   (h) 9727

2. Given only each of the following results, what, if anything, does Fermat's Little Theorem allow you to conclude about whether the modulus is prime?

   (a)* $2^{6600} = 1 \bmod 6601$

   (b) $3^{22606} = 3955 \bmod 22607$

   (c)* $4^{39996} = 17872 \bmod 39997$

   (d) $5^{52632} = 1 \bmod 52633$

3. Use binary exponentiation to find the following. Given only each of the results, what, if anything, does Fermat's Little Theorem allow you to conclude about whether the modulus is prime?

   (a)* $2^{118} \bmod 119$

   (b) $3^{90} \bmod 91$

   (c)* $4^{90} \bmod 91$

   (d) $5^{142} \bmod 143$

4. (a) Show that 15 is pseudoprime to the base 4, but not to the base 2.

   (b) Show that 9 is pseudoprime to the base 8, but not to the base 7.

   (c) Use binary exponentiation to show that 91 is pseudoprime to the base 4, but not to the base 2.

   (d) Use binary exponentiation to show that 124 is pseudoprime to the base 5, but not to the base 3.

5. Given that standard 12-point fonts typically produce on average 10 printed digits per inch, verify that the prime number $2^{77232917} - 1$ of length 23249425 digits would stretch for more than 36 miles if printed in a standard 12-point font.

6. Find some information about one or more other primality tests besides trial divisions and the test based on Fermat's Little Theorem and illustrated in Example 9.17, and write a summary of your findings.

7. Find some information about the *Great Internet Mersenne Prime Search* volunteer project, and write a summary of your findings.

9.9   Integer Factorization

Recall again that to break an RSA cipher, the only real difficulty that an outsider should face is in factoring $m = p \cdot q$ to find p and q. As with primality testing, the development of public-key ciphers like RSA has in part motivated much research over the past few decades in the area of integer factorization.

As we noted in Section 9.8, the Fundamental Theorem of Arithmetic guarantees that every nonprime greater than 1 can be expressed as a product of primes, with the primes in this product being unique. The *prime factorization* of a positive integer is simply a list of these prime factors, usually expressed as a product, often with the primes in increasing order.

As with primality testing, the most obvious technique for finding the prime factorization of a positive integer is to just do actual division. For a given positive integer n, we could do this systematically by checking to see if any prime number starting with 2 divides n evenly.[4] We will also call this technique for finding prime factorizations *trial divisions*. In Example 9.16, we essentially used trial divisions to find the prime factorization of 1073. Since the prime 29 divides 1073 evenly, and the result of this division is 37, which is also prime, the prime factorization of 1073 is $29 \cdot 37$.

Example 9.18 To use trial divisions to find the prime factorization of 985439, we begin by dividing this number by the primes 2, 3, and 5, each of which fails to divide 985439 evenly. The next prime, 7, does divide 985439 evenly, though, with quotient 140777. Thus, we have the following.

$$985439 \;=\; 7 \cdot 140777$$

Continuing now with 140777, note that we can skip dividing this number by the primes 2, 3, and 5, since if any of these primes divided 140777, they

---

[4] Also, recall as we noted in Section 9.8, if no prime less than or equal to $\sqrt{n}$ divides n evenly, then n is itself prime.

would also have divided 985439. So we can begin by dividing 140777 by 7, which divides 140777 evenly, with quotient 20111. Thus, we have the following.

$$985439 = 7 \cdot 7 \cdot 20111$$

Similarly, 7 divides 20111 evenly, with quotient 2873. Thus, we have the following.

$$985439 = 7 \cdot 7 \cdot 7 \cdot 2873$$

Next, the smallest prime that divides 2873 evenly is 13, with quotient 221. Thus, we have the following.

$$985439 = 7 \cdot 7 \cdot 7 \cdot 13 \cdot 221$$

Finally, the smallest prime that divides 221 evenly is 13, with quotient 17, which is also prime. Thus, the complete prime factorization of 985439 is the following.

$$985439 = 7 \cdot 7 \cdot 7 \cdot 13 \cdot 13 \cdot 17$$
$$= 7^3 \cdot 13^2 \cdot 17 \qquad \Box$$

As with primality testing, one problem with using trial divisions to find the prime factorization of a positive integer n is that if n had any extremely large prime factors, there would be a tremendously large number of primes that could need to be tested. More specifically, for a number containing two prime factors, each hundreds of digits in length, it could take millions of years to find the prime factorization by trial divisions, even if the process were programmed on a computer capable of millions of divisions per second.

One relatively simple and efficient technique for finding prime factorizations of certain given positive integers is called *Fermat factorization*, named for none other than Pierre de Fermat, who discovered the technique. Fermat factorization is a technique for finding prime factorizations of integers that are the product of two distinct primes that are relatively close together, and it is a useful technique even when both primes are extremely large.

In order to present the technique of Fermat factorization, suppose for a given value of m with $m = p \cdot q$ for some unknown pair of distinct prime numbers p and q, we wish to determine p and q. Symbolically, let $x = \frac{p+q}{2}$ and $y = \frac{p-q}{2}$. Note then that $x^2 - y^2$ can be simplified as follows.

$$
\begin{aligned}
x^2 - y^2 &= \frac{(p+q)^2}{4} - \frac{(p-q)^2}{4} \\
&= \frac{p^2 + 2 \cdot p \cdot q + q^2 - p^2 + 2 \cdot p \cdot q - q^2}{4} \\
&= \frac{4 \cdot p \cdot q}{4} = p \cdot q
\end{aligned}
$$

Thus, $p \cdot q = x^2 - y^2$, and so $m = p \cdot q = x^2 - y^2 = (x + y) \cdot (x - y)$. Since the prime factors of m are p and q, p and q must be $x + y$ and $x - y$, and so to determine p and q, we must only find x and y. To do this, we can begin by assuming x is the smallest integer greater than $\sqrt{m}$. Since $m = x^2 - y^2$, if this assumed value of x were correct, then $x^2 - m$ would be the perfect square y^2. On the other hand, if $x^2 - m$ were not a perfect square, then we would know that the assumed value of x was incorrect, and could simply increase x by one and repeat. The full Fermat factorization technique is just to repeat this process as many times as necessary, each time increasing x by one, until $x^2 - m$ is a perfect square.

Example 9.19 To use Fermat factorization to determine the prime factors of $m = p \cdot q = 103935290639$, we first find $\sqrt{103935290639} \approx 322389.97$. Since the smallest integer greater than 322389.97 is 322390, we begin with $x = 322390$. For this value of x, it follows that $x^2 - m = 21461$. However, $\sqrt{21461} \approx 146.50$, and so 21461 is not a perfect square. Thus, 322390 is not the correct value of x. Next, we try $x = 322391$. For this value of x, it follows that $x^2 - m = 666242$. However, $\sqrt{666242} \approx 816.24$, and so 666242 is not a perfect square. Thus, 322391 is also not the correct value of x. Next, we try $x = 322392$. For this value of x, it follows that $x^2 - m = 1311025$. Since $\sqrt{1311025} = 1145$, 322392 is the correct value of x, and the corresponding value of y is $y = 1145$. The prime factors of m are then $p = x + y = 323537$ and $q = x - y = 321247$. $\square$

Note that only if p and q were relatively close together, as they are in Example 9.19, would the number of choices for x required in the Fermat factorization process be relatively small. This is why we noted previously that Fermat factorization is a useful technique, but only for factoring integers that are the product of two distinct primes that are relatively close together. When the primes are not relatively close together, Fermat factorization fails to be efficient. On the other hand, when the primes are relatively close together, Fermat factorization is very efficient, even when both primes are extremely large. More specifically, for finding p and q from $m = p \cdot q$, Fermat factorization requires on average $|p - q|/2$ steps, a number that does not depend directly on the size of p and q, but rather on the distance between them.

   The existence of Fermat factorization means that for an RSA cipher to be secure from technologically savvy outsiders, not only must the primes chosen for the cipher be extremely large, but they must also be relatively far apart. In addition, we should note that in comparison with the problems of finding extremely large prime numbers p and q and factoring $m = p \cdot q$, finding p and q would in general be much less time consuming than factoring m. More precisely, the utility of RSA ciphers is based on the fact that it would in general be much less time consuming for the intended recipient of a message to choose p and q than for an outsider to factor $m = p \cdot q$. This is an obvious and necessary condition for any cipher to be useful; using a cipher should in general be much easier than breaking it.

9.9.1   Exercises

1. Use trial divisions to find the prime factors of the following integers.

    (a)* 1573

    (b) 2767

    (c)* 2773

    (d) 4187

    (e)* 8731

    (f) 21879

    (g)* 89175

    (h) 135541

2. Use Fermat factorization to find the prime factors of the following integers.

    (a)* $m = p \cdot q = 93343$

    (b) $m = p \cdot q = 321179$

    (c)* $m = p \cdot q = 4701041$

    (d) $m = p \cdot q = 6550477$

3. Find some information about one or more other techniques for finding prime factorizations besides trial divisions and Fermat factorization, and write a summary of your findings.

9.10   The RSA Factoring Challenges

When the details of their type of cipher became known in 1977, Rivest, Shamir, and Adleman presented a challenge to the public: break a specific RSA cipher, with a token cash prize of \$100 offered to anyone who could do so within five years. This RSA cipher used a modulus $m = p \cdot q$ that was 129 digits in length, and it was estimated that given the factorization techniques and technology available at the time, it would take 40 quadrillion years for m to be factored. Techniques and technology both improved, of course, and in 1994 a team of 600 people under the direction of Dutch mathematician Arjen Lenstra, working on 1600 computers for a little over six months, successfully factored m.

Rivest, Shamir, and Adleman did not just propose this challenge for fun, but also to encourage research in the area of integer factorization, for it was the general difficulty of this that made their type of cipher legitimate. By the end of the 1980s when the 129-digit value of m from their original challenge remained unfactored, RSA Laboratories, as the RSA Data Security corporation formed by Rivest, Shamir, and Adleman had come to be known, decided that the public needed a little more encouragement. So on March 18, 1991, RSA Laboratories presented a broader challenge, publishing a list of numbers that were each the product of a pair of distinct extremely large primes, and offering larger cash prizes for their prime factorizations. The smallest number published was 100 digits in length, and was factored within 14 days by Arjen Lenstra, earning him $1000. The largest number published was 617 digits in length, and was offered for a cash prize of $200,000.

In 2007, RSA Laboratories, citing advances in techniques and technology, declared the broader challenge inactive. Before this time a total of 12 of the published numbers were factored, the longest of which was 200 digits in length, earning a team under the direction of German mathematician Jens Franke of the University of Bonn $20,000 when it was factored in 2005. As of this writing, the longest of the published numbers to be factored was 232 digits in length, by a team under the direction of German mathematician Thorsten Kleinjung of the University of Bonn in 2009, too late for them to collect the retracted prize of $50,000.

9.10.1   Exercises

1. Find some information about the original factoring challenge presented by Rivest, Shamir, and Adleman in 1977, including the motivation for and specifics of the challenge, as well as attempts by researchers at solving it, and write a summary of your findings.

2. Find some information about the broader factoring challenge presented by RSA Laboratories in 1991, including the motivation for and specifics of the challenge, as well as attempts by researchers at solving parts of it, and write a summary of your findings.

Chapter 10

ElGamal Ciphers

Recall that when Whitfield Diffie and Martin Hellman first explained the idea of public-key ciphers, they did not suggest an actual type of public-key cipher. What they did was explain how mathematical operations that are easy to do but difficult to undo could be used to create public-key ciphers. Operations that are easy to do but difficult to undo are sometimes called *one-way functions*. We have already seen an example of a one-way function, the one that forms the basis of RSA ciphers: given a pair of very large prime numbers p and q, it is easy to form $p \cdot q$, but given only the result of this multiplication, it is in general very difficult to find p and q.

Although Diffie and Hellman did not suggest an actual type of public-key cipher in their original paper, they did propose a method through which two parties communicating only over an insecure communication line could secretly agree upon encryption keys for any cipher of their choice. This can be just as important to consider as public-key ciphers, because in practice, known public-key ciphers are in general much slower than known non-public-key ciphers when all are implemented properly, a fact anticipated by Diffie and Hellman before any actual types of public-key ciphers had been discovered. Diffie and Hellman conjectured that their key exchange method could be used by two parties communicating only over an insecure communication line to secretly agree upon encryption keys for any type of cipher, including non-public-key ciphers, thereby increasing their utility. We will begin this chapter by considering the details of this key exchange method, called the *Diffie-Hellman key exchange* in honor of its creators.[1]

---

[1]Malcolm Williamson, a British cryptologist, while working for the British intelligence agency GCHQ, is now credited as independently inventing the Diffie-Hellman key exchange around the same time as Diffie and Hellman. This discovery, however, was not revealed publicly until 1997, when its top-secret classification expired.

10.1   The Diffie-Hellman Key Exchange

Although the key exchange method proposed by Diffie and Hellman can be used for agreeing upon encryption keys for non-public-key ciphers, because the only type of modern cipher we have considered so far is RSA, and the general process of the Diffie-Hellman key exchange is independent of any specific type of cipher, in this section we will consider a way in which the Diffie-Hellman key exchange could be used with RSA ciphers.

For two parties communicating over an insecure line of communication and wishing to use an RSA cipher to exchange a message, it can be argued that it might be beneficial for the encryption exponent for the cipher to be kept secret from outsiders. The basic steps in using the Diffie-Hellman key exchange to accomplish this can be summarized as follows.

1. The intended recipient of the message initiates the process by choosing prime numbers p and q, and forming $m = p \cdot q$ and $f = (p-1) \cdot (q-1)$. The intended recipient then chooses an integer k between 1 and m with $\gcd(k, m) = 1$, and sends the values of k and m to the originator.

2. The intended recipient chooses an integer r between 1 and m, forms $k^r \bmod m$, and sends the result to the originator (but not r). Meanwhile, the originator chooses an integer s between 1 and m, forms $k^s \bmod m$, and sends the result to the intended recipient (but not s).

3. The intended recipient and originator both form the potential encryption exponent $e = k^{r \cdot s} \bmod m$, which the intended recipient can find as $(k^s \bmod m)^r \bmod m$, and the originator as $(k^r \bmod m)^s \bmod m$.

4. The intended recipient determines if e is an acceptable encryption exponent by checking whether $\gcd(e, f) = 1$. If e is not an acceptable encryption exponent, then the two parties repeat the process (from either step 1 or 2) as many times as necessary until an acceptable encryption exponent is obtained.[2]

After obtaining an acceptable encryption exponent e, the two parties could then proceed with the usual RSA encryption procedure with encryption exponent e and modulus m.

Example 10.1 Suppose you want to receive a secret message from a colleague over an insecure line of communication using an RSA cipher, and you wish to use the Diffie-Hellman key exchange as it is presented in this section in order to keep the encryption exponent for the cipher secret from outsiders.

---

[2] Although RSA encryption exponents are typically chosen less than f, if e happens to lie between f and m, it will still work as the encryption exponent in an RSA cipher provided $\gcd(e, f) = 1$.

1. You choose primes $p = 11$ and $q = 13$, and form $m = p \cdot q = 143$ and $f = (p - 1) \cdot (q - 1) = 120$. You then choose $k = 15$, and send the values of k and m to your colleague.

2. You choose $r = 4$, form $k^r \bmod m = 15^4 \bmod 143 = 3$, and send the result to your colleague while keeping r secret. Meanwhile, your colleague chooses $s = 5$, forms $k^s \bmod m = 15^5 \bmod 143 = 45$, and sends the result to you while keeping s secret.

3. You and your colleague both form the potential encryption exponent $e = 100$, which you can find as $45^4 \bmod 143 = 100$, and your colleague as $3^5 \bmod 143 = 100$.

4. You find that $\gcd(e, f) = \gcd(100, 120) \neq 1$, and so $e = 100$ is not an acceptable encryption exponent with $m = 143$. Thus, the process must be repeated.

For the second attempt, suppose you and your colleague repeat beginning from step 2.

2. You choose $r = 2$, form $k^r \bmod m = 15^2 \bmod 143 = 82$, and send the result to your colleague while keeping r secret. Meanwhile, your colleague chooses $s = 3$, forms $k^s \bmod m = 15^3 \bmod 143 = 86$, and sends the result to you while keeping s secret.

3. You and your colleague both form the potential encryption exponent $e = 103$, which you can find as $86^2 \bmod 143 = 103$, and your colleague as $82^3 \bmod 143 = 103$.

4. You find that $\gcd(e, f) = \gcd(103, 120) = 1$, and so $e = 103$ is an acceptable encryption exponent with $m = 143$.

You and your colleague could then proceed with the usual RSA encryption procedure with encryption exponent $e = 103$ and modulus $m = 143$.   □

Since the Diffie-Hellman key exchange as we described it indicates that the process may need to be repeated an unspecified number of times, it is natural to wonder the number of times one should expect to repeat the process before achieving success. To estimate this, we simulated the process as we described it 777,000 times, split as 111,000 for seven different maximum sizes of $m = p \cdot q$. These seven different maximum sizes of m were 5, 10, 20, 40, 80, 160, and 320 digits, and for each the success rate, with success being that the Diffie-Hellman key exchange with random k, r, s, and $m = p \cdot q$ resulted in a valid encryption exponent e for an RSA cipher with modulus m, ranged from a low of 30.016% to a high of 31.394%. As such, the probability of success on a single trial of the Diffie-Hellman key exchange as we described it seems to be around 30% and independent of the size of m.

10.1.1   Exercises

1.* Suppose you want to receive a secret message from a colleague over an insecure communication line using an RSA cipher, and you wish to use the Diffie-Hellman key exchange as it is presented in this section to keep the encryption exponent for the cipher secret from outsiders. You choose primes $p = 11$ and $q = 17$, and form m and f. You then choose $k = 13$, and send the values of k and m to your colleague.

   (a) Using $r = 4$, find the value of $k^r \bmod m$ you would send to your colleague.

   (b) Suppose you receive the value $k^s \bmod m = 135$ from your colleague. Using the value of r in part (a), find the resulting potential encryption exponent $e = k^{r \cdot s} \bmod m$.

   (c) Determine whether your answer to part (b) is an acceptable encryption exponent with m, and justify your answer.

2. Suppose you want to receive a secret message from a colleague over an insecure communication line using an RSA cipher, and you wish to use the Diffie-Hellman key exchange as it is presented in this section to keep the encryption exponent for the cipher secret from outsiders. You choose primes $p = 13$ and $q = 19$, and form m and f. You then choose $k = 21$, and send the values of k and m to your colleague.

   (a) Using $r = 5$, find the value of $k^r \bmod m$ you would send to your colleague.

   (b) Suppose you receive the value $k^s \bmod m = 70$ from your colleague. Using the value of r in part (a), find the resulting potential encryption exponent $e = k^{r \cdot s} \bmod m$.

   (c) Determine whether your answer to part (b) is an acceptable encryption exponent with m, and justify your answer.

3. Suppose you want to receive a secret message from a colleague over an insecure communication line using an RSA cipher, and you wish to use the Diffie-Hellman key exchange as it is presented in this section to keep the encryption exponent for the cipher secret from outsiders. You choose primes $p = 127$ and $q = 181$, and form m and f. You then choose $k = 215$, and send the values of k and m to your colleague.

   (a)* Using $r = 85$, find the value of $k^r \bmod m$ you would send to your colleague.

(b) Suppose you receive the value $k^s \bmod m = 799$ from your colleague. Using the value of r in part (a), find the resulting potential encryption exponent $e = k^{r \cdot s} \bmod m$.

(c) Determine whether your answer to part (b) is an acceptable encryption exponent with m, and justify your answer.

4.* Suppose you want to send a secret message to a colleague over an insecure communication line using an RSA cipher, and you wish to use the Diffie-Hellman key exchange as it is presented in this section to keep the encryption exponent for the cipher secret from outsiders. Your colleague chooses primes p and q, and forms $m = 161$ and f. Your colleague then chooses $k = 12$, and sends the values of k and m to you.

(a) Using $s = 5$, find the value of $k^s \bmod m$ you would send to your colleague.

(b) Suppose you receive the value $k^r \bmod m = 39$ from your colleague. Using the value of s in part (a), find the resulting potential encryption exponent $e = k^{r \cdot s} \bmod m$.

(c) Determine whether your answer to part (b) is an acceptable encryption exponent with m, and justify your answer.

5. Suppose you want to send a secret message to a colleague over an insecure communication line using an RSA cipher, and you wish to use the Diffie-Hellman key exchange as it is presented in this section to keep the encryption exponent for the cipher secret from outsiders. Your colleague chooses primes p and q, and forms $m = 209$ and f. Your colleague then chooses $k = 28$, and sends the values of k and m to you.

(a) Using $s = 4$, find the value of $k^s \bmod m$ you would send to your colleague.

(b) Suppose you receive the value $k^r \bmod m = 149$ from your colleague. Using the value of s in part (a), find the resulting potential encryption exponent $e = k^{r \cdot s} \bmod m$.

(c) Determine whether your answer to part (b) is an acceptable encryption exponent with m, and justify your answer.

6. Suppose you want to send a secret message to a colleague over an insecure communication line using an RSA cipher, and you wish to use the Diffie-Hellman key exchange as it is presented in this section to keep the encryption exponent for the cipher secret from outsiders. Your colleague chooses primes p and q, and forms $m = 2545$ and f.

Your colleague then chooses $k = 281$, and sends the values of k and m to you.

(a)* Using $s = 114$, find the value of k^s mod m you would send to your colleague.

(b) Suppose you receive the value k^r mod $m = 571$ from your colleague. Using the value of s in part (a), find the resulting potential encryption exponent $e = k^{r \cdot s}$ mod m.

(c) Determine whether your answer to part (b) is an acceptable encryption exponent with m, and justify your answer.

7. Find some information about the career in cryptology of Malcolm Williamson, and write a summary of your findings.

10.2 Discrete Logarithms

In the basic steps in the Diffie-Hellman key exchange summarized on page 324, note that an outsider could intercept the values of k, m, k^r mod m, and k^s mod m, since each was transmitted over an insecure communication line. In order for this key exchange method to be secure, it should be a difficult problem for an outsider to determine the candidate encryption exponent $e = k^{r \cdot s}$ mod m from the knowledge of k, m, k^r mod m, and k^s mod m. This is called the *Diffie-Hellman problem*.

The most direct way to solve the Diffie-Hellman problem would involve first finding r (or s). This would be an example of solving the *discrete logarithm problem*, specifically finding the *discrete logarithm* r from the knowledge of k, m, and k^r mod m (or s from the knowledge of k, m, and k^s mod m).

Example 10.2 Suppose $k = 13$ and $m = 187$. One way to solve the discrete logarithm problem of finding s for which k^s mod $m = 135$ is by trial multiplications. Specifically, we can form k^s mod m for positive integers starting with $s = 1$, continuing until we find a value of s for which k^s mod $m = 135$. The following table shows the results of these calculations, each of which after the first can be found by multiplying the previous result by k and reducing modulo m.

s	k^s mod m	s	k^s mod m	s	k^s mod m
1	13	7	106	13	30
2	169	8	69	14	16
3	140	9	149	15	21
4	137	10	67	16	86
5	98	11	123	17	183
6	152	12	103	18	135

Thus, $s = 18$ is a value for which $k^s \bmod m = 135$. Incidentally, there are other values of s for which $k^s \bmod m = 135$. For example, both $s = 38$ and $s = 58$ satisfy $k^s \bmod m = 135$. In fact, every twentieth positive integer starting with $s = 18$ satisfies $k^s \bmod m = 135$. However, to solve this discrete logarithm problem, it is only necessary to find one s for which $k^s \bmod m = 135$. $\qquad\square$

It is always possible to break a Diffie-Hellman key exchange by finding a discrete logarithm. To see this, suppose an outsider intercepts values of k, m, $k^r \bmod m$, and $k^s \bmod m$ in a Diffie-Hellman key exchange, and wishes to determine the candidate encryption exponent $e = k^{r \cdot s} \bmod m$. If the outsider could find the discrete logarithm r from the knowledge of k, m, and $k^r \bmod m$ (or s from the knowledge of k, m, and $k^s \bmod m$), then the outsider could determine the candidate encryption exponent by forming $(k^s \bmod m)^r \bmod m$ (or $(k^r \bmod m)^s \bmod m$).

Example 10.3 Suppose two parties perform a Diffie-Hellman key exchange, and you intercept their values of $k = 13$, $m = 187$, $k^r \bmod m = 137$, and $k^s \bmod m = 135$. You could determine the value of $e = k^{r \cdot s} \bmod m$ as follows. First, you solve the discrete logarithm problem as in Example 10.2 to find that $s = 18$ satisfies $k^s \bmod m = 135$. You then form $e = (k^r \bmod m)^s \bmod m = 137^{18} \bmod m = 103$. Also, recall from Example 10.2 that there are other values of s for which $k^s \bmod m = 135$. As it turns out, any s for which $k^s \bmod m = 135$ would result in the same value of $e = k^{r \cdot s} \bmod m$. For example, $s = 38$ also satisfies $k^s \bmod m = 135$, and with this value of s you could still find the value of $e = k^{r \cdot s} \bmod m$ by forming $e = (k^r \bmod m)^s \bmod m = 137^{38} \bmod m = 103$. $\qquad\square$

As we demonstrated in Example 10.3, it is always possible to break a Diffie-Hellman key exchange by finding a discrete logarithm. Does this mean that in general the Diffie-Hellman key exchange is not secure, since an outsider could always break it by finding a discrete logarithm? Only if discrete logarithms are in general easy to find. Of course, with a relatively small modulus, discrete logarithms are indeed easy to find, by trial multiplications, as we demonstrated in Example 10.2. However, as the modulus grows larger, the time it would take to find a discrete logarithm by trial multiplications increases very quickly. Techniques for finding discrete logarithms that are faster than trial multiplications do exist, but with a very large modulus, even the fastest known techniques for finding discrete logarithms would essentially take forever. More specifically, for a modulus hundreds of digits in length, the fastest known techniques for finding discrete logarithms would in general take millions of years to find a single discrete logarithm, even when programmed on a computer capable of millions of operations per second.

It is the general difficulty of finding discrete logarithms that gives the Diffie-Hellman key exchange its high level of security. Knowing k, m, $k^r \bmod m$, and $k^s \bmod m$ in a Diffie-Hellman key exchange is not enough to determine $e = k^{r \cdot s} \bmod m$. To determine e, an outsider would have to find either r or s, something that in general would be essentially impossible to do if m was very large. On the other hand, the difficulty of finding discrete logarithms would not pose a problem for the two parties performing the key exchange, since each would have chosen either r or s. In addition, with a modulus m that is not prime (as is the case in the Diffie-Hellman key exchange as it is presented in Section 10.1), it has been argued that finding discrete logarithms would require factoring m. Thus, if m were the product of two very large prime numbers, then the factorization problem that provides security to RSA ciphers would provide an additional equal amount of security to the Diffie-Hellman key exchange.

The general difficulty of finding discrete logarithms also leads to our second example of a one-way function. It would be relatively easy using the process of binary exponentiation to raise a base to a given power with modular arithmetic, even if the power and modulus were both very large. However, after a base was raised to a power with modular arithmetic, it would be extremely difficult to find the power if it and the modulus were both very large.

10.2.1   Exercises

1.* (a) For $k = 12$ and $m = 161$, find r for which $k^r \bmod m = 85$.

   (b) For $k = 12$ and $m = 161$, find r for which $k^r \bmod m = 87$.

   (c) Suppose two parties perform a Diffie-Hellman key exchange, and you intercept their values of $k = 12$, $m = 161$, $k^r \bmod m = 87$, and $k^s \bmod m = 39$. Use your answer to part (b) to determine the candidate RSA encryption exponent $e = k^{r \cdot s} \bmod m$.

2. (a) For $k = 21$ and $m = 247$, find s for which $k^s \bmod m = 60$.

   (b) For $k = 21$ and $m = 247$, find s for which $k^s \bmod m = 64$.

   (c) Suppose two parties perform a Diffie-Hellman key exchange, and you intercept their values of $k = 21$, $m = 247$, $k^r \bmod m = 34$, and $k^s \bmod m = 64$. Use your answer to part (b) to determine the candidate RSA encryption exponent $e = k^{r \cdot s} \bmod m$.

3. Suppose two parties perform a Diffie-Hellman key exchange, and you intercept the following values of k, m, $k^r \bmod m$, and $k^s \bmod m$. Suppose you then find the following discrete logarithm r or s, which

results in the given value of $k^r \bmod m$ or $k^s \bmod m$. Determine the candidate RSA encryption exponent $e = k^{r \cdot s} \bmod m$.

(a)* $k = 8$, $m = 203$, $k^r \bmod m = 71$, $k^s \bmod m = 43$, $r = 6$

(b) $k = 10$, $m = 203$, $k^r \bmod m = 67$, $k^s \bmod m = 53$, $s = 4$

(c)* $k = 41$, $m = 299$, $k^r \bmod m = 216$, $k^s \bmod m = 62$, $r = 87$

(d) $k = 58$, $m = 299$, $k^r \bmod m = 31$, $k^s \bmod m = 282$, $s = 112$

4. Suppose two parties perform a Diffie-Hellman key exchange, and you intercept the following values of k, m, $k^r \bmod m$, and $k^s \bmod m$. Determine the candidate RSA encryption exponent $e = k^{r \cdot s} \bmod m$.

(a)* $k = 35$, $m = 209$, $k^r \bmod m = 30$, $k^s \bmod m = 64$

(b) $k = 35$, $m = 209$, $k^r \bmod m = 175$, $k^s \bmod m = 5$

(c)* $k = 17$, $m = 253$, $k^r \bmod m = 74$, $k^s \bmod m = 106$

(d) $k = 17$, $m = 253$, $k^r \bmod m = 31$, $k^s \bmod m = 49$

10.3 ElGamal Ciphers

Recall again that in their 1976 paper *New Directions in Cryptography* [6], Whitfield Diffie and Martin Hellman explained how one-way functions could be used to create public-key ciphers, and any type of one-way function would suffice. Recall also, as we noted at the end of Section 10.2, the general difficulty of finding discrete logarithms leads to a one-way function. For the rest of this chapter, we will consider the details of a type of public-key cipher that results from this one-way function. These ciphers are called *ElGamal* ciphers in honor of their creator, Stanford-trained Egyptian computer scientist Taher Elgamal,[3] who first published the type of cipher in 1985. Elgamal went on to serve as chief scientist at Netscape and director of engineering at RSA Laboratories before founding his own company, Securify, in 1998.

Before presenting ElGamal ciphers, we need to consider one additional mathematical prerequisite. Suppose a is a positive integer less than a prime p, and consider the following list of numbers.

$$a, \ a^2 \bmod p, \ a^3 \bmod p, \ a^4 \bmod p, \ \ldots, \ a^{p-1} \bmod p$$

By Fermat's Little Theorem (Theorem 9.2 on page 306), we know that $a^{p-1} \bmod p = 1$, and so the last number in this list must equal 1. However,

---

[3]Elgamal prefers to spell his own surname with a lowercase g, to discourage its mispronunciation in English. However, the type of cipher that he published, as well as a system that he developed for obtaining digital signatures, are usually denoted using a capital G.

an earlier number in this list might equal 1 as well. That is, it could happen that there is a positive integer n less than $p - 1$ for which $a^n \bmod p = 1$. This does not have to happen, but it could. For some combinations of a and p, only the last number in this list will equal 1, and for other combinations an earlier number will equal 1 as well. If only the last number in this list equals 1, then we say that a is *primitive* relative to p. Whether a is primitive relative to p or not, if n is the smallest positive integer for which $a^n \bmod p = 1$, then we say that a *generates* n values modulo p.

Example 10.4 Suppose we wish to determine whether $a = 2$ is primitive relative to $p = 19$. For integers n from 1 through $p - 1 = 18$, the following table shows the values of $2^n \bmod 19$, each of which after the first can be found by multiplying the previous result by 2 and reducing modulo 19.

n	$2^n \bmod 19$	n	$2^n \bmod 19$	n	$2^n \bmod 19$
1	2	7	14	13	3
2	4	8	9	14	6
3	8	9	18	15	12
4	16	10	17	16	5
5	13	11	15	17	10
6	7	12	11	18	1

Since only the last value of $2^n \bmod 19$ equals 1, $a = 2$ is primitive relative to $p = 19$. Now suppose we wish to determine whether $a = 5$ is primitive relative to $p = 19$. For integers n from 1 through $p - 1 = 18$, the following table shows the values of $5^n \bmod 19$.

n	$5^n \bmod 19$	n	$5^n \bmod 19$	n	$5^n \bmod 19$
1	5	7	16	13	17
2	6	8	4	14	9
3	11	9	1	15	7
4	17	10	5	16	16
5	9	11	6	17	4
6	7	12	11	18	1

Since there is a positive integer n less than 18 for which $a^n \bmod p = 1$, namely $n = 9$, $a = 5$ is not primitive relative to $p = 19$. □

With ElGamal ciphers, we will again convert plaintext messages into numeric form using the ASCII correspondences given in Table 9.1 on page 302. Suppose the originator of a numeric plaintext message wishes to send the message to an intended recipient over an insecure communication line, and wants to use a cipher to disguise the message to protect it from outsiders who may observe it during transmission. The basic steps in using an ElGamal cipher to do this can be summarized as follows.

1. The intended recipient of the message initiates the process by choosing a prime number p. The intended recipient then chooses a positive integer a less than $p - 1$ that is primitive relative to p, and a positive integer n less than $p - 1$, forms $b = a^n \bmod p$, and sends the values of p, a, and b to the originator of the message over the insecure communication line.[4]

2. Suppose the numeric plaintext message is expressed as one or more positive integers x less than p. Then, for each plaintext integer x, the originator encrypts x by choosing a positive integer k less than $p - 1$, and forming the following pair of quantities y and z. For maximum security, a different k should be used for each plaintext integer.

$$y = a^k \bmod p$$
$$z = x \cdot b^k \bmod p$$

The originator then sends the resulting ciphertext pair (y, z) to the intended recipient over the insecure communication line.

3. The recipient can decrypt each ciphertext pair (y, z) by forming the following quantity, which results in the plaintext integer x from which z was formed.

$$x = z \cdot y^{p-1-n} \bmod p$$

For a ciphertext pair (y, z), it is by no means obvious that the calculation $z \cdot y^{p-1-n} \bmod p$ will result in the plaintext integer x from which z was formed. The reason this happens is as follows.

$$
\begin{aligned}
z \cdot y^{p-1-n} \bmod p &= x \cdot b^k \cdot (a^k)^{p-1-n} \bmod p \\
&= x \cdot b^k \cdot (a^k)^{p-1} \cdot (a^k)^{-n} \bmod p \\
&= x \cdot (a^n)^k \cdot 1 \cdot (a^k)^{-n} \bmod p \\
&= x \cdot (a^k)^n \cdot (a^k)^{-n} \bmod p \\
&= x \cdot (a^k)^0 \bmod p \\
&= x \cdot 1 \bmod p \\
&= x
\end{aligned}
$$

It is worth noting that $z \cdot y^{-n} \bmod p$ would also result in the plaintext integer x from which z was formed. The reason for forming $z \cdot y^{p-1-n} \bmod p$ instead of $z \cdot y^{-n} \bmod p$ is because the exponent $p - 1 - n$ on y would always be positive.

---

[4] ElGamal ciphers are often presented in the literature with the requirement that a generate a large number of values modulo p, as opposed to being primitive relative to p. Requiring that a be primitive relative to p maximizes the number of possibilities for b.

Example 10.5 Suppose you wish to send the secret message Queen to a colleague over an insecure communication line using an ElGamal cipher with the ASCII correspondences, under which the plaintext converts into the list of integers 81, 117, 101, 101, 110. Your colleague initiates the process by choosing prime $p = 131$. Your colleague then chooses $a = 2$ (which is primitive relative to 131) and $n = 14$, forms $b = a^n \bmod p = 9$, and sends the values of p, a, and b to you over the insecure communication line. Consider the plaintext integers denoted in order as $x_1 = 81$, $x_2 = 117$, $x_3 = 101$, $x_4 = 101$, $x_5 = 110$. With five plaintext numbers to encrypt, you choose five different values of k, which we will denote as $k_1 = 3$, $k_2 = 4$, $k_3 = 5$, $k_4 = 6$, $k_5 = 7$. Then, with the values $a = 2$ and $b = 9$ received from your colleague, you encrypt each plaintext integer x_i by using k_i to form the following pair of ciphertext integers y_i and z_i.

$$x_1 = 81 \rightarrow \begin{cases} y_1 = 2^3 \bmod 131 & = & 8 \\ z_1 = 81 \cdot 9^3 \bmod 131 & = & 99 \end{cases}$$

$$x_2 = 117 \rightarrow \begin{cases} y_2 = 2^4 \bmod 131 & = & 16 \\ z_2 = 117 \cdot 9^4 \bmod 131 & = & 108 \end{cases}$$

$$x_3 = 101 \rightarrow \begin{cases} y_3 = 2^5 \bmod 131 & = & 32 \\ z_3 = 101 \cdot 9^5 \bmod 131 & = & 43 \end{cases}$$

$$x_4 = 101 \rightarrow \begin{cases} y_4 = 2^6 \bmod 131 & = & 64 \\ z_4 = 101 \cdot 9^6 \bmod 131 & = & 125 \end{cases}$$

$$x_5 = 110 \rightarrow \begin{cases} y_5 = 2^7 \bmod 131 & = & 128 \\ z_5 = 110 \cdot 9^7 \bmod 131 & = & 67 \end{cases}$$

You then send the ciphertext pairs (y_i, z_i) in order to your colleague. That is, you send your colleague the list of ciphertext pairs $(8, 99)$, $(16, 108)$, $(32, 43)$, $(64, 125)$, $(128, 67)$. Your colleague can then use the values of $p = 131$ and $n = 14$ to decrypt the message as follows.

$$\left. \begin{array}{l} y_1 = 8 \\ z_1 = 99 \end{array} \right\} \rightarrow x_1 = 99 \cdot 8^{116} \bmod 131 = 81$$

$$\left. \begin{array}{l} y_2 = 16 \\ z_2 = 108 \end{array} \right\} \rightarrow x_2 = 108 \cdot 16^{116} \bmod 131 = 117$$

$$\left. \begin{array}{l} y_3 = 32 \\ z_3 = 43 \end{array} \right\} \rightarrow x_3 = 43 \cdot 32^{116} \bmod 131 = 101$$

$$\left. \begin{array}{l} y_4 = 64 \\ z_4 = 125 \end{array} \right\} \rightarrow x_4 = 125 \cdot 64^{116} \bmod 131 = 101$$

$$\left.\begin{array}{rcl} y_5 & = & 128 \\ z_5 & = & 67 \end{array}\right\} \quad \rightarrow \quad x_5 = 67 \cdot 128^{116} \bmod 131 = 110$$

Your colleague can then convert these plaintext numbers back into the original plaintext characters. □

As we noted in the basic steps in using an ElGamal cipher, for maximum security a different k should be used for each plaintext integer. A result of this (but not the reason for it) is that although the plaintext characters are all encrypted separately in Example 10.5, the ElGamal cipher in this example is not a substitution cipher. Even so, as was true of RSA ciphers in Chapter 9, ElGamal ciphers can be more secure if plaintext numbers are encrypted in groups. More specifically, the basic steps in using an ElGamal cipher listed on page 332 can be followed exactly as presented, even if the plaintext integers x result from grouping ASCII numbers into larger blocks, provided the blocks are all less than the prime p. This is how we will implement most of our ElGamal ciphers in the rest of this chapter, with the plaintext ASCII numbers grouped in order into blocks, using for each block the maximum possible collection of numbers such that the block remains less than p. As was also true of RSA ciphers, ElGamal ciphers that are used in industry generally require modular exponentiation with very large powers. Recall that this can always be done relatively easily using binary exponentiation.

Example 10.6 Suppose again that you wish to send the secret message Queen to a colleague over an insecure communication line using an ElGamal cipher with the ASCII correspondences, under which the plaintext converts into the list of integers 81, 117, 101, 101, 110. Your colleague initiates the process by choosing prime $p = 126127$. Your colleague then chooses $a = 2347$ (which is primitive relative to 126127) and $n = 6789$, forms $b = a^n \bmod p = 108329$, and sends the values of p, a, and b to you over the insecure communication line. With the value $p = 126127$ received from your colleague, you can group the first two plaintext numbers into the single block 81117, which remains less than p. Similarly, you can group the next two plaintext numbers into the block 101101, with only the last plaintext number 110 left to form the last block. That is, the full list of plaintext blocks to be encrypted can be $x_1 = 81117$, $x_2 = 101101$, $x_3 = 110$. With three plaintext numbers to encrypt, you choose three different values of k, which we will denote as $k_1 = 300$, $k_2 = 400$, $k_3 = 500$. Then, with the values $a = 2347$ and $b = 108329$ received from your colleague, you encrypt each plaintext integer x_i by using k_i to form the following pair of ciphertext integers y_i and z_i.

$$x_1 = 81117 \quad \rightarrow \quad \left\{\begin{array}{lll} y_1 & = a^{k_1} \bmod p & = 19803 \\ z_1 & = x_1 \cdot b^{k_1} \bmod p & = 86747 \end{array}\right.$$

$$x_2 = 101101 \quad \rightarrow \quad \begin{cases} y_2 = a^{k_2} \bmod p & = \quad 28521 \\ z_2 = x_2 \cdot b^{k_2} \bmod p & = \quad 26634 \end{cases}$$

$$x_3 = \quad 110 \quad \rightarrow \quad \begin{cases} y_3 = a^{k_3} \bmod p & = \quad 56401 \\ z_3 = x_3 \cdot b^{k_3} \bmod p & = \; 125508 \end{cases}$$

You then send the list of ciphertext pairs $(19803, 86747)$, $(28521, 26634)$, $(56401, 125508)$ to your colleague. Your colleague can use the values of $p = 126127$ and $n = 6789$ to decrypt the message as follows.

$$\left. \begin{array}{l} y_1 = \quad 19803 \\ z_1 = \quad 86747 \end{array} \right\} \quad \rightarrow \quad x_1 = z_1 \cdot y_1^{p-1-n} \bmod p = \quad 81117$$

$$\left. \begin{array}{l} y_2 = \quad 28521 \\ z_2 = \quad 26634 \end{array} \right\} \quad \rightarrow \quad x_2 = z_2 \cdot y_2^{p-1-n} \bmod p = \quad 101101$$

$$\left. \begin{array}{l} y_3 = \quad 56401 \\ z_3 = \quad 125508 \end{array} \right\} \quad \rightarrow \quad x_3 = z_3 \cdot y_3^{p-1-n} \bmod p = \quad 110$$

Your colleague can then split these blocks into numbers that correspond to a single ASCII character each, and convert these back into characters. □

Example 10.7 Suppose again that you wish to send the secret message **Queen** to a colleague over an insecure communication line using an ElGamal cipher with the ASCII correspondences, under which the plaintext converts into the list of integers 81, 117, 101, 101, 110. Your colleague initiates by choosing prime $p = 126126126126197$. Your colleague then chooses $a = 23456$ (which is primitive relative to p) and $n = 78901$, forms $b = a^n \bmod p = 69036247048399$, and sends the values of p, a, and b to you over the insecure communication line. With the value of p received from your colleague, you can group the plaintext numbers into the single block $x = 81117101101110$, which is less than p. With one plaintext number to encrypt, you choose the one value of $k = 30000$. Then, with the values of a and b received from your colleague, you encrypt the plaintext block x by using k to form the following pair of ciphertext integers y and z.

$$y = a^k \bmod p \quad = \quad 25772079736491$$
$$z = x \cdot b^k \bmod p = 114827885818362$$

You then send the ciphertext pair (y, z) to your colleague. Your colleague can then use the values of p and n to decrypt the message as follows.

$$x = z \cdot y^{p-1-n} \bmod p = 81117101101110$$

Your colleague can then split this plaintext block into numbers that correspond to a single ASCII character each, and convert these numbers back into the original plaintext characters. □

10.3.1 Exercises

1. For the following values of a and p, determine whether a is primitive relative to p, and justify your answer.

   (a)* $a = 2$, $p = 17$

   (b) $a = 3$, $p = 17$

   (c) $a = 4$, $p = 17$

   (d)* $a = 11$, $p = 29$

   (e) $a = 12$, $p = 29$

   (f) $a = 13$, $p = 29$

2. Consider an ElGamal cipher with prime $p = 137$ and the ASCII correspondences.

   (a) For this cipher with $a = 12$ (which is primitive relative to 137) and $n = 6$, verify that $b = 69$.

   (b)* Use this cipher with $a = 12$ and $n = 6$ (for which b is given in part (a)), $k_1 = 2$, $k_2 = 3$, $k_3 = 4$, and $k_4 = 5$ to encrypt Rick, with the plaintext numbers all encrypted separately.

   (c) Use this cipher with $a = 12$ and $n = 6$ (for which b is given in part (a)), $k_1 = 2$, $k_2 = 3$, $k_3 = 4$, and $k_4 = 5$ to encrypt T.C., with the plaintext numbers all encrypted separately.

   (d) Decrypt $(37, 61)$, $(33, 12)$, $(122, 87)$, $(94, 94)$, $(32, 98)$, which was formed using this cipher with $n = 133$, and in which the plaintext numbers were all encrypted separately.

   (e) For this cipher with $a = 20$ (which is primitive relative to 137) and $n = 47$, find b.

   (f)* Use this cipher with $a = 20$ and $n = 47$ (for which b is the answer to part (e)), $k_1 = 2$, $k_2 = 3$, $k_3 = 4$, $k_4 = 5$, and $k_5 = 6$ to encrypt Hutch, with the plaintext numbers all encrypted separately.

   (g) Use this cipher with $a = 20$ and $n = 47$ (for which b is the answer to part (e)), $k_1 = 2$, $k_2 = 3$, $k_3 = 4$, $k_4 = 5$, $k_5 = 6$, $k_6 = 7$, and $k_7 = 8$ to encrypt Starsky, with the plaintext numbers all encrypted separately.

   (h) Decrypt $(100, 72)$, $(82, 27)$, $(133, 136)$, $(57, 63)$, $(44, 18)$, which was formed using this cipher with $n = 120$, and in which the plaintext numbers were all encrypted separately.

3. Consider an ElGamal cipher with prime $p = 126131$ and the ASCII correspondences.

   (a) For this cipher with $a = 30$ (which is primitive relative to 126131) and $n = 5$, verify that $b = 82848$.

   (b)* Use this cipher with $a = 30$ and $n = 5$ (for which b is given in part (a)), $k_1 = 2$, and $k_2 = 3$ to encrypt Mike, with the plaintext numbers grouped into the largest possible blocks before being encrypted.

   (c) Use this cipher with $a = 30$ and $n = 5$ (for which b is given in part (a)), $k_1 = 2$, $k_2 = 3$, and $k_3 = 4$ to encrypt Sonny, with the plaintext numbers grouped into the largest possible blocks before being encrypted.

   (d) Decrypt $(12517, 63381)$, $(123248, 4513)$, $(39641, 9694)$, which was formed using this cipher with $n = 126126$, and in which the plaintext numbers were grouped into the largest possible blocks before being encrypted.

   (e) For this cipher with $a = 41$ (which is primitive relative to 126131) and $n = 161$, find b.

   (f)* Use this cipher with $a = 41$ and $n = 161$ (for which b is the answer to part (e)), $k_1 = 2$, and $k_2 = 3$ to encrypt Ruth, with the plaintext numbers grouped into the largest possible blocks before being encrypted.

   (g) Use this cipher with $a = 41$ and $n = 161$ (for which b is the answer to part (e)), $k_1 = 2$, $k_2 = 3$, and $k_3 = 4$ to encrypt Aaron, with the plaintext numbers grouped into the largest possible blocks before being encrypted.

   (h) Decrypt $(90465, 119990)$, $(51266, 81860)$, $(83810, 73028)$, which was formed using this cipher with $n = 126098$, and in which the plaintext numbers were grouped into the largest possible blocks before being encrypted.

4. Consider an ElGamal cipher with prime $p = 126126139$ and the ASCII correspondences.

   (a) For this cipher with $a = 50$ (which is primitive relative to p) and $n = 4$, verify that $b = 6250000$.

   (b)* Use this cipher with $a = 50$ and $n = 4$ (for which b is given in part (a)), $k_1 = 2$, and $k_2 = 3$ to encrypt Nash, with the plaintext numbers grouped into the largest possible blocks before being encrypted.

(c) Use this cipher with $a = 50$ and $n = 4$ (for which b is given in part (a)), $k_1 = 2$, and $k_2 = 3$ to encrypt `Stills`, with the plaintext numbers grouped into the largest possible blocks before being encrypted.

(d) Decrypt $(92682299, 1652699)$, $(93573946, 14892797)$, which was formed using this cipher with $n = 126126133$, and in which the plaintext numbers were grouped into the largest possible blocks before being encrypted.

(e) For this cipher with $a = 60$ (which is primitive relative to p) and $n = 683$, find b.

(f)* Use this cipher with $a = 60$ and $n = 683$ (for which b is the answer to part (e)), and $k = 2$ to encrypt `ELO`, with the plaintext numbers grouped into a single block before being encrypted.

(g) Use this cipher with $a = 60$ and $n = 683$ (for which b is the answer to part (e)), and $k = 2$ to encrypt `BTO`, with the plaintext numbers grouped into a single block before being encrypted.

(h) Decrypt $(115332886, 74068236)$, which was formed using this cipher with $n = 126126074$, and in which the plaintext numbers were grouped into a single block before being encrypted.

5. Find some information about the career in cryptology of Taher Elgamal, including his background, accomplishments, and importance in its history, and write a summary of your findings.

6. The cryptographic method *Pretty Good Privacy* (*PGP*) uses both RSA and ElGamal ciphers. Find some information about PGP, including the reasons for and history of its development, and one or more of its real-life uses, and write a summary of your findings.

7. Find some information about one or more real-life uses of ElGamal ciphers besides in PGP, and write a summary of your findings.

10.4 Cryptanalysis of ElGamal Ciphers

We would assume an outsider who intercepts a ciphertext formed using an ElGamal cipher would know that each ciphertext pair (y, z) was formed as $y = a^k \bmod p$ and $z = x \cdot b^k \bmod p$ for some plaintext block x and encryption keys a, b, k, and p, with p prime, a primitive relative to p, and $b = a^n \bmod p$ for some integer n. We would also assume the outsider knows the actual values of a, b, and p, since, noting the basic steps in using an ElGamal cipher listed on page 332, each was sent over an insecure communication line. For example, we would assume an outsider who intercepts the ciphertext pairs

in Example 10.6 on page 335 would know that each ciphertext pair (y_i, z_i)
was formed as $y_i = 2347^{k_i} \bmod 126127$ and $z_i = x_i \cdot 108329^{k_i} \bmod 126127$
for some integer k_i and plaintext block x_i. This implies that the cipher
in Example 10.6 is not mathematically secure, because an outsider who
intercepts the ciphertext pairs (y_i, z_i) and knows the values of a, b, and p
could use the following steps to break the cipher.

1. Find a discrete logarithm n that satisfies $b = a^n \bmod p$.

2. Recover the plaintext blocks x_i by forming $x_i = z_i \cdot y_i^{p-1-n} \bmod p$.

These steps are not unique to breaking the cipher in Example 10.6, of
course. Any ElGamal cipher could be broken using the same steps. In
addition, even with extremely large values for the ciphertext pairs (y_i, z_i)
and p, the second step could be done easily using binary exponentiation.
The first step, on the other hand, is very difficult to complete in general.
The general difficulty of finding discrete logarithms, the same problem we
noted in Section 10.2 that gives the Diffie-Hellman key exchange its high
level of security, gives ElGamal ciphers an equally high level of security.

This high level of security is dependent, though, upon the users of an
ElGamal cipher choosing a very large value of n. If the value of n were small
enough to be determined relatively quickly by trial multiplications, then the
cipher would be relatively easy to break. For example, a cryptanalyst using
a hand-held calculator could find the value of n in Example 10.6 ($n = 6789$)
by trial multiplications in no more than a few hours.

Also, recall that discrete logarithms are not in general unique, even
when required to be positive and less than a given modulus. For example,
as we noted in Example 10.2 on page 328, every twentieth positive integer
starting with $s = 18$ is a discrete logarithm with $13^s \bmod 187 = 135$. As
we further noted and illustrated in Example 10.3 on page 329, any discrete
logarithm for a particular base and modulus can be used to break a Diffie-
Hellman key exchange. The same is true for ElGamal ciphers, a fact which
underscores the importance of using a value of a in an ElGamal cipher that
is primitive relative to p. If a were primitive relative to p, then for any
possible value of b there would be only one positive integer n less than p
for which $b = a^n \bmod p$, a value of n that it would be necessary to find in
order to break the cipher using the two steps listed in the first paragraph in
this section. On the other hand, if a were not primitive relative to p, then
it could be the case that a discrete logarithm much smaller than n could
be used to break the cipher.

Example 10.8 Consider $(3904, 1286)$, $(3904, 4868)$, $(3904, 9658)$, a ciphertext formed using an ElGamal cipher with $p = 10111$, $a = 9$, and $b = 6051$.
Incidentally, this value of a is not primitive relative to p. Furthermore,

the exponent used to form b was 9109. That is, $9^{9109} \bmod 10111 = 6051$, although as a cryptanalyst we would not know that this exponent was used to form b. To break this cipher, we can begin by using trial multiplications to find a discrete logarithm n for which $9^n \bmod 10111 = 6051$. The following table shows the values of $9^n \bmod 10111$ for positive integers starting with $n = 1$, continuing until we find the first value of n for which $9^n \bmod 10111 = 6051$.

n	$9^n \bmod 10111$	n	$9^n \bmod 10111$
1	9	6	5669
2	81	7	466
3	729	8	4194
4	6561	9	7413
5	8494	10	6051

With the value $n = 10$, we can now decrypt the ciphertext as follows.

$$\left. \begin{array}{l} y_1 = 3904 \\ z_1 = 1286 \end{array} \right\} \quad \rightarrow \quad x_1 = z_1 \cdot y_1^{p-1-n} \bmod p = 6682$$

$$\left. \begin{array}{l} y_2 = 3904 \\ z_2 = 4868 \end{array} \right\} \quad \rightarrow \quad x_2 = z_2 \cdot y_2^{p-1-n} \bmod p = 4553$$

$$\left. \begin{array}{l} y_3 = 3904 \\ z_3 = 9658 \end{array} \right\} \quad \rightarrow \quad x_3 = z_3 \cdot y_3^{p-1-n} \bmod p = 5257$$

Splitting these plaintext blocks into numbers that correspond to a single ASCII character each, and converting these numbers back into the original plaintext characters, yields the plaintext BR-549. $\qquad\square$

In Example 10.8, the intended recipient of the message would not have to find a discrete logarithm to decrypt the ciphertext, since the intended recipient would have chosen $n = 9109$ in the initial setup of the cipher. The intended recipient would know $n = 9109$ satisfies $9^n \bmod 10111 = 6051$, and could thus decrypt the ciphertext using the same three final calculations as in Example 10.8, but with $n = 9109$ instead of $n = 10$. Since both $n = 10$ and $n = 9109$ yield the same value of $b = a^n \bmod p$, either can be used to decrypt the ciphertext. This illustrates why, for an ElGamal cipher with a given modulus p, it is optimal to choose a primitive relative to p. If a is chosen primitive relative to p, then among positive integers less than p, only the value of n chosen by the intended recipient could be used to break the cipher. For instance, suppose that in Example 10.8 the intended recipient had chosen $a = 12$, which is primitive relative to $p = 10111$, and formed $b = 12^{9109} \bmod 10111 = 1512$. Then in breaking the cipher, finding a discrete logarithm n for which $12^{9109} \bmod 10111 = 1512$ would require trial multiplications through $n = 9109$.

As we have just shown, given a list of ciphertext pairs (y_i, z_i) formed using an ElGamal cipher with $y_i = a^{k_i} \bmod p$ and $z_i = x_i \cdot b^{k_i} \bmod p$ for unknown plaintext blocks x_i, an outsider could recover all of the plaintext blocks x_i by finding a single discrete logarithm n with $b = a^n \bmod p$. Of course, for any specific ciphertext pair (y_i, z_i), an outsider could also recover x_i by finding a discrete logarithm k_i with $y_i = a^{k_i} \bmod p$, and then forming $x_i = z_i \cdot (b^{k_i})^{-1} \bmod p$. This, by the way, indicates the necessity for users of an ElGamal cipher to also choose very large values of k.

Finding a discrete logarithm k_i that satisfies $y_i = a^{k_i} \bmod p$ for each ciphertext pair is not typically an effective way to break an ElGamal cipher, though, since assuming that a different value of k was used to form each ciphertext pair, an outsider would have to find a different discrete logarithm for each ciphertext pair. However, this is not the reason why we noted in the basic steps in using an ElGamal cipher on page 332 that a different k should be used to form each ciphertext pair. The reason why we noted this is because, as it turns out, if an outsider somehow had a crib (that is, a small known part of the plaintext) giving the plaintext block corresponding to one ciphertext pair, the outsider could decrypt any other ciphertext pair formed using the same k without having to find a discrete logarithm. More specifically, consider, for example, an ElGamal cipher with plaintext blocks x_1 and x_2, and corresponding ciphertext values $z_1 = x_1 \cdot b^k \bmod p$ and

Elliptic Curve Cryptography

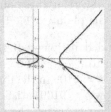

In practice ElGamal ciphers are typically implemented using elliptic curves, because this can give a high level of security with less computation. One type of elliptic curve contains the ordered pairs of solutions to $y^2 = x^3 + ax + b \bmod p$ for prime $p > 3$ and integers a and b satisfying $4a^3 + 27b^2 \neq 0 \bmod p$. For example, $(3,6)$ is one element in the elliptic curve satisfying $y^2 = x^3 + x + 6 \bmod 19$, since this equation is true when $x = 3$ and $y = 6$. Using a special operation for combining elements, an elliptic curve forms a mathematical structure called a *group*. The encryption and decryption procedures for an ElGamal cipher can be generalized to work with any group.

Besides cryptology, elliptic curves have many other applications in mathematics. They have proven useful in primality testing and integer factorization, and played an important role in Andrew Wiles' celebrated proof of Fermat's Last Theorem.

$z_2 = x_2 \cdot b^k \mod p$ formed using the same k. If an outsider somehow knew x_1, the outsider could determine x_2 by forming $x_2 = x_1 \cdot z_2 \cdot (z_1)^{-1} \mod p$.

More generally, consider an ElGamal cipher with any two plaintext blocks x_i and x_j, and corresponding ciphertext values z_i and z_j formed using the same k. If an outsider somehow knew x_i, the outsider could determine x_j using the following formula.[5]

$$x_j = x_i \cdot z_j \cdot (z_i)^{-1} \mod p$$

Using this formula requires finding the modular inverse $(z_i)^{-1} \mod p$. Recall that this can always be done relatively easily using the Euclidean algorithm. Using this formula also requires somehow having a crib giving x_i. As we have noted, it is not unreasonable to suppose an outsider in possession of a ciphertext might know or be able to correctly guess a small part of the corresponding plaintext. For example, the outsider may know where or from whom the message originated, and correctly guess that the first several letters in the plaintext are a time or location stamp or that the last few letters in the plaintext are the originator's name.

Example 10.9 Consider again the ciphertext $(3904, 1286)$, $(3904, 4868)$, $(3904, 9658)$ from Example 10.8, which was formed using an ElGamal cipher with $p = 10111$. These ciphertext pairs all have a common first component because each was formed using the same k. Suppose an outsider somehow knew that the first two letters in the plaintext were BR, and thus that the first plaintext block was $x_1 = 6682$. Then, the outsider could determine x_2 and x_3 by first using the Euclidean algorithm to find that $(z_1)^{-1} \mod p = 1286^{-1} \mod 10111 = 3656$, and forming the following.

$$
\begin{aligned}
x_2 &= x_1 \cdot z_2 \cdot (z_1)^{-1} \mod p \\
&= 6682 \cdot 4868 \cdot 3656 \mod 10111 \\
&= 4553
\end{aligned}
$$

$$
\begin{aligned}
x_3 &= x_1 \cdot z_3 \cdot (z_1)^{-1} \mod p \\
&= 6682 \cdot 9658 \cdot 3656 \mod 10111 \\
&= 5257 \qquad \square
\end{aligned}
$$

10.4.1 Exercises

1.[*] Decrypt the following ciphertexts, which were formed using ElGamal ciphers with the given values of p, a, and b.

---

[5]This formula works because $z_i \cdot (x_i)^{-1} \mod p$ would be the same as $z_j \cdot (x_j)^{-1} \mod p$, both being equal to $b^k \mod p$. That is, $z_i \cdot (x_i)^{-1} = z_j \cdot (x_j)^{-1} \mod p$. Solving this equation for x_j gives the formula.

(a) $p = 131$; $a = 2$; $b = 41$;
ciphertext $= (112, 30)$, $(93, 39)$, $(55, 67)$

(b) $p = 131$; $a = 6$; $b = 59$;
ciphertext $= (112, 39)$, $(17, 6)$, $(102, 17)$

(c) $p = 137$; $a = 6$; $b = 133$;
ciphertext $= (18, 48)$, $(108, 15)$, $(100, 85)$

(d) $p = 137$; $a = 12$; $b = 6$;
ciphertext $= (61, 30)$, $(47, 42)$, $(16, 78)$

2. Decrypt the following ciphertexts, which were formed using ElGamal ciphers with the given values of p, b, and k.

(a) $p = 131$; $b = 84$; $k_1 = 2$; $k_2 = 3$; $k_3 = 4$;
ciphertext $= (64, 50)$, $(119, 19)$, $(35, 106)$

(b) $p = 131$; $b = 64$; $k_1 = 2$; $k_2 = 12$; $k_3 = 112$;
ciphertext $= (100, 40)$, $(34, 67)$, $(65, 65)$

3.* Decrypt the following ciphertexts, which were formed using ElGamal ciphers with the given values of p, a, and b.

(a) $p = 137$; $a = 13$; $b = 14$;
ciphertext $= (65, 34)$, $(23, 80)$, $(25, 104)$

(b) $p = 137$; $a = 20$; $b = 102$;
ciphertext $= (121, 133)$, $(120, 56)$, $(30, 13)$

4. Decrypt the following ciphertexts, which were formed using ElGamal ciphers with the given value of p and crib, and for which each ciphertext pair was formed using the same k.

(a) $p = 131$; crib $=$ N, giving the first plaintext block;
ciphertext $= (84, 90)$, $(84, 12)$, $(84, 37)$

(b) $p = 131$; crib $=$ A, giving the second plaintext block;
ciphertext $= (112, 73)$, $(112, 9)$, $(112, 119)$

(c) $p = 137$; crib $=$ A, giving the first plaintext block;
ciphertext $= (122, 49)$, $(122, 57)$, $(122, 32)$

(d) $p = 137$; crib $=$ X, giving the third plaintext block;
ciphertext $= (130, 116)$, $(130, 77)$, $(130, 18)$

5. Find some information about how the size of the prime p and/or the number of values generated by a believed to be necessary for ElGamal ciphers to be secure has increased over the years, including whatever the current industry standards or opinions are, and write a summary of your findings.

Chapter 11

The Advanced Encryption Standard

As we have noted, in practice, known public-key ciphers are in general much slower than known non-public-key ciphers when all are implemented properly. Because of this, non-public-key ciphers continue to play a practical role in modern society. Of course, as we have also noted, the progress in technology that our society achieved over the last several decades has made the types of ciphers we considered before Chapter 9 obsolete, at least from a security perspective. However, non-public-key ciphers do exist that are useful in modern society. Such ciphers are typically called *symmetric-key* ciphers, since users must usually agree upon encryption and decryption keys that are clearly related, often identical, and which must both be kept secret from outsiders. Technically, each of the types of ciphers we considered before Chapter 9 are symmetric-key, but these types of ciphers are more commonly called *classical* ciphers, since they are not actually useful for transmitting sensitive information in modern society.

In this chapter, we will consider some details related to the use of symmetric-key ciphers in modern society. Special attention will be given to the Advanced Encryption Standard, a type of symmetric-key cipher secure enough that it was selected by the National Institute of Standards and Technology in 2001 as a Federal Information Processing Standard, a function it continues to serve as of this writing.

11.1 Representations of Numbers

Since modern cryptographic methods are almost exclusively automated on computers, we will begin by briefly describing some basics about how computers store information and manipulate data, particularly as numbers expressed in binary and hexadecimal formats.

The numbers we see and use in our everyday lives are typically expressed in decimal, or base 10, format. A number is expressed in *decimal* format if the digits used to represent the number (reading from left to right) would be the coefficients if the number were written as a sum of nonnegative integer multiples of nonnegative integer powers of 10 (with decreasing powers of 10). For example, the decimal number 40217 uses the digits **4-0-2-1-7** from left to right, because if we write 40217 as a sum of multiples of powers of 10, we obtain the following.

$$
\begin{aligned}
40217 &= 40000 + 200 + 10 + 7 \\
&= \mathbf{4} \cdot 10000 + \mathbf{0} \cdot 1000 + \mathbf{2} \cdot 100 + \mathbf{1} \cdot 10 + \mathbf{7} \cdot 1 \\
&= \mathbf{4} \cdot 10^4 + \mathbf{0} \cdot 10^3 + \mathbf{2} \cdot 10^2 + \mathbf{1} \cdot 10^1 + \mathbf{7} \cdot 10^0
\end{aligned}
$$

However, computers usually do not store information and manipulate data as numbers expressed in decimal format, but instead use binary or hexadecimal format. These formats work similarly to decimal; they just use sums of multiples of powers of different bases.

11.1.1 Binary

A number is expressed in *binary*, or base 2, format if the digits used to represent the number (reading from left to right) would be the coefficients if the decimal expression of the number were written as a sum of nonnegative integer multiples of nonnegative integer powers of 2 (with decreasing powers of 2). For example, to express the decimal number 217 in binary, we form the following sum.

$$
\begin{aligned}
217 &= 128 + 64 + 16 + 8 + 1 \\
&= 1 \cdot 2^7 + 1 \cdot 2^6 + 0 \cdot 2^5 + 1 \cdot 2^4 + 1 \cdot 2^3 + 0 \cdot 2^2 + 0 \cdot 2^1 + 1 \cdot 2^0
\end{aligned}
$$

The coefficients in this sum give the binary representation of the decimal number 217. That is, the binary representation of the decimal number 217 is 11011001.

Expressing a decimal number in binary can be done more easily as follows. We begin by dividing the decimal number by 2, obtaining quotient q_1 and remainder r_1. We then divide q_1 by 2, obtaining quotient q_2 and remainder r_2. Next, we divide q_2 by 2, obtaining quotient q_3 and remainder r_3, and continue this process of repeated division, stopping the first time we obtain a quotient of 0. Listing the resulting remainders in order, starting with the last remainder (corresponding to the first quotient of 0) and ending with r_1, gives the binary representation of the decimal number with which we started.

Example 11.1 To express the decimal number 217 in binary, we perform the following divisions.

$$217 \div 2 \quad \rightarrow \quad q_1 = 108, \, r_1 = 1$$
$$108 \div 2 \quad \rightarrow \quad q_2 = 54, \, r_2 = 0$$
$$54 \div 2 \quad \rightarrow \quad q_3 = 27, \, r_3 = 0$$
$$27 \div 2 \quad \rightarrow \quad q_4 = 13, \, r_4 = 1$$
$$13 \div 2 \quad \rightarrow \quad q_5 = 6, \, r_5 = 1$$
$$6 \div 2 \quad \rightarrow \quad q_6 = 3, \, r_6 = 0$$
$$3 \div 2 \quad \rightarrow \quad q_7 = 1, \, r_7 = 1$$
$$1 \div 2 \quad \rightarrow \quad q_8 = 0, \, r_8 = 1$$

The list of remainders $r_8 r_7 r_6 r_5 r_4 r_3 r_2 r_1 = 11011001$ in order is the binary representation of the decimal number 217. □

Example 11.2 To express the decimal number 85 in binary, we perform the following divisions.

$$85 \div 2 \quad \rightarrow \quad q_1 = 42, \, r_1 = 1$$
$$42 \div 2 \quad \rightarrow \quad q_2 = 21, \, r_2 = 0$$
$$21 \div 2 \quad \rightarrow \quad q_3 = 10, \, r_3 = 1$$
$$10 \div 2 \quad \rightarrow \quad q_4 = 5, \, r_4 = 0$$
$$5 \div 2 \quad \rightarrow \quad q_5 = 2, \, r_5 = 1$$
$$2 \div 2 \quad \rightarrow \quad q_6 = 1, \, r_6 = 0$$
$$1 \div 2 \quad \rightarrow \quad q_7 = 0, \, r_7 = 1$$

Thus, the binary representation of the decimal number 85 is 1010101. □

For a given binary representation of a number, to express the number in decimal, it is easiest to just form a sum of multiples of powers of 2 using the digits in the binary representation as coefficients, and then combine the terms in this sum.

Example 11.3 To express the binary number 11011001 in decimal, we form the following sum.

$$\mathbf{1} \cdot 2^7 + \mathbf{1} \cdot 2^6 + \mathbf{0} \cdot 2^5 + \mathbf{1} \cdot 2^4 + \mathbf{1} \cdot 2^3 + \mathbf{0} \cdot 2^2 + \mathbf{0} \cdot 2^1 + \mathbf{1} \cdot 2^0$$
$$= \quad 128 + 64 + 16 + 8 + 1$$
$$= \quad 217$$

Thus, the decimal representation of the binary number 11011001 is 217. □

Example 11.4 To express the binary number 1010101 in decimal, we form the following sum.

$$1 \cdot 2^6 + 0 \cdot 2^5 + 1 \cdot 2^4 + 0 \cdot 2^3 + 1 \cdot 2^2 + 0 \cdot 2^1 + 1 \cdot 2^0$$
$$= \quad 64 + 16 + 4 + 1$$
$$= \quad 85$$

Thus, the decimal representation of the binary number 1010101 is 85.   □

When expressing a decimal number in binary, whether using a sum of multiples of powers of 2 or repeated division by 2, the only possible digits that could result are 0s and 1s. That is, the binary representation of a number will always consist only of 0s and 1s. These individual 0s and 1s are usually referred to as *bits*. A string of eight bits is called a *byte*. For example, the binary representation 11011001 of the decimal number 217, which we found in Example 11.1, is a byte. The binary representation 1010101 of the decimal number 85, which we found in Example 11.2, is not a byte, since it is only contains seven bits. A bit string shorter than eight bits can always be expanded to a full byte by padding the string on the left with 0s. For example, the bit string 1010101 can be expanded to the full byte 01010101. Similarly, the bit string 10101 can be expanded to the full byte 00010101.

Computers typically store information as bytes. More specifically, computers typically store information by using the ASCII correspondences to convert characters into decimal numbers, and then converting these decimal numbers into bytes.

Example 11.5 To convert the list of characters UB40 into a list of bytes using the ASCII correspondences given in Table 9.1 on page 302, we begin by converting these characters into the following corresponding ASCII decimal numbers.

$$85,\ 66,\ 52,\ 48$$

Next, we express each of these decimal numbers in binary.

$$1010101,\ 1000010,\ 110100,\ 110000$$

Finally, we expand each of these binary numbers to a full byte.

$$01010101,\ 01000010,\ 00110100,\ 00110000$$ □

Binary numbers can be combined using arithmetic operations such as addition, subtraction, and multiplication. One important operation for combining binary numbers that we will use later in this chapter is the *exclusive or* operation, denoted XOR, or, more commonly, with the symbol $\oplus$. The

XOR operation combines pairs of binary digits by adding the digits and reducing the result modulo 2. That is, $0 \oplus 0 = 0$, $1 \oplus 0 = 1$, $0 \oplus 1 = 1$, and $1 \oplus 1 = 0$. A pair of binary numbers of the same length can be combined using the XOR operation, by applying the operation to each pair of corresponding bits. A pair of binary numbers of different lengths can also be combined using the XOR operation, by padding the shorter number on the left with 0s until it is the same length as the longer number, and then applying the operation to each pair of corresponding bits.

Example 11.6 To find $11011001 \oplus 1010101$, we begin by padding the second term on the left with a single 0, so that the numbers are the same length. We then form $11011001 \oplus 01010101$ by applying the XOR operation to each pair of corresponding bits. For example, since the first bit in the first term is 1, and the first bit in the second term is 0, then the first bit in the sum is $1 \oplus 0 = 1$. Similarly, since the second bit in the first term is 1, and the second bit in the second term is 1, then the second bit in the sum is $1 \oplus 1 = 0$. The full process is sometimes easier to follow if the numbers are written with the second term below the first, so that corresponding bits form columns.

$$
\begin{array}{c}
1\ 1\ 0\ 1\ 1\ 0\ 0\ 1 \\
\oplus\ \ 0\ 1\ 0\ 1\ 0\ 1\ 0\ 1 \\
\hline
1\ 0\ 0\ 0\ 1\ 1\ 0\ 0
\end{array}
$$

Thus, $11011001 \oplus 01010101 = 10001100$. $\qquad\square$

11.1.2 Hexadecimal

A number is expressed in *hexadecimal*, or base 16, format if the digits used to represent the number (reading from left to right) would be the coefficients if the decimal expression of the number were written as a sum of nonnegative integer multiples of nonnegative integer powers of 16 (with decreasing powers of 16).

Just as binary numbers require only two possible digits (0 and 1), and decimal numbers require 10 possible digits (0–9), hexadecimal numbers require 16 possible digits. This requires six new digits, for which it is most common to use the letters A–F, in order. That is, the most common list of digits used in hexadecimal numbers is shown in Table 11.1 on page 350.

Expressing a decimal number in hexadecimal can be done via the procedure illustrated for binary in Examples 11.1 and 11.2, except with repeated division by 16 instead of 2. That is, to express a decimal number in hexadecimal, we begin by dividing the decimal number by 16, obtaining quotient q_1 and remainder r_1. We then divide q_1 by 16, obtaining quotient q_2 and

Decimal	Hexadecimal	Decimal	Hexadecimal
0	0	8	8
1	1	9	9
2	2	10	A
3	3	11	B
4	4	12	C
5	5	13	D
6	6	14	E
7	7	15	F

Table 11.1   Correspondences between decimal and hexadecimal.

remainder r_2, and continue this process of repeated division, stopping the first time we obtain a quotient of 0. Listing the resulting remainders in order, starting with the last remainder (corresponding to the first quotient of 0) and ending with r_1, gives the hexadecimal representation of the decimal number with which we started.

Example 11.7 To express the decimal number 217 in hexadecimal, we perform the following divisions.

$$217 \div 16 \quad \rightarrow \quad q_1 = 13, r_1 = 9$$
$$13 \div 16 \quad \rightarrow \quad q_2 = 0, r_2 = 13 \text{ (D)}$$

Thus, the hexadecimal representation of the decimal number 217 is D9. □

Example 11.8 To express the decimal number 114151 in hexadecimal, we perform the following divisions.

$$114151 \div 16 \quad \rightarrow \quad q_1 = 7134, r_1 = 7$$
$$7134 \div 16 \quad \rightarrow \quad q_2 = 445, r_2 = 14 \text{ (E)}$$
$$445 \div 16 \quad \rightarrow \quad q_3 = 27, r_3 = 13 \text{ (D)}$$
$$27 \div 16 \quad \rightarrow \quad q_4 = 1, r_4 = 11 \text{ (B)}$$
$$1 \div 16 \quad \rightarrow \quad q_5 = 0, r_5 = 1$$

Thus, the hexadecimal representation of the decimal number 114151 is 1BDE7. □

As with binary, for a given hexadecimal representation of a number, to express the number in decimal, it is easiest to just form a sum of multiples of powers of 16 using the digits in the hexadecimal representation as coefficients, and then combine the terms in this sum.

Example 11.9 To express the hexadecimal number D9 in decimal, we form the following sum.

$$
\begin{aligned}
\mathbf{D} \cdot 16^1 + \mathbf{9} \cdot 16^0 &= \mathbf{13} \cdot 16^1 + \mathbf{9} \cdot 16^0 \\
&= 208 + 9 \\
&= 217
\end{aligned}
$$

Thus, the decimal representation of the hexadecimal number D9 is 217. □

Example 11.10 To express the hexadecimal number 1BDE7 in decimal, we form the following sum.

$$
\begin{aligned}
\mathbf{1} \cdot 16^4 + \mathbf{B} \cdot 16^3 + \mathbf{D} \cdot 16^2 &+ \mathbf{E} \cdot 16^1 + \mathbf{7} \cdot 16^0 \\
&= \mathbf{1} \cdot 16^4 + \mathbf{11} \cdot 16^3 + \mathbf{13} \cdot 16^2 + \mathbf{14} \cdot 16^1 + \mathbf{7} \cdot 16^0 \\
&= 65536 + 45056 + 3328 + 224 + 7 \\
&= 114151
\end{aligned}
$$

Thus, the decimal representation of the hexadecimal number 1BDE7 is 114151. □

It is always possible to express a given binary number in hexadecimal by first converting the number from binary into decimal, and then converting the result from decimal into hexadecimal. However, expressing a given binary number in hexadecimal can be done more easily using the correspondences between binary and hexadecimal shown in Table 11.2.

Binary	Hexadecimal	Binary	Hexadecimal
0000	0	1000	8
0001	1	1001	9
0010	2	1010	A
0011	3	1011	B
0100	4	1100	C
0101	5	1101	D
0110	6	1110	E
0111	7	1111	F

Table 11.2 Correspondences between binary and hexadecimal.

To express a given binary number in hexadecimal, we can split the number into blocks of four bits starting from the right, padding the leftmost block on the left with 0s if necessary so that it contains exactly four bits, and then convert each block of four bits into the hexadecimal digit with which it corresponds in Table 11.2.

Example 11.11 To express the binary number 11011001 in hexadecimal, we split the number into the following blocks of four bits, with the hexadecimal digit with which each block corresponds in Table 11.2 written below the block.

$$
\begin{array}{cc}
1101 & 1001 \\
D & 9
\end{array}
$$

Thus, the hexadecimal representation of the binary number 11011001 is D9.   □

Example 11.12 To express the binary number 11011110111100111 in hexadecimal, we split the number into the following blocks of four bits, padding the leftmost block on the left with three 0s.

$$
\begin{array}{ccccc}
0001 & 1011 & 1101 & 1110 & 0111 \\
1 & B & D & E & 7
\end{array}
$$

Thus, the hexadecimal representation of the binary number 1101111011110 0111 is 1BDE7.   □

To express a given hexadecimal number in binary, it is easiest to just reverse the process illustrated in Examples 11.11 and 11.12.

Example 11.13 To express the hexadecimal number 1BDE7 in binary, we write the digits in the number below, with the block of four bits with which each digit corresponds in Table 11.2 written below the digit.

$$
\begin{array}{ccccc}
1 & B & D & E & 7 \\
0001 & 1011 & 1101 & 1110 & 0111
\end{array}
$$

With the 0s truncated from the left of the leftmost block, the binary representation of the hexadecimal number 1BDE7 is 11011110111100111.   □

11.1.3   Exercises

1. Express the following decimal numbers in binary.

   (a)* 49

   (b) 157

   (c)* 197

   (d) 881

2. Express the following binary numbers in decimal.

   (a)* 101011

   (b) 11001010

(c)* 11101011

(d) 1100101010

3. Convert the following lists of characters into lists of bytes using the ASCII correspondences given in Table 9.1 on page 302.

(a)* Jets

(b) Giants

4. Convert the following lists of bytes into lists of characters using the ASCII correspondences given in Table 9.1 on page 302.

(a) 01001110, 01100101, 01110100, 01110011

(b) 01001011, 01101110, 01101001, 01100011, 01101011, 01110011

5. Find the following.

(a)* $101011 \oplus 110010$

(b) $10010111 \oplus 11100011$

(c)* $11010110 \oplus 100011$

(d) $10101011 \oplus 1100100011$

6. Express the following decimal numbers in hexadecimal.

(a)* 1954

(b) 35275

(c)* 840110

(d) 7986063

7. Express the following hexadecimal numbers in decimal.

(a)* 9CF

(b) 51AB

(c)* 4EF07

(d) 49D113

8. Express the following binary numbers in hexadecimal.

(a)* 11110011

(b) 110011010101

(c)* 110101101101110

(d) 1011000100101011100010

9. Express the following hexadecimal numbers in binary.

   (a)* 4B

   (b) 2E7

   (c)* A17F

   (d) D18C8

10. Write a summary of how a base 8 number system could work. Include in your summary at least one example of expressing a nontrivial decimal number in base 8, and at least one example of expressing a nontrivial base 8 number in decimal.

11. Write a summary of how a base 12 number system could work. Include in your summary at least one example of expressing a nontrivial decimal number in base 12, and at least one example of expressing a nontrivial base 12 number in decimal.

12. Find some information about the base 20 number system used by the ancient Mayans, and write a summary of your findings.

13. Find some information about the base 60 number system used by the ancient Babylonians, and write a summary of your findings.

11.2   Stream Ciphers

A *stream* cipher is a symmetric-key cipher in which plaintexts are encrypted one character at a time. Technically, all monoalphabetic and polyalphabetic substitution ciphers are stream ciphers, although the term *stream cipher* is usually only used to refer to ciphers in which plaintext characters are expressed as bytes, and then encrypted one bit at a time using a binary key and the XOR operation. The key for a stream cipher is often as long as the plaintext itself.

Example 11.14 Consider the message Mets, which we will encrypt using a stream cipher. We begin by using the ASCII correspondences given in Table 9.1 on page 302 to convert this list of plaintext characters into the corresponding list of decimal numbers 77, 101, 116, 115. Next, we express each of these decimal numbers in binary, expanding each to a full byte, which yields the following list of bytes.

$$01001101, \ 01100101, \ 01110100, \ 01110011$$

We will encrypt this string of 32 plaintext bits using the following 32 key bits with the XOR operation.

01101100, 10011001, 01110000, 00010110

That is, we will encrypt the message as follows.

plain   0 1 0 0 1 1 0 1   0 1 1 0 0 1 0 1   0 1 1 1 0 1 0 0   0 1 1 1 0 0 1 1
key $\oplus$ 0 1 1 0 1 1 0 0   1 0 0 1 1 0 0 1   0 1 1 1 0 0 0 0   0 0 0 1 0 1 1 0
_____

cipher   0 0 1 0 0 0 0 1   1 1 1 1 1 1 0 0   0 0 0 0 0 1 0 0   0 1 1 0 0 1 0 1

Since the decimal numbers that are equivalent to these ciphertext bytes are not all small enough to have corresponding ASCII characters, this ciphertext must be left in numeric form. Also, since addition and subtraction are identical operations with modulo 2 arithmetic, this string of 32 ciphertext bits can be decrypted using the same 32 key bits with the XOR operation. That is, this ciphertext can be decrypted as follows.

cipher   0 0 1 0 0 0 0 1   1 1 1 1 1 1 0 0   0 0 0 0 0 1 0 0   0 1 1 0 0 1 0 1
key $\oplus$ 0 1 1 0 1 1 0 0   1 0 0 1 1 0 0 1   0 1 1 1 0 0 0 0   0 0 0 1 0 1 1 0
_____

plain   0 1 0 0 1 1 0 1   0 1 1 0 0 1 0 1   0 1 1 1 0 1 0 0   0 1 1 1 0 0 1 1

Using the ASCII correspondences to convert these plaintext bytes back into characters gives the original message. $\Box$

Ideally the key for a stream cipher should be a random string of bits. However, generating a string of bits that is truly random can be very difficult. Thus, typically the key for a stream cipher is a string of bits that is *pseudorandom*, meaning the bits appear random, even though they are not. Many software packages have predefined pseudorandom number generators. To be secure (i.e., to produce bits that appear random, even though they are not), a pseudorandom bit generator should produce strings of bits in which later bits are not predictable based upon earlier bits.

One popular pseudorandom bit generator that is generally believed to be secure is the *Blum Blum Shub* generator, or *BBS* for short, named for Lenore Blum, Manuel Blum, and Michael Shub, who proposed it in 1986. The basic steps in using BBS to generate a pseudorandom string of bits can be summarized as follows.

1. Choose prime numbers p and q with $p = 3 \bmod 4$ and $q = 3 \bmod 4$, and form $m = p \cdot q$.

2. Choose an integer x between 1 and m with $\gcd(x, m) = 1$, and form the following *initial seed* x_0.

$$x_0 = x^2 \bmod m$$

3. Form the sequence of numbers $x_1, x_2, x_3, \ldots$ using the following formula.

$$x_k = (x_{k-1})^2 \bmod m$$

4. Form the pseudorandom string of bits $b_1 b_2 b_3 \cdots$ using the following formula.

$$b_k = \begin{cases} 0 & \text{if } x_k \text{ is even} \\ 1 & \text{if } x_k \text{ is odd} \end{cases}$$

Example 11.15 Suppose we wish to use BBS to generate a pseudorandom string of eight bits. We first choose primes $p = 47$ and $q = 59$, and form $m = p \cdot q = 2773$. We then choose $x = 11$, and form initial seed $x_0 = 11^2 \bmod 2773 = 121$. Next, we form $x_1, x_2, \ldots, x_8$ as follows.

$$
\begin{array}{ccccccc}
x_1 & = & (x_0)^2 \bmod 2773 & = & 121^2 \bmod 2773 & = & 776 \\
x_2 & = & (x_1)^2 \bmod 2773 & = & 776^2 \bmod 2773 & = & 435 \\
x_3 & = & (x_2)^2 \bmod 2773 & = & 435^2 \bmod 2773 & = & 661 \\
x_4 & = & (x_3)^2 \bmod 2773 & = & 661^2 \bmod 2773 & = & 1560 \\
x_5 & = & (x_4)^2 \bmod 2773 & = & 1560^2 \bmod 2773 & = & 1679 \\
x_6 & = & (x_5)^2 \bmod 2773 & = & 1679^2 \bmod 2773 & = & 1673 \\
x_7 & = & (x_6)^2 \bmod 2773 & = & 1673^2 \bmod 2773 & = & 972 \\
x_8 & = & (x_7)^2 \bmod 2773 & = & 972^2 \bmod 2773 & = & 1964 \\
\end{array}
$$

Finally, we form the string of bits $b_1 b_2 b_3 \cdots b_8 = 01101100$, in which each b_k is 0 if x_k is even, and 1 if x_k is odd. (Incidentally, if we continued to form $x_9, x_{10}, \ldots, x_{32}$, the complete resulting string of bits $b_1 b_2 b_3 \cdots b_{32}$ would be the 32 key bits in Example 11.14.)   □

The security of BBS is, as we similarly saw for RSA ciphers in Chapter 9, due to the general difficulty of factoring m. A problem with BBS, though, is that it can be very time consuming to use. One way to help alleviate this problem is by having each x_k produce more than one bit, which can be done by simply converting x_k to binary, and then taking more than one bit from the end of this binary representation.[1] If speed is of the essence and some security can be sacrificed, there are many alternatives to BBS for generating pseudorandom strings of bits. One particularly interesting technique involves the use of a *linear feedback shift register*. We will leave this to the reader to investigate further.

Stream ciphers are used extensively in modern society, most notably in electronic devices in which memory and processing resources are limited,

---

[1]This is consistent with BBS as it is described on page 356, in which each x_k produces only one bit, since if x_k were even, the last bit in its binary representation would be 0, and if x_k were odd, the last bit in its binary representation would be 1.

due to their simplicity in implementation. For a ciphertext formed using a stream cipher in which the key is as long as the corresponding plaintext, truly random, never reused, and kept secret, it has been proved that the cipher would be impossible to break. Such ciphers are called *one-time pads*, and were initially described in part in 1917 by Gilbert Vernam, an engineer at AT&T Bell Labs who invented the concept of a stream cipher. The notion of a one-time pad was completed shortly thereafter by U.S. Army Major Joseph Mauborgne, head of research and engineering at the Signal Corps who went on to become its chief. Vernam and Mauborgne were each awarded patents for their ideas. In [14], the National Security Agency describes Vernam's patent as "one of the most important in the history of cryptography." Even more strongly, in *The Codebreakers* [13], David Kahn describes it as "the most important in the history of cryptology." Kahn goes on to describe Vernam as "the man who had automated cryptography," since his patent included the concept for a teletype machine that encrypted characters using the XOR operation.

To better understand the importance of the requirement in a one-time pad that the key be truly random, consider again Example 11.14. At the end of this example, we decrypted a string of 32 ciphertext bits as follows.

cipher	00100001	11111100	00000100	01100101
key ⊕	01101100	10011001	01110000	00010110
plain	01001101	01100101	01110100	01110011

Using the ASCII correspondences to convert these bytes back into characters yields the plaintext Mets. However, were the key in the cipher truly random, the 32 bits used for the key above would be no more likely the correct key than any other string of 32 bits. For example, they would be no more likely the correct key than the string of 32 bits used for the key in the following decryption.

cipher	00100001	11111100	00000100	01100101
key ⊕	01100010	10001001	01100110	00010110
plain	01000011	01110101	01100010	01110011

Using the ASCII correspondences to convert these bytes back into characters yields the plaintext Cubs. Similarly, the string of 32 bits used for the key in the following decryption would be just as likely the correct key.

cipher	00100001	11111100	00000100	01100101
key ⊕	01110011	10011001	01100000	00010110
plain	01010010	01100101	01100100	01110011

Using the ASCII correspondences to convert these bytes back into characters yields the plaintext Reds. The point is that were the key in the cipher truly random, any sequence of four ASCII characters would be just as likely as any other to be the correct plaintext. In other words, were the key in the cipher truly random, the ciphertext would give no information about the plaintext. Further, even if part of the plaintext were somehow determined, and part of the key thus revealed, nothing about the rest of the key could be deduced. The fact is that one-time pads, when implemented correctly, are *information-theoretically secure*, meaning they are unbreakable even by outsiders who have unlimited computing power. This absolute security of one-time pads was first proved by famed Russian information theorist Vladimir Kotelnikov, and then proved independently a short time later by American Claude Shannon, the Father of Information Theory.

One-time pads have been used in some important historical capacities. For example, the Moscow–Washington hotline, which was established in 1963 after the Cuban Missile Crisis and allows for direct communication between the leaders of Russia and the United States, has used one-time pads. However, the disadvantages of one-time pads have limited their use. One disadvantage is that often the key must be very long, and thus time consuming to create and difficult to transmit securely. Also, it would be a mistake to ever reuse a key, since any knowledge of one plaintext on which the key were used would give information about any other plaintext on which it were used. In addition, as we noted previously, generating a string of bits that is truly random can be very difficult. One-time pads also typically provide no direct assurance to the intended recipient of a message that it actually came from the originator claiming to have sent it. As a result of these and other disadvantages, encryption methods which make key distribution easier and are known or at least strongly believed to be secure against attacks involving current and anticipated technologies are usually preferred over one-time pads, despite their proven invincibility.

11.2.1   Exercises

1. Use a stream cipher with the given key to encrypt the following plaintexts.

   (a)* Key = 11011000, 11110110, 11001001;
   plaintext = Avs

   (b) Key = 11011000, 11110110, 11001001, 01101011;
   plaintext = Habs   .

   (c)* Key = 11011000, 11110110, 11001001, 01101011, 11111001;
   plaintext = Canes

(d) Key = 11011000, 11110110, 11001001, 01101011, 11111001, 00110111;
plaintext = Lanche

2. Decrypt the following ciphertexts, which were formed using a stream cipher with the given key.

(a) Key = 11011000, 11110110, 11001001;
ciphertext = 10001011, 10011001, 10110001

(b) Key = 11011000, 11110110, 11001001, 01101011;
ciphertext = 10010110, 10010111, 10111101, 00011000

(c) Key = 11011000, 11110110, 11001001, 01101011, 11111001;
ciphertext = 10001000, 10011110, 10100000, 00000111, 10001010

(d) Key = 11011000, 11110110, 11001001, 01101011, 11111001, 00110111;
ciphertext = 10011010, 10000011, 10101010, 00001000, 10010110, 01000100

3. Use BBS as it is described on page 356 with the given values of p, q, and x to generate a pseudorandom string of eight bits.

(a)* $p = 7$, $q = 19$, $x = 5$

(b) $p = 67$, $q = 103$, $x = 50$

(c)* $p = 139$, $q = 307$, $x = 500$

(d) $p = 1019$, $q = 4003$, $x = 5000$

4.* Recall we mentioned in this section that BBS is more efficient if each x_k produces more than one bit, which can be done by simply converting x_k to binary, and then taking more than one bit from the end of this binary representation. Use this variation of BBS with the values of p, q, and x given in Exercise 3 and each x_k producing two bits to generate a pseudorandom string of eight bits.

5. Find some information about linear feedback shift registers and their use in generating pseudorandom strings of bits, and write a summary of your findings.

6. Find some information about one or more other methods for generating pseudorandom strings of bits in addition to BBS and linear feedback shift registers, and write a summary of your findings.

7. Find some information about a teletype machine with encryption for which Gilbert Vernam was awarded a patent, and write a summary of your findings.

8. Find some information about Claude Shannon, including some of his work in cryptology, and write a summary of your findings.

9. Find some information about the cryptographic method RC4, including the history of its development and one or more of its real-life uses, and write a summary of your findings.

10. Find some information about one or more other real-life uses of stream ciphers in addition to the cryptographic method RC4, and write a summary of your findings.

11. Find some information about one or more other real-life uses of one-time pads in addition to the Moscow–Washington hotline, and write a summary of your findings.

11.3   AES Preliminaries

From 1977 until 2001, the *Data Encryption Standard*, or *DES* for short, was a Federal Information Processing Standard for encryption. Throughout this time, DES was used extensively in electronic commerce and the banking industry both in the U.S. and internationally. However, as predicted, over time successful attacks against DES were developed. In 1997, the National Institute of Standards and Technology made an open request for candidates to replace DES. Fifteen designs were submitted and studied, from which five were selected in 1999 for further and more intense scrutiny. From these five finalists the algorithm *Rijndael*, named for its creators, Belgian cryptologists Joan Daemen and Vincent Rijmen, was chosen as the winner. In November, 2001, Rijndael was adopted as the *Advanced Encryption Standard*, or *AES* for short, becoming a new (and current, as of this writing) Federal Information Processing Standard for encryption.

AES ciphers are symmetric-key. They are also block ciphers, meaning encryption occurs in groups, or blocks, of more than one character at a time. More specifically, encryption in AES ciphers occurs in blocks of 16 characters, or, equivalently, 128 bits, at a time. AES ciphers use a variety of different steps and types of operations for encryption (which we will describe in detail in Section 11.4), and include features of substitution, transposition, shift, Hill, and stream ciphers.

In practice, AES, like DES, is implemented primarily using binary arithmetic. However, unlike DES, AES has a precise mathematical description that can be outlined using for the most part ideas that we presented previously in this book. This is our goal for the remainder of this chapter—to precisely describe the Advanced Encryption Standard mathematically using ideas that we presented previously in this book.

11.3.1 Plaintext Format

With AES ciphers, we will again convert plaintext messages into numeric form using the ASCII correspondences given in Table 9.1 on page 302. In addition, we will convert these decimal ASCII numbers into binary, as we demonstrated in Example 11.2 on page 347, expanding each to a full byte, as we demonstrated in Example 11.5 on page 348. For example, the character U, with corresponding ASCII decimal number 85, would be represented as the byte 01010101.

Encryption in AES ciphers occurs in blocks of 16 characters, or, equivalently, 128 bits, at a time. A plaintext containing 16 or fewer characters is formatted for an AES cipher as a single 4×4 matrix, called the *plaintext matrix*, which we will denote as P. To form P, we first pad the plaintext at the end if necessary so that it contains exactly 16 characters. Consider these 16 characters labeled in order as follows.

$$x_{1,1}\ x_{2,1}\ x_{3,1}\ x_{4,1}\ x_{1,2}\ x_{2,2}\ x_{3,2}\ x_{4,2}\ x_{1,3}\ x_{2,3}\ x_{3,3}\ x_{4,3}\ x_{1,4}\ x_{2,4}\ x_{3,4}\ x_{4,4}$$

Next, we convert each plaintext character $x_{i,j}$ into its corresponding ASCII decimal number $y_{i,j}$, yielding the following list.

$$y_{1,1}\ y_{2,1}\ y_{3,1}\ y_{4,1}\ y_{1,2}\ y_{2,2}\ y_{3,2}\ y_{4,2}\ y_{1,3}\ y_{2,3}\ y_{3,3}\ y_{4,3}\ y_{1,4}\ y_{2,4}\ y_{3,4}\ y_{4,4}$$

We then take these decimal numbers in order in groups of four, and place these groups as columns from left to right in a 4×4 matrix. That is, we form the following 4×4 matrix.

$$\begin{bmatrix} y_{1,1} & y_{1,2} & y_{1,3} & y_{1,4} \\ y_{2,1} & y_{2,2} & y_{2,3} & y_{2,4} \\ y_{3,1} & y_{3,2} & y_{3,3} & y_{3,4} \\ y_{4,1} & y_{4,2} & y_{4,3} & y_{4,4} \end{bmatrix}$$

Finally, we convert each decimal number $y_{i,j}$ in this matrix into its binary representation $p_{i,j}$, expanding each to a full byte. Thus, for the 16-character plaintext with which we began, the following is the plaintext matrix P.

$$P = \begin{bmatrix} p_{1,1} & p_{1,2} & p_{1,3} & p_{1,4} \\ p_{2,1} & p_{2,2} & p_{2,3} & p_{2,4} \\ p_{3,1} & p_{3,2} & p_{3,3} & p_{3,4} \\ p_{4,1} & p_{4,2} & p_{4,3} & p_{4,4} \end{bmatrix}$$

For a plaintext containing more than 16 characters, the plaintext must be split into blocks containing exactly 16 characters, with the last block (and only the last block) padded at the end if necessary so that it contains exactly 16 characters. Each block of 16 characters is then formatted as a 4×4 plaintext matrix, and encrypted separately using the AES encryption process (which we will describe in detail in Section 11.4).

Example 11.16 With the plaintext `Friday the 13th`, to form the plaintext matrix P, we begin by padding the plaintext at the end with an extra space so that it contains exactly 16 characters. Next, we convert each plaintext character into its corresponding ASCII decimal number.

F	r	i	d	a	y		t	h	e		1	3	t	h	
70	114	105	100	97	121	32	116	104	101	32	49	51	116	104	32

We then place these decimal numbers in order in groups of four as columns from left to right in a 4×4 matrix.

$$\begin{bmatrix} 70 & 97 & 104 & 51 \\ 114 & 121 & 101 & 116 \\ 105 & 32 & 32 & 104 \\ 100 & 116 & 49 & 32 \end{bmatrix}$$

Finally, we convert each decimal number in this matrix into its binary representation, expanding each to a full byte.

$$P = \begin{bmatrix} 01000110 & 01100001 & 01101000 & 00110011 \\ 01110010 & 01111001 & 01100101 & 01110100 \\ 01101001 & 00100000 & 00100000 & 01101000 \\ 01100100 & 01110100 & 00110001 & 00100000 \end{bmatrix}$$

This plaintext matrix P is now ready for the AES encryption process. We will describe this encryption process and demonstrate beginning with this plaintext matrix later in this chapter.   □

11.3.2   The S-Box

The *S-box* is used in the Advanced Encryption Standard to transform a given byte into another byte. This is done using Table 11.3 on page 363, which is itself called the *AES S-box*.

The AES S-box contains 16 rows and 16 columns, each labeled with the hexadecimal representations of the decimal numbers 0 through 15, in order. To use the S-box to transform a given input byte into an output byte, we first convert the input byte from binary into an input two-digit hexadecimal number (with a leading digit of 0, if only a single digit would normally be required). We then find an output two-digit hexadecimal number as the entry in the S-box in the row labeled with the first digit of the input hexadecimal number and the column labeled with the second digit of the input hexadecimal number. Finally, we convert this output two-digit hexadecimal number into binary, expanding to a full byte if necessary. The result is the output byte.

	0	1	2	3	4	5	6	7	8	9	A	B	C	D	E	F
0	63	7C	77	7B	F2	6B	6F	C5	30	01	67	2B	FE	D7	AB	76
1	CA	82	C9	7D	FA	59	47	F0	AD	D4	A2	AF	9C	A4	72	C0
2	B7	FD	93	26	36	3F	F7	CC	34	A5	E5	F1	71	D8	31	15
3	04	C7	23	C3	18	96	05	9A	07	12	80	E2	EB	27	B2	75
4	09	83	2C	1A	1B	6E	5A	A0	52	3B	D6	B3	29	E3	2F	84
5	53	D1	00	ED	20	FC	B1	5B	6A	CB	BE	39	4A	4C	58	CF
6	D0	EF	AA	FB	43	4D	33	85	45	F9	02	7F	50	3C	9F	A8
7	51	A3	40	8F	92	9D	38	F5	BC	B6	DA	21	10	FF	F3	D2
8	CD	0C	13	EC	5F	97	44	17	C4	A7	7E	3D	64	5D	19	73
9	60	81	4F	DC	22	2A	90	88	46	EE	B8	14	DE	5E	0B	DB
A	E0	32	3A	0A	49	06	24	5C	C2	D3	AC	62	91	95	E4	79
B	E7	C8	37	6D	8D	D5	4E	A9	6C	56	F4	EA	65	7A	AE	08
C	BA	78	25	2E	1C	A6	B4	C6	E8	DD	74	1F	4B	BD	8B	8A
D	70	3E	B5	66	48	03	F6	0E	61	35	57	B9	86	C1	1D	9E
E	E1	F8	98	11	69	D9	8E	94	9B	1E	87	E9	CE	55	28	DF
F	8C	A1	89	0D	BF	E6	42	68	41	99	2D	0F	B0	54	BB	16

Table 11.3   The AES S-box.

Example 11.17 To use the S-box to transform the input byte 01100101 into an output byte, we first use Table 11.2 on page 351 to convert 01100101 from binary into the hexadecimal number 65. The entry in the S-box in the row labeled 6, and the column labeled 5 is 4D. Using Table 11.2 to convert 4D from hexadecimal into binary gives the output byte 01001101.   □

With AES ciphers, we will need to be able to not only apply the S-box, but invert it as well. That is, in addition to finding the output byte into which the S-box would transform a given input byte, we will also need to be able to find the input byte that would be transformed by the S-box into a given output byte. This can be done using another table similar to Table 11.3, although it is virtually as easy to just reverse the steps for applying the S-box. More specifically, to find the input byte that would be transformed by the S-box into a given output byte, we first convert the output byte from binary into an output two-digit hexadecimal number. We then find an input two-digit hexadecimal number by writing together the row and column labels (in that order) of the position in the S-box in which the output two-digit hexadecimal number is located. Finally, we convert this input two-digit hexadecimal number into binary, expanding to a full byte if necessary. The result is the input byte.

Example 11.18 To find the input byte that would be transformed by the S-box into the output byte 01100101, we first use Table 11.2 on page 351 to convert 01100101 from binary into the hexadecimal number 65. The entry 65 is located in the S-box in the row labeled B, and the column labeled C. Using Table 11.2 to convert BC from hexadecimal into binary gives the input byte 10111100.   □

11.3.3   Key Format and Generation

Encryption in AES ciphers involves several repeated steps, or *rounds*, the number of which depends on the size of an initial key. Possible sizes for the initial key are 128, 192, or 256 bits, with respective number of encryption rounds 10, 12, or 14. A recursive key generation process, called the *key schedule*, is used to generate keys for the encryption rounds. In this section, we will describe the initial key format and key schedule.

Initial Key Format

The initial key for an AES cipher is formatted similarly to how a plaintext is formatted, as a matrix with four rows, with each entry consisting of one byte. Like a plaintext, an initial key is taken in order in groups of four bytes, with these groups placed as columns from left to right in a matrix. This matrix is called the *initial key matrix*, which we will denote as K. The size of the initial key matrix depends on the size of the initial key. Specifically, a 128-bit initial key yields a 4×4 initial key matrix, a 192-bit initial key a 4×6 initial key matrix, and a 256-bit initial key a 4×8 initial key matrix. For simplicity, we will limit our general discussion, examples, and computational exercises to 128-bit initial keys.

Initial keys can be formed using characters in a keyword, or a pseudo-random bit generator like BBS (see Section 11.2). For simplicity, we will only consider initial keys formed using keywords. For a 128-bit initial key, a 16-character keyword is required, and K is formed identically to how a plaintext matrix would be formed. That is, we first pad a keyword containing fewer than 16 characters at the end so that it contains exactly 16 characters. Consider these 16 characters labeled in order as follows.

$$x_{1,1}\ x_{2,1}\ x_{3,1}\ x_{4,1}\ x_{1,2}\ x_{2,2}\ x_{3,2}\ x_{4,2}\ x_{1,3}\ x_{2,3}\ x_{3,3}\ x_{4,3}\ x_{1,4}\ x_{2,4}\ x_{3,4}\ x_{4,4}$$

Next, we convert each keyword character $x_{i,j}$ into its corresponding ASCII decimal number $y_{i,j}$, yielding the following list.

$$y_{1,1}\ y_{2,1}\ y_{3,1}\ y_{4,1}\ y_{1,2}\ y_{2,2}\ y_{3,2}\ y_{4,2}\ y_{1,3}\ y_{2,3}\ y_{3,3}\ y_{4,3}\ y_{1,4}\ y_{2,4}\ y_{3,4}\ y_{4,4}$$

We then take these decimal numbers in order in groups of four, and place these groups as columns from left to right in a 4×4 matrix. That is, we form the following 4×4 matrix.

$$\begin{bmatrix} y_{1,1} & y_{1,2} & y_{1,3} & y_{1,4} \\ y_{2,1} & y_{2,2} & y_{2,3} & y_{2,4} \\ y_{3,1} & y_{3,2} & y_{3,3} & y_{3,4} \\ y_{4,1} & y_{4,2} & y_{4,3} & y_{4,4} \end{bmatrix}$$

Finally, we convert each decimal number $y_{i,j}$ in this matrix into its binary representation $k_{i,j}$, expanding each to a full byte. Thus, for the 16-character keyword with which we began, the following is the initial key matrix K.

$$K = \begin{bmatrix} k_{1,1} & k_{1,2} & k_{1,3} & k_{1,4} \\ k_{2,1} & k_{2,2} & k_{2,3} & k_{2,4} \\ k_{3,1} & k_{3,2} & k_{3,3} & k_{3,4} \\ k_{4,1} & k_{4,2} & k_{4,3} & k_{4,4} \end{bmatrix}$$

Example 11.19 With the keyword Jason Voorhees, to form the initial key matrix K, we begin by padding the keyword at the end with two extra spaces so that it contains exactly 16 characters. Next, we convert each keyword character into its corresponding ASCII decimal number.

J   a   s   o   n   V   o   o   r   h   e   e   s
74 97 115 111 110 32 86 111 111 114 104 101 101 115 32 32

We then place these decimal numbers in order in groups of four as columns from left to right in a 4×4 matrix.

$$\begin{bmatrix} 74 & 110 & 111 & 101 \\ 97 & 32 & 114 & 115 \\ 115 & 86 & 104 & 32 \\ 111 & 111 & 101 & 32 \end{bmatrix}$$

Finally, we convert each decimal number in this matrix into its binary representation, expanding each to a full byte.

$$K = \begin{bmatrix} 01001010 & 01101110 & 01101111 & 01100101 \\ 01100001 & 00100000 & 01110010 & 01110011 \\ 01110011 & 01010110 & 01101000 & 00100000 \\ 01101111 & 01101111 & 01100101 & 00100000 \end{bmatrix}$$

This initial key matrix K is now ready for the key schedule generation process. We will describe this key generation process and demonstrate beginning with this initial key matrix next.   □

Key Schedule

Recall that with a 128-bit initial key, the AES encryption process involves 10 rounds. To use the key schedule to generate the keys for these rounds, we begin by labeling the columns in the initial key matrix K as $\mathbf{k}_1$–$\mathbf{k}_4$. That is, we begin by considering the initial key matrix with columns labeled as follows.

$$K = \begin{bmatrix} \mathbf{k}_1 & \mathbf{k}_2 & \mathbf{k}_3 & \mathbf{k}_4 \end{bmatrix}$$

We then construct new columns $\mathbf{k}_5, \mathbf{k}_6, \ldots, \mathbf{k}_{44}$ recursively as follows. Given $\mathbf{k}_1, \mathbf{k}_2, \ldots, \mathbf{k}_j$, with $4 \leq j \leq 43$, we construct $\mathbf{k}_{j+1}$ using the following formula, which involves a transformation T that we will describe next.

$$\mathbf{k}_{j+1} = \mathbf{k}_{j-3} \oplus T(\mathbf{k}_j)$$

To describe the transformation T in this formula, consider the entries in $\mathbf{k}_j$ denoted as follows.

$$\mathbf{k}_j = \begin{bmatrix} k_{1,j} \\ k_{2,j} \\ k_{3,j} \\ k_{4,j} \end{bmatrix}$$

The result of $T(\mathbf{k}_j)$ depends on whether the subscript j is an integer multiple of the number of columns in the initial key matrix K, which is 4. If j is not an integer multiple of 4, then $T(\mathbf{k}_j) = \mathbf{k}_j$. If j is an integer multiple of 4, then $T(\mathbf{k}_j)$ is found using the following three steps.

1. The entries in $\mathbf{k}_j$ are each shifted up one position, with the entry at the top wrapping to the bottom. This yields the following new column.

$$\begin{bmatrix} k_{2,j} \\ k_{3,j} \\ k_{4,j} \\ k_{1,j} \end{bmatrix}$$

2. Next, the S-box is applied to each of the entries in the column that results from the first step. This yields the following new column.

$$\begin{bmatrix} s_{2,j} \\ s_{3,j} \\ s_{4,j} \\ s_{1,j} \end{bmatrix}$$

3. Finally, to the first entry (and only the first entry) in the column that results from the second step, a byte r_j called the *round constant* is added using the XOR operation. This yields the following result.

$$T(\mathbf{k}_j) = \begin{bmatrix} s_{2,j} \oplus r_j \\ s_{3,j} \\ s_{4,j} \\ s_{1,j} \end{bmatrix}$$

The round constant r_j depends on j. The value of r_j for each j for which a round constant is needed in the key schedule is shown in Table 11.4 on page 367.

j	r_j
4	00000001
8	00000010
12	00000100
16	00001000
20	00010000
24	00100000
28	01000000
32	10000000
36	00011011
40	00110110

Table 11.4 Round constants for the AES key schedule.

Recall again that with a 128-bit initial key, the AES encryption process involves 10 rounds. Each round uses a key matrix of size 4×4 formed from the columns resulting from the key schedule, taken in order four at a time starting with $\mathbf{k}_5$. That is, the key matrix used in the first round consists of columns $\mathbf{k}_5$–$\mathbf{k}_8$, the key matrix used in the second round consists of columns $\mathbf{k}_9$–$\mathbf{k}_{12}$, and so on, through the key matrix used in the tenth round, which consists of columns $\mathbf{k}_{41}$–$\mathbf{k}_{44}$. We will denote these key matrices by K_1, K_2, and so on, through K_{10}. In addition, the initial key matrix K is used in the encryption process before the first round begins. Since the initial key matrix consists of columns $\mathbf{k}_1$–$\mathbf{k}_4$, for consistency we will denote the initial key matrix by K_0. In summary, the key matrices K_i for $i = 0, 1, \ldots, 10$ that are used before and during the AES encryption rounds are defined as follows.

$$K_i = \begin{bmatrix} \mathbf{k}_{4i+1} & \mathbf{k}_{4i+2} & \mathbf{k}_{4i+3} & \mathbf{k}_{4i+4} \end{bmatrix}$$

Example 11.20 For the initial key matrix $K_0 = K$ in Example 11.19, we will construct the key matrix K_1 that would be used in the first round of the AES encryption process. We begin by labeling the columns in K_0 as $\mathbf{k}_1$–$\mathbf{k}_4$, and we must then form the new columns $\mathbf{k}_5$–$\mathbf{k}_8$. We have first that $\mathbf{k}_5 = \mathbf{k}_1 \oplus T(\mathbf{k}_4)$. To find $T(\mathbf{k}_4)$, since the subscript in $T(\mathbf{k}_4)$ is an integer multiple of 4, we must use the three steps listed on page 366. Note that $\mathbf{k}_4$ is the following column.

$$\mathbf{k}_4 = \begin{bmatrix} 01100101 \\ 01110011 \\ 00100000 \\ 00100000 \end{bmatrix}$$

For the first step, we shift each of the entries in $\mathbf{k}_4$ up one position, with the entry at the top wrapping to the bottom. This yields the following column.

$$\begin{bmatrix} 01110011 \\ 00100000 \\ 00100000 \\ 01100101 \end{bmatrix}$$

Next, we apply the S-box to each of the entries in the preceding column. This yields the following column.

$$\begin{bmatrix} 10001111 \\ 10110111 \\ 10110111 \\ 01001101 \end{bmatrix} \quad \text{(see Example 11.17)}$$

We then add to the first entry in the preceding column the round constant 00000001 for $j = 4$ from Table 11.4. This yields the following column.

$$T(\mathbf{k}_4) = \begin{bmatrix} 10001111 \oplus 00000001 \\ 10110111 \\ 10110111 \\ 01001101 \end{bmatrix}$$

$$= \begin{bmatrix} 10001110 \\ 10110111 \\ 10110111 \\ 01001101 \end{bmatrix}$$

Finally, we form $\mathbf{k}_5$ as follows.

$$\mathbf{k}_5 = \mathbf{k}_1 \oplus T(\mathbf{k}_4)$$

$$= \begin{bmatrix} 01001010 \\ 01100001 \\ 01110011 \\ 01101111 \end{bmatrix} \oplus \begin{bmatrix} 10001110 \\ 10110111 \\ 10110111 \\ 01001101 \end{bmatrix}$$

$$= \begin{bmatrix} 01001010 \oplus 10001110 \\ 01100001 \oplus 10110111 \\ 01110011 \oplus 10110111 \\ 01101111 \oplus 01001101 \end{bmatrix}$$

$$= \begin{bmatrix} 11000100 \\ 11010110 \\ 11000100 \\ 00100010 \end{bmatrix}$$

Next, we have that $\mathbf{k}_6 = \mathbf{k}_2 \oplus T(\mathbf{k}_5)$. To find $T(\mathbf{k}_5)$, since the subscript in $T(\mathbf{k}_5)$ is not an integer multiple of 4, then $T(\mathbf{k}_5) = \mathbf{k}_5$. Thus, we have the following.

$$
\begin{aligned}
\mathbf{k}_6 &= \mathbf{k}_2 \oplus T(\mathbf{k}_5) \\
&= \mathbf{k}_2 \oplus \mathbf{k}_5 \\
&= \begin{bmatrix} 01101110 \\ 00100000 \\ 01010110 \\ 01101111 \end{bmatrix} \oplus \begin{bmatrix} 11000100 \\ 11010110 \\ 11000100 \\ 00100010 \end{bmatrix} \\
&= \begin{bmatrix} 01101110 \oplus 11000100 \\ 00100000 \oplus 11010110 \\ 01010110 \oplus 11000100 \\ 01101111 \oplus 00100010 \end{bmatrix} \\
&= \begin{bmatrix} 10101010 \\ 11110110 \\ 10010010 \\ 01001101 \end{bmatrix}
\end{aligned}
$$

The columns $\mathbf{k}_7$ and $\mathbf{k}_8$ are constructed similarly to $\mathbf{k}_6$, and are as follows.

$$
\mathbf{k}_7 = \begin{bmatrix} 11000101 \\ 10000100 \\ 11111010 \\ 00101000 \end{bmatrix}
$$

$$
\mathbf{k}_8 = \begin{bmatrix} 10100000 \\ 11110111 \\ 11011010 \\ 00001000 \end{bmatrix}
$$

Now with $\mathbf{k}_5$–$\mathbf{k}_8$, we can form the key matrix K_1 that would be used in the first round of the AES encryption process as follows.

$$
\begin{aligned}
K_1 &= \begin{bmatrix} \mathbf{k}_5 & \mathbf{k}_6 & \mathbf{k}_7 & \mathbf{k}_8 \end{bmatrix} \\
&= \begin{bmatrix} 11000100 & 10101010 & 11000101 & 10100000 \\ 11010110 & 11110110 & 10000100 & 11110111 \\ 11000100 & 10010010 & 11111010 & 11011010 \\ 00100010 & 01001101 & 00101000 & 00001000 \end{bmatrix}
\end{aligned}
$$

We will describe the rounds in the AES encryption process and demonstrate with this key matrix later in this chapter. □

11.3.4   Exercises

1. For the following plaintexts, find the plaintext matrix or matrices for the AES encryption process, padding with space characters when necessary.

    (a)* Halloween

    (b) Child's Play

    (c) Dawn of the Dead

    (d)* A Nightmare on Elm Street

    (e) The Texas Chain Saw Massacre

2. For each of the following input bytes, use the AES S-box to transform the input byte into an output byte.

    (a)* 11100010

    (b) 01111001

    (c) 01101111

    (d)* 01110010

    (e) 10111001

3. For each of the following output bytes, find the input byte that would be transformed by the AES S-box into the output byte.

    (a)* 11100010

    (b) 01111001

    (c) 01101111

    (d)* 01110010

    (e) 10111001

4. For the following keywords with a 128-bit initial key, find the initial key matrix for the AES encryption process, padding with space characters when necessary.

    (a)* Michael Myers

    (b) Chucky Lee Ray

    (c) George A. Romero

    (d)* Freddy Krueger

    (e) Drayton Sawyer

5.* For each part of Exercise 4, use the AES key schedule to find the key matrix K_1 that would be used in the first round of the AES encryption process.

6.* For each part of Exercise 5, use the AES key schedule to find the key matrix K_2 that would be used in the second round of the AES encryption process.

7.* As a continuation of Example 11.20, use the AES key schedule to find the key matrix K_2 that would be used in the second round of the AES encryption process.

8. As a continuation of Exercise 7, use the AES key schedule to find the key matrix K_3 that would be used in the third round of the AES encryption process.

9. Recall that AES was created as a replacement for DES. Find some information about DES, including its development, how it operates, and some of the reasons why a replacement was necessary, and write a summary of your findings.

10. Find some information about the controversy surrounding the design of the S-boxes used in DES, and write a summary of your findings.

11. Recall that the algorithm Rijndael was chosen from among five finalists to be the Advanced Encryption Standard. Find some information about one or more of the other finalists, including their creators, how they operate, and some of the reasons why they were not chosen, and write a summary of your findings.

12. Recall that the Advanced Encryption Standard can use initial keys of sizes 128, 192, or 256 bits. The general discussion, examples, and computational exercises in this book are limited to 128-bit initial keys. Find some information about how the AES initial key format and key schedule differ for initial keys of sizes 192 and/or 256 bits (when compared to 128 bits), and write a summary of your findings. Include at least one example for illustration.

11.4 AES Encryption

Recall that AES ciphers use a variety of different steps and types of operations for encryption. In this section, we will give an overview of the full AES encryption process, and describe in detail the steps and types of operations it entails.

11.4.1 Overview

The AES encryption process includes the following four types of operations, each of which we will describe in detail in Section 11.4.2.

- *ByteSub* (*BS*): The entries in an input matrix are transformed using the S-box.

- *ShiftRow* (*SR*): The entries in an input matrix are shifted to the left by zero, one, two, or three positions, with the entries at the left wrapping to the right.

- *MixColumn* (*MC*): An input matrix is multiplied on the left by a fixed matrix.

- *AddRoundKey* (*ARK*): To an input matrix, a changing key matrix is added using the XOR operation.

Recall that plaintexts are formatted for AES ciphers in blocks of 128 bits, arranged one byte at a time in a 4×4 plaintext matrix. Recall also that with a 128-bit initial key, encryption in AES ciphers involves 10 repeated rounds. Assuming a 4×4 plaintext matrix P and a 4×4 initial key matrix K_0 resulting from a 128-bit initial key, the basic steps in the full AES encryption process are as follows.

1. Before round 1 begins, ARK is applied to P, using the initial key matrix K_0. For consistency, this initial application of ARK is often referred to as *round 0*.

2. For rounds 1–9, BS, SR, MC, and ARK are applied in order, each to the output of the previous operation, using for ARK in round i the key matrix K_i given by the key schedule.

3. For round 10, BS, SR, and ARK are applied in order, each to the output of the previous operation, using for ARK the key matrix K_{10} given by the key schedule.

Note that MixColumn is not included in round 10.[2] So in summary, the full AES encryption process with a 128-bit initial key is as follows.

$$
\begin{array}{ll}
\text{Round} \ \ 0: & \text{ARK} \\
\text{Round} \ \ 1: & \text{BS, SR, MC, ARK} \\
\text{Round} \ \ 2: & \text{BS, SR, MC, ARK} \\
\vdots & \\
\text{Round} \ \ 9: & \text{BS, SR, MC, ARK} \\
\text{Round} \ 10: & \text{BS, SR, ARK}
\end{array}
$$

---

[2]We will explain the reason for this in Section 11.5.

The encryption process ends with the conclusion of round 10, with the resulting 4×4 matrix being the final *ciphertext matrix*. The bytes that form the entries in this ciphertext matrix, when strung together one at a time reading down the columns taken from left to right, make the final 128-bit ciphertext. For convenience, when writing ciphertexts we will express these 16 bytes using their equivalent hexadecimal representations.

11.4.2   The Operations

In this section, we will describe the AES encryption operations in detail. We will also demonstrate these operations in rounds 0 and 1 of the AES encryption process as a continuation of Examples 11.16 on page 362 and 11.20 on page 367.

Suppose we wish to use the AES encryption process to encrypt a 128-bit plaintext using a 128-bit initial key. We would begin by forming a 4×4 plaintext matrix P as described in Section 11.3.1, and a 4×4 initial key matrix $K_0 = K$ as described in Section 11.3.3. Then, for round 0 in the process, we would apply AddRoundKey to P using the key matrix K_0, resulting in the following matrix A_0.

$$A_0 = K_0 \oplus P$$

Example 11.21 Suppose we wish to use the AES encryption process to encrypt the plaintext Friday the 13th using a 128-bit initial key with the keyword Jason Voorhees. Using space characters to pad the plaintext and keyword yields the plaintext matrix P in Example 11.16 and initial key matrix $K_0 = K$ in Example 11.19 on page 365. Then, for round 0 in the process, we form A_0 as follows.

$$
\begin{aligned}
A_0 \; = \; & K_0 \oplus P \\[6pt]
= \; & \begin{bmatrix}
01001010 & 01101110 & 01101111 & 01100101 \\
01100001 & 00100000 & 01110010 & 01110011 \\
01110011 & 01010110 & 01101000 & 00100000 \\
01101111 & 01101111 & 01100101 & 00100000
\end{bmatrix} \\[6pt]
\oplus \; & \begin{bmatrix}
01000110 & 01100001 & 01101000 & 00110011 \\
01110010 & 01111001 & 01100101 & 01110100 \\
01101001 & 00100000 & 00100000 & 01101000 \\
01100100 & 01110100 & 00110001 & 00100000
\end{bmatrix} \\[6pt]
= \; & \begin{bmatrix}
00001100 & 00001111 & 00000111 & 01010110 \\
00010011 & 01011001 & 00010111 & 00000111 \\
00011010 & 01110110 & 01001000 & 01001000 \\
00001011 & 00011011 & 01010100 & 00000000
\end{bmatrix}
\end{aligned}
$$

We are now ready to begin round 1 in the AES encryption process, with A_0 as the input matrix for the first operation.   □

ByteSub

For rounds 1–10 in the AES encryption process, each round i begins with a matrix A_{i-1} output at the end of the previous round. The first operation in the round is ByteSub, which transforms each entry in A_{i-1} into another entry using the S-box, and results in a matrix labeled B_i. In order to describe the next operation in the process, it will be convenient to denote the entries in this matrix as follows.

$$B_i = \begin{bmatrix} b_{i:1,1} & b_{i:1,2} & b_{i:1,3} & b_{i:1,4} \\ b_{i:2,1} & b_{i:2,2} & b_{i:2,3} & b_{i:2,4} \\ b_{i:3,1} & b_{i:3,2} & b_{i:3,3} & b_{i:3,4} \\ b_{i:4,1} & b_{i:4,2} & b_{i:4,3} & b_{i:4,4} \end{bmatrix}$$

Example 11.22 As a continuation of Example 11.21, we would begin round 1 in the AES encryption process by applying ByteSub to A_0. This requires transforming each entry in A_0 into another entry using the S-box, and results in the following matrix B_1.

$$B_1 = \begin{bmatrix} 11111110 & 01110110 & 11000101 & 10110001 \\ 01111101 & 11001011 & 11110000 & 11000101 \\ 10100010 & 00111000 & 01010010 & 01010010 \\ 00101011 & 10101111 & 00100000 & 01100011 \end{bmatrix}$$

We are now ready to continue round 1 in the AES encryption process, with B_1 as the input matrix for the next operation.   □

ShiftRow

For rounds 1–10 in the AES encryption process, after ByteSub, the next operation in the round is ShiftRow, which in each round i is applied to the matrix B_i output by ByteSub. ShiftRow leaves the entries in the first row of B_i unchanged, shifts the entries in the second row to the left by one position, shifts the entries in the third row to the left by two positions, and shifts the entries in the fourth row to the left by three positions, with the entries at the left of the last three rows wrapping to the right. This results in a matrix labeled C_i, which can be expressed using the same notation as in B_i as follows.

$$C_i = \begin{bmatrix} b_{i:1,1} & b_{i:1,2} & b_{i:1,3} & b_{i:1,4} \\ b_{i:2,2} & b_{i:2,3} & b_{i:2,4} & b_{i:2,1} \\ b_{i:3,3} & b_{i:3,4} & b_{i:3,1} & b_{i:3,2} \\ b_{i:4,4} & b_{i:4,1} & b_{i:4,2} & b_{i:4,3} \end{bmatrix}$$

Example 11.23 As a continuation of Example 11.22, we would continue round 1 in the AES encryption process by applying ShiftRow to B_1, which results in the following matrix C_1.

$$C_1 = \begin{bmatrix} 11111110 & 01110110 & 11000101 & 10110001 \\ 11001011 & 11110000 & 11000101 & 01111101 \\ 01010010 & 01010010 & 10100010 & 00111000 \\ 01100011 & 00101011 & 10101111 & 00100000 \end{bmatrix}$$

We are now ready to continue round 1 in the AES encryption process, with C_1 as the input matrix for the next operation. $\qquad\square$

MixColumn

For rounds 1–9 in the AES encryption process, after ShiftRow, the next operation in the round is MixColumn, which in each round i is applied to the matrix C_i output by ShiftRow. MixColumn multiplies C_i on the left by the following matrix M, which is expressed with entries in decimal form.

$$M = \begin{bmatrix} 2 & 3 & 1 & 1 \\ 1 & 2 & 3 & 1 \\ 1 & 1 & 2 & 3 \\ 3 & 1 & 1 & 2 \end{bmatrix}$$

The result is the following matrix D_i.

$$D_i = MC_i$$

While the matrix multiplication defining D_i is usual matrix multiplication, the addition and multiplication involving bit strings within this matrix multiplication are not usual addition and multiplication. The addition involving bit strings is the XOR operation. The multiplication involving bit strings, which we will denote with the symbol $\otimes$, depends on whether the decimal number in the multiplication is 1, 2, or 3. Multiplying a bit string by 1 is trivial, as it just returns the bit string. For example, $1 \otimes 01010010 = 01010010$. To multiply a bit string by 2 or 3, we convert the 2 or 3 into its binary representation 10 or 11, respectively, and form a long multiplication tableau, using the XOR operation for the addition at the end. This results in a bit string of length nine; however, we want the result of this operation to be a byte. To obtain a byte, if the leftmost bit in the bit string of length nine is 0, we discard this leftmost bit and take the remaining eight bits. If the leftmost bit in the bit string of length nine is 1, then we add 100011011 to the bit string of length nine using the XOR operation; the leftmost bit in the result will then be 0, which we discard and take the remaining eight bits.

Example 11.24 To find $2 \otimes 01110110$, we convert 2 into its binary representation 10, and form the following long multiplication tableau.

$$
\begin{array}{r}
0\ 1\ 1\ 1\ 0\ 1\ 1\ 0 \\
\otimes \qquad\qquad 1\ 0 \\
\hline
0\ 0\ 0\ 0\ 0\ 0\ 0\ 0 \\
\oplus\ \ 0\ 1\ 1\ 1\ 0\ 1\ 1\ 0 \\
\hline
0\ 1\ 1\ 1\ 0\ 1\ 1\ 0\ 0
\end{array}
$$

Since the leftmost bit in the result is 0, we discard it to get the final answer 11101100. That is, $2 \otimes 01110110 = 11101100$. $\square$

Example 11.25 To find $3 \otimes 11110000$, we convert 3 into its binary representation 11, and form the following long multiplication tableau.

$$
\begin{array}{r}
1\ 1\ 1\ 1\ 0\ 0\ 0\ 0 \\
\otimes \qquad\qquad 1\ 1 \\
\hline
1\ 1\ 1\ 1\ 0\ 0\ 0\ 0 \\
\oplus\ \ 1\ 1\ 1\ 1\ 0\ 0\ 0\ 0 \\
\hline
1\ 0\ 0\ 0\ 1\ 0\ 0\ 0\ 0
\end{array}
$$

Since the leftmost bit in the result is 1, we compute the following.

$$
\begin{array}{r}
1\ 0\ 0\ 0\ 1\ 0\ 0\ 0\ 0 \\
\oplus\ \ 1\ 0\ 0\ 0\ 1\ 1\ 0\ 1\ 1 \\
\hline
0\ 0\ 0\ 0\ 0\ 1\ 0\ 1\ 1
\end{array}
$$

Discarding the 0 in the leftmost position of the result gives the final answer 00001011. That is, $3 \otimes 11110000 = 00001011$. $\square$

For convenience, the results of multiplying every possible byte by the decimal numbers 2 and 3 using the $\otimes$ operation are shown in Tables 11.5 and 11.6 on page 377. So that these tables will fit in these pages, the information in the tables is condensed into hexadecimal format. To demonstrate how the tables can be used, consider $2 \otimes 01110110$, which we found in Example 11.24. To find this using Table 11.5, we begin by converting 01110110 into its hexadecimal representation 76. Then, with this two-digit hexadecimal number, we find the entry in the table in the row labeled with the first digit 7 and the column labeled with the second digit 6. This entry is EC. Converting EC from hexadecimal into binary gives the final answer 11101100. Similarly, consider $3 \otimes 11110000$, which we found in Example 11.25. To

2⊗	0	1	2	3	4	5	6	7	8	9	A	B	C	D	E	F
0	00	02	04	06	08	0A	0C	0E	10	12	14	16	18	1A	1C	1E
1	20	22	24	26	28	2A	2C	2E	30	32	34	36	38	3A	3C	3E
2	40	42	44	46	48	4A	4C	4E	50	52	54	56	58	5A	5C	5E
3	60	62	64	66	68	6A	6C	6E	70	72	74	76	78	7A	7C	7E
4	80	82	84	86	88	8A	8C	8E	90	92	94	96	98	9A	9C	9E
5	A0	A2	A4	A6	A8	AA	AC	AE	B0	B2	B4	B6	B8	BA	BC	BE
6	C0	C2	C4	C6	C8	CA	CC	CE	D0	D2	D4	D6	D8	DA	DC	DE
7	E0	E2	E4	E6	E8	EA	EC	EE	F0	F2	F4	F6	F8	FA	FC	FE
8	1B	19	1F	1D	13	11	17	15	0B	09	0F	0D	03	01	07	05
9	3B	39	3F	3D	33	31	37	35	2B	29	2F	2D	23	21	27	25
A	5B	59	5F	5D	53	51	57	55	4B	49	4F	4D	43	41	47	45
B	7B	79	7F	7D	73	71	77	75	6B	69	6F	6D	63	61	67	65
C	9B	99	9F	9D	93	91	97	95	8B	89	8F	8D	83	81	87	85
D	BB	B9	BF	BD	B3	B1	B7	B5	AB	A9	AF	AD	A3	A1	A7	A5
E	DB	D9	DF	DD	D3	D1	D7	D5	CB	C9	CF	CD	C3	C1	C7	C5
F	FB	F9	FF	FD	F3	F1	F7	F5	EB	E9	EF	ED	E3	E1	E7	E5

Table 11.5   Results of multiplying by the decimal number 2 using ⊗.

3⊗	0	1	2	3	4	5	6	7	8	9	A	B	C	D	E	F
0	00	03	06	05	0C	0F	0A	09	18	1B	1E	1D	14	17	12	11
1	30	33	36	35	3C	3F	3A	39	28	2B	2E	2D	24	27	22	21
2	60	63	66	65	6C	6F	6A	69	78	7B	7E	7D	74	77	72	71
3	50	53	56	55	5C	5F	5A	59	48	4B	4E	4D	44	47	42	41
4	C0	C3	C6	C5	CC	CF	CA	C9	D8	DB	DE	DD	D4	D7	D2	D1
5	F0	F3	F6	F5	FC	FF	FA	F9	E8	EB	EE	ED	E4	E7	E2	E1
6	A0	A3	A6	A5	AC	AF	AA	A9	B8	BB	BE	BD	B4	B7	B2	B1
7	90	93	96	95	9C	9F	9A	99	88	8B	8E	8D	84	87	82	81
8	9B	98	9D	9E	97	94	91	92	83	80	85	86	8F	8C	89	8A
9	AB	A8	AD	AE	A7	A4	A1	A2	B3	B0	B5	B6	BF	BC	B9	BA
A	FB	F8	FD	FE	F7	F4	F1	F2	E3	E0	E5	E6	EF	EC	E9	EA
B	CB	C8	CD	CE	C7	C4	C1	C2	D3	D0	D5	D6	DF	DC	D9	DA
C	5B	58	5D	5E	57	54	51	52	43	40	45	46	4F	4C	49	4A
D	6B	68	6D	6E	67	64	61	62	73	70	75	76	7F	7C	79	7A
E	3B	38	3D	3E	37	34	31	32	23	20	25	26	2F	2C	29	2A
F	0B	08	0D	0E	07	04	01	02	13	10	15	16	1F	1C	19	1A

Table 11.6   Results of multiplying by the decimal number 3 using ⊗.

find this using Table 11.6, we convert 11110000 into its hexadecimal representation F0, and find the entry in the table in the row labeled F and the column labeled 0, which is 0B. Converting 0B from hexadecimal into binary gives the final answer 00001011.

Example 11.26 As a continuation of Example 11.23, we would continue round 1 in the AES encryption process by applying MixColumn to C_1. To make Tables 11.5 and 11.6 easier to use, we will first convert the entries in C_1 into hexadecimal. This yields the following representation of C_1.

$$C_1 = \begin{bmatrix} FE & 76 & C5 & B1 \\ CB & F0 & C5 & 7D \\ 52 & 52 & A2 & 38 \\ 63 & 2B & AF & 20 \end{bmatrix}$$

Then the formula defining D_i on page 375 gives the following.

$$D_1 = MC_1$$

$$= \begin{bmatrix} 2 & 3 & 1 & 1 \\ 1 & 2 & 3 & 1 \\ 1 & 1 & 2 & 3 \\ 3 & 1 & 1 & 2 \end{bmatrix} \begin{bmatrix} FE & 76 & C5 & B1 \\ CB & F0 & C5 & 7D \\ 52 & 52 & A2 & 38 \\ 63 & 2B & AF & 20 \end{bmatrix}$$

$$= \begin{bmatrix} 10010000 & 10011110 & 11001000 & 11100110 \\ 11100110 & 01010000 & 00000110 & 00100011 \\ 00110100 & 01011111 & 10110101 & 11011100 \\ 01000110 & 01101110 & 01110110 & 11001101 \end{bmatrix}$$

As a demonstration of the matrix product MC_1 that results in D_1, consider the entry in the first row and second column of D_1. To find this entry, we would multiply the first row of M times the second column of C_1 as follows.

$$(2 \otimes 76) \oplus (3 \otimes F0) \oplus (1 \otimes 52) \oplus (1 \otimes 2B)$$

$$= EC \oplus 0B \oplus 52 \oplus 2B$$

$$= 11101100 \oplus 00001011 \oplus 01010010 \oplus 00101011$$

$$= 10011110$$

We are now ready to continue round 1 in the AES encryption process, with D_1 as the input matrix for the next operation.   □

AddRoundKey

For rounds 1–9 in the AES encryption process, after MixColumn, the final operation in the round is AddRoundKey, which in each round i is applied to the matrix D_i output by MixColumn. AddRoundKey uses the XOR operation to add to D_i the key matrix K_i given by the key schedule, resulting in the following matrix A_i.

$$A_i = K_i \oplus D_i$$

Example 11.27 As a continuation of Example 11.26, we would continue round 1 in the AES encryption process by applying AddRoundKey to D_1, using the key matrix K_1 in Example 11.20 on page 367. This results in the following matrix A_1.

$$
\begin{aligned}
A_1 &= K_1 \oplus D_1 \\[6pt]
&= \begin{bmatrix}
11000100 & 10101010 & 11000101 & 10100000 \\
11010110 & 11110110 & 10000100 & 11110111 \\
11000100 & 10010010 & 11111010 & 11011010 \\
00100010 & 01001101 & 00101000 & 00001000
\end{bmatrix} \\[6pt]
&\oplus \begin{bmatrix}
10010000 & 10011110 & 11001000 & 11100110 \\
11100110 & 01010000 & 00000110 & 00100011 \\
00110100 & 01011111 & 10110101 & 11011100 \\
01000110 & 01101110 & 01110110 & 11001101
\end{bmatrix} \\[6pt]
&= \begin{bmatrix}
01010100 & 00110100 & 00001101 & 01000110 \\
00110000 & 10100110 & 10000010 & 11010100 \\
11110000 & 11001101 & 01001111 & 00000110 \\
01100100 & 00100011 & 01011110 & 11000101
\end{bmatrix}
\end{aligned}
$$

This completes round 1 in the AES encryption process, and we are ready to begin round 2, with A_1 as the input matrix for the first operation. $\quad\square$

Recall that the AES encryption process with a 128-bit initial key involves 10 rounds, but MixColumn is not included in the last round. Thus, for round 10 (and round 10 only), after ShiftRow, the next operation in the round, which is also the final operation in the full AES encryption process, is AddRoundKey, applied to the matrix C_{10} output by ShiftRow. This results in the following final ciphertext matrix A_{10}.

$$A_{10} = K_{10} \oplus C_{10}$$

Example 11.28 The full AES encryption process encrypts the plaintext `Friday the 13th` using a 128-bit initial key with the keyword `Jason Voorhees` to the following ciphertext matrix A_{10}.

$$
A_{10} = \begin{bmatrix}
10001110 & 11001001 & 10100100 & 10101110 \\
11111111 & 00100111 & 11101010 & 10001110 \\
01101100 & 10011111 & 11101110 & 11100011 \\
01111011 & 01100010 & 01010100 & 10011110
\end{bmatrix}
$$

Recall that the bytes that form the entries in this ciphertext matrix, when strung together one at a time reading down the columns taken from left to right, make the final 128-bit ciphertext. Recall also that for convenience, we will express these 16 bytes using their equivalent hexadecimal representations. Converting the entries in A_{10} into hexadecimal yields the following representation of A_{10}.

$$A_{10} = \begin{bmatrix} 8E & C9 & A4 & AE \\ FF & 27 & EA & 8E \\ 6C & 9F & EE & E3 \\ 7B & 62 & 54 & 9E \end{bmatrix}$$

Thus, the final ciphertext is 8E FF 6C 7B C9 27 9F 62 A4 EA EE 54 AE 8E E3 9E. $\Box$

11.4.3   Exercises

1. For each of the following bytes, find the result of multiplying the byte by the decimal number 2 using the $\otimes$ operation.

   (a)* 01101011

   (b) 01100011

   (c) 10111110

   (d)* 11000101

   (e) 11001010

2. For each of the following bytes, find the result of multiplying the byte by the decimal number 3 using the $\otimes$ operation.

   (a)* 01100011

   (b) 01011000

   (c) 11110010

   (d)* 11000101

   (e) 10101111

3.* Suppose you wish to use the AES encryption process to encrypt the plaintext Halloween using a 128-bit initial key with the keyword Michael Myers, padding both with space characters. (This is the plaintext used in Exercise 1a in Section 11.3, and the keyword used in Exercises 4a, 5a, and 6a in Section 11.3.)

   (a) Find the matrix A_0 that would be output at the end of round 0.

    (b) Find the matrix B_1 that would be output by ByteSub during round 1.

    (c) Find the matrix C_1 that would be output by ShiftRow during round 1.

    (d) Find the matrix D_1 that would be output by MixColumn during round 1.

    (e) Find the matrix A_1 that would be output at the end of round 1.

    (f) Find the matrix A_2 that would be output at the end of round 2.

4. Suppose you wish to use the AES encryption process to encrypt the plaintext Child's Play using a 128-bit initial key with the keyword Chucky Lee Ray, padding both with space characters. (This is the plaintext used in Exercise 1b in Section 11.3, and the keyword used in Exercises 4b, 5b, and 6b in Section 11.3.)

    (a) Find the matrix A_0 that would be output at the end of round 0.

    (b) Find the matrix B_1 that would be output by ByteSub during round 1.

    (c) Find the matrix C_1 that would be output by ShiftRow during round 1.

    (d) Find the matrix D_1 that would be output by MixColumn during round 1.

    (e) Find the matrix A_1 that would be output at the end of round 1.

    (f) Find the matrix A_2 that would be output at the end of round 2.

5. Suppose you wish to use the AES encryption process to encrypt the plaintext Dawn of the Dead using a 128-bit initial key with the keyword George A. Romero. (This is the plaintext used in Exercise 1c in Section 11.3, and the keyword used in Exercises 4c, 5c, and 6c in Section 11.3.)

    (a) Find the matrix A_0 that would be output at the end of round 0.

    (b) Find the matrix B_1 that would be output by ByteSub during round 1.

    (c) Find the matrix C_1 that would be output by ShiftRow during round 1.

    (d) Find the matrix D_1 that would be output by MixColumn during round 1.

    (e) Find the matrix A_1 that would be output at the end of round 1.

    (f) Find the matrix A_2 that would be output at the end of round 2.

6.* Suppose you wish to use the AES encryption process to encrypt the plaintext A Nightmare on E using a 128-bit initial key with the keyword Freddy Krueger, padding the keyword with space characters. (This is the first 16 characters in the plaintext used in Exercise 1d in Section 11.3, and the keyword used in Exercises 4d, 5d, and 6d in Section 11.3.)

   (a) Find the matrix A_0 that would be output at the end of round 0.

   (b) Find the matrix B_1 that would be output by ByteSub during round 1.

   (c) Find the matrix C_1 that would be output by ShiftRow during round 1.

   (d) Find the matrix D_1 that would be output by MixColumn during round 1.

   (e) Find the matrix A_1 that would be output at the end of round 1.

   (f) Find the matrix A_2 that would be output at the end of round 2.

7. Suppose you wish to use the AES encryption process to encrypt the plaintext The Texas Chain using a 128-bit initial key with the keyword Drayton Sawyer, padding both with space characters. (This is the first 15 characters in the plaintext used in Exercise 1e in Section 11.3, and the keyword used in Exercises 4e, 5e, and 6e in Section 11.3.)

   (a) Find the matrix A_0 that would be output at the end of round 0.

   (b) Find the matrix B_1 that would be output by ByteSub during round 1.

   (c) Find the matrix C_1 that would be output by ShiftRow during round 1.

   (d) Find the matrix D_1 that would be output by MixColumn during round 1.

   (e) Find the matrix A_1 that would be output at the end of round 1.

   (f) Find the matrix A_2 that would be output at the end of round 2.

8.* This exercise is a continuation of Example 11.27.

   (a) Find the matrix B_2 that would be output by ByteSub during round 2.

   (b) Find the matrix C_2 that would be output by ShiftRow during round 2.

   (c) Find the matrix D_2 that would be output by MixColumn during round 2.

   (d) Find the matrix A_2 that would be output at the end of round 2.

9. As a continuation of Exercise 8, find the matrix A_3 that would be output at the end of round 3.

11.5   AES Decryption

To decrypt a ciphertext that was formed using the AES encryption process, we must use the inverses of the encryption operations. The inverses of the operations ByteSub, ShiftRow, MixColumn, and AddRoundKey, respectively, can be described as follows.

- *InvByteSub* (*IBS*): The entries in an input matrix are transformed using the inverse of the S-box.

- *InvShiftRow* (*ISR*): The entries in an input matrix are shifted to the right by zero, one, two, or three positions, with the entries at the right wrapping to the left.

- *InvMixColumn* (*IMC*): An input matrix is multiplied on the left by a fixed matrix.

- AddRoundKey is its own inverse.

More specifically, in a matrix of bytes, InvByteSub considers each entry as output by the S-box, finds the input byte that would be transformed by the S-box into this output byte, and replaces the output byte with this input byte. InvShiftRow leaves the entries in the first row of a matrix unchanged, shifts the entries in the second row to the right by one position, shifts the entries in the third row to the right by two positions, and shifts the entries in the fourth row to the right by three positions, with the entries at the right of the last three rows wrapping to the left. InvMixColumn multiplies a matrix of bytes on the left by the following matrix M^{-1}, which is expressed with entries in decimal form, and is the inverse of the matrix M on page 375 (with respect to usual matrix multiplication, using the XOR and $\otimes$ operations for the addition and multiplication within this matrix multiplication).

$$M^{-1} = \begin{bmatrix} 14 & 11 & 13 & 9 \\ 9 & 14 & 11 & 13 \\ 13 & 9 & 14 & 11 \\ 11 & 13 & 9 & 14 \end{bmatrix}$$

As with MixColumn, the matrix multiplication involving M^{-1} in InvMixColumn is the usual matrix multiplication, and the addition involving bit strings within this matrix multiplication is the XOR operation. The multiplication involving bit strings within this matrix multiplication is an extension of the same operation in MixColumn, and we will also denote it with the symbol $\otimes$. To multiply a bit string by 9, 11, 13, or 14, we convert the 9, 11, 13, or 14 into its binary representation 1001, 1011, 1101, or 1110, respectively, and form a long multiplication tableau, using the XOR operation for the addition at the end. This results in a bit string of length eleven; however, we want the result of this operation to be a byte. To obtain a byte, we use the following three steps.

1. We first turn the bit string of length eleven into a bit string of length ten as follows. If the leftmost bit in the bit string of length eleven is 0, we discard this leftmost bit and take the remaining ten bits. If the leftmost bit in the bit string of length eleven is 1, then we add 10001101100 to the bit string of length eleven using the XOR operation; the leftmost bit in the result will then be 0, which we discard and take the remaining ten bits.

2. Next, we turn the bit string of length ten into a bit string of length nine as follows. If the leftmost bit in the bit string of length ten is 0, we discard this leftmost bit and take the remaining nine bits. If the leftmost bit in the bit string of length ten is 1, then we add 1000110110 to the bit string of length ten using the XOR operation; the leftmost bit in the result will then be 0, which we discard and take the remaining nine bits.

3. Finally, we turn the bit string of length nine into a byte exactly as in MixColumn. That is, if the leftmost bit in the bit string of length nine is 0, we discard this leftmost bit and take the remaining eight bits. If the leftmost bit in the bit string of length nine is 1, then we add 100011011 to the bit string of length nine using the XOR operation; the leftmost bit in the result will then be 0, which we discard and take the remaining eight bits.

Example 11.29 To find $9 \otimes 10100110$, we convert 9 into its binary representation 1001, and form the following long multiplication tableau.

$$
\begin{array}{r}
1\,0\,1\,0\,0\,1\,1\,0 \\
\otimes \quad\quad\quad 1\,0\,0\,1 \\
\hline
1\,0\,1\,0\,0\,1\,1\,0 \\
0\,0\,0\,0\,0\,0\,0\,0 \\
0\,0\,0\,0\,0\,0\,0\,0 \\
\oplus\; 1\,0\,1\,0\,0\,1\,1\,0 \\
\hline
1\,0\,1\,1\,0\,0\,1\,0\,1\,1\,0
\end{array}
$$

Since the leftmost bit in the result is 1, we compute the following.

$$1\ 0\ 1\ 1\ 0\ 0\ 1\ 0\ 1\ 1\ 0$$
$$\oplus\ 1\ 0\ 0\ 0\ 1\ 1\ 0\ 1\ 1\ 0\ 0$$

$$0\ 0\ 1\ 1\ 1\ 1\ 1\ 1\ 0\ 1\ 0$$

Discarding the 0 in the leftmost position of the result gives the bit string 0111111010 of length ten. Since the leftmost bit in this bit string of length ten is 0, we discard it to obtain the bit string 111111010 of length nine. Since the leftmost bit in this bit string of length nine is 1, we compute the following.

$$1\ 1\ 1\ 1\ 1\ 1\ 0\ 1\ 0$$
$$\oplus\ 1\ 0\ 0\ 0\ 1\ 1\ 0\ 1\ 1$$

$$0\ 1\ 1\ 1\ 0\ 0\ 0\ 0\ 1$$

Discarding the 0 in the leftmost position of the result gives the final answer 11100001; that is, $9 \otimes 10100110 = 11100001$. □

For convenience, the results of multiplying every possible byte by the decimal numbers 9 and 11 using the $\otimes$ operation are shown in Tables 11.7 and 11.8 on page 386, and by the decimal numbers 13 and 14 in Tables 11.9 and 11.10 on page 387. These tables are formatted and can be used in the same way as Tables 11.5 and 11.6 on page 377, which we used previously for multiplying bytes by the decimal numbers 2 and 3. For example, consider $9 \otimes 10100110$, which we found in Example 11.29. To find this using Table 11.7, we begin by converting 10100110 into its hexadecimal representation A6. Then, with this two-digit hexadecimal number, we find the entry in the table in the row labeled with the first digit A and the column labeled with the second digit 6. This entry is E1. Converting E1 from hexadecimal into binary gives the final answer 11100001.

Recall that the full AES encryption process with a 128-bit initial key can be summarized as follows.

Encryption Round 0:	ARK
Encryption Round 1:	BS, SR, MC, ARK
Encryption Round 2:	BS, SR, MC, ARK
$\vdots$	
Encryption Round 9:	BS, SR, MC, ARK
Encryption Round 10:	BS, SR, ARK

To decrypt a ciphertext that was formed using this encryption process, we must apply the inverses of these operations in the reverse order starting at the bottom. Thus, the full AES decryption process with a 128-bit initial key can be summarized as follows.

9⊗	0	1	2	3	4	5	6	7	8	9	A	B	C	D	E	F
0	00	09	12	1B	24	2D	36	3F	48	41	5A	53	6C	65	7E	77
1	90	99	82	8B	B4	BD	A6	AF	D8	D1	CA	C3	FC	F5	EE	E7
2	3B	32	29	20	1F	16	0D	04	73	7A	61	68	57	5E	45	4C
3	AB	A2	B9	B0	8F	86	9D	94	E3	EA	F1	F8	C7	CE	D5	DC
4	76	7F	64	6D	52	5B	40	49	3E	37	2C	25	1A	13	08	01
5	E6	EF	F4	FD	C2	CB	D0	D9	AE	A7	BC	B5	8A	83	98	91
6	4D	44	5F	56	69	60	7B	72	05	0C	17	1E	21	28	33	3A
7	DD	D4	CF	C6	F9	F0	EB	E2	95	9C	87	8E	B1	B8	A3	AA
8	EC	E5	FE	F7	C8	C1	DA	D3	A4	AD	B6	BF	80	89	92	9B
9	7C	75	6E	67	58	51	4A	43	34	3D	26	2F	10	19	02	0B
A	D7	DE	C5	CC	F3	FA	E1	E8	9F	96	8D	84	BB	B2	A9	A0
B	47	4E	55	5C	63	6A	71	78	0F	06	1D	14	2B	22	39	30
C	9A	93	88	81	BE	B7	AC	A5	D2	DB	C0	C9	F6	FF	E4	ED
D	0A	03	18	11	2E	27	3C	35	42	4B	50	59	66	6F	74	7D
E	A1	A8	B3	BA	85	8C	97	9E	E9	E0	FB	F2	CD	C4	DF	D6
F	31	38	23	2A	15	1C	07	0E	79	70	6B	62	5D	54	4F	46

Table 11.7   Results of multiplying by the decimal number 9 using ⊗.

11⊗	0	1	2	3	4	5	6	7	8	9	A	B	C	D	E	F
0	00	0B	16	1D	2C	27	3A	31	58	53	4E	45	74	7F	62	69
1	B0	BB	A6	AD	9C	97	8A	81	E8	E3	FE	F5	C4	CF	D2	D9
2	7B	70	6D	66	57	5C	41	4A	23	28	35	3E	0F	04	19	12
3	CB	C0	DD	D6	E7	EC	F1	FA	93	98	85	8E	BF	B4	A9	A2
4	F6	FD	E0	EB	DA	D1	CC	C7	AE	A5	B8	B3	82	89	94	9F
5	46	4D	50	5B	6A	61	7C	77	1E	15	08	03	32	39	24	2F
6	8D	86	9B	90	A1	AA	B7	BC	D5	DE	C3	C8	F9	F2	EF	E4
7	3D	36	2B	20	11	1A	07	0C	65	6E	73	78	49	42	5F	54
8	F7	FC	E1	EA	DB	D0	CD	C6	AF	A4	B9	B2	83	88	95	9E
9	47	4C	51	5A	6B	60	7D	76	1F	14	09	02	33	38	25	2E
A	8C	87	9A	91	A0	AB	B6	BD	D4	DF	C2	C9	F8	F3	EE	E5
B	3C	37	2A	21	10	1B	06	0D	64	6F	72	79	48	43	5E	55
C	01	0A	17	1C	2D	26	3B	30	59	52	4F	44	75	7E	63	68
D	B1	BA	A7	AC	9D	96	8B	80	E9	E2	FF	F4	C5	CE	D3	D8
E	7A	71	6C	67	56	5D	40	4B	22	29	34	3F	0E	05	18	13
F	CA	C1	DC	D7	E6	ED	F0	FB	92	99	84	8F	BE	B5	A8	A3

Table 11.8   Results of multiplying by the decimal number 11 using ⊗.

13⊗	0	1	2	3	4	5	6	7	8	9	A	B	C	D	E	F
0	00	0D	1A	17	34	39	2E	23	68	65	72	7F	5C	51	46	4B
1	D0	DD	CA	C7	E4	E9	FE	F3	B8	B5	A2	AF	8C	81	96	9B
2	BB	B6	A1	AC	8F	82	95	98	D3	DE	C9	C4	E7	EA	FD	F0
3	6B	66	71	7C	5F	52	45	48	03	0E	19	14	37	3A	2D	20
4	6D	60	77	7A	59	54	43	4E	05	08	1F	12	31	3C	2B	26
5	BD	B0	A7	AA	89	84	93	9E	D5	D8	CF	C2	E1	EC	FB	F6
6	D6	DB	CC	C1	E2	EF	F8	F5	BE	B3	A4	A9	8A	87	90	9D
7	06	0B	1C	11	32	3F	28	25	6E	63	74	79	5A	57	40	4D
8	DA	D7	C0	CD	EE	E3	F4	F9	B2	BF	A8	A5	86	8B	9C	91
9	0A	07	10	1D	3E	33	24	29	62	6F	78	75	56	5B	4C	41
A	61	6C	7B	76	55	58	4F	42	09	04	13	1E	3D	30	27	2A
B	B1	BC	AB	A6	85	88	9F	92	D9	D4	C3	CE	ED	E0	F7	FA
C	B7	BA	AD	A0	83	8E	99	94	DF	D2	C5	C8	EB	E6	F1	FC
D	67	6A	7D	70	53	5E	49	44	0F	02	15	18	3B	36	21	2C
E	0C	01	16	1B	38	35	22	2F	64	69	7E	73	50	5D	4A	47
F	DC	D1	C6	CB	E8	E5	F2	FF	B4	B9	AE	A3	80	8D	9A	97

Table 11.9   Results of multiplying by the decimal number 13 using ⊗.

14⊗	0	1	2	3	4	5	6	7	8	9	A	B	C	D	E	F
0	00	0E	1C	12	38	36	24	2A	70	7E	6C	62	48	46	54	5A
1	E0	EE	FC	F2	D8	D6	C4	CA	90	9E	8C	82	A8	A6	B4	BA
2	DB	D5	C7	C9	E3	ED	FF	F1	AB	A5	B7	B9	93	9D	8F	81
3	3B	35	27	29	03	0D	1F	11	4B	45	57	59	73	7D	6F	61
4	AD	A3	B1	BF	95	9B	89	87	DD	D3	C1	CF	E5	EB	F9	F7
5	4D	43	51	5F	75	7B	69	67	3D	33	21	2F	05	0B	19	17
6	76	78	6A	64	4E	40	52	5C	06	08	1A	14	3E	30	22	2C
7	96	98	8A	84	AE	A0	B2	BC	E6	E8	FA	F4	DE	D0	C2	CC
8	41	4F	5D	53	79	77	65	6B	31	3F	2D	23	09	07	15	1B
9	A1	AF	BD	B3	99	97	85	8B	D1	DF	CD	C3	E9	E7	F5	FB
A	9A	94	86	88	A2	AC	BE	B0	EA	E4	F6	F8	D2	DC	CE	C0
B	7A	74	66	68	42	4C	5E	50	0A	04	16	18	32	3C	2E	20
C	EC	E2	F0	FE	D4	DA	C8	C6	9C	92	80	8E	A4	AA	B8	B6
D	0C	02	10	1E	34	3A	28	26	7C	72	60	6E	44	4A	58	56
E	37	39	2B	25	0F	01	13	1D	47	49	5B	55	7F	71	63	6D
F	D7	D9	CB	C5	EF	E1	F3	FD	A7	A9	BB	B5	9F	91	83	8D

Table 11.10   Results of multiplying by the decimal number 14 using ⊗.

Decryption Round 0:   ARK, ISR, IBS
Decryption Round 1:   ARK, IMC, ISR, IBS
Decryption Round 2:   ARK, IMC, ISR, IBS

$\vdots$

Decryption Round 9:   ARK, IMC, ISR, IBS
Decryption Round 10:   ARK

However, because the Advanced Encryption Standard was never intended to be done by hand, and in practice is exclusively automated, it would be desirable if the steps in the AES decryption process could be written to more closely resemble the steps in the encryption process. That way, the automation necessary to decrypt a ciphertext would more closely resemble the automation necessary to form the ciphertext.

Fortunately, it is possible to write the steps in the AES decryption process so that they more closely resemble the steps in the encryption process. To see this, note first that since ByteSub and ShiftRow operate on specific entries in an input matrix, their order could be reversed in the encryption process without changing the process itself. Similarly, the order of InvShiftRow and InvByteSub can be reversed in the decryption process without changing the process itself. It is not possible, though, to directly reverse the order of AddRoundKey and InvMixColumn in the decryption process without changing the process itself. However, through a little matrix algebra we can identify an alternative way to reverse their order. For rounds 1–9 in the encryption process, consider the matrices C_i output by ShiftRow, D_i by MixColumn, K_i by the key schedule, and A_i by AddRoundKey. Then the formulas defining A_i on page 379 and D_i on page 375 can be combined, resulting in the following equation.

$$A_i = K_i \oplus MC_i$$

Solving for C_i in this equation yields the following.

$$MC_i = K_i \oplus A_i$$
$$C_i = M^{-1}(K_i \oplus A_i)$$
$$C_i = M^{-1}K_i \oplus M^{-1}A_i$$

The term $M^{-1}A_i$ in this last equation represents InvMixColumn applied to A_i. Thus, if we define *InvAddRoundKey* (*IARK*) to be addition of the decryption key matrix $M^{-1}K_i$ to the input matrix $M^{-1}A_i$ using the XOR operation, then we can invert the encryption operations "MC, ARK" with "IMC, IARK." This change, along with replacing "ISR, IBS" with "IBS, ISR," which we noted above is allowable, and moving "IBS, ISR" from the end of each decryption round 1–9 to the start of the following round 2–10, yields the following equivalent summary of the steps in the full AES decryption process with a 128-bit initial key.

Decryption Round 0: ARK
Decryption Round 1: IBS, ISR, IMC, IARK
Decryption Round 2: IBS, ISR, IMC, IARK
$$\vdots$$
Decryption Round 9: IBS, ISR, IMC, IARK
Decryption Round 10: IBS, ISR, ARK

In comparison with the steps in the full AES encryption process with a 128-bit initial key, note that these decryption steps do indeed very closely resemble the encryption steps, with ByteSub, ShiftRow, and MixColumn replaced by their inverses, and AddRoundKey replaced by InvAddRound-Key in rounds 1–9.[3] Also, of course, in the decryption process the key matrices given by the key schedule would be used in the reverse of the order in which they were used in the encryption process. That is, assuming a 4×4 ciphertext matrix A_{10} and a 4×4 initial key matrix K_0 resulting from a 128-bit initial key, the basic steps in the full AES decryption process can be described as follows.

1. For round 0, ARK is applied to A_{10}, using the key matrix K_{10} given by the key schedule.

2. For rounds 1–9, IBS, ISR, IMC, and IARK are applied in order, each to the output of the previous operation, using for IARK in decryption round i the decryption key matrix $M^{-1}K_{10-i}$, with K_{10-i} given by the key schedule.

3. For round 10, IBS, ISR, and ARK are applied in order, each to the output of the previous operation, using for ARK the initial key matrix K_0.

The decryption process ends with the conclusion of round 10, with the resulting 4×4 matrix being the plaintext matrix P.

We will demonstrate the operations in the AES decryption process in the following examples. For simplicity, we will begin with a matrix A_1 formed from a plaintext using rounds 0 and 1 in the encryption process, and show how the last part of the decryption process can be used to recover the plaintext. Since rounds 0 and 1 in the encryption process include five operations, namely ARK, BS, SR, MC, ARK, in order, we will recover the plaintext by applying the last five operations in the decryption process,

---

[3]Our discussion in this section indicates why MixColumn is not included in round 10 of the AES encryption process. If MixColumn were included in this round, then encryption round 10 would be "BS, SR, MC, ARK." This would make decryption round 0, after reversing operation orders and moving IBS and ISR to decryption round 1, be "IMC, IARK," which would leave an extra InvMixColumn at the beginning of the decryption process.

namely IMC, IARK, IBS, ISR, ARK, in order. We will do this one operation at a time in the following five examples.

Example 11.30 Consider the following matrix A_1 from Example 11.27 on page 379.

$$A_1 = \begin{bmatrix} 01010100 & 00110100 & 00001101 & 01000110 \\ 00110000 & 10100110 & 10000010 & 11010100 \\ 11110000 & 11001101 & 01001111 & 00000110 \\ 01100100 & 00100011 & 01011110 & 11000101 \end{bmatrix}$$

In this example, we will begin the process of recovering the plaintext by applying InvMixColumn to A_1. To make Tables 11.7–11.10 easier to use, we first convert the entries in A_1 into hexadecimal. This yields the following representation of A_1.

$$A_1 = \begin{bmatrix} 54 & 34 & 0D & 46 \\ 30 & A6 & 82 & D4 \\ F0 & CD & 4F & 06 \\ 64 & 23 & 5E & C5 \end{bmatrix}$$

We can then apply InvMixColumn to A_1 as follows.

$$M^{-1}A_1 = \begin{bmatrix} 14 & 11 & 13 & 9 \\ 9 & 14 & 11 & 13 \\ 13 & 9 & 14 & 11 \\ 11 & 13 & 9 & 14 \end{bmatrix} \begin{bmatrix} 54 & 34 & 0D & 46 \\ 30 & A6 & 82 & D4 \\ F0 & CD & 4F & 06 \\ 64 & 23 & 5E & C5 \end{bmatrix}$$

$$= \begin{bmatrix} 00001011 & 01110011 & 00011001 & 10001101 \\ 11010001 & 11100011 & 01011100 & 11000000 \\ 01010100 & 01110010 & 01111100 & 01101111 \\ 01111110 & 10011110 & 10100111 & 01110011 \end{bmatrix}$$

As a demonstration of this matrix product $M^{-1}A_1$, consider the entry in the third row and second column of the result. To find this entry, we would multiply the third row of M^{-1} by the second column of A_1 as follows.

$$(13 \otimes 34) \oplus (9 \otimes A6) \oplus (14 \otimes CD) \oplus (11 \otimes 23)$$
$$= 5F \oplus E1 \oplus AA \oplus 66$$
$$= 01011111 \oplus 11100001 \oplus 10101010 \oplus 01100110$$
$$= 01110010$$

We are now ready to continue the process of recovering the plaintext, with $M^{-1}A_1$ as the input matrix for the next operation. □

Example 11.31 As a continuation of Example 11.30, we would continue the process of recovering the plaintext by applying InvAddRoundKey to $M^{-1}A_1$, using the decryption key matrix $M^{-1}K_1$, with K_1 from Example 11.20 on page 367. Again, to make Tables 11.7–11.10 easier to use, we first convert the entries in K_1 into hexadecimal. This yields the following representation of K_1.

$$K_1 = \begin{bmatrix} C4 & AA & C5 & A0 \\ D6 & F6 & 84 & F7 \\ C4 & 92 & FA & DA \\ 22 & 4D & 28 & 08 \end{bmatrix}$$

We can then form $M^{-1}K_1$ as follows.

$$M^{-1}K_1 = \begin{bmatrix} 14 & 11 & 13 & 9 \\ 9 & 14 & 11 & 13 \\ 13 & 9 & 14 & 11 \\ 11 & 13 & 9 & 14 \end{bmatrix} \begin{bmatrix} C4 & AA & C5 & A0 \\ D6 & F6 & 84 & F7 \\ C4 & 92 & FA & DA \\ 22 & 4D & 28 & 08 \end{bmatrix}$$

$$= \begin{bmatrix} 11110101 & 00000101 & 11011100 & 00111100 \\ 00011010 & 00010011 & 10011001 & 10111101 \\ 00000110 & 00100000 & 11011110 & 01010111 \\ 00011101 & 10110101 & 00001000 & 01010011 \end{bmatrix}$$

We can then apply InvAddRoundKey to $M^{-1}A_1$ as follows. Note that the result is the matrix C_1 in Example 11.23 on page 375.

$$C_1 = M^{-1}K_1 \oplus M^{-1}A_1$$

$$= \begin{bmatrix} 11110101 & 00000101 & 11011100 & 00111100 \\ 00011010 & 00010011 & 10011001 & 10111101 \\ 00000110 & 00100000 & 11011110 & 01010111 \\ 00011101 & 10110101 & 00001000 & 01010011 \end{bmatrix}$$

$$\oplus \begin{bmatrix} 00001011 & 01110011 & 00011001 & 10001101 \\ 11010001 & 11100011 & 01011100 & 11000000 \\ 01010100 & 01110010 & 01111100 & 01101111 \\ 01111110 & 10011110 & 10100111 & 01110011 \end{bmatrix}$$

$$= \begin{bmatrix} 11111110 & 01110110 & 11000101 & 10110001 \\ 11001011 & 11110000 & 11000101 & 01111101 \\ 01010010 & 01010010 & 10100010 & 00111000 \\ 01100011 & 00101011 & 10101111 & 00100000 \end{bmatrix}$$

We are now ready to continue the process of recovering the plaintext, with C_1 as the input matrix for the next operation. $\square$

Example 11.32 As a continuation of Example 11.31, we would continue the process of recovering the plaintext by applying InvByteSub to C_1, resulting in the following, which we will denote IBS(C_1).

$$\text{IBS}(C_1) = \begin{bmatrix} 00001100 & 00001111 & 00000111 & 01010110 \\ 01011001 & 00010111 & 00000111 & 00010011 \\ 01001000 & 01001000 & 00011010 & 01110110 \\ 00000000 & 00001011 & 00011011 & 01010100 \end{bmatrix}$$

We are now ready to continue the process of recovering the plaintext, with this as the input matrix for the next operation. □

Example 11.33 As a continuation of Example 11.32, we would continue the process of recovering the plaintext by applying InvShiftRow to IBS(C_1), which results in the following. Note that the result is the matrix A_0 in Example 11.21 on page 373.

$$A_0 = \begin{bmatrix} 00001100 & 00001111 & 00000111 & 01010110 \\ 00010011 & 01011001 & 00010111 & 00000111 \\ 00011010 & 01110110 & 01001000 & 01001000 \\ 00001011 & 00011011 & 01010100 & 00000000 \end{bmatrix}$$

We are now ready to continue the process of recovering the plaintext, with A_0 as the input matrix for the last operation. □

Example 11.34 As a continuation of Example 11.33, we would complete the process of recovering the plaintext by first applying AddRoundKey to A_0, using the initial key matrix $K_0 = K$ in Example 11.19 on page 365, which results in the following. Note that the result is the plaintext matrix P in Example 11.16 on page 362.

$$
\begin{aligned}
P &= K_0 \oplus A_0 \\[4pt]
&= \begin{bmatrix} 01001010 & 01101110 & 01101111 & 01100101 \\ 01100001 & 00100000 & 01110010 & 01110011 \\ 01110011 & 01010110 & 01101000 & 00100000 \\ 01101111 & 01101111 & 01100101 & 00100000 \end{bmatrix} \\[4pt]
&\oplus \begin{bmatrix} 00001100 & 00001111 & 00000111 & 01010110 \\ 00010011 & 01011001 & 00010111 & 00000111 \\ 00011010 & 01110110 & 01001000 & 01001000 \\ 00001011 & 00011011 & 01010100 & 00000000 \end{bmatrix} \\[4pt]
&= \begin{bmatrix} 01000110 & 01100001 & 01101000 & 00110011 \\ 01110010 & 01111001 & 01100101 & 01110100 \\ 01101001 & 00100000 & 00100000 & 01101000 \\ 01100100 & 01110100 & 00110001 & 00100000 \end{bmatrix}
\end{aligned}
$$

We then convert the entries in P into decimal form, which yields the following.

$$\begin{bmatrix} 70 & 97 & 104 & 51 \\ 114 & 121 & 101 & 116 \\ 105 & 32 & 32 & 104 \\ 100 & 116 & 49 & 32 \end{bmatrix}$$

These decimal numbers, when strung together one at a time reading down the columns taken from left to right, yields 70 114 105 100 97 121 32 116 104 101 32 49 51 116 104 32. Converting each decimal number into its corresponding ASCII character finally gives the plaintext: `Friday the 13th.` □

11.5.1 Exercises

1. For each of the following bytes, find the result of multiplying the byte by the decimal number 9 using the $\otimes$ operation.

   (a)* 01011110

   (b) 00111110

   (c) 10011111

2. For each of the following bytes, find the result of multiplying the byte by the decimal number 11 using the $\otimes$ operation.

   (a)* 01000110

   (b) 11101101

   (c) 10000011

3. For each of the following bytes, find the result of multiplying the byte by the decimal number 13 using the $\otimes$ operation.

   (a)* 01001001

   (b) 01110010

   (c) 10011000

4. For each of the following bytes, find the result of multiplying the byte by the decimal number 14 using the $\otimes$ operation.

   (a)* 10010101

   (b) 00110010

   (c) 11110010

5. Consider the following matrix A_1, which was formed from a 128-bit plaintext by applying rounds 0 and 1 in the AES encryption process, using a 128-bit initial key with the keyword **Sean Connery**, padded with space characters, for which the following matrices K_0 and K_1 are the initial key matrix and first key matrix given by the key schedule.

$$A_1 = \begin{bmatrix} 10010101 & 11011111 & 10011101 & 01011001 \\ 01000110 & 01001010 & 00000011 & 11001001 \\ 01001001 & 11000000 & 11110011 & 01011010 \\ 01011110 & 10010110 & 01100101 & 10010101 \end{bmatrix}$$

$$K_0 = \begin{bmatrix} 01010011 & 00100000 & 01101110 & 00100000 \\ 01100101 & 01000011 & 01100101 & 00100000 \\ 01100001 & 01101111 & 01110010 & 00100000 \\ 01101110 & 01101110 & 01111001 & 00100000 \end{bmatrix}$$

$$K_1 = \begin{bmatrix} 11100101 & 11000101 & 10101011 & 10001011 \\ 11010010 & 10010001 & 11110100 & 11010100 \\ 11010110 & 10111001 & 11001011 & 11101011 \\ 11011001 & 10110111 & 11001110 & 11101110 \end{bmatrix}$$

(a)* Find the matrix $M^{-1}A_1$ that results from applying InvMixColumn to A_1.

(b) Find the decryption key matrix $M^{-1}K_1$.

(c)* Find the matrix C_1 that results from applying InvAddRoundKey to $M^{-1}A_1$.

(d) Find the matrix $\text{IBS}(C_1)$ that results from applying InvByteSub to C_1.

(e)* Find the matrix A_0 that results from applying InvShiftRow to $\text{IBS}(C_1)$.

(f) Find the plaintext matrix P that results from applying AddRoundKey to A_0, and then find the plaintext.

6. Consider the following matrix A_1, which was formed from a 128-bit plaintext by applying rounds 0 and 1 in the AES encryption process, using a 128-bit initial key with the keyword **Roger Moore**, padded with space characters, for which the following matrices K_0 and K_1 are the initial key matrix and first key matrix given by the key schedule.

$$A_1 = \begin{bmatrix} 00110010 & 10111001 & 01101110 & 11100110 \\ 11101101 & 11100001 & 10111110 & 00101101 \\ 01110010 & 01100010 & 00011100 & 00000011 \\ 00111110 & 10100000 & 00000110 & 10000010 \end{bmatrix}$$

$$K_0 = \begin{bmatrix} 01010010 & 01110010 & 01101111 & 00100000 \\ 01101111 & 00100000 & 01110010 & 00100000 \\ 01100111 & 01001101 & 01100101 & 00100000 \\ 01100101 & 01101111 & 00100000 & 00100000 \end{bmatrix}$$

$$K_1 = \begin{bmatrix} 11100100 & 10010110 & 11111001 & 11011001 \\ 11011000 & 11111000 & 10001010 & 10101010 \\ 11010000 & 10011101 & 11111000 & 11011000 \\ 11010010 & 10111101 & 10011101 & 10111101 \end{bmatrix}$$

(a)* Find the matrix $M^{-1}A_1$ that results from applying InvMixColumn to A_1.

(b) Find the decryption key matrix $M^{-1}K_1$.

(c)* Find the matrix C_1 that results from applying InvAddRoundKey to $M^{-1}A_1$.

(d) Find the matrix IBS(C_1) that results from applying InvByteSub to C_1.

(e)* Find the matrix A_0 that results from applying InvShiftRow to IBS(C_1).

(f) Find the plaintext matrix P that results from applying AddRoundKey to A_0, and then find the plaintext.

7. Consider the following matrix A_1, which was formed from a 128-bit plaintext by applying rounds 0 and 1 in the AES encryption process, using a 128-bit initial key with the keyword **Pierce Brosnan**, padded with space characters, for which the following matrices K_0 and K_1 are the initial key matrix and first key matrix given by the key schedule.

$$A_1 = \begin{bmatrix} 11110010 & 01111011 & 10011011 & 00101011 \\ 10000011 & 10000110 & 01100000 & 10000001 \\ 10011000 & 10100000 & 00111000 & 10011110 \\ 10011111 & 00100010 & 01011101 & 01101111 \end{bmatrix}$$

$$K_0 = \begin{bmatrix} 01010000 & 01100011 & 01110010 & 01100001 \\ 01101001 & 01100101 & 01101111 & 01101110 \\ 01100101 & 00100000 & 01110011 & 00100000 \\ 01110010 & 01000010 & 01101110 & 00100000 \end{bmatrix}$$

$$K_1 = \begin{bmatrix} 11001110 & 10101101 & 11011111 & 10111110 \\ 11011110 & 10111011 & 11010100 & 10111010 \\ 11010010 & 11110010 & 10000001 & 10100001 \\ 10011101 & 11011111 & 10110001 & 10010001 \end{bmatrix}$$

(a) Find the matrix $M^{-1}A_1$ that results from applying InvMixColumn to A_1.

(b) Find the decryption key matrix $M^{-1}K_1$.

(c) Find the matrix C_1 that results from applying InvAddRoundKey to $M^{-1}A_1$.

(d) Find the matrix IBS(C_1) that results from applying InvByteSub to C_1.

(e) Find the matrix A_0 that results from applying InvShiftRow to IBS(C_1).

(f) Find the plaintext matrix P that results from applying Add-RoundKey to A_0, and then find the plaintext.

8.* Consider the following matrix A_2, which was formed from a 128-bit plaintext by applying rounds 0–2 in the AES encryption process, using a 128-bit initial key with the keyword Michael Myers, padded with space characters. (This is the keyword used in Exercises 4a, 5a, and 6a in Section 11.3.) Find the plaintext.

$$A_2 = \begin{bmatrix} 10101101 & 11001010 & 00001101 & 10011000 \\ 00101100 & 11001000 & 01011100 & 10110001 \\ 11000110 & 00010010 & 10101000 & 11101000 \\ 10110101 & 10010110 & 01110000 & 11110000 \end{bmatrix}$$

9. Consider the following matrix A_2, which was formed from a 128-bit plaintext by applying rounds 0–2 in the AES encryption process, using a 128-bit initial key with the keyword George A. Romero. (This is the keyword used in Exercises 4c, 5c, and 6c in Section 11.3.) Find the plaintext.

$$A_2 = \begin{bmatrix} 00001101 & 00000010 & 11001000 & 10101110 \\ 00001011 & 10011001 & 10011000 & 01111010 \\ 10000010 & 01101001 & 10100101 & 10101100 \\ 10010110 & 10001010 & 00001011 & 00011000 \end{bmatrix}$$

10. Show that the matrix M^{-1} on page 383 is the inverse of the matrix M on page 375 (with respect to usual matrix multiplication, using the XOR and $\otimes$ operations for the addition and multiplication within this matrix multiplication).

11.6   AES Security

When the algorithm Rijndael was adopted as the Advanced Encryption Standard, the number of encryption rounds was chosen to be at least ten, because beyond six, no attacks against the algorithm were known that would be faster than brute force. It was believed at the time that at least four extra encryption rounds beyond six would successfully prevent all possible attacks against the system for many years. Also, it was shown at the time that the machines used to break DES would take trillions of years to break Rijndael, even if Rijndael were implemented using the smallest initial key size of 128 bits. In addition, it would be very easy to strengthen Rijndael by doing nothing more than increasing the number of encryption rounds. Even with expected advances in technology, AES has the potential to remain secure well past the twenty years that spanned the utility of DES.

We should also emphasize that AES ciphers are symmetric-key, and thus have the deficiency that users must have a secure way to exchange the initial key. There are effective methods for overcoming this, though. For example, two parties wishing to communicate secretly over an insecure communication line using an AES cipher could first use a public-key cipher to exchange the initial key for the AES cipher, and then proceed with the AES cipher as usual. However, why then would the parties not just use a public-key cipher for all their communication? The answer to this lies in the first sentence of this chapter—in practice known public-key ciphers are in general much slower than known non-public-key ciphers. This is why it is common, for parties needing to exchange a large amount of information, to first use a slower public-key cipher to exchange the initial key for an AES cipher, and then proceed with the faster AES cipher to actually exchange the information.

11.6.1 Exercises

1. Find some information about one or more cryptanalytic attacks against AES, and write a summary of your findings.

2. Find some information about one or more techniques for securely exchanging AES initial keys, and write a summary of your findings.

3. Find some information about one or more real-life uses of AES, and write a summary of your findings.

4. The ByteSub, ShiftRow, and MixColumn operations, as well as the format of the key schedule, were each part of the Rijndael algorithm for specific reasons. Find some information about one or more of these reasons, and write a summary of your findings.

Chapter 12

Message Authentication

As we have noted, public and symmetric-key ciphers each have advantages over the other. One important advantage of symmetric-key ciphers is that they are less susceptible to exploitation related to message authentication. For example, when Whitfield Diffie and Martin Hellman first explained their idea of public-key ciphers, the way they envisioned public-key ciphers could be used most effectively was by a group of people who all wished to be able to communicate with each other spontaneously across a series of insecure communication lines. By way of illustration, suppose the group would like to use RSA ciphers to encrypt their messages. Each member of the group could choose their own personal encryption keys e and m, and then make these keys public knowledge by, say, publishing them online. Then, whenever one member of the group wanted to send another a secret message, they could use the intended recipient's public encryption keys to encrypt the message. A problem could result, though—the intended recipient may have no way of verifying that the received message was really sent by the person claiming to have sent it, and not by an impostor posing as the person claiming to have sent it. This problem can be overcome through the use of a *digital signature*, a method for authenticating, or "signing," a message to verify it was really sent by the person claiming to have sent it.

Another problem related to message authentication in public-key ciphers is that the originator of a message may have no way to verify that the public keys published by the intended recipient were really published by the intended recipient, and not by an impostor posing as the intended recipient. This problem can be overcome through the use of a well-defined *public-key infrastructure*, a method for defining procedures for generating and publishing public keys that binds them to particular individuals.

In this chapter, we will consider some specific ways in which these potential problems can be overcome. Message authentication has been a very significant area of cryptologic research and study for many years, and has only increased in significance over the past few decades with the advent of public-key ciphers. In fact, it is interesting to note that for the paper [20] in which Ron Rivest, Adi Shamir, and Len Adleman published the first specific type of public-key cipher, they chose the title *A Method for Obtaining Digital Signatures and Public-Key Cryptosystems*, giving the notion of a digital signature precedence over that of a public-key cipher.

12.1   RSA Signatures

Suppose you wish to send a numeric message x electronically to a colleague, and while you are not actually concerned with keeping the message itself confidential (i.e., x can be sent unencrypted, or *in the clear*), your colleague would like to have some assurance that the message is really from you, and not from an impostor posing as you. One way to do this is as follows. Suppose you have published RSA encryption exponent e_o and modulus m_o (with m_o greater than x), while keeping the corresponding decryption exponent d_o secret. You could then use d_o and m_o to "sign" the message by forming $s = x^{d_o} \bmod m_o$, and send both x and the "signature" s to your colleague. Upon receipt, your colleague could obviously read the message x, as it was not encrypted. However, using s your colleague could also obtain some assurance that the message is really from you. Just as your secret d_o undoes what the publicly known e_o does in an RSA cipher, e_o dually undoes what d_o does. That is, your colleague can use the publicly known e_o and m_o to form $\overline{x} = s^{e_o} \bmod m_o$, and check to see if x and $\overline{x}$ match. Only if $x \neq \overline{x}$ should your colleague be suspicious. If $x = \overline{x}$, and your RSA keys are secure, your colleague can be sure that the message is really from you.

The description above of how an RSA cipher can be used to authenticate a message is called the *RSA signature* scheme. In summary, the following are the basic steps in the RSA signature scheme.

1. The originator of a numeric message x publishes RSA encryption exponent e_o and modulus m_o, and keeps the corresponding decryption exponent d_o secret.

2. Suppose x is expressed as one or more positive integers x_i less than m_o. Then, for each message integer x_i, the originator forms signature s_i with the following calculation.

$$s_i = x_i^{d_o} \bmod m_o$$

The originator then sends the message integer(s) x_i and corresponding signature(s) s_i to the intended recipient.

3. For each signature s_i, the recipient verifies the signature with the following calculation.
$$\overline{x}_i = s_i^{e_o} \bmod m_o$$
If $x_i = \overline{x}_i$, and the originator's RSA keys are secure, the recipient can be sure that the message integer x_i is from the originator.

Example 12.1 Suppose you wish to send the message PIN # 9089 electronically to a colleague in the clear, but you and your colleague would like to use the RSA signature scheme to give some assurance to your colleague that the message is really from you, and not from an impostor posing as you. Suppose also that you have published the RSA encryption exponent $e_o = 67$ and modulus $m_o = 9169$, and kept the corresponding decryption exponent $d_o = 267$ secret. With the ASCII correspondences given in Table 9.1 on page 302, your message converts into the list of integers 80, 73, 78, 32, 35, 32, 57, 48, 56, 57, or, equivalently, the numeric message $x = 80737832353257485657$. With the value $m_o = 9169$, you can split x into the message integers $x_1 = 8073$, $x_2 = 7832$, $x_3 = 3532$, $x_4 = 5748$, and $x_5 = 5657$. Then, with the values $d_o = 267$ and $m_o = 9169$, you sign each message integer x_i by forming the following signature s_i.

$$
\begin{aligned}
x_1 = 8073 &\quad\rightarrow\quad s_1 = 8073^{267} \bmod 9169 = 5465 \\
x_2 = 7832 &\quad\rightarrow\quad s_2 = 7832^{267} \bmod 9169 = 4036 \\
x_3 = 3532 &\quad\rightarrow\quad s_3 = 3532^{267} \bmod 9169 = 5644 \\
x_4 = 5748 &\quad\rightarrow\quad s_4 = 5748^{267} \bmod 9169 = 7827 \\
x_5 = 5657 &\quad\rightarrow\quad s_5 = 5657^{267} \bmod 9169 = 2489
\end{aligned}
$$

You then send the message integers x_i and corresponding signatures s_i to your colleague. With the values $e_o = 67$ and $m_o = 9169$, your colleague verifies each signature s_i by forming the following quantity $\overline{x}_i$.

$$
\begin{aligned}
s_1 = 5465 &\quad\rightarrow\quad \overline{x}_1 = 5465^{67} \bmod 9169 = 8073 \\
s_2 = 4036 &\quad\rightarrow\quad \overline{x}_2 = 4036^{67} \bmod 9169 = 7832 \\
s_3 = 5644 &\quad\rightarrow\quad \overline{x}_3 = 5644^{67} \bmod 9169 = 3532 \\
s_4 = 7827 &\quad\rightarrow\quad \overline{x}_4 = 7827^{67} \bmod 9169 = 5748 \\
s_5 = 2489 &\quad\rightarrow\quad \overline{x}_5 = 2489^{67} \bmod 9169 = 5657
\end{aligned}
$$

Since each x_i and $\overline{x}_i$ match, assuming (for the sake of illustration) your RSA keys are secure, your colleague can be sure that the message integers x_i are from you. □

Note that in Example 12.1, the values of your RSA modulus m_o, each x_i, and each $s_i = x_i^{d_o} \bmod m_o$ would obviously be known to your colleague, but would also have to be assumed to be known publicly if they were transmitted over an insecure communication line. Would this reveal the value of your decryption exponent d_o? Only if your colleague or someone intercepting your transmissions could solve the discrete logarithm problem. The general difficulty of finding discrete logarithms, the same problem that we noted in Sections 10.2 and 10.4 gives the Diffie-Hellman key exchange and ElGamal ciphers their high level of security, protects your decryption exponent in this example, and the originator's decryption exponent in the RSA signature scheme in general.

It is obviously possible to modify the RSA signature scheme to include encryption. The following steps summarize one way in which this can be done, with both the message and signature(s) encrypted. We will call this the RSA signature scheme *with encryption*.

1. The originator of a numeric message x publishes RSA encryption exponent e_o and modulus m_o, and keeps the corresponding decryption exponent d_o secret. Meanwhile, the intended recipient publishes RSA encryption exponent e_r and modulus m_r, and keeps the corresponding decryption exponent d_r secret. Suppose also that m_r is greater than m_o.

2. Suppose x is expressed as one or more positive integers x_i less than m_o. Then, for each plaintext integer x_i, the originator forms ciphertext integer y_i with the following calculation.

$$y_i = x_i^{e_r} \bmod m_r$$

Next, the originator forms encrypted signature z_i with the following pair of calculations.

$$s_i = x_i^{d_o} \bmod m_o$$
$$z_i = s_i^{e_r} \bmod m_r$$

The originator then sends the ciphertext integer(s) y_i and corresponding encrypted signature(s) z_i to the intended recipient.

3. For each ciphertext integer y_i, the recipient decrypts y_i with the following calculation.
$$x_i = y_i^{d_r} \bmod m_r$$

Then, for each encrypted signature z_i, the recipient decrypts and verifies the signature with the following pair of calculations.

$$s_i = z_i^{d_r} \bmod m_r$$
$$\overline{x}_i = s_i^{e_o} \bmod m_o$$

If $x_i = \overline{x}_i$, and both sets of RSA keys are secure, the recipient can be sure that the plaintext integer x_i is from the originator.

Note that in the RSA signature scheme with encryption as it is described on page 402, it is necessary for a signature s_i formed in step 2 to be encrypted as z_i before being transmitted over an insecure communication line. Were a signature transmitted as s_i, then an intruder who intercepts s_i could recover the corresponding plaintext integer x_i by calculating $x_i = s_i^{e_o} \bmod m_o$.

Example 12.2 Suppose again you wish to send the message PIN # 9089 electronically to a colleague, but you and your colleague would like to use the RSA signature scheme with encryption to keep the message secret and give some assurance to your colleague that the message is really from you, and not from an impostor posing as you. Suppose also that you have published the RSA encryption exponent $e_o = 67$ and modulus $m_o = 9169$, and kept the corresponding decryption exponent $d_o = 267$ secret, and your colleague has published the RSA encryption exponent $e_r = 91$ and modulus $m_r = 10921$ (which, note, is greater than m_o), and kept the corresponding decryption exponent $d_r = 235$ secret. As in Example 12.1, your message is equivalent to the numeric message $x = 80737832353257485657$, which, with the value $m_o = 9169$, you can split into the plaintext integers $x_1 = 8073$, $x_2 = 7832$, $x_3 = 3532$, $x_4 = 5748$, and $x_5 = 5657$. Then, with the values $e_r = 91$ and $m_r = 10921$, you encrypt each plaintext integer x_i by forming the following ciphertext integer y_i.

$$x_1 = 8073 \quad \rightarrow \quad y_1 = 8073^{91} \bmod 10921 = \quad 2869$$
$$x_2 = 7832 \quad \rightarrow \quad y_2 = 7832^{91} \bmod 10921 = \quad 7473$$
$$x_3 = 3532 \quad \rightarrow \quad y_3 = 3532^{91} \bmod 10921 = \quad 4636$$
$$x_4 = 5748 \quad \rightarrow \quad y_4 = 5748^{91} \bmod 10921 = 10522$$
$$x_5 = 5657 \quad \rightarrow \quad y_5 = 5657^{91} \bmod 10921 = \quad 3647$$

Next, with the values $d_o = 267$ and $m_o = 9169$, you sign each plaintext integer x_i by forming the signatures $s_1 = 5465$, $s_2 = 4036$, $s_3 = 5644$, $s_4 = 7827$, and $s_5 = 2489$ as in Example 12.1. Then, with the values $e_r = 91$ and $m_r = 10921$, you encrypt each signature s_i by forming the following quantity z_i.

$$s_1 = 5465 \quad \rightarrow \quad z_1 = 5465^{91} \bmod 10921 = \quad 9552$$
$$s_2 = 4036 \quad \rightarrow \quad z_2 = 4036^{91} \bmod 10921 = \quad 2879$$
$$s_3 = 5644 \quad \rightarrow \quad z_3 = 5644^{91} \bmod 10921 = \quad 7837$$
$$s_4 = 7827 \quad \rightarrow \quad z_4 = 7827^{91} \bmod 10921 = 10389$$
$$s_5 = 2489 \quad \rightarrow \quad z_5 = 2489^{91} \bmod 10921 = \quad 7319$$

You then send the ciphertext integers y_i and corresponding encrypted signatures z_i to your colleague. With the values $d_r = 235$ and $m_r = 10921$, your colleague decrypts each ciphertext integer y_i as follows.

$$y_1 = 2869 \;\rightarrow\; x_1 = 2869^{235} \bmod 10921 = 8073$$
$$y_2 = 7473 \;\rightarrow\; x_2 = 7473^{235} \bmod 10921 = 7832$$
$$y_3 = 4636 \;\rightarrow\; x_3 = 4636^{235} \bmod 10921 = 3532$$
$$y_4 = 10522 \;\rightarrow\; x_4 = 10522^{235} \bmod 10921 = 5748$$
$$y_5 = 3647 \;\rightarrow\; x_5 = 3647^{235} \bmod 10921 = 5657$$

Next, with the values $d_r = 235$ and $m_r = 10921$, your colleague decrypts each encrypted signature z_i as follows.

$$z_1 = 9552 \;\rightarrow\; s_1 = 9552^{235} \bmod 10921 = 5465$$
$$z_2 = 2879 \;\rightarrow\; s_2 = 2879^{235} \bmod 10921 = 4036$$
$$z_3 = 7837 \;\rightarrow\; s_3 = 7837^{235} \bmod 10921 = 5644$$
$$z_4 = 10389 \;\rightarrow\; s_4 = 10389^{235} \bmod 10921 = 7827$$
$$z_5 = 7319 \;\rightarrow\; s_5 = 7319^{235} \bmod 10921 = 2489$$

Finally, with the values $e_o = 67$ and $m_o = 9169$, your colleague verifies each signature s_i by forming the quantities $\overline{x}_1 = 8073$, $\overline{x}_2 = 7832$, $\overline{x}_3 = 3532$, $\overline{x}_4 = 5748$, and $\overline{x}_5 = 5657$ as in Example 12.1. Since each x_i and $\overline{x}_i$ match, assuming (again for the sake of illustration) your and your colleague's RSA keys are secure, your colleague can be sure that the plaintext integers x_i are from you. $\square$

Recall that in the first step in the RSA signature scheme with encryption as it is described on page 402, we supposed m_r was greater than m_o. This assumption dictated the order in which the signature formation and encryption calculations were done, with the signature formation $s_i = x_i^{d_o} \bmod m_o$ done first, and the signature encryption $z_i = s_i^{e_r} \bmod m_r$ done second. For signature decryption and verification, these calculations must be undone in the reverse order, with the signature decryption $s_i = z_i^{d_r} \bmod m_r$ done first. However, if m_r were less than m_o, then the scheme as it is described on page 402 can fail. Specifically, if m_r were less than m_o, and the signature formation $s_i = x_i^{d_o} \bmod m_o$ resulted in a value of s_i between m_r and m_o, then it would not be possible for the signature decryption calculation $z_i^{d_r} \bmod m_r$ to return s_i, since $z_i^{d_r} \bmod m_r$ must be less than m_r.

It is easy to modify the RSA signature scheme with encryption, though, so that it cannot fail when m_r is less than m_o. If m_r is less than m_o, then the signature formation and encryption calculations must be done in the reverse order, with $s_i = x_i^{e_r} \bmod m_r$ done first, and $z_i = s_i^{d_o} \bmod m_o$ done second. That is, when m_r is less than m_o, the following steps summarize the RSA signature scheme with encryption.

1. The originator of a numeric message x publishes RSA encryption exponent e_o and modulus m_o, and keeps the corresponding decryption exponent d_o secret. Meanwhile, the intended recipient publishes RSA encryption exponent e_r and modulus m_r, and keeps the corresponding decryption exponent d_r secret. Suppose also that m_r is less than m_o.

2. Suppose x is expressed as one or more positive integers x_i less than m_r. Then, for each plaintext integer x_i, the originator forms ciphertext integer y_i with the following calculation.

$$y_i = x_i^{e_r} \bmod m_r$$

Next, the originator forms encrypted signature z_i with the following pair of calculations.

$$s_i = x_i^{e_r} \bmod m_r$$
$$z_i = s_i^{d_o} \bmod m_o$$

The originator then sends the ciphertext integer(s) y_i and corresponding encrypted signature(s) z_i to the intended recipient.

3. For each ciphertext integer y_i, the recipient decrypts y_i with the following calculation.

$$x_i = y_i^{d_r} \bmod m_r$$

Then, for each encrypted signature z_i, the recipient decrypts and verifies the signature with the following pair of calculations.

$$s_i = z_i^{e_o} \bmod m_o$$
$$\overline{x}_i = s_i^{d_r} \bmod m_r$$

If $x_i = \overline{x}_i$, and both sets of RSA keys are secure, the recipient can be sure that the plaintext integer x_i is from the originator.

Incidentally, note that the RSA signature scheme with encryption as it is described above for when m_r is less than m_o includes some duplicate calculations. Specifically, the calculations forming y_i and s_i in step 2 are identical, and thus the calculations forming x_i and $\overline{x}_i$ in step 3 are also identical. As a result, the scheme in this case effectively operates by signing the ciphertext, and then leaving this signature unencrypted. However, this is only true of the RSA signature scheme with encryption when m_r is less than m_o. The scheme as it is described on page 402 for when m_r is greater than m_o does not contain duplicate calculations. Also, we have deliberately written the steps in the scheme for when m_r is less than m_o to most closely resemble the steps in the scheme for when m_r is greater than m_o, not just for consistency, but also so that the scheme can be most easily modified to include hashing, which we will do in Section 12.3.

The RSA signature scheme as it is presented in this section is an example of a signature scheme *with appendix*, since signatures are separate entities from messages. It is not always necessary for messages and signatures to be separate entities, though. For example, the RSA signature scheme with encryption can be modified as a *message recovery* scheme, in which signatures serve to also encrypt messages. However, even this does not necessarily overcome the problem that when signatures are formed that are as large as the messages being signed, twice as much data must be formed. We can actually see this in Example 12.2, in which the ciphertext and signature integers together constitute twice as much data as the ciphertext integers alone. The solution to this problem is hashing, which, as we will see in the next two sections, gives the means for forming signatures that are significantly smaller than the messages being signed.

12.1.1   Exercises

1. Suppose you have published the RSA encryption exponent $e_o = 173$ and modulus $m_o = 247$, for which the corresponding decryption exponent is $d_o = 5$. For the following ASCII characters with corresponding ASCII numbers taken one at a time, use the RSA signature scheme (without encryption) to form the signatures you would send to your colleague.

    (a)* VA

    (b) NC

2. Suppose you have published the RSA encryption exponent $e_o = 80529$ and modulus $m_o = 129163$, for which the corresponding decryption exponent is $d_o = 161$. For the following pairs of ASCII characters with corresponding ASCII numbers grouped as a single integer, use the RSA signature scheme (without encryption) to form the signature you would send to your colleague.

    (a)* VA

    (b) NC

3. Suppose your colleague has published the RSA encryption exponent $e_o = 3$ and modulus $m_o = 391$, and uses the RSA signature scheme (without encryption) to form signatures from ASCII characters with corresponding ASCII numbers taken one at a time. Assuming these

RSA keys are secure, for the following ASCII characters and corresponding signatures received from your colleague, determine whether you can be sure the characters are from your colleague.

(a)* Characters = VT; signatures = 222, 67

(b) Characters = WF; signatures = 377, 300

(c)* Characters = GT; signatures = 5, 299

(d) Characters = BC; signatures = 264, 237

4. Suppose your colleague has published the RSA encryption exponent $e_o = 683$ and modulus $m_o = 1010189$, and uses the RSA signature scheme (without encryption) to form signatures from pairs of ASCII characters with corresponding ASCII numbers grouped as a single integer. Assuming these RSA keys are secure, for the following pairs of ASCII characters and corresponding signatures received from your colleague, determine whether you can be sure the characters are from your colleague.

(a)* Characters = VT; signature = 551011

(b) Characters = WF; signature = 435749

(c)* Characters = GT; signature = 325046

(d) Characters = BC; signature = 762157

5. Suppose you have published the RSA encryption exponent $e_o = 173$ and modulus $m_o = 247$, for which the corresponding decryption exponent is $d_o = 5$. Meanwhile, your colleague has published the RSA encryption exponent $e_r = 3$ and modulus $m_r = 391$. For the following ASCII characters with corresponding ASCII numbers taken one at a time, use the RSA signature scheme with encryption to form the ciphertext and encrypted signatures you would send to your colleague.

(a)* JAG

(b) NCIS

6. Suppose you have published the RSA encryption exponent $e_r = 173$ and modulus $m_r = 323$, for which the corresponding decryption exponent is $d_r = 5$. Meanwhile, your colleague has published the RSA encryption exponent $e_o = 7$ and modulus $m_o = 143$, and uses the RSA signature scheme with encryption to form ciphertexts and signatures from ASCII characters with corresponding ASCII numbers taken one at a time. Assuming these RSA keys are secure, for the following ciphertexts and encrypted signatures received from your colleague,

decrypt the ciphertext and determine whether you can be sure the resulting characters are from your colleague.

(a)* Ciphertext = 103, 226;
    encrypted signatures = 226, 225

(b) Ciphertext = 33, 87, 275;
    encrypted signatures = 21, 275, 95

(c)* Ciphertext = 229, 107, 87, 174;
    encrypted signatures = 246, 238, 300, 5

(d) Ciphertext = 33, 174, 318, 320, 115;
    encrypted signatures = 299, 99, 37, 196, 282

7. Suppose you have published the RSA encryption exponent $e_o = 387$ and modulus $m_o = 649$, for which the corresponding decryption exponent is $d_o = 3$. Meanwhile, your colleague has published the RSA encryption exponent $e_r = 5$ and modulus $m_r = 381$. For the following ASCII characters with corresponding ASCII numbers taken one at a time, use the RSA signature scheme with encryption to form the ciphertext and encrypted signatures you would send to your colleague.

    (a)* STL

    (b) ARI

8. Suppose you have published the RSA encryption exponent $e_r = 197$ and modulus $m_r = 415$, for which the corresponding decryption exponent is $d_r = 5$. Meanwhile, your colleague has published the RSA encryption exponent $e_o = 7$ and modulus $m_o = 553$, and uses the RSA signature scheme with encryption to form ciphertexts and signatures from ASCII characters with corresponding ASCII numbers taken one at a time. Assuming these RSA keys are secure, for the following ciphertexts and encrypted signatures received from your colleague, decrypt the ciphertext and determine whether you can be sure the resulting characters are from your colleague.

    (a)* Ciphertext = 93, 395, 136;
        encrypted signatures = 351, 16, 161

    (b) Ciphertext = 123, 329, 296;
        encrypted signatures = 130, 490, 359

    (c)* Ciphertext = 145, 307, 18;
        encrypted signatures = 537, 48, 403

    (d) Ciphertext = 227, 395, 83;
        encrypted signatures = 273, 179, 311

9.* Write examples of two different specific discrete logarithm problems (containing actual numbers) that your colleague or someone intercepting your transmissions in Example 12.1 could solve in order to find the value of your decryption exponent d_o.

10.* Consider the RSA signature scheme with encryption as it is described on page 402 for when m_r is greater than m_o.

   (a) Write a general discrete logarithm problem (containing variables) that the recipient could solve in order to find the value of the originator's decryption exponent d_o.

   (b) Explain why someone intercepting the originator's transmissions, in order to find the value of the originator's decryption exponent d_o, would have to first break an RSA cipher before solving a discrete logarithm problem.

11.* Write examples of two different specific discrete logarithm problems (containing actual numbers) that your colleague in Example 12.2 could solve in order to find the value of your decryption exponent d_o.

12. Consider the RSA signature scheme with encryption as it is described on page 405 for when m_r is less than m_o.

   (a) Write a general discrete logarithm problem (containing variables) that the recipient could solve in order to find the value of the originator's decryption exponent d_o.

   (b) Explain how someone intercepting the originator's transmissions, in order to find the value of the originator's decryption exponent d_o, could solve a discrete logarithm problem without having to first break an RSA cipher.

12.2   Hash Functions

Recall that in the RSA signature scheme as it is described in Section 12.1, signatures are formed that can be as large as the messages being signed. In practice, the RSA signature scheme is typically not employed in this manner, since it is usually not desirable for users to have to create, store, and transmit as much signature data as message data. What is typically done in practice is that users first apply a hash function to a message. This reduces the size of the message, perhaps considerably, and then only the smaller hashed message is signed.

A *hash function* can be thought of as a procedure that takes a stream of message data of arbitrary length, and converts, or "hashes," it into a stream of fixed length called the message *digest*. The idea behind using hash functions with digital signatures is to take a message that is perhaps very long, hash it into a shorter fixed-length digest, and then sign just the digest in a way such that the signature indicates that the full message was really sent by the person claiming to have sent it.

Consider the following very primitive example of a hash function h. Given a numeric message x and modulus m, suppose we split x into message integers $x_1, x_2, \ldots, x_n$, with each x_i less than m, and then form message digest $h(x)$ as follows.

$$h(x) = (x_1 + x_2 + \cdots + x_n) \bmod m$$

We will refer to this type of hash function as a *modular* hash function.

Example 12.3 Consider again the message PIN # 9089. As in Example 12.1 on page 401, this message is equivalent to the numeric message $x = 80737832353257485657$. To hash this message using a modular hash function h with modulus $m = 9001$, we can split x into the message integers $x_1 = 8073$, $x_2 = 7832$, $x_3 = 3532$, $x_4 = 5748$, and $x_5 = 5657$, and form message digest $h(x)$ as follows.

$$\begin{aligned} h(x) &= (8073 + 7832 + 3532 + 5748 + 5657) \bmod 9001 \\ &= 3839 \end{aligned}$$

Thus, for the message $x = 80737832353257485657$, a modular hash function h with modulus $m = 9001$ gives message digest $h(x) = 3839$. □

Message digests produced by hash functions are usually expressed in hexadecimal format. For example, the digest $h(x) = 3839$ in Example 12.3 would be expressed in hexadecimal format as EFF. Also, message digests are usually understood to be of *exactly* a fixed length, as opposed to *at most* a fixed length. For a modular hash function, this can be done by requiring digests to be represented using the same number of hexadecimal digits as the hexadecimal representation of the modulus. For example, since the modulus $m = 9001$ in Example 12.3 has hexadecimal representation 2329, with four digits, then message digests produced by h would be expressed with four hexadecimal digits, despite this sometimes requiring one or more zeros to be written on the left of a number, such as in the hexadecimal representation 0EFF of the digest $h(x) = 3839$ in Example 12.3.

There are m possible numbers that can result from a calculation modulo m, and for a modular hash function with a modulus chosen randomly, each of these m numbers are equally likely to result from a large input chosen randomly. Thus, for a given message digest produced by a modular hash function with modulus m, the probability that a large input chosen randomly would produce the same digest is $\frac{1}{m}$. For example, for the hash function in Example 12.3, the probability that a large input chosen randomly would produce the same digest as x is $\frac{1}{9001} = 0.00011$. This suggests a method for using a modular hash function to verify that a received message was most likely sent by the person claiming to have sent it. For example, if the intended recipient of the message in Example 12.3 was expecting the received message to have digest 0EFF, and an unaware impostor posing as the originator sent a false message, there is only a 0.00011 probability that the false message would produce the expected digest.

Despite this, a modular hash function, even one with a very large modulus, would not be useful in practice. It would defeat the whole purpose of using a public-key cipher if the communicating parties had to have a secret way to identify message digests. Hash functions are usually assumed to be public. Also, like all cryptographic methods, hash functions must be resistant to all types of cryptanalytic attacks. As a result, it is generally considered that for a hash function to be useful in practice, in addition to being fast and easy to use, it must satisfy each of the following conditions.

1. A hash function must be a one-way function. Given a hash function h and message digest c, it should be extremely difficult to find a message x for which $h(x) = c$. In other words, given a hash function and digest, it should be extremely difficult to find a message that hashes to the digest.

2. A hash function must be *weakly collision resistant*. Given a hash function h and message x with corresponding message digest $h(x) = c$, it should be extremely difficult to find a message y for which $x \neq y$ and $h(y) = c$. In other words, given a hash function and message with corresponding digest, it should be extremely difficult to find a different message that hashes to the same digest.

3. A hash function must be *strongly collision resistant*. Given a hash function h, it should be extremely difficult to find a pair of messages x and y for which $x \neq y$ and $h(x) = h(y)$. In other words, given a hash function, it should be extremely difficult to find two different messages that hash to the same digest.

In addition, the set of possible input messages into a hash function should be much larger than the set of possible output message digests. In fact, ideally there should be infinitely many possible input messages, but finitely many possible output digests. As a result, there should be numerous examples of different messages with identical digests. However, the properties above do not say that such examples should not exist, just that they should be extremely difficult to find.

Example 12.4 For the modular hash function with modulus $m = 9001$ used in Example 12.3, consider the following.

1. For any message digest $c < 9001$, it is true that $h(c) = c$. For example, $h(3839) = 3839$, and so for the digest $c = 3839$, the message $x = 3839$ gives $h(x) = c$. Thus, this hash function fails to be a one-way function.

2. As we saw in Example 12.3, the message PIN # 9089 has message digest 3839. However, it is easy to verify (and should be obvious) that PIN # 8990 also has digest 3839. Thus, this hash function fails to be weakly collision resistant.

3. It is easy to verify (and should be obvious) that the messages ITEM and EMIT have identical message digests. Thus, this hash function fails to be strongly collision resistant.

Thus, this hash function fails to satisfy each of the three conditions of useful hash functions listed on page 411. □

While modular hash functions are not useful in practice, they do clearly demonstrate the foundational idea that a hash function should use an entire message of arbitrary length to create a message digest of fixed length. Also, while given a message and corresponding digest produced by a modular hash function, it is not difficult to find a different message that hashes to the same digest, it may be difficult to find one that is legible, especially if the modulus were large and the size of the integers into which messages were split were varied. Even with a public modular hash function, message digests indicate with some measure of likelihood that a received message was really sent by the person claiming to have sent it. An industrial-strength public hash function, on the other hand, can increase this likelihood up to essential certainty, which a modular hash function cannot do.

One family of hash functions that have been widely used are the *MD* functions, which were developed by Ron Rivest. The most important MD functions were MD2, MD4, and MD5, published between 1989 and 1991. In particular, MD5, which for an input message of arbitrary length gives a 128-bit digest expressed as 32 hexadecimal digits, has been widely used worldwide, even after some serious flaws were discovered in it over a period of several years beginning in 1995. For instance, several examples have been produced showing that MD5 is not strongly collision resistant. Despite this, MD5 has been widely used to provide assurance that digital files have not been corrupted during online transfer. Specifically, websites often publish MD5 digests of files uploaded for transfer. Once a file has been downloaded, the digest of the downloaded file is compared to the published digest of the uploaded file, and if these digests match, it is assumed that the downloaded file matches the file that was uploaded.

MD5 has also been widely used for password storage and confirmation. Specifically, password-protected systems often store MD5 digests of passwords rather than the passwords in the clear, since storing passwords in the clear would make the system vulnerable to an intruder who could identify their location. Once a password has been entered by a user, the digest of the entered password is compared to the stored digest of the user's password, and if these digests match, access to the system is granted. Typically for this type of system using MD5, which can produce 2^{128} possible digests, it would be extremely unlikely for an intruder to successfully guess a password whose digest matched the digest of an actual password, with probability of success for a specific guess of around $\frac{1}{2^{128}} \approx 2.94 \times 10^{-39}$.

Another common family of hash functions that have been widely used are the *Secure Hash Algorithm*, or *SHA*, functions, for which the initial function was developed by the National Security Agency and selected by the National Institute of Standards and Technology (NIST) in 1994 as a Federal Information Processing Standard. The first three SHA functions were SHA-0, SHA-1, and SHA-2, published between 1993 and 2001. Of these, SHA-1, which for an input message of arbitrary length gives a 160-bit digest expressed as 40 hexadecimal digits, was the most widely used. Typically for a system using SHA-1, which can produce 2^{160} possible digests, the probability of success for a specific guess in trying to match the

Quantum Computing: The Next Change in Direction for Cryptology?

Because of the difficulty of factoring numbers and finding discrete logarithms, it seems that RSA and ElGamal ciphers are unbreakable. However, it may be possible for these types of ciphers to be broken through applying quantum mechanics to computing. Known as *quantum computing*, this would allow for computations to be done simultaneously, instead of one at a time like in traditional computing. A procedure called *Shor's algorithm* designed for factoring numbers using quantum computers is astronomically faster than the current procedures available for traditional computers.

If quantum computing becomes a reality, will we still be able to transmit information securely? More than likely, yes. Through the use of quantum channels, it should be possible for keys to be determined through the use of *qubits*, which are the equivalent in quantum computing of bits in traditional computing. Keys for quantum computers formed using qubits would be equivalent to one-time pads, which are unbreakable even with unlimited computing power.

Quantum computing is a technology that could still be years away from being useful. If it is ever perfected though, it could lead to the next great change in direction for cryptology.

digest of an actual message is around $\frac{1}{2^{160}} \approx 6.84 \times 10^{-49}$. This makes SHA-1 more secure than MD5. The trade-off is that SHA-1 is slower than MD5. For security reasons, NIST has strongly recommended that users stop using SHA-1, and in 2007 opened a competition for the development of a new SHA-3. Ron Rivest submitted an MD6 function to this competition, but it failed to advance to the second round. In 2012 NIST announced the winner of this competition, an algorithm submitted by a team that included Joan Daemen, which it approved in 2015 as the SHA-3 Federal Information Processing Standard.

Technical descriptions of the MD and SHA hash functions are beyond the scope of this book. Interested readers should be able to find numerous such descriptions on the Internet.

12.2.1   Exercises

1. Use a modular hash function with the given modulus m to form the digest of the given message. Express your answer in both decimal and hexadecimal formats.

    (a)* $m = 9169$;

    message = DOOMSDAY, split into integers corresponding to two characters each

    (b) $m = 10921$;

    message = GRITZ BLITZ, split into integers corresponding to two characters each, padded at the end with a space character

    (c)* $m = 1010189$;

    message = ORANGE CRUSH, split into integers corresponding to three characters each

    (d) $m = 1358237$;

    message = STEEL CURTAIN, split into integers corresponding to three characters each, padded at the end with two space characters

2. For the modular hash function in Exercise 1b, find a message different from the message in Exercise 1b that results in the same digest as the message in Exercise 1b.

3.* Repeat Exercise 2, but use the modular hash function and message in Exercise 1c.

4. Consider a password-protected system which stores modular hash function digests of passwords rather than the passwords in the clear. Suppose the system uses a modular hash function with the given modulus m, a user enters the given potential password, and the system

has the given digest stored for the user. Determine whether the user should be granted access to the system.

(a)* $m = 9169$;

potential password = LILLY, split into integers corresponding to two characters each, padded at the end with a space character; digest = 5943 in decimal, which is 1737 in hexadecimal

(b) $m = 10921$;

potential password = BREZINA, split into integers corresponding to two characters each, padded at the end with a space character; digest = 1042 in decimal, which is 0412 in hexadecimal

(c)* $m = 1010189$;

potential password = GRADISHAR, split into integers corresponding to three characters each; digest = 304380 in decimal, which is 4A4FC in hexadecimal

(d) $m = 1358237$;

potential password = GREENWOOD, split into integers corresponding to three characters each; digest = 855887 in decimal, which is 0D0F4F in hexadecimal

5.* Consider the modular hash function in Exercise 4b, and assume each of the possible digests that can be produced by the function are equally likely to result from a large input chosen randomly.

(a) Suppose an intruder tries a random potential password in the hope that its digest will match the digest stored for a known user. Find the probability of success for a specific random guess.

(b) Suppose an intruder somehow knows the digest stored for the user in Exercise 4b. Find a potential password the intruder could enter that would produce this digest.

6. Repeat Exercise 5, but use the modular hash function and digest in Exercise 4c.

7. Let f be a function that takes as input an arbitrary string of eight bits, and gives as output the same string with each bit shifted to the left by one position and the leftmost bit wrapped to the right. That is, for bits $b_1, b_2, \ldots, b_8$, suppose f operates as follows.

$$f(b_1 b_2 b_3 b_4 b_5 b_6 b_7 b_8) = b_2 b_3 b_4 b_5 b_6 b_7 b_8 b_1$$

Now suppose x is a numeric message in binary format, padded at the end with zeros if necessary so that its length is a multiple of eight, and then split into bytes $x_1, x_2, \ldots, x_k$. From the bytes x_i, construct the following bytes $c_1, c_2, \ldots, c_k$ using the function f and the XOR operation $\oplus$.

$$c_1 = f(x_1)$$
$$c_2 = f(c_1 \oplus x_2)$$
$$c_3 = f(c_2 \oplus x_3)$$
$$\vdots$$
$$c_k = f(c_{k-1} \oplus x_k)$$

Finally, let h be the hash function that takes as input the message x, and gives as output the byte c_k. That is, suppose h is the hash function defined by $h(x) = c_k$.

(a)* For the message $x =$ 01000100 00101101 01000100, find $h(x)$. Express your answer in both binary and hexadecimal formats.

(b) For the message $y =$ 11000100 00011010 01011111 00010000, find $h(y)$. Express your answer in both binary and hexadecimal formats.

(c)* Find the number of possible digests that can be produced by h.

(d) Assuming each of the possible digests that can be produced by h are equally likely to result from a large input chosen randomly, find the probability that a large input chosen randomly would produce a given digest.

(e)* Find a message different from the message in part (a) that results in the same digest as the message in part (a).

(f) Find a message different from the message in part (b) that results in the same digest as the message in part (b).

8. Find some information about one or more of the MD hash functions, including how they operate, and write a summary of your findings.

9. Find some information about one or more real-life uses of the MD hash functions, and write a summary of your findings.

10. Find some information about one or more of the SHA hash functions, including how they operate, and write a summary of your findings.

11. Find some information about one or more real-life uses of the SHA hash functions, and write a summary of your findings.

12. Find some information about the NIST competition for the development of the SHA-3 hash function, and write a summary of your findings. Include in your summary some details about one or more of the five finalists in the competition, with some information about how they operate and their creators.

12.3 RSA Signatures with Hashing

Recall that in the RSA signature scheme as it is presented in Section 12.1, signatures are formed that are as large as the messages being signed. This can severely limit the efficiency of the scheme, especially with longer messages, since it requires twice as much data to be formed. However, if the scheme is implemented with a hash function, then the necessary computation and data storage can be greatly reduced.

The following steps summarize one way in which the RSA signature scheme (without encryption) can be implemented with a hash function. We will call this the RSA signature scheme *with hashing*.

1. The originator of a numeric message x publishes RSA encryption exponent e_o and modulus m_o, and keeps the corresponding decryption exponent d_o secret. In addition, the originator and intended recipient agree (publicly, if necessary) upon a hash function h that produces message digests that are less than m_o.

2. Next, the originator applies h to x, resulting in message digest $h(x)$, and forms signature s with the following calculation.

$$s = h(x)^{d_o} \bmod m_o$$

The originator then sends the message x and signature s to the intended recipient.

3. The recipient applies h to x, resulting in message digest $h(x)$, and then verifies the signature with the following calculation.

$$\overline{h(x)} = s^{e_o} \bmod m_o$$

If $h(x) = \overline{h(x)}$, and the originator's RSA keys and the hash function are secure, the recipient can be sure that the message x is from the originator.

Example 12.5 Suppose again you wish to send the message PIN # 9089 electronically to a colleague in the clear, but you and your colleague would like to use the RSA signature scheme with hashing to give some assurance to your colleague that the message is really from you, and not from an impostor posing as you. Suppose also that you have published the RSA encryption exponent $e_o = 67$ and modulus $m_o = 9169$, and kept the corresponding decryption exponent $d_o = 267$ secret. In addition, suppose you and your colleague agree to use a modular hash function h with modulus $m = 9001$. As in Example 12.1 on page 401, your message is equivalent to the numeric message $x = 80737832353257485657$, which, with the value $m = 9001$, you can form the message digest $h(x) = 3839$ as in Example 12.3 on page 410. Next, with the values $d_o = 267$ and $m_o = 9169$, you sign $h(x)$ by forming the following signature s.

$$s = 3839^{267} \bmod 9169 = 9146$$

You then send the message x and signature s to your colleague. Your colleague applies h to x to find $h(x) = 3839$. With the values $e_o = 67$ and $m_o = 9169$, your colleague then verifies the signature s by forming the following quantity $\overline{h(x)}$.

$$\overline{h(x)} = 9146^{67} \bmod 9169 = 3839$$

Since $h(x)$ and $\overline{h(x)}$ match, assuming (again for the sake of illustration) your RSA keys and the hash function are secure, your colleague can be sure that the message x is from you. □

Comparing Examples 12.1 and 12.5 makes the benefit of using a hash function in the RSA signature scheme transparent. Example 12.5 requires only 20% of the computation and data storage in Example 12.1, in exchange for only the very limited computation and data storage in Example 12.3.

It is obviously possible to modify the RSA signature scheme with hashing to include encryption. The following steps summarize one way in which this can be done, with both the message and signature encrypted. We will call this the RSA signature scheme with hashing *and encryption*.

1. The originator of a numeric message x publishes RSA encryption exponent e_o and modulus m_o, and keeps the corresponding decryption exponent d_o secret. Meanwhile, the intended recipient publishes RSA encryption exponent e_r and modulus m_r, and keeps the corresponding decryption exponent d_r secret. Suppose also that m_r is greater than m_o. In addition, the originator and intended recipient agree (publicly, if necessary) upon a hash function h that produces message digests that are less than m_o.

2. Suppose x is expressed as one or more positive integers x_i less than m_r. Then, for each plaintext integer x_i, the originator forms ciphertext integer y_i with the following calculation.

$$y_i = x_i^{e_r} \bmod m_r$$

Next, the originator applies h to x, resulting in message digest $h(x)$, and forms encrypted signature z with the following pair of calculations.

$$s = \check{h}(x)^{d_o} \bmod m_o$$
$$z = s^{e_r} \bmod m_r$$

The originator then sends the ciphertext integer(s) y_i and encrypted signature z to the intended recipient.

3. For each ciphertext integer y_i, the recipient decrypts y_i with the following calculation.

$$x_i = y_i^{d_r} \bmod m_r$$

The recipient then decrypts and verifies the signature with the following pair of calculations.

$$s = z^{d_r} \bmod m_r$$
$$\overline{h(x)} = s^{e_o} \bmod m_o$$

If $h(x) = \overline{h(x)}$, and both sets of RSA keys and the hash function are secure, the recipient can be sure that the plaintext integers x_i are from the originator.

Example 12.6 Suppose again you wish to send the message PIN # 9089 electronically to a colleague, but you and your colleague would like to use the RSA signature scheme with hashing and encryption to keep the message secret and give some assurance to your colleague that the message is really from you, and not from an impostor posing as you. Suppose also that you have published the RSA encryption exponent $e_o = 67$ and modulus $m_o = 9169$, and kept the corresponding decryption exponent $d_o = 267$ secret, and your colleague has published the RSA encryption exponent $e_r = 91$ and modulus $m_r = 10921$ (which, note, is greater than m_o), and kept the corresponding decryption exponent $d_r = 235$ secret. In addition, suppose you and your colleague agree to use a modular hash function h with modulus $m = 9001$. As in Example 12.1 on page 401, your message is equivalent to the numeric message $x = 80737832353257485657$, which, with the value $m_r = 10921$, you can split into the plaintext integers $x_1 = 8073$, $x_2 = 7832$, $x_3 = 3532$, $x_4 = 5748$, and $x_5 = 5657$. Then, with the values $e_r = 91$ and $m_r = 10921$, you encrypt the plaintext integers x_i by forming the ciphertext integers $y_1 = 2869$, $y_2 = 7473$, $y_3 = 4636$, $y_4 = 10522$, and $y_5 = 3647$ as in Example 12.2 on page 403. Next, with the value $m = 9001$, you can form the message digest $h(x) = 3839$ as in Example 12.3 on page 410. Then, with the values $d_o = 267$ and $m_o = 9169$, you form signature

$s = 9146$ as in Example 12.5, and, with the values $e_r = 91$ and $m_r = 10921$, encrypt s by forming the following encrypted signature z.

$$z = 9146^{91} \bmod 10921 = 10464$$

You then send the ciphertext integers y_i and encrypted signature z to your colleague. With the values $d_r = 235$ and $m_r = 10921$, your colleague decrypts the ciphertext integers y_i by forming the plaintext integers $x_1 = 8073$, $x_2 = 7832$, $x_3 = 3532$, $x_4 = 5748$, and $x_5 = 5657$ as in Example 12.2. Next, with the values $d_r = 235$ and $m_r = 10921$, your colleague decrypts the encrypted signature z as follows.

$$s = 10464^{235} \bmod 10921 = 9146$$

Finally, with the values $e_o = 67$ and $m_o = 9169$, your colleague verifies the signature s by forming $\overline{h(x)} = 3839$ as in Example 12.5. Since $h(x)$ and $\overline{h(x)}$ match, assuming (again for the sake of illustration) your and your colleague's RSA keys and the hash function are secure, your colleague can be sure that the message x is from you. □

There are other digital signature schemes that can be implemented with hash functions, including the *ElGamal signature scheme*, whose security is based on the difficulty of finding discrete logarithms, and the *Digital Signature Algorithm*, one of several variants of the ElGamal signature scheme. The Digital Signature Algorithm was developed by the National Security Agency and selected by the National Institute of Standards and Technology in 1993 as a Federal Information Processing Standard, to serve with the Secure Hash Algorithm in the *Digital Signature Standard*. Descriptions of the ElGamal signature scheme and Digital Signature Algorithm are beyond the scope of this book, but remain open to the interested reader for investigation.

12.3.1　Exercises

1. Suppose you have published the RSA encryption exponent $e_o = 173$ and modulus $m_o = 247$, for which the corresponding decryption exponent is $d_o = 5$. For the following ASCII characters, use the RSA signature scheme with hashing (but without encryption) to form the signature you would send to your colleague. For hashing, use a modular hash function with modulus $m = 241$ and ASCII numbers taken one at a time.

   (a)* TIGERS
   (b) WILDCATS

2. Suppose you have published the RSA encryption exponent $e_o = 80529$ and modulus $m_o = 129163$, for which the corresponding decryption exponent is $d_o = 161$. For the following ASCII characters, use the RSA signature scheme with hashing (but without encryption) to form the signature you would send to your colleague. For hashing, use a modular hash function with modulus $m = 129127$ and ASCII numbers grouped into integers corresponding to two characters each.

   (a)* TIGERS

   (b) WILDCATS

3. Suppose your colleague has published the RSA encryption exponent $e_o = 3$ and modulus $m_o = 391$, and uses the RSA signature scheme with hashing (but without encryption) to form signatures from ASCII characters using a modular hash function with modulus $m = 389$ and ASCII numbers taken one at a time. Assuming these RSA keys and this hash function are secure, for the following ASCII characters and signatures received from your colleague, determine whether you can be sure the characters are really from your colleague.

   (a)* Characters = TIDE; signature = 348

   (b) Characters = GATORS; signature = 246

   (c)* Characters = BULLDOGS; signature = 301

   (d) Characters = VOLUNTEERS; signature = 225

4. Suppose your colleague has published the RSA encryption exponent $e_o = 683$ and modulus $m_o = 1010189$, and uses the RSA signature scheme with hashing (but without encryption) to form signatures from ASCII characters using a modular hash function with modulus $m = 1010179$ and ASCII numbers grouped into integers corresponding to two characters each. Assuming these RSA keys and this hash function are secure, for the following ASCII characters and signatures received from your colleague, determine whether you can be sure the characters are really from your colleague.

   (a)* Characters = TIDE; signature = 204665

   (b) Characters = GATORS; signature = 395577

   (c)* Characters = BULLDOGS; signature = 73385

   (d) Characters = VOLUNTEERS; signature = 144684

5. Suppose you have published the RSA encryption exponent $e_o = 173$ and modulus $m_o = 247$, for which the corresponding decryption exponent is $d_o = 5$. Meanwhile, your colleague has published the RSA

encryption exponent $e_r = 3$ and modulus $m_r = 391$. For the following ASCII characters, use the RSA signature scheme with hashing and encryption to form the ciphertext and encrypted signature you would send to your colleague. For hashing, use a modular hash function with modulus $m = 241$ and ASCII numbers taken one at a time.

(a)* SMOKE

(b) JUMPERS

6. Suppose you have published the RSA encryption exponent $e_r = 173$ and modulus $m_r = 323$, for which the corresponding decryption exponent is $d_r = 5$. Meanwhile, your colleague has published the RSA encryption exponent $e_o = 7$ and modulus $m_o = 143$, and uses the RSA signature scheme with hashing and encryption to form ciphertexts and signatures from ASCII characters using a modular hash function with modulus $m = 139$ and ASCII numbers taken one at a time. Assuming these RSA keys and this hash function are secure, for the following ciphertexts and encrypted signatures received from your colleague, decrypt the ciphertext and determine whether you can be sure the resulting characters are really from your colleague.

(a)* Ciphertext $= 103, 226, 275, 33$;
encrypted signature $= 222$

(b) Ciphertext $= 102, 226, 107, 229, 107$;
encrypted signature $= 5$

(c)* Ciphertext $= 50, 119, 226, 50, 247, 103$;
encrypted signature $= 299$

(d) Ciphertext $= 52, 275, 300, 33, 103, 300, 50$;
encrypted signature $= 318$

7. Recall that in the first step in the RSA signature scheme with hashing and encryption as it is described on page 418, we supposed m_r was greater than m_o.

(a)* Explain how this scheme could fail if m_r were less than m_o.

(b) Explain how this scheme can be modified so that it cannot fail if m_r were less than m_o.

8. Recall that in the first step in the RSA signature scheme with hashing as it is described on page 417, we supposed h produced message digests less than m_o. Explain how this scheme could fail if h could produce message digests greater than m_o.

9. Find some information about the *ElGamal signature scheme*, and write a summary of your findings.

10. Find some information about the *Digital Signature Algorithm*, and write a summary of your findings.

12.4 The Man-in-the-Middle Attack

As we have seen, digital signatures provide a method through which the recipient of an electronic message can verify with any desired level of certainty that the message was really sent by the person claiming to have sent it. Digital signatures are not the solution to all problems related to message authentication, though. In this section, we will describe another way in which public-key ciphers are susceptible to exploitation, through the *man-in-the-middle* attack.

We will specifically describe how the man-in-the-middle attack could be used against users of an RSA cipher. To this end, suppose you want to make a purchase from a small business remotely using a credit card, and you wish to send your credit card number to the manager of the business over an insecure communication line. In order that your credit card number will not need to be transmitted in the clear, you and the manager decide to use an RSA cipher to encrypt the number. First, the manager chooses RSA encryption exponent e_r and modulus m_r, with corresponding decryption exponent d_r, and sends the values of e_r and m_r to you. However, an outsider monitoring your communication intercepts and stops this transmission. The outsider then chooses RSA encryption exponent e_c and modulus m_c, with corresponding decryption exponent d_c, and, posing as the manager, sends the values of e_c and m_c on to you. You, having no reason to suspect any clandestine behavior, use e_c and m_c to encrypt your credit card number, and send the ciphertext to the manager. The outsider intercepts and stops this transmission as well, and uses d_c and m_c to decrypt the ciphertext. The outsider then uses e_r and m_r to encrypt your credit card number, and, posing as you, sends the ciphertext on to the manager. The manager, also having no reason to suspect any clandestine behavior, uses d_r and m_r to decrypt the ciphertext. Note that through this process, the outsider gains possession of your credit card number in the clear, without your or the manager's knowledge, and never has to break an RSA cipher.

Example 12.7 Suppose you wish to send the message PIN # 9089 to your spouse over an insecure communication line. In order that your message will not need to be transmitted in the clear, you and your spouse decide to use an RSA cipher to encrypt the message. First, your spouse chooses the RSA encryption exponent $e_r = 91$ and modulus $m_r = 10921$, for which the corresponding decryption exponent is $d_r = 235$, and sends the values of e_r and m_r to you. However, an outsider monitoring your communication intercepts and stops this transmission. The outsider then chooses the RSA encryption exponent $e_c = 55$ and modulus $m_c = 9379$, for which the corresponding decryption exponent is $d_c = 167$, and, posing as your spouse, sends the values of e_c and m_c on to you. As in Example 12.1 on page 401, your message is equivalent to the numeric message $x = 80737832353257485657$, which, with the value $m_c = 9379$, you can split into the plaintext integers $x_1 = 8073$, $x_2 = 7832$, $x_3 = 3532$, $x_4 = 5748$, and $x_5 = 5657$. Then, with the values $e_c = 55$ and $m_c = 9379$, you encrypt each plaintext integer x_i by forming the following ciphertext integer y_i.

$$x_1 = 8073 \quad \rightarrow \quad y_1 = 8073^{55} \bmod 9379 = 2425$$
$$x_2 = 7832 \quad \rightarrow \quad y_2 = 7832^{55} \bmod 9379 = 9111$$
$$x_3 = 3532 \quad \rightarrow \quad y_3 = 3532^{55} \bmod 9379 = 3012$$
$$x_4 = 5748 \quad \rightarrow \quad y_4 = 5748^{55} \bmod 9379 = 3179$$
$$x_5 = 5657 \quad \rightarrow \quad y_5 = 5657^{55} \bmod 9379 = 3374$$

You then send the ciphertext integers y_i to your spouse. The outsider intercepts and stops this transmission as well, and, with the values $d_c = 167$ and $m_c = 9379$, decrypts each ciphertext integer y_i as follows.

$$y_1 = 2425 \quad \rightarrow \quad x_1 = 2425^{167} \bmod 9379 = 8073$$
$$y_2 = 9111 \quad \rightarrow \quad x_2 = 9111^{167} \bmod 9379 = 7832$$
$$y_3 = 3012 \quad \rightarrow \quad x_3 = 3012^{167} \bmod 9379 = 3532$$
$$y_4 = 3179 \quad \rightarrow \quad x_4 = 3179^{167} \bmod 9379 = 5748$$
$$y_5 = 3374 \quad \rightarrow \quad x_5 = 3374^{167} \bmod 9379 = 5657$$

The outsider converts these ASCII numbers back into characters and reads your message. Then, with the values $e_r = 91$ and $m_r = 10921$, the outsider encrypts each plaintext integer x_i by forming the following ciphertext integer $\overline{y}_i$.

$$x_1 = 8073 \quad \rightarrow \quad \overline{y}_1 = 8073^{91} \bmod 10921 = 2869$$
$$x_2 = 7832 \quad \rightarrow \quad \overline{y}_2 = 7832^{91} \bmod 10921 = 7473$$
$$x_3 = 3532 \quad \rightarrow \quad \overline{y}_3 = 3532^{91} \bmod 10921 = 4636$$
$$x_4 = 5748 \quad \rightarrow \quad \overline{y}_4 = 5748^{91} \bmod 10921 = 10522$$
$$x_5 = 5657 \quad \rightarrow \quad \overline{y}_5 = 5657^{91} \bmod 10921 = 3647$$

The outsider then, posing as you, sends the ciphertext integers $\overline{y}_i$ on to your spouse. With the values $d_r = 235$ and $m_r = 10921$, your spouse decrypts each ciphertext integer $\overline{y}_i$ as follows.

$$\overline{y}_1 = 2869 \quad \rightarrow \quad x_1 = 2869^{235} \bmod 10921 = 8073$$
$$\overline{y}_2 = 7473 \quad \rightarrow \quad x_2 = 7473^{235} \bmod 10921 = 7832$$
$$\overline{y}_3 = 4636 \quad \rightarrow \quad x_3 = 4636^{235} \bmod 10921 = 3532$$
$$\overline{y}_4 = 10522 \quad \rightarrow \quad x_4 = 10522^{235} \bmod 10921 = 5748$$
$$\overline{y}_5 = 3647 \quad \rightarrow \quad x_5 = 3647^{235} \bmod 10921 = 5657$$

Your spouse converts these ASCII numbers back into characters and reads your message, not knowing that your message has also been read by an outsider who did not have to break an RSA cipher.   □

The man-in-the-middle attack illustrates why the common terminology for public-key ciphers with regard to the exchange of public keys is that users "publish" them or make them "public knowledge," as opposed to transmitting them. Even then, the problem of verifying that keys were really published by the person claiming to have published them, and not by an impostor posing as the person claiming to have published them, remains. We will consider how this verification can be obtained in Section 12.5.

12.4.1   Exercises

1. Suppose you are monitoring communication between your arch enemy and his colleague as they communicate over an insecure communication line. In addition, suppose your arch enemy wishes to send a message electronically to his colleague using an RSA cipher, and you decide to use a man-in-the-middle attack to secretly read the message. First, your arch enemy's colleague chooses the RSA encryption exponent $e_r = 3$ and modulus $m_r = 391$, and sends the values of e_r and m_r to your arch enemy. However, you intercept and stop this transmission. You then choose the RSA encryption exponent $e_c = 173$ and modulus $m_c = 247$, for which the corresponding decryption exponent is $d_c = 5$, and, posing as your arch enemy's colleague, send the values of e_c and m_c on to your arch enemy. Your arch enemy encrypts his message, and sends the resulting ciphertext to his colleague, but you intercept and stop this transmission as well. For the following intercepted ciphertexts, (i) find the corresponding plaintext ASCII characters. Then, with the plaintext ASCII numbers taken one at a time, (ii) form the ciphertext you would, posing as your arch enemy, send on to his colleague.

    (a)* Intercepted ciphertext = 221, 228, 221, 91

    (b) Intercepted ciphertext = 145, 24, 36, 47, 91, 184

2. Suppose you are monitoring communication between your arch enemy and his colleague as they communicate over an insecure communication line. In addition, suppose your arch enemy wishes to send a message electronically to his colleague using an RSA cipher, and you decide to use a man-in-the-middle attack to secretly read the message. First, your arch enemy's colleague chooses the RSA encryption exponent $e_r = 683$ and modulus $m_r = 1010189$, and sends the values of e_r and m_r to your arch enemy. However, you intercept and stop this transmission. You then choose the RSA encryption exponent $e_c = 80529$ and modulus $m_c = 129163$, for which the corresponding decryption exponent is $d_c = 161$, and, posing as your arch enemy's colleague, send the values of e_c and m_c on to your arch enemy. Your arch enemy encrypts his message, and sends the resulting ciphertext to his colleague, but you intercept and stop this transmission as well. For the following intercepted ciphertexts, (i) find the corresponding plaintext ASCII characters. Then, with the plaintext ASCII numbers grouped into blocks corresponding to two characters, (ii) form the ciphertext you would, posing as your arch enemy, send on to his colleague.

   (a)* Intercepted ciphertext = 101370, 104005, 118149

   (b) Intercepted ciphertext = 14180, 88435, 99196, 50631

3.* Describe a way in which the man-in-the-middle attack could be used to exploit the RSA signature scheme with encryption as it is described on page 402.

4. Describe a way in which the man-in-the-middle attack could be used to exploit ElGamal ciphers as they are described in Section 10.3.

5. Find some information about how the *interlock protocol* can be used to prevent the man-in-the-middle attack, and write a summary of your findings.

12.5   Public-Key Infrastructures

As the man-in-the-middle attack illustrates, for the originator of a message, verifying that the public keys belonging to the intended recipient really belong to the intended recipient, and not to an impostor posing as the intended recipient, can be as critical as actually encrypting the message securely. One method through which such verification is sometimes possible is to have the intended recipient post his or her public keys on a personal web page provided by a known organization with which the recipient is affiliated, since the originator may be able to externally confirm this affiliation. This method may not be efficient, though, if the information to be transmitted is extensive. It may also not be exceptionally secure, since an outsider might be able to create a web page that appears virtually identical to the intended recipient's web page.

However, efficient and exceptionally secure methods for binding public keys to particular individuals or organizations do exist. These methods are examples of *public-key infrastructures*. Many public-key infrastructures are too complex to be included in this book, but we will present two in this section which are readily accessible.

12.5.1 Key Formation

Any public-key infrastructure begins with the formation of keys. This can be done directly by a user, or by a security officer within the organization with which the user is affiliated. In the former approach, the user must possess and trust a copy of the key formation software, while in the latter, the user must trust the security officer, and a secure way must exist for the private key (e.g., the decryption exponent for an RSA cipher) to be transferred from the security officer to the user.

Regarding actual key formation, we will briefly consider some details specific to RSA ciphers. Typically, RSA ciphers used in practice employ a modulus that when expressed in binary contain a specified minimum number of bits, often 1024 or 2048. If the binary representation of a particular modulus m contains b bits, then the decimal representation of m will satisfy $2^{b-1} \leq m < 2^b$. Recall also that an RSA modulus m is the product of a pair of primes p and q. In practice, these primes are usually chosen of roughly the same length, since this maximizes the difficulty of factoring m.[1] Therefore, assuming b is even, in order for m to contain around b bits, the binary representations of p and q should each contain around $b/2$ bits. Thus, the decimal representations of p and q should satisfy $2^{b/2-1} \leq p < 2^{b/2}$ and $2^{b/2-1} \leq q < 2^{b/2}$.

Example 12.8 Suppose we wish to form RSA keys with a modulus m whose binary representation contains around $b = 16$ bits. To do this using primes p and q of roughly the same length, the binary representations of p and q should each contain around $b/2 = 8$ bits. Thus, since $2^7 = 128$ and $2^8 = 256$, the decimal representations of p and q should satisfy $128 \leq p < 256$ and $128 \leq q < 256$. For example, for $p = 199$, which has binary representation 11000111, and $q = 229$, which has binary representation 11100101, we have $m = p \cdot q = 45571$, which has binary representation

---

[1]Recall, as we demonstrated in Section 9.9, Fermat factorization poses a potential security risk if p and q are extremely close together. However, the probability that a pair of randomly chosen primes of roughly the same length would be close enough for Fermat factorization to be effective is negligible.

1011001000000011. To complete the key formation process, we would then form $f = (p-1) \cdot (q-1) = 45144$, choose an encryption exponent e for which $\gcd(e, f) = 1$, for example, $e = 101$, and use the Euclidean algorithm to find the corresponding decryption exponent $d = e^{-1} \bmod f = 15197$. □

When forming RSA keys for use in actual practice, it is also helpful to have access to a good method for generating pseudorandom numbers. One such method is the Blum Blum Shub generator, which we presented in Section 11.2. In addition, for an RSA cipher in actual practice, the encryption exponent e is usually much smaller than the corresponding decryption exponent d. This is not only because it has been shown that with a small decryption exponent the RSA algorithm can be insecure, but also because it is then faster for a signature to be verified than formed. This increases the overall efficiency of the system, since signatures are typically formed only once, but must frequently be verified more than once.

12.5.2　Web of Trust

Public-key infrastructures often distribute keys through electronic documents called *certificates* that use digital signatures to bind the keys to their owners. Certificates are sometimes signed by organizations called *certificate authorities* that specialize in forming keys and verifying that they belong to particular entities, but they can also be signed and verified through a *web of trust* scheme, with signatures included for the owner (yielding a *self-signed* certificate) and/or one or more other users called *endorsers*. In a web of trust scheme, no matter who signs a certificate, the signatures serve as an attestation by the signers that the keys belong to the entity claiming ownership.

In a public-key system protected by a web of trust scheme, each user has a certificate, and trust in a particular user's certificate is provided by the other users in the system. An individual user can assign varying levels of trust to the other users in the system, including complete trust, partial trust, no trust, and no information. For example, suppose Trixie, Sophie, and Allie are users in a public-key system protected by a web of trust scheme, and Allie wishes to send Sophie an encrypted message. However, suppose Allie does not know Sophie, and is concerned that Sophie's published keys may not actually belong to Sophie. If Trixie, on the other hand, knows Sophie well, and completely trusts that Sophie's published keys do indeed belong to Sophie, then Trixie could sign Sophie's certificate as an attestation that Sophie's published keys do indeed belong to Sophie. Then if Allie knows and trusts Trixie with some level of trust, with Trixie's signature on Sophie's certificate, Allie can have a corresponding level of confidence in sending Sophie an encrypted message.

One way in which a public-key system protected by a web of trust scheme can be strengthened is through an event called a *key signing party*. At key signing parties, users in the system present their public keys to the other users in the system, and sign the certificates of users whose public keys they trust as belonging to their claimed owners. Through key signing parties and other means, each user can create and maintain a file called a *keyring*, which contains the various levels of trust the user has in other users' certificates.

The idea of a web of trust was originally expressed by Phil Zimmermann, who is also the creator of the cryptographic method *Pretty Good Privacy* (*PGP*). PGP, which uses both symmetric and public-key encryption, was initially available as a free download on the Internet, although it is now privately owned by the Symantec corporation and available for purchase at https://www.symantec.com/products/encryption.

12.5.3 X.509 Certificates

For a certificate signed by a certificate authority (*CA*), the CA typically requires some type of proof of identification from the user. The type of proof of identification acceptable to a CA varies, and can include a driver's license, a notarized certificate, or even the fingerprints of the user. Knowing the type of proof of identification acceptable to a CA allows other users to form a level of confidence that the certificate issued by the CA can be trusted to correctly bind keys with their claimed owner. Companies that can serve as certificate authorities include VeriSign, RSA Laboratories, and Entrust Technologies, and are normally assumed to be trustworthy. If a user requests a certificate from a CA, the CA will produce it using the user's public keys, and then either sign it so that it can be verified by potential users, or, if more requests are received than can be filled, authorize a *registration authority* (*RA*) to sign it. In order that users can trust an RA, the RA will themselves have a certificate signed by the CA, giving them the authority to sign certificates. This can create a *certification hierarchy*, in which the verification of a certificate is certified by an RA, which is then certified by a CA.

One of the most common types of certificates issued are *X.509* certificates. An X.509 certificate contains the following standard parameters.

- Version: This gives the version of the certificate. Three versions of X.509 certificates have been used, with the most recent Version 3 first used in 1997.

- Serial Number: This gives the unique serial number, often expressed in hexadecimal format, issued by the CA and included on each certificate.

- Signature Algorithm: This gives the hash function used to hash the information on the certificate, and the encryption method used to sign this hashed information.

- Issuer: This gives the name of the CA and RA (if applicable) signing the certificate. Identifiers for this parameter can include CN for common name (which could be an individual or entity sponsoring the certificate), O for organization, OU for organizational unit, L for locality (usually a city), S for state, and C for country.

- Validity: This gives the time period for which the certificate is valid, which is always finite, and identified by the starting date and time and ending date and time.

- Subject: This gives the name of the individual or entity to whom the keys belong. The same identifiers used for the issuer can be used for this parameter.

- Subject Public Key Info: This gives the type of cipher for which the keys are designed.

- Subject's Public Key: This gives the actual keys, often expressed in hexadecimal format.

- Signature Information: This gives the result of first hashing the preceding information on the certificate, and then signing this hashed information using the CA's private decryption key, often expressed in hexadecimal format.

Examples of X.509 certificates can be found in [22] and also online at https://en.wikipedia.org/wiki/X.509.

Example 12.9 The following shows the general format of the information on an X.509 certificate.

```
Certificate
  Version: 3
  Serial Number: 15 4C
  Signature Algorithm: MODULAR (mh = 128) With RSA Encryption
  Issuer: CN = Certificate Company/Key Sign, Inc.
          O = Cryptography Division
          OU = Key Generation
          L = Washington
          S = DC
          C = USA
  Validity
```

```
          Not Before: Sep 14 12:00:00 AM 2011 EST
          Not After : Dec 31 11:59:59 PM 2021 EST
    Subject: CN = Juanita Sigmon
             O = Art Secrets
             OU = Painting
             L = Newton
             S = NC
             C = USA
    Subject Public Key Info: RSA Encryption
    Subject's Public Key:
      Modulus:
        B2 03
      Exponent:
        65
    Signature Algorithm: MODULAR (mh = 128) With RSA Encryption
    Signature Value:
        03 6A
```

The serial number of this certificate is 5452, which appears in hexadecimal format as 154C. It is next indicated on the certificate that the RSA signature scheme with hashing and encryption was used in forming the signature on the certificate, with a modular hash function with modulus $m_h = 128$. For the issuer, the common name on the certificate identifies that the CA issuing the certificate is Certificate Company, with RA Key Sign, Inc. It is later indicated on the certificate that the keys on the certificate, which are designed for RSA encryption, are modulus $m_r = 45571$, which appears in hexadecimal format as B203, and encryption exponent $e_r = 101$, which appears in hexadecimal format as 65. The certificate ends with the signature value 874, expressed in hexadecimal format as 36A. To form this signature, the modular hash function with modulus $m_h = 128$ is applied to the ASCII numbers corresponding to the characters on the certificate, starting with the characters Certificate at the top and ending with the characters 65 giving the subject's encryption exponent, ignoring any spaces that precede the first printed character on a line and the carriage return that follows each line. That is, the modular hash function is applied to the ASCII numbers corresponding to the characters CertificateVersion: 3Serial Number: 15 4CSignature Algorithm: MODULAR (mh = 128) With RSA EncryptionIssuer: CN = Certificate Company/Key Sign, Inc.O = Cryptography DivisionOU = Key GenerationL = WashingtonS = DCC = USAValidityNot Before: Sep 14 12:00:00 AM 2011 ESTNot After : Dec 31 11:59:59 PM 2021 ESTSubject: CN = Juanita SigmonO = Art SecretsOU = PaintingL = NewtonS = NCC = USASubject Public Key Info: RSA EncryptionSubject's Public Key:Modulus:B2 03Exponent:65. This results in digest 100, which is then signed using the issuer's decryption exponent and modulus, $d_i = 211$ and $m_i = 2537$, which, of course, do not appear on the certificate. This gives signature value $100^{211} \bmod 2537 = 874$, which appears on the certificate in hexadecimal format as 36A. □

To check the validity of the sample X.509 certificate in Example 12.9, we would first hash the certificate to find the digest value 100. We would then use the issuer's public encryption exponent and modulus, which are $e_i = 127$ and $m_i = 2537$, and the signature at the bottom of the certificate expressed in decimal format as 874, to decrypt the signature by forming $874^{127} \bmod 2537 = 100$. Since this decrypted signature matches the digest, assuming we trust the issuer of the certificate, we can trust the subject's keys on the certificate. If we needed to verify our trust in the issuer, we could use the CA's public keys to check the certificate issued by the CA to the RA. However, typically verifying trust in an issuer is unnecessary, as CAs are financially motivated to be trustworthy.

Certificates for individuals are not uncommon, although they are also often generated and verified for particular websites. Internet browsers regularly come prepackaged with certificates of common websites that are created by companies, such as VeriSign, that specialize in generating certificates. Details of these certificates can sometimes be found by searching under the Tools option at the top of the browser window.

12.5.4　Exercises

1. Use primes p and q of the same length to form an RSA modulus m whose binary representation contains the given number of bits.

   (a)* 8

   (b) 12

   (c)* 20

   (d) 24

2. Construct a Version 1 X.509 certificate with decimal serial number 1500 and the following information. The certificate was issued by the CA Certificate Generators (and no RA) of Raleigh, North Carolina, which is part of the organizational unit Public Keys in the organization Security Firm, and issued to Vicky Klima of Boone, North Carolina, who is part of the organizational unit Canine in the organization Sophie's Good Pets. The certificate became valid at 12:00:00 AM EST on January 1, 2012, and was set to expire at 11:59:59 PM EST on December 31, 2019. The subject's public keys, which were

designed for RSA encryption, are modulus $m_r = 667$ and encryption exponent $e_r = 19$. The certificate's hexadecimal signature value, formed using the hash function MD5 with RSA encryption, is 29 75 92 71 BB 86 2C 20 99 DD CB 9C 64 51 44 ED C3 76 BD CB B0 4D 31 F2.

3. Construct a Version 2 X.509 certificate with decimal serial number 9999 and the following information. The certificate was issued by the CA Rocket Flyers, with RA Secure Transport, Inc., of Cape Canaveral, Florida, which is part of the organizational unit Launch Codes in the organization Space Flight, and was issued to Neil Armstrong of Houston, Texas, who is part of the organizational unit Apollo in the organization Flight Heroes. The certificate became valid at 8:00:15 AM CST on May 1, 2017, and expired at 9:01:30 AM CST on May 1, 2017. The subject's public keys, which were designed for RSA encryption, are modulus $m_r = 28459$ and encryption exponent $e_r = 5537$. The certificate's hexadecimal signature value, which was formed using the hash function SHA-1 with RSA encryption, is 0B 5A 6D C8 5C C5 3B 6A 34 A5 46 37 3E 49 06 69 0E 52 69 43 21 F0 2C 29 68 3B 3D 5E.

4.* Construct a Version 3 X.509 certificate with decimal serial number 99999 and the following information. The certificate was issued by the CA Todd Harkrader (and no RA) of Atlanta, Georgia, who is part of the organizational unit Military Heroes in the organization Armed Forces Security, and issued to Mandy Sigmon of Christiansburg, Virginia, who is part of the organizational unit Fighter Planes in the organization Rudy's Metalworks. The certificate became valid at 8:00:05 AM EST on March 1, 2015, and was set to expire at 8:00:05 AM EST on March 1, 2025. The subject's public keys, which were designed for RSA encryption, are modulus $m_r = 2716454479$ and encryption exponent $e_r = 65537$. The certificate's hexadecimal signature value, which was formed using the RSA signature scheme with hashing and encryption with a modular hash function with modulus $m_h = 1207$, is 4A CC 75 59 70 48 1F 5E 39 B0 72 C3 CE 98 77 D7.

5. Find some information about the *Kerberos* computer network authentication protocol, and write a summary of your findings.

6. Find some information about the *Transport Layer Security* protocol, and write a summary of your findings.

7. Find some information about the career in cryptology of Phil Zimmermann, and write a summary of your findings.

Bibliography

[1] T. Barr. *Invitation to Cryptology*. Prentice Hall, Upper Saddle River, NJ, 2002.

[2] C. Bauer. *Secret History: The Story of Cryptology*. Chapman & Hall/CRC, Boca Raton, FL, 2013.

[3] F. Bauer. *Decrypted Secrets: Methods and Maxims of Cryptology, Fourth Edition*. Springer, New York, NY, 2007.

[4] J. Brawley. In memory of Jack Levine (1907-2005). *Cryptologia*, 30(2):83–97, 2006.

[5] F. Carter. *The Turing Bombe*. Bletchley Park Trust, Milton Keynes, UK, 2008.

[6] W. Diffie and M. Hellman. New directions in cryptography. *IEEE Transactions in Information Theory*, 22:644–654, 1976.

[7] G. Ellsbury. The Enigma and the Bombe. Available at http://www.ellsbury.com/enigmabombe.htm, 2007.

[8] W. Friedman. *Elements of Cryptanalysis*. Aegean Park Press, Walnut Creek, CA, 1976.

[9] W. Friedman. *The Index of Coincidence and Its Applications in Cryptography*. Aegean Park Press, Walnut Creek, CA, 1998.

[10] L. Hill. Cryptography in an algebraic alphabet. *American Mathematical Monthly*, 36:306–312, 1929.

[11] L. Hill. Concerning certain linear transformation apparatus of cryptography. *American Mathematical Monthly*, 38:135–154, 1931.

[12] D. Kahn. *Seizing the Enigma: The Race to Break the German U-Boat Codes, 1939-1943*. Houghton Mifflin Harcourt, Boston, MA, 1991.

[13] D. Kahn. *The Codebreakers: The Comprehensive History of Secret Communication from Ancient Times to the Internet.* Scribner, New York, NY, 1996.

[14] M. Klein. *Securing Record Communications: The TSEC/KW-26.* Center for Cryptologic History, National Security Agency, Ft. Meade, MD, 2003.

[15] J. Levine. Variable matrix substitution in algebraic cryptography. *American Mathematical Monthly*, 65:170–179, 1958.

[16] A. Miller. The cryptographic mathematics of Enigma. *Cryptologia*, 19:65–80, 1995.

[17] Naval History and Heritage Command. Navajo Code Talkers' Dictionary. Available at https://www.history.navy.mil/research/library/online-reading-room/title-list-alphabetically/n/navajo-code-talker-dictionary.html, 2017.

[18] C. Nez. *Code Talker: The First and Only Memoir by One of the Original Navajo Code Talkers of WWII.* Penguin Publishing Group, London, UK, 2012.

[19] D. Rijmenants. The German Enigma Cipher Machine. Available at http://users.telenet.be/d.rijmenants/en/enigma.htm, 2017.

[20] R. Rivest, A. Shamir, and L. Adleman. A method for obtaining digital signatures and public-key cryptosystems. *Communications of the ACM*, 21(2):120–126, 1978.

[21] S. Singh. *The Code Book.* Anchor Books, New York, NY, 2000.

[22] W. Trappe and L. Washington. *Introduction to Cryptography with Coding Theory, Second Edition.* Prentice Hall, Upper Saddle River, NJ, 2005.

[23] D. Turing. *Demystifying the Bombe.* The History Press, Stroud, UK, 2015.

[24] U.S. Army Field Manual 34-40-2. Basic Cryptanalysis. Available at http://www.contestcen.com/ArmyFieldManual.pdf, 1990.

[25] P. Wiedman. Cryptopop's Hints. Available at http://www.freewebs.com/gidusko/cryptopop/, 2007.

[26] J. Wilcox. *Solving the Enigma: History of the Cryptanalytic Bombe.* Center for Cryptologic History, National Security Agency, Ft. Meade, MD, 2004.

Hints and Answers for Selected Exercises

2.1.3 Exercises

1. (a) VZUWS WSD AXGDR RDGDAOUAD

4. (a) G TDRNN DOUR TOUR

7. (a), DMKIZAMN GWREK WG AYARC UWKN

2.2.1 Exercises

1. (a) The two most common letters in the ciphertext correspond to the two most common letters in ordinary English.

   (b) The two most common letters in the ciphertext correspond to the two most common letters in ordinary English.

   (c) The most common letter in the ciphertext corresponds to the plaintext letter A.

2. (a) The first word in the plaintext is THE, and there are three keywords, the second of which is NEW.

   (b) The second most common letter in the ciphertext corresponds to the plaintext letter N, and the third and fourth most common letters in the ciphertext correspond to the plaintext letters R and S.

   (c) The six most common letters in the ciphertext correspond to the six most common letters in ordinary English.

   (d) The most common letter in the ciphertext corresponds to the most common letter in ordinary English, and there are two keywords.

3. The two most common letters in the ciphertext each correspond to
   vowels in the plaintext, and the first two words in the plaintext each
   have two letters.

4. The two most common characters in the ciphertext correspond to the
   two most common letters in ordinary English.

5. (a) A MWJ AZ UKI SGGY

2.3.1 Exercises

1. (a) QMIEHNAVBVLBXI

2. (a) FCZOZGOMQZSFELABFIFSIU

2.4.1 Exercises

1. One possible ciphertext is CHA-GEE GINI KLESH WOL-LA-CHEE A-CHIN
   BA-AH-NE-DI-TININ CHA-GEE BESH-LO, with literal English transla-
   tion BLUE JAY CHICKEN HAWK SNAKE ANT NOSE KEY BLUE JAY IRON
   FISH.

2. One possible ciphertext is BESH-LEGAI-NAH-KIH TSE-NILL TSE-GAH
   WOL-LA-CHEE SHUSH JEHA TSE-NILL A-KEH-DI-GLINI AH-NAH CHA-
   GEE BE-EH-HO-ZINI, with literal English translation TWO SILVER BARS
   AXE HAIR ANT BEAR GUM AXE VICTOR EYE BLUE JAY ORDER.

3. (a) One possible ciphertext is YE-TSAN CHA TLO-CHIN CLA-GI-AIH
   AH-NAH, with literal English translation RUN AWAY FROM HAT
   ONION PANT EYE.

4. (a) The literal English translation is MOUSE APPLE DEER ELK ITCH
   NEEDLE BETWEEN WATERS, and the plaintext is MADE IN BRITAIN.

5. (a) One possible ciphertext is CHAY-DA-GAHI BEH-ELI-DOH-BE-CAH-
   ALI-TAS-AI CHINDI DZEH BI-SO-DIH AH-JAD DZEH D-AH AH-NAH
   CHINDI, with literal English translation TORTOISE AMMUNITION
   DEVIL ELK PIG LEG ELK TEA EYE DEVIL.

6. (a) The literal English translation is ATTACK SHEEP UNCLE RABBIT
   ICE BADGER ANT CAT HAIR ITCH AT DAWN, and the plaintext is
   ATTACK SURIBACHI AT DAWN.

3.1.3 Exercises

1. (a) LLIPE OAASS NRKCI TEKEI NAWPD UTY

6. (a) HIOIA OLAFW RBTNN IAUEH ATRHT MNODT SRRCX YHSAR AXEFW
   NYRIG SELYR X

3.2.3 Exercises

2. (a) LBMNT EKASR LNLPD IGKCN DTEEO UINIA AAOSN OYID

3. (a) There are five letters in the keyword(s).

   (c) There are fewer than eleven letters in the keyword(s), and the
   correct TE digraph is the one in the third ciphertext block.

   (d) There are fewer than nine letters in the keyword(s).

3.3.1 Exercises

1. (a) FAAFX DDFXF GAGAA FDFFX DFFFF D

2. (a) FDDAX AGGFF DFXAA GAAXD FFXAA GXGAF DDAGA XAGFD DXGGD
   XFXFA DGXGF AAFAA XDXDF FDFAG FFDXA FXDFX AGGXF AXAAD
   AGFFG XGAGX FDDF

4. (a) DXXFV VFAAX XAXVV GVXAA DGDDA FGGXV DXGAA XAFGF GDFF

5. (a) ADXDD DXDGD DGGGD GGAAV DFXDV VDDVD DDXAV VAAGD DVVDG
   GGXVD VDAXD VXXDD DGFDX FDVAG DGDVX GGA

4.1.1 Exercises

2. (a) S

   (c) V

3. (a) B

4. (a) 3

   (c) 9

5. (a) U

6. (a) Y

7. (a) W

8. (a) BELJ, BEMK

9. The first two letters in the ciphertext are NO, and the entries in the
    following table summarize the encryption process for these letters.

Summary of encryption of plaintext GERMAN using a Kriegsmarine M4 Enigma with plugboard connections GY, NS, and RT, left-to-right rotors γ, II, VIII, IV with ring settings 2, 22, 16, 7 and initial window letters BELI, and reflector C.						
Input letters	G	E	R	M	A	N
Window letters	BELJ	BEMK				
Rotor IV offset	3	4				
Rotor VIII offset	22	23				
Rotor II offset	9	9				
Rotor γ offset	0	0				
Plugboard	Y	E				
Add rotor IV offset	B	I				
Rotor IV from right	S	Y				
Subtract rotor IV offset	P	U				
Add rotor VIII offset	L	R				
Rotor VIII from right	S	M				
Subtract rotor VIII offset	W	P				
Add rotor II offset	F	Y				
Rotor II from right	I	O				
Subtract rotor II offset	Z	F				
Add rotor γ offset	Z	F				
Rotor γ from right	D	N				
Subtract rotor γ offset	D	N				
Reflector C	B	F				
Add rotor γ offset	B	F				
Rotor γ from left	L	A				
Subtract rotor γ offset	L	A				
Add rotor II offset	U	J				
Rotor II from left	H	B				
Subtract rotor II offset	Y	S				
Add rotor VIII offset	U	P				
Rotor VIII from left	W	M				
Subtract rotor VIII offset	A	P				
Add rotor IV offset	D	T				
Rotor IV from left	V	S				
Subtract rotor IV offset	S	O				
Plugboard	N	O				
Output letters	N	O				

4.2.4   Exercises

1. (a) 20

2. (a) 10,000

3. (a) 10,000,000

5. (a) 29,654,190,720

6. (a) 455

7. (a) 40,320

9. (a) 3960

10. (a) 4060

11. (a) 75,287,520

13. $25! \approx 1.5511 \times 10^{25}$

15. 325

17. $26! \approx 4.0329 \times 10^{26}$

18. Starting with A, how many possible letters could A be paired with? Then, for the next letter in alphabetical order that has not yet been paired, how many possible letters could this letter be paired with? Continue in this manner until all letters have been paired.

19. 60

20. 120

23. 17,576

25. From a collection of n objects, consider the experiment of choosing t of the objects, and then arranging these t objects in an ordered list. As a whole, the number of outcomes for this experiment is $P(n,t)$. Alternatively, thinking of the two parts of this experiment separately, the number of outcomes for the first part is $C(n,t)$, and the number of outcomes for the second part is $t!$. The multiplication principle then gives that the number of outcomes for this experiment is $C(n,t) \cdot t!$.

4.3.3　Exercises

1. (a) 10,767,019,638,375
   (c) 205,552,193,096,250

6. (a) $399,132,267,215,502,300,480,000 \approx 3.9913 \times 10^{23}$
   (c) $7,619,797,828,659,589,372,800,000 \approx 7.6198 \times 10^{24}$

7. (a) $3,021,910,221,542,011,017,394,176,000 \approx 3.0219 \times 10^{27}$
   (d) $57,691,013,320,347,483,059,343,360,000 \approx 5.7691 \times 10^{28}$

8. For a ciphertext output letter to match the corresponding plaintext input letter, the current traveling through the machine would have to follow the same path through the rotors from left to right and from right to left. Think about why this was impossible in German Enigmas.

5.1.1 Exercises

1. (a) PARLH EWCNI CGIR and RLHEW CNICG IRGK

   (c) CWGBT GHGDD THDH

3. (a) MUTVM UUPWF FTSF

   (b) Consider the following alignment of crib and ciphertext letters with position numbers.

Position:	1	2	3	4	5	6	7	8	9	10	11	12	13	14
Crib:	G	O	O	D	P	E	T	O	F	T	E	N	I	S
Cipher:	M	U	T	V	M	U	U	P	W	F	F	T	S	F

   The following is a menu that expresses the crib/ciphertext pairs in this alignment.

   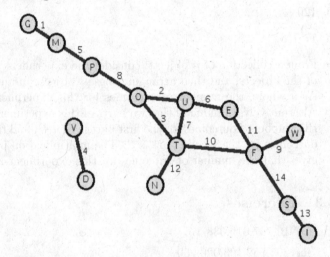

5. Consider the following alignment of crib and ciphertext letters with position numbers.

Position:	1	2	3	4	5	6	7	8	9	10	11	12	13	14
Crib:	G	O	R	D	O	N	W	E	L	C	H	M	A	N
Cipher:	E	M	H	S	S	M	B	N	S	X	N	L	W	W

The following is a menu that expresses the crib/ciphertext pairs in this alignment.

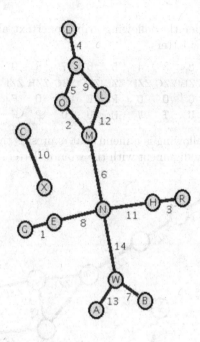

5.2.1 Exercises

2. (b) The closed loop T → U → O → T results from the machine in the sequence of positions 7, 2, 3; the closed loop T → U → E → F → T results from the machine in the sequence of positions 7, 6, 11, 10; and the closed loop T → O → U → E → F → T results from the machine in the sequence of positions 3, 2, 6, 11, 10.

   (d) The closed loop M → O → S → L → M results from the machine in the sequence of positions 2, 5, 9, 12.

5.3.1 Exercises

1. (a) 6, 1, 15
   (c) 14, 19, 5

2. (a) 20, 25, 11
   (c) 12, 7, 21

3. (a) D

   (c) E

5. (a) Consider the following crib/ciphertext alignment notated with window letters.

1	2	3	4	5	6	7	8	9	10	11	12	13	14
ZZA	ZZB	ZZC	ZZD	ZZE	ZZF	ZZG	ZZH	ZZI	ZZJ	ZZK	ZZL	ZZM	ZZN
G	O	O	D	P	E	T	O	F	T	E	N	I	S
M	U	T	V	M	U	U	P	W	F	F	T	S	F

The following is a menu that expresses the crib/ciphertext pairs in this alignment with the window letters shown along each link.

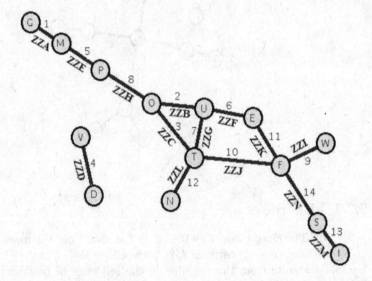

(b) For example, for the menu loop T → U → O → T, the choice K for the plugboard partner of T at the start of the loop results in the following.

Drum Setting:	ZZG	ZZB	ZZC
Menu Letter:	T ⟶ U ⟶	O ⟶	T
Plug Partner:	K ⟶ B ⟶	C ⟶	K

(c) The menu has three loops, which can be represented as follows.

   i. T → U → O → T

   ii. T → U → E → F → T

   iii. T → O → U → E → F → T

The complete list of cycles that result for each loop is given in the following table.

Loop	Cycles
i	$C_1 = $ (AIGDJWSE), $C_2 = $ (BMXKYLFPCHVROUT), $C_3 = $ (N), $C_4 = $ (Q), $C_5 = $ (Z)
ii	$D_1 = $ (AX), $D_2 = $ (BKOFPNGZV), $D_3 = $ (CTRSILJUMHY), $D_4 = $ (D), $D_5 = $ (EW), $D_6 = $ (Q)
iii	$E_1 = $ (AWUFJDZVYOSEIXHTMK), $E_2 = $ (BR), $E_3 = $ (CNGL), $E_4 = $ (P), $E_5 = $ (Q)

Only Q cannot be eliminated as the plugboard partner of the central letter.

(d) With the menu loops labeled as in part (c), the complete list of cycles that result for each loop is given in the following table.

Loop	Cycles
i	$C_1 = $ (ABFKZLG), $C_2 = $ (CRV), $C_3 = $ (D), $C_4 = $ (EYMTQOUP), $C_5 = $ (HXNWSJI)
ii	$D_1 = $ (AYPQSUZ), $D_2 = $ (BFKWINJ), $D_3 = $ (CER), $D_4 = $ (DXLG), $D_5 = $ (HTOM), $D_6 = $ (V)
iii	$E_1 = $ (ADXTHNL), $E_2 = $ (BYREQOSI), $E_3 = $ (CV), $E_4 = $ (F), $E_5 = $ (G), $E_6 = $ (JUMPZW), $E_7 = $ (K)

Every letter can be eliminated as the plugboard partner of the central letter.

7. For menus with 1 loop, the expected number of stops in a bombe run would be 676.

5.4.1 Exercises

1. (a) Yes
   (c) No

2. (a) Consider the following crib/ciphertext alignment notated with window letters.

1	2	3	4	5	6	7	8	9	10	11	12	13	14
ZZA	ZZB	ZZC	ZZD	ZZE	ZZF	ZZG	ZZH	ZZI	ZZJ	ZZK	ZZL	ZZM	ZZN
G	O	R	D	O	N	W	E	L	C	H	M	A	N
E	M	H	S	S	M	B	N	S	X	N	L	W	W

The following is a menu that expresses the crib/ciphertext pairs
in this alignment with the window letters shown along each link.

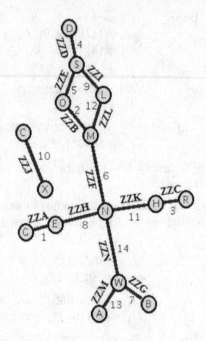

(b) The complete list of cycles that result for the menu loop M →
O → S → L → M is (ALSCIKQTUHNVMDOGERYWXJ)(B)(FP)(Z).

(c) For example, traveling in the menu from M to G results in the
following.

Drum Setting:	ZZF	ZZH	ZZA
Menu Letter:	M ⟶	N ⟶	E ⟶ G
Plug Partner:	A ⟶	U ⟶	H ⟶ L

A complete list of the plugboard pairs that would activate the
diagonal board is M/A, O/N, S/E, L/D, E/H, G/L, H/R, and A/D. A
complete list of the plugboard pairs that would not activate the
diagonal board is D/Q, N/U, R/Q, W/Q, and B/K.

(d) For example, for the plugboard pair E/H, exchanging the menu
letter and plugboard partner and following the links in the menu
from H back to M results in the following.

Drum Setting:	ZZK	ZZF
Menu Letter:	H ⟶	N ⟶ M
Plug Partner:	E ⟶	J ⟶ Z

    (e) Only B cannot be eliminated as the plugboard partner of the central letter.

4. (a) 45

5.5.1   Exercises

1. (a) Yes

   (c) No

2. (b) The first plaintext letter is A.

4. (b) The second letter in the plaintext after the end of the crib is R.

5. For example, traveling in the menu from T to I results in the following.

$$\begin{array}{lcccc}
\textbf{Drum Setting:} & \text{ZZJ} & \text{ZZN} & \text{ZZM} \\
\textbf{Menu Letter:} & \text{T} \longrightarrow \text{F} \longrightarrow \text{S} \longrightarrow \text{I} \\
\textbf{Plug Partner:} & \text{K} \longrightarrow \text{F} \longrightarrow \text{X} \longrightarrow \text{K}
\end{array}$$

6. (a) The plugboard pairs that follow from the menu are T/Q, N/R, L/U, F/Z, E/S, B/P, and G/V. The letters left unconnected in the plugboard that follow from the menu are D, I, M, O, and W.

   (b) The remaining plugboard pairs are A/C, H/Y, and J/K. The remaining letter left unconnected in the plugboard is X.

9. (a) The plugboard pairs that follow from the menu are M/B, O/U, I/S, D/V, L/W, A/G, E/Y, and H/Q. The letters left unconnected in the plugboard that follow from the menu are N and R.

   (b) The remaining plugboard pairs are C/T and P/X. The remaining letters left unconnected in the plugboard are F, J, K, and Z.

5.6.1   Exercises

4. The middle rotor rotated during encryption at the window letter position ZAV.

6. The middle rotor rotated during encryption at the window letter position ZAX.

5.7.1   Exercises

1. (a) M

(c) Z

2. (a) 9

    (c) 16

6. (a) A turnover of the middle rotor first occurred at the 23rd letter in the message.

    (b) Since a turnover of the middle rotor first occurred at the 23rd letter in the message, then after 22 turnovers of the rightmost rotor during encryption, the notch letter E would have been the rightmost window letter. Going backwards on the circle of letters around the rotor 22 positions from E gives I.

    (c) For the rightmost initial window letter Y and ring setting 19, since Y is the 25th letter in the alphabet, the rotor core starting position would be $25 - 19 = 6$. Thus, since the actual rightmost initial window letter I is the 9th letter in the alphabet, the actual initial ring setting would be $9 - 6 = 3$.

    (d) The initial window letters and ring settings that worked in the cryptanalysis process were ZZY and 26, 23, 19. Since ZZY are the alphabet letters in positions 26, 26, 25, the ring settings 26, 23, 19 with these initial window letters result in the rotor core starting positions $26 - 26 = 0$, $26 - 23 = 3$, $25 - 19 = 6$.

    (e) The actual initial window letters were TMI, and the actual ring settings were 20, 10, 3.

8. (a) A turnover of the middle rotor first occurred at the 24th letter in the message.

    (b) Since a turnover of the middle rotor first occurred at the 24th letter in the message, then after 23 turnovers of the rightmost rotor during encryption, the notch letter V would have been the rightmost window letter. Going backwards on the circle of letters around the rotor 23 positions from V gives Y.

    (c) For the rightmost initial window letter Z and ring setting 9, since Z is the 26th letter in the alphabet, the rotor core starting position would be $26 - 9 = 17$. Thus, since the actual rightmost initial window letter Y is the 25th letter in the alphabet, the actual initial ring setting would be $25 - 17 = 8$.

    (d) The initial window letters and ring settings that worked in the cryptanalysis process were ZZZ and 26, 11, 9. Since ZZZ are the alphabet letters in positions 26, 26, 26, the ring settings 26, 11, 9 with these initial window letters result in the rotor core starting positions $26 - 26 = 0$, $26 - 11 = 15$, $26 - 9 = 17$.

(e) The actual initial window letters were ALY, and the actual ring settings were 1, 23, 8.

5.8.1 Exercises

1. 40 hours

3. One of the 32 possible rotor arrangements that could be indicated in the codebook for the next day would be **II, I, IV**.

4. 21.333 hours

6.1.1 Exercises

1. (a) $q = 5, r - 3$
   (d) $q = -4, r = 4$

2. (a) 3
   (d) 4

3. (a) 9 o'clock
   (d) 11 o'clock

4. One possible congruence class is $\{\ldots, -21, -14, -7, 0, 7, 14, 21, \ldots\}$.

5. One possible congruence class is $\{\ldots, -30, -20, -10, 0, 10, 20, 30, \ldots\}$.

6. (a) 3
   (b) 8

7. (a) 21
   (d) There is no such value of x.

8. (a) Are relatively prime, since $\gcd(11, 26) = 1$

9. (a) Has multiplicative inverse 7
   (b) Does not have a multiplicative inverse, since $\gcd(5, 10) \neq 1$

10. (a) 1, 3, 7, 9

11. (a) One is its own inverse, and 5 is its own inverse.

12. (b) $a = 23, b = 10$

6.2.1   Exercises

1. (a) HWWXE UXWH

2. (a) OBOFN PNZNA B

3. (b) ZNKVX UHRKS COZNN GBOTM

   (d) $x = (y + 20) \bmod 26$

4. (a) JXUJH EKRBU MYJXT EYDW

5. (a) NBYXI QHMCX YIZVY CHA

6.3.1   Exercises

4. (a) ZOYZV OGRYC KIXOO NVOCC DYCKI

6.4.1   Exercises

1. (a) QMRBT SGK

   (c) $x = 15(y + 8) \bmod 26$

2. (a) XCPED MITK

3. (a) KPYJK SUEYE GREKR TUHU

4. (a) WNNWF WAXMW JMGVX EGEAY JPFWR IEGVW GKLEX MJPWR MHYJW
   GVWVE

6.5.1   Exercises

1. (a) The encryption formula is $y = (15x + 12) \bmod 26$.

   (c) The encryption formula is $y = (5x + 24) \bmod 26$.

2. (a) The most common letter in the ciphertext corresponds to the
   third most common letter in ordinary English, and one of the
   second most common letters in the ciphertext corresponds to the
   plaintext letter R.

   (b) The most common letter in the ciphertext corresponds to the
   most common letter in ordinary English, and one of the second
   most common letters in the ciphertext corresponds to the second
   most common letter in ordinary English.

(c) Two of the second most common letters in the ciphertext correspond to the two most common letters in ordinary English, and the most common letter in the ciphertext corresponds to the plaintext letter O.

(d) The most common letter in the ciphertext corresponds to the most common letter in ordinary English, and one of the second most common letters in the ciphertext corresponds to the second most common letter in ordinary English.

3. (a) GAQNK GJUPO C

4. 7.775 hours

5. (a) There are 72 possible values of a.

7.1.1 Exercises

1. (a) Gcevl ayss& d
   (c) Gcexa &lvyl tM&&t hevtn kib

2. (a) Ganlp dyefh iMqxx aaNqv syod& ndhv

7.2.3 Exercises

1. (a) VKVFZ LMAQC RUDNB

2. (a) OEAVR ANFKO VFAVJ KBQHV

4. (a) NAPYR LIITR IE

5. (a) EOHVO WGNXY TZOYM

6. (a) VTRXU IKLDV MWVRK LZEUV YHTSI AZMGY KIVKO TUII

7.3.1 Exercises

1. (a) $\frac{5}{6} \approx 0.833$
   (c) $\frac{5}{6} \approx 0.833$
   (e) $\frac{2}{5} = 0.4$

2. (a) $\frac{1}{36} \approx 0.028$
   (c) $\frac{1}{36} \approx 0.028$
   (e) $\frac{6}{36} \approx 0.167$

3. (a) No, since it is not possible for the sum of the results of the two
       rolls to be 14.

   (c) $\frac{2}{36} \approx 0.056$

   (e) $\frac{1}{6} \approx 0.167$

4. (a) $\frac{48}{52} \approx 0.923$

   (c) $\frac{8}{52} \approx 0.154$

   (e) $\frac{13}{39} \approx 0.333$

5. (a) $\frac{16}{2652} \approx 0.006$

   (c) $\frac{11}{51} \approx 0.216$

   (e) $\frac{3}{51} \approx 0.059$

6. (a) $\frac{16}{2704} \approx 0.006$

   (c) $\frac{12}{52} \approx 0.231$

   (e) $\frac{4}{52} \approx 0.077$

7. (a) $\frac{40}{48} \approx 0.833$

   (c) $\frac{16}{48} \approx 0.333$

   (e) $\frac{12}{36} \approx 0.333$

8. (a) $\frac{64}{2256} \approx 0.028$

   (c) $\frac{23}{47} \approx 0.489$

   (e) $\frac{7}{47} \approx 0.149$

9. (a) $\frac{64}{2304} \approx 0.028$

   (c) $\frac{24}{48} = 0.5$

   (e) $\frac{8}{48} \approx 0.167$

10. (a) $\frac{90}{100} = 0.9$

    (c) $\frac{35}{100} = 0.35$

    (e) $\frac{20}{90} \approx 0.222$

11. (a) $\frac{200}{9900} \approx 0.020$

    (c) $\frac{6320}{9900} \approx 0.638$

    (e) 1

12. (a) $\frac{200}{10000} = 0.02$

    (c) $\frac{6400}{10000} = 0.64$

    (e) $\frac{99}{100} = 0.99$

7.4.3 Exercises

1. (a) Approximately 0.034

3. (a) The index of coincidence is approximately 0.0629, and the cipher is more likely to be monoalphabetic.

   (b) The estimate for the length of the keyword is approximately 1.1061, and so the most likely length of the keyword is one letter.

4. (a) The estimate for the length of the keyword is approximately 2.7985, and so the most likely length of the keyword is three letters.

5. (a) n

7.5.1 Exercises

1. 6

3. (a) The longest repeated group of letters is LVV, and the most likely length of the keyword is four letters.

   (c) The longest repeated group of letters is WICD, and the most likely length of the keyword is six letters.

7.6.3 Exercises

1. (a) 4

   (b) Coset 1: LESKWVKWLJDFKVKSWVSUWVWELWJFSWJWLAWAJ
       Coset 2: PMOQVNBVWMTLBPQBAAEICBBQPGLOVGLLPOZX
       Coset 3: OYKTJUAZSLEKGUMAGIRTYUKTKCRZJCAOKKKN
       Coset 4: FFVGREQFBHHEAJAEAEFORQERXBRUXBFAIAPR

3. (a) SIGN

5. (a) Ciphertext A was formed using a monoalphabetic cipher.

   (c) The most likely length of the keyword is five letters.

6. (b) The keyword is seven letters long, and is a common word in ordinary English.

7. (a) TSUXM VFIGV KZVRM

8.1.5   Exercises

1. (a) $\begin{bmatrix} 7 & 10 \\ -2 & -2 \\ 4 & 8 \end{bmatrix}$

    (c) $\begin{bmatrix} 15 & 30 \\ -18 & 21 \\ 3 & 0 \end{bmatrix}$

2. (a) $\begin{bmatrix} 7 & 0 \\ 8 & 8 \\ 4 & 8 \end{bmatrix}$

    (c) $\begin{bmatrix} 5 & 0 \\ 2 & 1 \\ 3 & 0 \end{bmatrix}$

3. (a) $\begin{bmatrix} 13 & 9 \end{bmatrix}$

    (c) $\begin{bmatrix} 38 & 22 & 40 \end{bmatrix}$

4. (a) $\begin{bmatrix} 13 & 9 \end{bmatrix}$

    (c) $\begin{bmatrix} 12 & 22 & 14 \end{bmatrix}$

5. (a) $AB = \begin{bmatrix} 19 & 22 \\ 43 & 50 \end{bmatrix}$, $BA = \begin{bmatrix} 23 & 34 \\ 31 & 46 \end{bmatrix}$

    (c) $AB = \begin{bmatrix} 38 & 22 & 40 \\ -20 & -32 & -60 \\ -32 & 9 & 13 \end{bmatrix}$, $BA = \begin{bmatrix} 18 & -30 & -1 \\ 36 & -2 & 0 \\ 30 & -40 & 3 \end{bmatrix}$

6. (a) $AB = \begin{bmatrix} 19 & 22 \\ 17 & 24 \end{bmatrix}$, $BA = \begin{bmatrix} 23 & 8 \\ 5 & 20 \end{bmatrix}$

    (c) $AB = \begin{bmatrix} 12 & 22 & 14 \\ 6 & 20 & 18 \\ 20 & 9 & 13 \end{bmatrix}$, $BA = \begin{bmatrix} 18 & 22 & 25 \\ 10 & 24 & 0 \\ 4 & 12 & 3 \end{bmatrix}$

7. (a) $AB = \begin{bmatrix} 19 & 22 & 25 \\ 43 & 50 & 57 \end{bmatrix}$, BA does not exist

    (c) AB does not exist, $BA = \begin{bmatrix} 8 & 14 \\ 0 & 50 \\ 18 & 44 \end{bmatrix}$

8. (a) $AB = \begin{bmatrix} 19 & 22 & 25 \\ 17 & 24 & 5 \end{bmatrix}$, BA does not exist

(c) AB does not exist, $BA = \begin{bmatrix} 8 & 14 \\ 0 & 24 \\ 18 & 18 \end{bmatrix}$

9. **Ab** does not exist, but **bA** does.

10. **Ab** mod 26 does not exist, but **bA** mod 26 does.

11. Both **ab** and **ba** exist.

12. Both **ab** mod 26 and **ba** mod 26 exist.

13. (a) -23

    (c) 34

14. (a) 3

    (c) 8

15. (a) $\begin{bmatrix} -3/23 & 4/23 \\ 17/23 & -15/23 \end{bmatrix}$

    (c) $\begin{bmatrix} 15/34 & -7/34 \\ -8/34 & 6/34 \end{bmatrix}$

16. (a) $\begin{bmatrix} 1 & 16 \\ 3 & 5 \end{bmatrix}$

    (c) Does not exist

17. (a) $B = A^{-1} \bmod 26$

19. (a) 25

20. (a) $\begin{bmatrix} 2 & 9 & 16 \\ 25 & 7 & 18 \\ 24 & 9 & 19 \end{bmatrix}$

8.2.1 Exercises

1. (a) The plaintext WHEN YOU PLAY, PLAY HARD A encrypts to XQTEQ MDWKL JAKLN SLEYD.

2. (a) EUHER OOZ

    (c) $\begin{bmatrix} 1 & 16 \\ 3 & 5 \end{bmatrix}$

3. (a) SEZCM KBZMB DX

4. (a) The plaintext NOBODY CARES HOW MUCH YOU KNOW AA encrypts to
        LQXCP GOMNE AJAUI YWRQE WEVYI CU.

5. (a) MYLPT K

6. (a) IQANK ODTJ

10. Decrypting the ciphertext yields I WOSL BE LATE A, which contains
    errors in both the third and fourth letters. Since plaintext letters were
    encrypted in pairs to form the ciphertext in Example 8.17, incorrectly
    transcribing one ciphertext letter can also cause the ciphertext letter
    with which it was formed to decrypt to the wrong plaintext letter.

8.3.1   Exercises

1. The key matrix is $\begin{bmatrix} 11 & 6 \\ 0 & 3 \end{bmatrix}$.

3. The key matrix is $\begin{bmatrix} 21 & 6 \\ 3 & 5 \end{bmatrix}$.

5. You may find it useful to know that for the matrix $\begin{bmatrix} 6 & 14 & 13 \\ 0 & 21 & 24 \\ 1 & 4 & 0 \end{bmatrix}$,
    the inverse modulo 26 is $\begin{bmatrix} 16 & 0 & 9 \\ 22 & 13 & 24 \\ 23 & 6 & 18 \end{bmatrix}$.

6. You may find it useful to know that for the matrix $\begin{bmatrix} 19 & 7 & 4 \\ 0 & 19 & 19 \\ 22 & 8 & 11 \end{bmatrix}$,
    the inverse modulo 26 is $\begin{bmatrix} 7 & 15 & 7 \\ 8 & 3 & 25 \\ 18 & 8 & 1 \end{bmatrix}$.

7. (a) CWOVO ZSBTI

8. 793.361 days

10. (a) The plaintext ANNAPOLIS A encrypts to OPOCW AQSLW.

11. (a) VWHAH J

    (f) The plaintext YEPPERS A encrypts to DIENQ XXJ.

9.2.1   Exercises

1. (a) 1, 20, 20, 7, 16

9.3.1   Exercises

1. (a) 180
   (c) 8
   (e) 1

2. (a) $s = 1, t = -1$
   (c) $s = -91, t = 334$
   (e) $s = 1727, t = -4934$

3. (a) 49
   (c) Does not exist
   (e) 2964

4. (a) 37
   (c) 1663
   (e) Does not exist

9.4.1   Exercises

1. (a) 6578
   (c) 769
   (e) 5447

2. (a) 8
   (c) 11
   (e) 16

3. (a) 6578
   (c) Does not exist
   (e) 5447

9.6.1   Exercises

1. (a) 246, 275, 207, 376, 237

(e) 42, 189, 189, 151

(g) 15

2. (a) 15146, 17934, 113431

(e) 76516, 80882

(g) 80529

3. (a) 31796559, 38332181, 49221620

(e) 108088274

9.7.1 Exercises

2. (a) 110, 90, 59, 136

9.8.1 Exercises

1. (a) Prime

(c) Not prime

(e) Prime

(g) Prime

2. (a) May be prime

(c) Definitely not prime

3. (a) $2^{118} = 30 \bmod 119$, and the modulus is definitely not prime

(c) $4^{90} = 1 \bmod 91$, and the modulus may be prime

9.9.1 Exercises

1. (a) $11^2 \cdot 13$

(c) $47 \cdot 59$

(e) 8731

(g) $3 \cdot 5^2 \cdot 29 \cdot 41$

2. (a) $347 \cdot 269$

(c) $2347 \cdot 2003$

10.1.1   Exercises

1. (a) 137

   (b) 103

   (c) Acceptable encryption exponent, since $\gcd(e, f) = 1$

3. (a) 2939

4. (a) 87

   (b) 121

   (c) Not an acceptable encryption exponent, since $\gcd(e, f) \neq 1$

6. (a) 1936

10.2.1   Exercises

1. (a) One possible answer is $r = 18$.

   (b) One possible answer is $r = 5$.

   (c) 121

3. (a) 92

   (c) 220

4. (a) 58

   (c) 171

10.3.1   Exercises

1. (a) Not primitive

   (d) Primitive

2. (b) $(7, 89)$, $(84, 133)$, $(49, 49)$, $(40, 59)$

   (f) $(126, 11)$, $(54, 98)$, $(121, 47)$, $(91, 92)$, $(39, 42)$

3. (b) $(900, 83197)$, $(27000, 32036)$

   (f) $(1681, 77682)$, $(68921, 67904)$

4. (b) $(2500, 22896383)$, $(125000, 123687952)$

   (f) $(3600, 39986954)$

10.4.1   Exercises

1. (a) The value of n is greater than 120, and $2^{120} \bmod 131 = 60$.

    (b) The value of n is greater than 120, and $6^{120} \bmod 131 = 80$.

    (c) The value of n is less than 10.

    (d) The value of n is less than 10.

3. (a) The values of k_1, k_2, and k_3 are all less than 10.

    (b) The values of k_1 and k_2 are both less than 15. The value of k_3 is greater than 110, and $20^{110} \bmod 137 = 118$.

11.1.3   Exercises

1. (a) 110001

    (c) 11000101

2. (a) 43

    (c) 235

3. (a) 01001010, 01100101, 01110100, 01110011

5. (a) 011001

    (c) 11110101

6. (a) 7A2

    (c) CD1AE

7. (a) 2511

    (c) 323335

8. (a) F3

    (c) 6B6E

9. (a) 1001011

    (c) 1010000101111111

11.2.1   Exercises

1. (a) 10011001, 10000000, 10111010

    (c) 10011011, 10010111, 10100111, 00001110, 10001010

3. (a) 10010110
   (c) 01011101

4. (a) 01000011
   (c) 10111001

11.3.4 Exercises

1. (a)
$$\begin{bmatrix} 01001000 & 01101111 & 01101110 & 00100000 \\ 01100001 & 01110111 & 00100000 & 00100000 \\ 01101100 & 01100101 & 00100000 & 00100000 \\ 01101100 & 01100101 & 00100000 & 00100000 \end{bmatrix}$$

(d)
$$\begin{bmatrix} 01000001 & 01100111 & 01100001 & 01101111 \\ 00100000 & 01101000 & 01110010 & 01101110 \\ 01001110 & 01110100 & 01100101 & 00100000 \\ 01101001 & 01101101 & 00100000 & 01000101 \end{bmatrix},$$

$$\begin{bmatrix} 01101100 & 01110100 & 01110100 & 00100000 \\ 01101101 & 01110010 & 00100000 & 00100000 \\ 00100000 & 01100101 & 00100000 & 00100000 \\ 01010011 & 01100101 & 00100000 & 00100000 \end{bmatrix}$$

2. (a) 10011000
   (d) 01000000

3. (a) 00111011
   (d) 00011110

4. (a)
$$\begin{bmatrix} 01001101 & 01100001 & 01001101 & 01110011 \\ 01101001 & 01100101 & 01111001 & 00100000 \\ 01100011 & 01101100 & 01100101 & 00100000 \\ 01101000 & 00100000 & 01110010 & 00100000 \end{bmatrix}$$

(d)
$$\begin{bmatrix} 01000110 & 01100100 & 01110010 & 01100101 \\ 01110010 & 01111001 & 01110101 & 01110010 \\ 01100101 & 00100000 & 01100101 & 00100000 \\ 01100100 & 01001011 & 01100111 & 00100000 \end{bmatrix}$$

5. (a)
$$\begin{bmatrix} 11111011 & 10011010 & 11010111 & 10100100 \\ 11011110 & 10111011 & 11000010 & 11100010 \\ 11010100 & 10111000 & 11011101 & 11111101 \\ 11100111 & 11000111 & 10110101 & 10010101 \end{bmatrix}$$

(d) $\begin{bmatrix} 00000111 & 01100011 & 00010001 & 01110100 \\ 11000101 & 10111100 & 11001001 & 10111011 \\ 11010010 & 11110010 & 10010111 & 10110111 \\ 00101001 & 01100010 & 00000101 & 00100101 \end{bmatrix}$

6. (a) $\begin{bmatrix} 01100001 & 11111011 & 00101100 & 10001000 \\ 10001010 & 00110001 & 11110011 & 00010001 \\ 11111110 & 01000110 & 10011011 & 01100110 \\ 10101110 & 01101001 & 11011100 & 01001001 \end{bmatrix}$

(d) $\begin{bmatrix} 11101111 & 10001100 & 10011101 & 11101001 \\ 01101100 & 11010000 & 00011001 & 10100010 \\ 11101101 & 00011111 & 10001000 & 00111111 \\ 10111011 & 11011001 & 11011100 & 11111001 \end{bmatrix}$

7. $\begin{bmatrix} 10101110 & 00000100 & 11000001 & 01100001 \\ 10000001 & 01110111 & 11110011 & 00000100 \\ 11110100 & 01100110 & 10011100 & 01000110 \\ 11000010 & 10001111 & 10100111 & 10101111 \end{bmatrix}$

11.4.3 Exercises

1. (a) 11010110
   (d) 10010001

2. (a) 10100101
   (d) 01010100

3. (a) $\begin{bmatrix} 00000101 & 00001110 & 00100011 & 01010011 \\ 00001000 & 00010010 & 01011001 & 00000000 \\ 00001111 & 00001001 & 01000101 & 00000000 \\ 00000100 & 01000101 & 01010010 & 00000000 \end{bmatrix}$

(b) $\begin{bmatrix} 01101011 & 10101011 & 00100110 & 11101101 \\ 00110000 & 11001001 & 11001011 & 01100011 \\ 01110110 & 00000001 & 01101110 & 01100011 \\ 11110010 & 01101110 & 00000000 & 01100011 \end{bmatrix}$

(c) $\begin{bmatrix} 01101011 & 10101011 & 00100110 & 11101101 \\ 11001001 & 11001011 & 01100011 & 00110000 \\ 01101110 & 01100011 & 01110110 & 00000001 \\ 01100011 & 11110010 & 01101110 & 00000000 \end{bmatrix}$

(d) $\begin{bmatrix} 10011011 & 10011010 & 11110001 & 10010000 \\ 00110011 & 01110001 & 00010100 & 10001110 \\ 11011011 & 10101011 & 00011011 & 11011111 \\ 11011100 & 10110001 & 10100011 & 00011101 \end{bmatrix}$

(e)
$$\begin{bmatrix} 01100000 & 00000000 & 00100110 & 00110100 \\ 11101101 & 11001010 & 11010110 & 01101100 \\ 00001111 & 00010011 & 11000110 & 00100010 \\ 00111011 & 01110110 & 00010110 & 10001000 \end{bmatrix}$$

(f)
$$\begin{bmatrix} 00110110 & 01001101 & 01100111 & 01111101 \\ 10110001 & 11101001 & 00000110 & 01100011 \\ 01111110 & 11010011 & 10011000 & 00011000 \\ 10010110 & 01110110 & 10001000 & 11000111 \end{bmatrix}$$

6. (a)
$$\begin{bmatrix} 00000111 & 00000011 & 00010011 & 00001010 \\ 01010010 & 00010001 & 00000111 & 00011100 \\ 00101011 & 01010100 & 00000000 & 00000000 \\ 00001101 & 00100110 & 01000111 & 01100101 \end{bmatrix}$$

(b)
$$\begin{bmatrix} 11000101 & 01111011 & 01111101 & 01100111 \\ 00000000 & 10000010 & 11000101 & 10011100 \\ 11110001 & 00100000 & 01100011 & 01100011 \\ 11010111 & 11110111 & 10100000 & 01001101 \end{bmatrix}$$

(c)
$$\begin{bmatrix} 11000101 & 01111011 & 01111101 & 01100111 \\ 10000010 & 11000101 & 10011100 & 00000000 \\ 01100011 & 01100011 & 11110001 & 00100000 \\ 01001101 & 11010111 & 11110111 & 10100000 \end{bmatrix}$$

(d)
$$\begin{bmatrix} 00100010 & 00010110 & 01000011 & 01001110 \\ 00110010 & 10011000 & 10100001 & 10100111 \\ 01010110 & 00011010 & 00011010 & 11011100 \\ 00101111 & 10011110 & 00011111 & 11010010 \end{bmatrix}$$

(e)
$$\begin{bmatrix} 00100101 & 01110101 & 01010010 & 00111010 \\ 11110111 & 00100100 & 01101000 & 00011100 \\ 10000100 & 11101000 & 10001101 & 01101011 \\ 00000110 & 11111100 & 00011010 & 11110111 \end{bmatrix}$$

(f)
$$\begin{bmatrix} 11111110 & 01110010 & 11001101 & 01110011 \\ 10110000 & 00101001 & 01101011 & 11100110 \\ 11100110 & 10001000 & 01100001 & 00000111 \\ 01000001 & 10000001 & 01100100 & 11001110 \end{bmatrix}$$

8. (a)
$$\begin{bmatrix} 00100000 & 00011000 & 11010111 & 01011010 \\ 00000100 & 00100100 & 00010011 & 01001000 \\ 10001100 & 10111101 & 10000100 & 01101111 \\ 01000011 & 00100110 & 01011000 & 10100110 \end{bmatrix}$$

(b)
$$\begin{bmatrix} 00100000 & 00011000 & 11010111 & 01011010 \\ 00100100 & 00010011 & 01001000 & 00000100 \\ 10000100 & 01101111 & 10001100 & 10111101 \\ 10100110 & 01000011 & 00100110 & 01011000 \end{bmatrix}$$

(c)
$$\begin{bmatrix} 00001110 & 00101001 & 11000111 & 01011101 \\ 01011001 & 11001100 & 11101110 & 11010110 \\ 11100110 & 00010000 & 11110110 & 11010111 \\ 10010111 & 11010010 & 11101010 & 11100111 \end{bmatrix}$$

(d)
$$\begin{bmatrix} 10100000 & 00101101 & 00000110 & 00111100 \\ 11011000 & 10111011 & 00011101 & 11010010 \\ 00010010 & 01110110 & 01101010 & 10010001 \\ 01010101 & 01011101 & 01001101 & 01001000 \end{bmatrix}$$

11.5.1   Exercises

1. (a) 10011000

2. (a) 11001100

3. (a) 00001000

4. (a) 10010111

5. (a)
$$\begin{bmatrix} 11001011 & 00010011 & 01010001 & 11111111 \\ 10000110 & 10011001 & 00110011 & 00001110 \\ 10000100 & 10010001 & 00101111 & 01000010 \\ 00001101 & 11011000 & 01000101 & 11101100 \end{bmatrix}$$

(c)
$$\begin{bmatrix} 01101111 & 00101001 & 01100011 & 11101101 \\ 10010011 & 01111100 & 11101101 & 11110000 \\ 01100011 & 01100011 & 11001001 & 10000100 \\ 01100011 & 10101111 & 00010101 & 10011100 \end{bmatrix}$$

(e)
$$\begin{bmatrix} 00000110 & 01001100 & 00000000 & 01010011 \\ 00010111 & 00100010 & 00000001 & 01010011 \\ 00010010 & 01001111 & 00000000 & 00000000 \\ 00011011 & 00101111 & 00011100 & 00000000 \end{bmatrix}$$

6. (a)
$$\begin{bmatrix} 11101011 & 01101110 & 11000110 & 11111110 \\ 11001110 & 11000101 & 11110111 & 11010111 \\ 10010110 & 10011010 & 00111011 & 10001111 \\ 00100000 & 10101011 & 11000000 & 11101100 \end{bmatrix}$$

(c)
$$\begin{bmatrix} 01110010 & 00000000 & 01111011 & 01100011 \\ 11111011 & 11110000 & 01100011 & 01100011 \\ 01000111 & 01100011 & 10101011 & 00111111 \\ 01100011 & 01000111 & 01101111 & 01100011 \end{bmatrix}$$

(e)
$$\begin{bmatrix} 00011110 & 01010010 & 00000011 & 00000000 \\ 00000000 & 01100011 & 00010111 & 00000000 \\ 00001110 & 00100101 & 00010110 & 00000000 \\ 00010110 & 00000110 & 00000000 & 00000000 \end{bmatrix}$$

8. $A_1 = \begin{bmatrix} 01111111 & 00000000 & 00100110 & 11010110 \\ 11001100 & 11001010 & 11010110 & 10001110 \\ 00110001 & 00010011 & 11000110 & 00011111 \\ 00100100 & 01110110 & 00010110 & 01010111 \end{bmatrix}$

12.1.1 Exercises

1. (a) 60, 221

2. (a) 107013

3. (a) Yes

   (c) No

4. (a) No

   (c) Yes

5. (a) Ciphertext = 148, 143, 146; encrypted signatures = 171, 306, 84

6. (a) Plaintext = ER; yes

   (c) Plaintext = MASH; no

7. (a) Ciphertext = 134, 153, 94; encrypted signatures = 261, 395, 513

8. (a) Plaintext = DAL; no

   (c) Plaintext = PHI; yes

9. One possibility is to find the value of d_o from the knowledge that $8073^{d_o} \bmod 9169 = 5465$.

10. (a) Find d_o from the knowledge of x_i, m_o, and $s_i = x_i^{d_o} \bmod m_o$.

    (b) Someone intercepting the originator's transmissions would have to first break an RSA cipher in order to find s_i.

11. One possibility is to find the value of d_o from the knowledge that $5657^{d_o} \bmod 9169 = 2489$.

12.2.1 Exercises

1. (a) 2306 in decimal, which is 902 in hexadecimal

   (c) 750210 in decimal, which is B7282 in hexadecimal

3. One possible answer is NGEORA CRUSH.

4. (a) Yes

    (c) No

5. (a) $\frac{1}{10921} \approx 0.000092$

    (b) One possible answer is NOBIS, split into integers corresponding to two characters each, padded at the end with a space character.

7. (a) 00011110 in binary, which is 1E in hexadecimal

    (c) 256

    (e) One possible answer is 00001111.

12.3.1 Exercises

1. (a) 65

2. (a) 75881

3. (a) No

    (c) Yes

4. (a) No

    (c) Yes

5. (a) Ciphertext = 145, 236, 379, 377, 69; encrypted signature = 366

6. (a) Plaintext = ERIC; yes

    (c) Plaintext = TURTLE; no

7. (a) If the signature formation $s = h(x)^{d_o} \bmod m_o$ resulted in a value of s between m_r and m_o, then it would not be possible for the signature decryption calculation $z^{d_r} \bmod m_r$ to return s, since $z^{d_r} \bmod m_r$ must be less than m_r.

12.4.1 Exercises

1. (a) Ciphertext for colleague = 143, 274, 143, 269

2. (a) Ciphertext for colleague = 335781, 579475, 968458

3. The following is one possible answer. Suppose the originator of a numeric message wishes to use the RSA signature scheme with encryption to send the message to an intended recipient over an insecure communication line. First, the originator chooses RSA encryption exponent e_o and modulus m_o, with corresponding decryption exponent d_o, and sends the values of e_o and m_o to the intended recipient. However, an outsider monitoring the communication intercepts and stops this transmission. Meanwhile, the intended recipient chooses RSA encryption exponent e_r and modulus m_r, with corresponding decryption exponent d_r, and sends the values of e_r and m_r to the intended recipient, but the outsider intercepts and stops this transmission as well. Suppose also that m_r is greater than m_o. The outsider then chooses RSA encryption exponent e_c and modulus m_c, with corresponding decryption exponent d_c, and, posing as the originator or intended recipient, sends the values of e_c and m_c on to both the originator and recipient. Suppose also that m_c is between m_o and m_r. The originator uses e_c and m_c to encrypt the message, uses d_o, m_o, e_c, and m_c to sign the message, and sends the resulting ciphertext and signature(s) to the intended recipient. The outsider intercepts and stops this transmission as well, and uses d_c and m_c to decrypt the ciphertext. The outsider then uses e_r and m_r to encrypt the message, uses d_c, m_c, e_r, and m_r to sign the message, and, posing as the originator, sends the resulting ciphertext and signature(s) on to the intended recipient.

12.5.4 Exercises

1. (a) One possible answer is $m = 143$.

   (c) One possible answer is $m = 770977$.

4. Certificate
   Version: 3
   Serial Number: 01 86 9F
   Signature Algorithm: MODULAR (mh = 1207) With RSA
   Encryption
   Issuer: CN = Todd Harkrader
           O = Armed Forces Security
           OU = Military Heroes
           L = Atlanta
           S = GA
           C = USA
   Validity

```
          Not Before: Mar 1 8:00:05 AM 2015 EST
          Not After : Mar 1 8:00:05 AM 2025 EST
Subject: CN = Mandy Sigmon
          O = Rudy's Metalworks
          OU = Fighter Planes
          L = Christiansburg
          S = VA
          C = USA
Subject Public Key Info: RSA Encryption
Subject's Public Key:
   Modulus:
          A1 E9 CE 4F
   Exponent:
          01 00 01
Signature Algorithm: MODULAR (mh = 1207) With RSA
Encryption
Signature Value:
          4A CC 75 59 70 48 1F 5E 39 B0 72 C3 CE 98 77 D7
```

Index

Printed in the United States
by Baker & Taylor Publisher Services